Solution of Problems
in Strength of Materials
and Mechanics of Solids

Solution of Problems in Strength of Materials and Mechanics of Solids

S. A. URRY, BSc(Eng), CEng, AFRAeS
Professor of Building Technology, Brunel University

P. J. TURNER BTech, CEng, MIMechE
School of Engineering, Brunel University

SI Metric Edition

PITMAN PUBLISHING

First published 1974

SIR ISAAC PITMAN AND SONS LTD
Pitman House, Parker Street, Kingsway, London WC2B 5PB
P.O. Box 46038, Banda Street, Nairobi, Kenya

SIR ISAAC PITMAN (AUST.) PTY LTD
Pitman House, 158 Bouverie Street, Carlton, Victoria 3053, Australia

PITMAN PUBLISHING CORPORATION
6 East 43rd Street, New York, N.Y. 10017, U.S.A.

SIR ISAAC PITMAN (CANADA) LTD
495 Wellington St West, Toronto 135, Canada

THE COPP CLARK PUBLISHING COMPANY
517 Wellington St West, Toronto 135, Canada

Cased edition ISBN 0273 36189 9
Paperback edition ISBN 0 273 36190 2

Text set in 10/11 pt. Monotype Times New Roman, printed by letterpress, and bound in Great Britain at The Pitman Press, Bath
G4 (T 1264/1353: 76)

Preface

This book is intended for engineering students in universities, polytechnics and colleges who are taking courses in Strength of Materials, Stress Analysis and Mechanics of Solids. Its standard is that of University and CNAA degrees and it should also prove suitable for courses leading to Higher National Certificates and Diplomas, and the examinations of the Council of Engineering Institutions. At the same time there is extensive coverage of the more elementary topics and it is hoped that students will find the book useful from the beginning of their courses.

SI units are used exclusively and an explanatory note on the system is given on page ix. Many of the problems first appeared in Imperial units in earlier editions of the book and, in the conversion to SI, numerical values have generally been rounded off.

We have taken the opportunity to make other changes; symbols and abbreviations have been brought into line with current practice and some new topics have been covered by the inclusion of additional examples. The new material includes laterally and eccentrically loaded struts, rotating discs, the bending of circular plates, curved beams, unsymmetrical bending, and plastic deformation.

The following abbreviations are used to indicate the sources of examination questions:

(U.L.) University of London
(I.Mech.E.) Institution of Mechanical Engineers
(I.Struct.E.) Institution of Structural Engineers
(R.Ae.S.) Royal Aeronautical Society

For permission to reproduce these questions we are indebted to the respective authorities but the responsibility for the conversion to SI units and for the solutions is entirely ours. The examinations of the various professional engineering institutions have now been replaced by those of the Council of Engineering Institutions.

Our thanks are due to Mrs Felicity Lefevre for typing the entire manuscript; to the Publishers for their care in the production of the book; and to our wives for their patience and encouragement.

Despite careful checking some errors may remain and any criticism or correction will be gratefully acknowledged.

S. A. Urry
P. J. Turner

Contents

Table of SI Units

Quantity	*Derivation*	*Name of Unit*	*Symbol*
Mass	—	kilogramme	kg
Length	—	metre	m
Time	—	second	s
Area	$(\text{length})^2$	—	m^2
Volume	$(\text{length})^3$	—	m^3
First moment of area	area × length	—	m^3
Second moment of area	area × $(\text{length})^2$	—	m^4
Velocity	length ÷ time	—	m/s
Acceleration	velocity ÷ time	—	m/s^2
Force	mass × acceleration	newton	N (= kg-m/s^2)
Stress and pressure	force ÷ area	—	N/m^2
Torque and moment	force × distance	—	N-m
Work and energy	force × distance	joule	J (=N-m)
Power	work ÷ time	watt	W (=N-m/s)

Multiples and Sub-multiples

Factor	*Unit prefix*	*Symbol*
$1000000000 = 10^9$	giga-	G
$1000000 = 10^6$	mega-	M
$1000 = 10^3$	kilo-	k
$0{\cdot}01 = \frac{1}{10^2} = 10^{-2}$	centi-	c
$0{\cdot}001 = \frac{1}{10^3} = 10^{-3}$	milli-	m
$0{\cdot}000001 = \frac{1}{10^6} = 10^{-6}$	micro-	μ

Note on SI Units

The tables opposite list the units used in this book. They conform to the SI (Système Internationale) scheme which is based on the three fundamental units: *kilogramme* (kg) for mass, *metre* (m) for length and *second* (s) for time. The units of other quantities are derived from these. For example, areas are measured in *square metres*, denoted by m^2, and volumes in *cubic metres* (m^3). Velocity is measured in *metres per second* (m/s) and acceleration in *metres per second per second* (m/s^2).

Force and stress

The units of force may be derived from the relationship force = mass × acceleration and, substituting the SI units of mass and acceleration, we obtain

$$\text{unit force} = 1\ \text{kg} \times 1\ \text{m/s}^2 = 1\ \text{kg-m/s}^2$$

Force is a quantity which occurs so frequently that this unit has been given a separate name, the *newton*, symbol N.

In most questions in this book, loads and forces are given in newtons but there are a few examples in which the load acting is the weight of a body whose mass is given. Suppose, for example, a beam supports a body whose mass is 200 kg, the gravitational acceleration being 9·81 m/s^2. Its weight—and therefore the load on the beam—is the product of its mass and the acceleration due to gravity. Thus:

$$\begin{aligned}\text{Load} &= 200\ \text{kg} \times 9{\cdot}81\ \text{m/s}^2 \\ &= 1962\ \text{kg-m/s}^2 \text{ or } 1962\ \text{N}\end{aligned}$$

Stress is the ratio force/area and its units are therefore *newtons per square metre*, abbreviated to N/m^2. The name *pascal* has been adopted for this unit but N/m^2 has the advantage of emphasizing the relationship between stress, force and area. Pressure has the same units as stress (N/m^2); practical values are sometimes given in terms of the unit *bar* which is 100000 N/m^2 and is roughly equal to atmospheric pressure at sea-level.

Work, energy and power

Work and energy have the units of force × distance, i.e. *newton metres* (N-m). This unit has been given the name *joule*, denoted by J. Moment or torque is also the product of force and length with the same units, N-m (but not the joule). Power is the rate of doing work and is therefore expressed in *joules per second* (J/s) a unit which has the name *watt* (W). Thus 1 watt is 1 newton metre per second.

Multiples of units

Practical values of stress and other physical quantities often lead to very large or very small numbers and there are two methods of writing values concisely. One is to express the numbers in terms of powers of 10 such as 10^3 (a thousand) or 10^6 (a million). Alternatively, the units themselves can be modified by similar factors using prefixes. For example 10^3 is denoted by kilo- (k) and 10^6 by mega- (M).

The *preferred* multiples are those in which the power of 10 is a multiple of 3, e.g. 10^6, 10^3, 10^{-3}, etc. The multiples used in this book are listed on page viii.

To illustrate the notation, suppose the stress in a steel bar is 84 500 000 N/m² and the modulus of elasticity for the material is 204 000 000 000 N/m². Then,

$$\text{Stress} = 84\,500\,000\ \text{N/m}^2 = 84{\cdot}5 \times 10^6\ \text{N/m}^2 \quad \text{or } 84{\cdot}5\ \text{MN/m}^2$$

$$\begin{aligned}\text{Modulus of elasticity} &= 204\,000\,000\,000\ \text{N/m}^2\\ &= 204 \times 10^9\ \text{N/m}^2\\ &\quad \text{or } 204\ \text{GN/m}^2\end{aligned}$$

In some books stress is expressed in *newtons per square millimetre* (N/mm²) and numerical values will then be the same as in MN/m².

A special problem arises with the quantity *second moment of area.* In many examples its value in m^4 leads to very small numbers but in mm^4 it becomes very large. It is often convenient to use cm^4 and it should be noted that $1\ \text{m}^4 = 10^8\ \text{cm}^4$. In the same way, areas are often expressed in *square centimetres* (cm²). Apart from these two quantities the prefix centi- has been avoided in this book.

Numerical examples

It is recommended that, in solving numerical problems, the prefixes to units are replaced by factors which are powers of 10. At the end of the working the prefix form can be used for the answer.

Example. The central deflection δ of a centrally-loaded uniform beam is given by $\delta = WL^3/48\,EI$. Calculate δ when $W = 144$ kN, $L = 4$ m, $E = 200$ GN/m² and $I = 8000$ cm⁴. The work may be set out as follows:

$$
\begin{aligned}
\delta &= \frac{144 \text{ kN} \times (4 \text{ m})^3}{48 \times 200 \text{ GN/m}^2 \times 8000 \text{ cm}^4} \\
&= \frac{(144 \times 10^3 \text{N}) \times (4 \text{ m})^3}{48 \times (200 \times 10^9 \text{ N/m}^2) \times (8000 \times 10^{-8} \text{ m}^4)} \\
&= \frac{144 \times 4^3}{48 \times 200 \times 8000} \times \frac{10^3}{10^9 \times 10^{-8}} \frac{\text{N} \times \text{m}^3}{(\text{N/m}^2) \times \text{m}^4} \\
&= \frac{12}{100 \times 1000} \times \frac{10^3 \times 10^8}{10^9} \text{ m} \\
&= 12 \times 10^{-3} \text{ m} \\
&= 12 \text{ mm}
\end{aligned}
$$

This layout may seem elaborate but it helps to avoid errors and it can be shortened as confidence in the method is gained.

Chapter 1

Simple Stress and Strain; Elasticity

The stress, strain, and proportionality between them for direct or for shear forces acting on a body are defined as follows:

For a direct force (tension or compression)

$$\text{Stress } \sigma = \frac{\text{force } P}{\text{area } A}$$

$$\text{Strain } \epsilon = \frac{\text{change in length}}{\text{original length}}$$

$$\text{Young's Modulus of Elasticity } E = \frac{\text{direct stress } \sigma}{\text{direct strain } \epsilon}$$

For a shear force

$$\text{Stress } \tau = \frac{\text{force } F}{\text{area resisting shear } A}$$

and

Shear strain γ = angular displacement (in radians) produced by the shear stress.

$$\text{Shear modulus or modulus of rigidity } G = \frac{\text{shear stress } \tau}{\text{shear strain } \gamma}$$

"Free" temperature expansion of a bar $= \alpha l t$

where

α = coefficient of linear expansion,
l = length of the bar,
t° = rise in temperature.

For a fall in temperature of $t°$ the same expression will give the "free" contraction of the bar.

WORKED EXAMPLES

1.1. Explain the terms "stress" and "strain" as applied to a bar in tension. What is the relation between these two quantities? A steel rod, 25 mm dia. and 6 m long, extends 6 mm under a pull of 100 kN. Calculate the stress and strain in the rod.

Solution. A bar in tension is one which is subjected to a pair of equal and opposite forces which tend to stretch it (Fig. 1.1). At any section, such as XX, there must be internal forces in the material which balance the applied force P.

The *stress* is defined as the total load (P) on the section divided by the cross-sectional area. Stress is usually denoted by σ and, if A is the cross-sectional area

$$\text{Stress } \sigma = \frac{\text{force } P}{\text{area } A}$$

(The same definition applies when the external forces are compressive, i.e. when they tend to shorten the bar.)

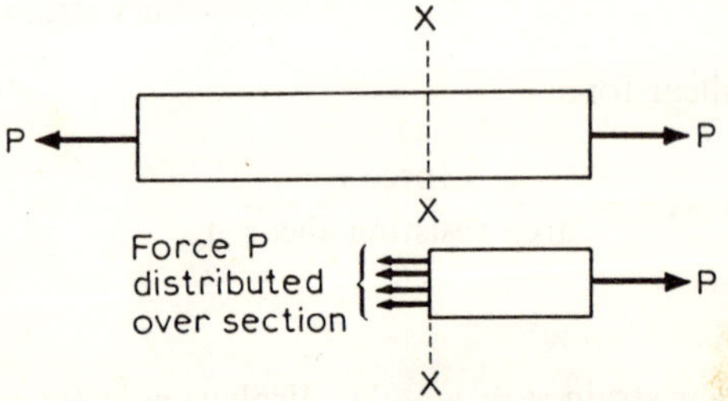

Fig. 1.1

The SI unit of area is the *square metre* (abbreviated to m^2) but for many structural and machine components the square centimetre (cm^2) or square millimetre (mm^2) is more convenient. Note that $1\ m^2 = 10^4\ cm^2 = 10^6\ mm^2$.

The SI unit of force is the *newton* (N) but for many practical purposes kilonewtons (kN) and meganewtons (MN) are used. Hence the units of stress are *newtons per square metre* (N/m^2) and practical values are usually expressed in kN/m^2 or MN/m^2. (N/mm^2 is used as an alternative to MN/m^2.) The name *pascal* (denoted by Pa) has been given to the unit N/m^2. Stress has the same units as pressure which is sometimes measured in *bars*. A bar is defined as $10^5\ N/m^2$.

The *strain* in a bar in tension is defined as the extension of the bar divided by its original length. It is denoted by ϵ and thus

$$\text{Strain } \epsilon = \frac{\text{increase in length}}{\text{original length}}$$

Since strain is a ratio of two lengths, it has no units.

Stress and strain are related by Hooke's Law, which states that *stress is proportional to strain*. This is almost exactly true for many engineering materials within a certain limit of stress called the *limit of proportionality*. Unless otherwise stated, it should be assumed in examples that the limit of proportionality is not exceeded and that Hooke's Law applies.

For the steel rod in the question,

$$\begin{aligned}\text{Cross-sectional area} &= \tfrac{1}{4}\pi \times (\text{diameter})^2\\ &= \tfrac{1}{4}\pi \times (25 \text{ mm})^2 = 491 \text{ mm}^2\end{aligned}$$

$$\begin{aligned}\text{Stress} &= \frac{\text{force or load}}{\text{cross-sectional area}}\\ &= \frac{100 \text{ kN}}{491 \text{ mm}^2} = \frac{100 \times 10^3 \text{ N}}{491 \times 10^{-6} \text{ m}^2}\\ &= 204 \times 10^6 \text{ N/m}^2\\ &= 204 \text{ MN/m}^2 \qquad (Ans)\end{aligned}$$

$$\begin{aligned}\text{Strain} &= \frac{\text{increase in length}}{\text{original length}}\\ &= \frac{6 \text{ mm}}{6 \text{ m}} = \frac{6 \times 10^{-3} \text{ m}}{6 \text{ m}} = 0{\cdot}001 \qquad (Ans)\end{aligned}$$

1.2. What is an "elastic material"? Define Young's modulus of elasticity. A load of 2 kN is to be raised at the end of a steel wire. If the stress in the wire must not exceed 80 MN/m² what is the minimum diameter required? What will be the extension of a 3 m length of the wire in this case? (Take $E = 206$ GN/m².)

Solution. An *elastic material* may be defined as one which obeys Hooke's Law and in which the strain disappears when the stress causing it is removed (see also page 430). Within the limit of proportionality, many engineering materials exhibit almost perfect elasticity. (The term "elastic limit" is often used to mean "limit of proportionality" though a distinction can be made.) Since stress is proportional to strain (Hooke's Law) then stress divided by strain is a constant. For tensile (and compressive) stresses this constant is called *Young's Modulus of Elasticity* and is denoted by E, that is

$$\text{Young's modulus } E = \frac{\text{stress } \sigma}{\text{strain } \epsilon}$$

Strain has no units and therefore Young's modulus has the same units as stress, namely newtons per square metre. In these units many engineering materials have very high numerical values of Young's modulus and it is convenient to express E in giganewtons per square metre (GN/m^2).

Typical values of E are

for steel, $206 \times 10^9\ N/m^2 = 206\ GN/m^2$
for copper, $103 \times 10^9\ N/m^2 = 103\ GN/m^2$
for timber, $10 \times 10^9\ N/m^2 = 10\ GN/m^2$

(In all examples where it is required, the value of E will be given.)

For the wire in question, since stress = load/area, then

$$\text{Minimum area required} = \frac{\text{load}}{\text{stress}} = \frac{2\ \text{kN}}{80\ \text{MN/m}^2}$$
$$= \frac{2 \times 10^3\ \text{N}}{80 \times 10^6\ \text{N/m}^2}$$
$$= 25 \times 10^{-6}\ \text{m}^2$$

But the area $= \frac{1}{4}\pi \times (\text{diameter})^2$, or $(\text{diameter})^2 = \frac{4}{\pi} \times \text{area}$.

Therefore

$$\text{Minimum diameter} = \sqrt{\left(\frac{4}{\pi} \times \text{area}\right)}$$
$$= \sqrt{\left(\frac{4}{\pi} \times 25 \times 10^{-6}\right)\ \text{m}^2} = 0{\cdot}00565\ \text{m}$$
$$= 5{\cdot}65\ \text{mm} \qquad (\textit{Ans})$$

Also

$$E = \frac{\text{stress}}{\text{strain}} = \text{stress} \div \frac{\text{extension}}{\text{original length}}$$
$$= \frac{\text{stress} \times \text{original length}}{\text{extension}}$$

Rearranging,

$$\text{Extension} = \frac{\text{stress} \times \text{original length}}{E}$$

$$= \frac{80\ \text{MN/m}^2 \times 3\ \text{m}}{206\ \text{GN/m}^2}$$

$$= \frac{80 \times 10^6\ \text{N/m}^2 \times 3\ \text{m}}{206 \times 10^9\ \text{N/m}^2}$$

$$= 0{\cdot}001\,17\ \text{m} = 1{\cdot}17\ \text{mm} \qquad (Ans)$$

1.3. The round bar shown in Fig. 1.2 is subjected to a tensile load of 150 kN. What must be the diameter of the middle portion if the stress there is to be 215 MN/m²?

What must be the length of the middle portion if the total extension of the bar under the given load is to be 0·2 mm? Take $E = 206$ GN/m².

Solution

Area of middle portion = load/stress

$$= \frac{150\ \text{kN}}{215\ \text{MN/m}^2} = \frac{150 \times 10^3\ \text{N}}{215 \times 10^6\ \text{N/m}^2}$$

$$= 0{\cdot}698 \times 10^{-3}\ \text{m}^2$$

Required diameter = $\sqrt{[(4/\pi) \times \text{area}]}$

$$= \sqrt{\left(\frac{4}{\pi} \times 0{\cdot}698 \times 10^{-3}\right)\ \text{m}^2}$$

$$= 0{\cdot}029\,8\ \text{m} = 29{\cdot}8\ \text{mm} \qquad (Ans)$$

If l m is the length of the middle portion, then $(0{\cdot}25 - l)$ m is the total length of the 50 mm diameter portions.

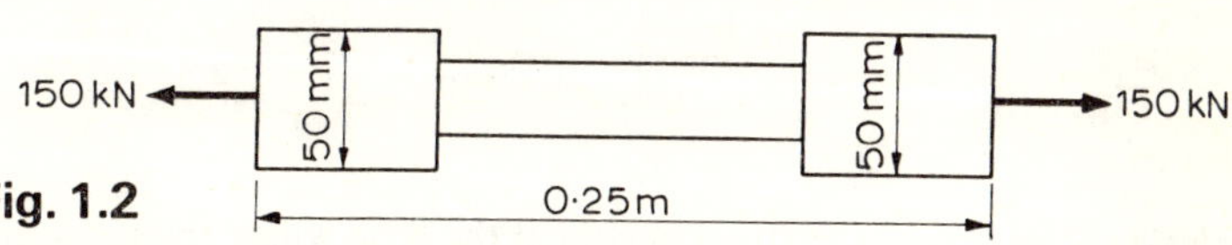

Fig. 1.2

Extension of the middle portion = $l \times$ strain

$$= l \times \frac{\text{stress}}{E} = l\ \text{m} \times \frac{215\ \text{MN/m}^2}{206\ \text{GN/m}^2}$$

$$= l\ \text{m} \times \frac{215 \times 10^6\ \text{N/m}^2}{206 \times 10^9\ \text{N/m}^2} = 1{\cdot}04\, l \times 10^{-3}\ \text{m}$$

Total extension of the two end portions

$$= (0{\cdot}25 - l) \times \frac{\text{stress}}{E} = (0{\cdot}25 - l) \times \frac{\text{load}}{\text{area} \times E}$$

$$= \frac{(0{\cdot}25 - l)\text{ m} \times 150 \times 10^3\text{ N}}{\frac{1}{4}\pi \times (5 \times 10^{-2}\text{ m})^2 \times 206 \times 10^9\text{ N/m}^2}$$

$$= 0{\cdot}371\,(0{\cdot}25 - l) \times 10^{-3}\text{ m}$$

Total extension of bar = extension of middle portion + extension of two end portions

or

$$0{\cdot}0002\text{ m} = 1{\cdot}04l \times 10^{-3}\text{ m} + 0{\cdot}371\,(0{\cdot}25 - l) \times 10^{-3}\text{ m}$$

from which

$$0{\cdot}669l = 0{\cdot}2 - 0{\cdot}093 = 0{\cdot}107$$

$$l = 0{\cdot}160\text{ m}$$

The middle portion of the bar should be 160 mm long. (*Ans*)

1.4. A cast-iron column of the section shown in Fig. 1.3 is 2 m high and supports a load of 20 kN, in addition to its own weight. What is the maximum compressive stress in the column? The density of cast iron is 7200 kg/m³. Take 1 kgf = 9·81 N.

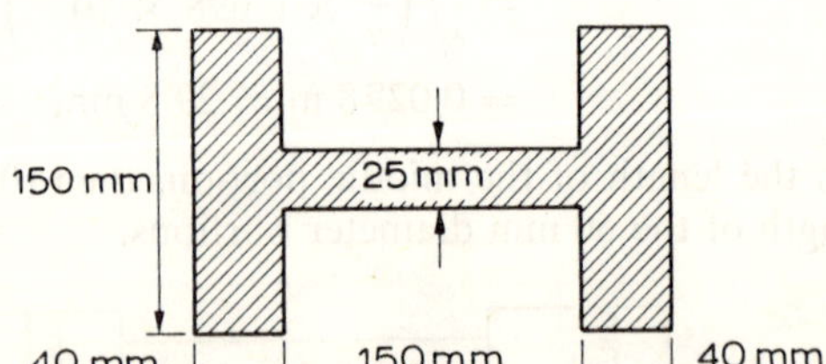

Fig. 1.3

Solution. The maximum stress occurs at the base where the total load of the section is the sum of the external load and the weight of the column.

Cross-sectional area = area of web + area of two flanges

$$= (0{\cdot}15 \times 0{\cdot}025)\text{ m}^2 + 2(0{\cdot}04 \times 0{\cdot}15)\text{ m}^2$$

$$= 15{\cdot}75 \times 10^{-3}\text{ m}^2$$

Volume of the column = cross-sectional area × height

$= 15{\cdot}75 \times 10^{-3}\ \text{m}^2 \times 2\ \text{m}$

$= 0{\cdot}0315\ \text{m}^3$

Mass of the column = volume × density

$= 0{\cdot}0315\ \text{m}^3 \times 7200\ \text{kg/m}^3$

$= 227\ \text{kg}$

The weight of one kilogramme is one kilogramme-force (1 kgf) and with the conversion factor given in the question

Weight of the column $= 227 \times 9{\cdot}81 = 2240\ \text{N} = 2{\cdot}24\ \text{kN}$

Total load on base section $= 2{\cdot}24\ \text{kN} + 20\ \text{kN}$

$= 22{\cdot}24\ \text{kN}$

Maximum compressive stress = total load/area

$$= \frac{22{\cdot}24\ \text{kN}}{0{\cdot}01575\ \text{m}^2} = 1{\cdot}41 \times 10^6\ \text{N/m}^2$$

$= 1{\cdot}41\ \text{MN/m}^2$ (*Ans*)

1.5. A rectangular base-plate is fixed at each of its four corners by a 20 mm bolt and nut as shown in Fig. 1.4. The plate rests on washers 22 mm internal diameter and 50 mm external diameter. The upper washers between the nuts and the plate are 22 mm internal diameter and 40 mm external diameter.

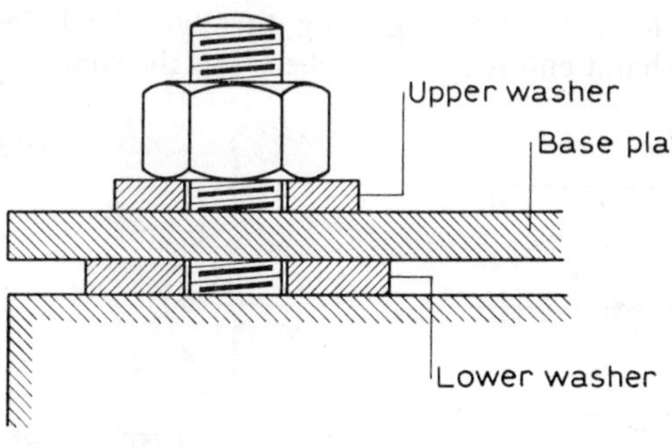

Fig. 1.4

If the base-plate carries a load of 160 kN (including its own weight) which is equally distributed between the four corners, find the stress (assumed uniform) in the lower washers before the nuts are tightened. What will be the stresses in the upper and lower washers when the nuts are tightened until there is a tension of 5 kN in each bolt?

Solution

$$\text{Area of lower washer} = \tfrac{1}{4}\pi[(50\text{ mm})^2 - (22\text{ mm})^2] = 1\,582\text{ mm}^2$$

Stress in lower washer due to load of 40 kN

$$= \text{load/area} = \frac{40 \times 10^3\text{ N}}{1\,582 \times 10^{-6}\text{ m}^2} = 25{\cdot}3\text{ MN/m}^2 \qquad (Ans)$$

After tightening, there is a compressive load in the upper washers equal to the tensile load in the bolt.

$$\text{Area of upper washer} = \tfrac{1}{4}\pi[(40\text{ mm})^2 - (22\text{ mm})^2] = 877\text{ mm}^2$$

$$\text{Stress in upper washer} = \text{load/area} = \frac{5\text{ kN}}{877\text{ mm}^2} = 5{\cdot}7\text{ MN/m}^2 \qquad (Ans)$$

The lower washer is also subjected to this compressive load, so that the total compressive load acting on it is 45 kN.

$$\text{Stress in lower washers} = \frac{45\text{ kN}}{1\,582\text{ mm}^2} = 28{\cdot}4\text{ MN/m}^2 \qquad (Ans)$$

1.6. Define "shear stress", "shear strain" and "modulus of rigidity". Calculate the shear stress in the tension member shown in Fig. 1.5 which is loaded by pins passing through the holes. The force at the right-hand end is trying to shear out the shaded part.

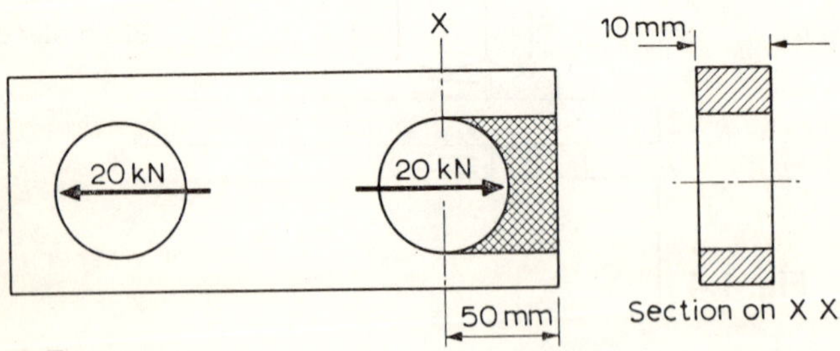

Fig. 1.5

Solution. Consider a rectangular block ABCD rigidly fixed at its base AD as shown in Fig. 1.6, and subjected to a force F acting along its top face.

At any section, such as EE, there is a tendency for sliding to take place between the upper and lower parts. This is resisted by internal forces in the material at the section.

The *shear stress* at any section, such as EE, is the shear force (F) divided by the area of the section, and is denoted by τ, i.e.

$$\text{Shear stress } (\tau) = \frac{\text{shear force } (F)}{\text{area } (A)}$$

(The reader should note that the area resisting shear is parallel to the line of action of the force.)

Due to the shearing force (F) the block distorts to a position such as AB′C′D (the distortion is exaggerated in the diagram).

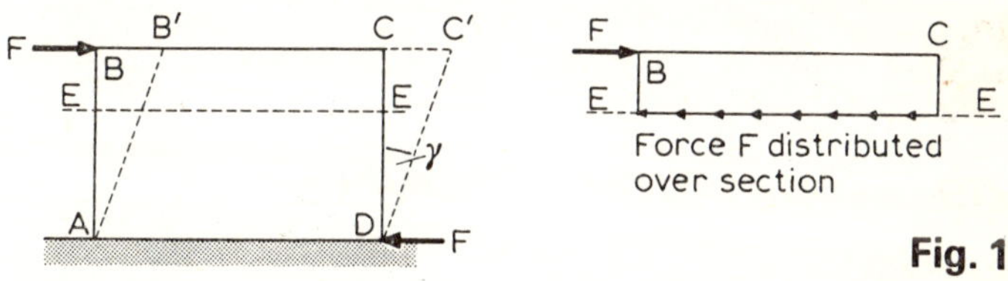

Fig. 1.6

The *shear strain* is the angle γ (in radians). Since this is always a small angle it is substantially equal to the ratio CC′/CD.

The *modulus of rigidity* (or shear modulus) is the ratio of shear stress to shear strain and it is denoted by G (sometimes C). Thus,

$$\text{Modulus of rigidity } (G) = \frac{\text{shear stress } (\tau)}{\text{shear strain } (\gamma)}$$

The units of shear stress and modulus of rigidity are N/m^2, the same as for direct stresses, and the multiples MN/m^2 and GN/m^2 are commonly used. The value of G is from one-third to two-fifths that of E for many metals.

In the problem given, the force of 20 kN tends to push out of the bar a block of the shape shown in Fig. 1.7 (p. 10). The areas in shear are two rectangles each 50 mm by 10 mm. Hence the total area resisting shear is $2 \times 500\ mm^2$, i.e. $1000\ mm^2$.

$$\begin{aligned} \text{Shear stress} &= \text{shear force/area} \\ &= \frac{20 \times 10^3\ N}{1000 \times 10^{-6}\ m^2} \\ &= 20\ MN/m^2 \qquad (Ans) \end{aligned}$$

1.7. A solid circular shaft and collar is forged in one piece (Fig. 1.8). If the maximum permissible shearing stress is 50 MN/m^2 what is the greatest compressive load P which may act on the shaft? What is the compressive stress in the shaft in this case?

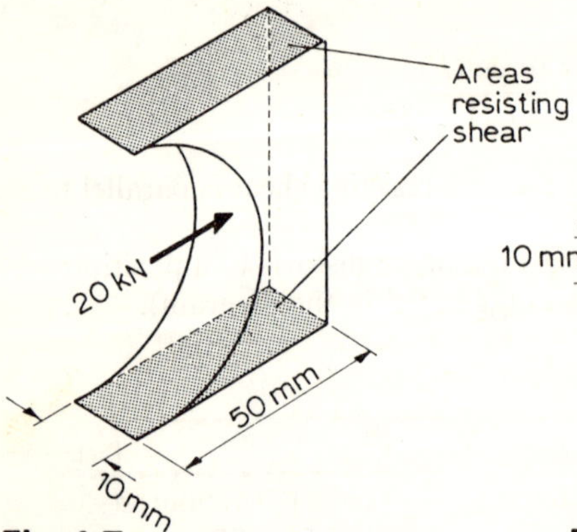

Fig. 1.7

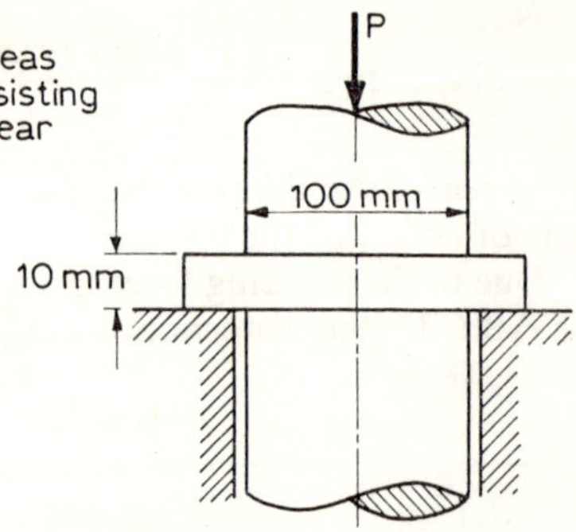

Fig. 1.8

Solution. Since the collar tends to slide along the shaft, the area resisting shear = (circumference of shaft) × (thickness of collar), i.e.

$$(\pi \times 100)\ \text{mm} \times 10\ \text{mm} = 3{\cdot}14 \times 10^{-3}\ \text{m}^2$$

$$\text{Force } P = \text{area} \times \text{shear stress} = (3{\cdot}14 \times 10^{-3}\ \text{m}^2) \times 50\ \text{MN/m}^2$$

$$= 157\ \text{kN} \qquad (Ans)$$

$$\text{Cross-sectional area of the shaft} = \tfrac{1}{4}\pi \times (100\ \text{mm})^2$$

$$= 7{\cdot}85 \times 10^{-3}\ \text{m}^2$$

Hence,

$$\text{Compressive stress in shaft} = \text{load } P/\text{area}$$

$$= \frac{157\ \text{kN}}{7{\cdot}85 \times 10^{-3}\ \text{m}^2}$$

$$= 20\ \text{MN/m}^2 \qquad (Ans)$$

1.8. Distinguish between single shear and double shear.

Two rails are joined by fishplates, as shown in Fig. 1.9, with one bolt to each rail at each joint. Due to temperature contraction, there is a tension in each rail of 40 kN. What is the shear stress in each bolt, diameter 50 mm, if

(*a*) two fishplates are used per joint (as shown),
(*b*) one fishplate is used per joint?

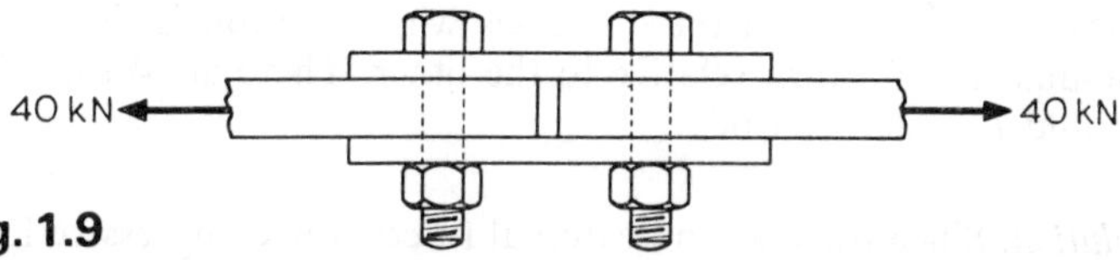

Fig. 1.9

Solution. A bolt, rivet or pin is said to be in single shear (Fig. 1.10a) when the tendency for shearing to take place occurs at one section only (such as AB). It is in double shear (Fig. 1.10b) when this tendency occurs at two sections simultaneously (such as XY and PQ). The area resisting shear is twice as much in the case of double shear as in the case of single shear, other things being equal.

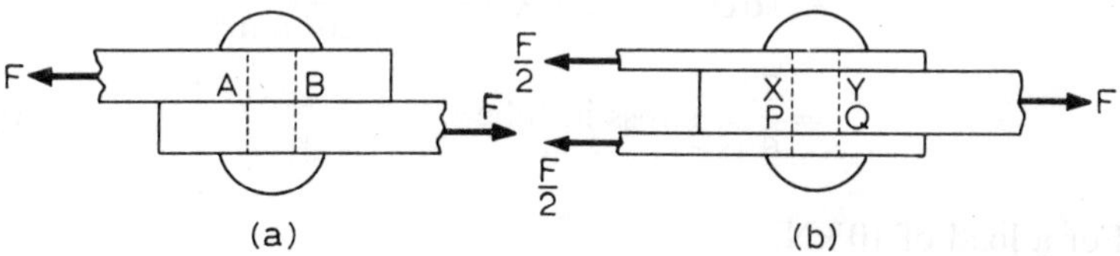

Fig. 1.10

(*a*) In this case each bolt transmits a force of 40 kN in double shear.

Total area resisting shear = 2 × cross-sectional area of bolt

$$= 2 \times \tfrac{1}{4}\pi \times (50\ \text{mm})^2 = 3{\cdot}93 \times 10^{-3}\ \text{m}^2$$

$$\text{Shear stress} = \text{shear force/area} = \frac{40\ \text{kN}}{3{\cdot}93 \times 10^{-3}\ \text{m}^2}$$

$$= 10{\cdot}2\ \text{MN/m}^2 \qquad (\textit{Ans})$$

(*b*) If there is only one fishplate per joint, the bolts are in single shear and the area resisting shear is the cross-sectional area, i.e. $1{\cdot}96 \times 10^{-3}\ \text{m}^2$. Hence

$$\text{Shear stress in bolts} = \frac{40\ \text{kN}}{1{\cdot}96 \times 10^{-3}\ \text{m}^2}$$

$$= 20{\cdot}4\ \text{MN/m}^2 \qquad (\textit{Ans})$$

1.9. A steel rod 20 mm diameter, passes centrally through a steel tube 30 mm internal diameter and 40 mm external diameter. The tube is 800 mm long and is closed by rigid washers of negligible thickness which are fastened by nuts threaded on the rod. The nuts are tightened until the compressive load on the tube is 10 kN. Calculate the stresses in the tube and rod.

Find the increase in these stresses when one nut is tightened by one-quarter of a turn relative to the other. There are 4 threads per cm and $E = 206\ \text{GN/m}^2$.

Solution. Since there are no external forces, the compressive load in the tube equals the tensile load in the rod. Hence

$$\text{Stress in rod} \times \text{Area of rod} = \text{Stress in tube} \times \text{Area of tube}$$

or

$$\begin{aligned}\text{Stress in rod} &= \text{stress in tube} \times \frac{\text{area of tube}}{\text{area of rod}} \\ &= \text{stress in tube} \times \frac{\frac{1}{4}\pi(40^2 - 30^2)\ \text{mm}^2}{\frac{1}{4}\pi \times 20^2\ \text{mm}^2} \\ &= \frac{7}{4} \times \text{stress in tube} \qquad \text{(i)}\end{aligned}$$

For a load of 10 kN,

$$\begin{aligned}\text{Compressive stress in tube} &= \text{load/area} \\ &= \frac{10\ \text{kN}}{\frac{1}{4}\pi(40^2 - 30^2)\ \text{mm}^2} \\ &= \frac{10 \times 10^3\ \text{N}}{\frac{1}{4}\pi(40^2 - 30^2) \times 10^{-6}\ \text{m}^2} \\ &= 18{\cdot}2\ \text{MN/m}^2 \qquad (\textit{Ans})\end{aligned}$$

By (i),

$$\begin{aligned}\text{Tensile stress in rod} &= 7/4 \times 18{\cdot}2\ \text{MN/m}^2 \\ &= 31{\cdot}8\ \text{MN/m}^2 \qquad (\textit{Ans})\end{aligned}$$

Let σ N/m² be the stress in tube due to the tightening of the nut. Then $7\sigma/4$ N/m² is the stress in rod due to the tightening of the nut.

Reduction in length of tube = length × (stress/E)

$$= \frac{0{\cdot}8 \text{ m} \times \sigma \text{ N/m}^2}{206 \times 10^9 \text{ N/m}^2} = \frac{0{\cdot}8\,\sigma}{206 \times 10^9} \text{ m}$$

$$\text{Extension of the rod} = \frac{0{\cdot}8 \text{ m} \times 7\sigma/4 \text{ N/m}^2}{206 \times 10^9 \text{ N/m}^2} = \frac{1{\cdot}4\sigma}{206 \times 10^9} \text{ m}$$

With 4 threads per cm, the pitch is 2·5 mm and, for one-quarter of a turn, the nut moves 2·5/4 mm. But the axial advance of the nut = contraction of tube + extension of rod.

i.e.

$$\frac{2{\cdot}5 \times 10^{-3}}{4} \text{ m} = \left(\frac{0{\cdot}8\sigma}{206 \times 10^9}\right) \text{m} + \left(\frac{1{\cdot}4\sigma}{206 \times 10^9}\right) \text{m}$$

Multiplying through by 206 × 10⁹ and rearranging,

$$2{\cdot}2\sigma = 206 \times (2{\cdot}5/4) \times 10^6 = 128{\cdot}8 \times 10^6$$

and therefore

$$\sigma = 58{\cdot}5 \times 10^6 \text{ MN/m}^2$$

Due to a quarter of a turn of the nut, therefore,

Stress in tube = 58·5 MN/m² (tensile) (*Ans*)

The corresponding stress in the rod is, by (i),

$$\frac{7}{4} \times 58{\cdot}5 \text{ MN/m}^2 = 102{\cdot}3 \text{ MN/m}^2 \text{ (compressive)} \qquad (\textit{Ans})$$

1.10. A weight of 300 kN is supported by a short concrete column 250 mm square. The column is strengthened by four steel bars in the corners of total cross-sectional area 50 cm². If the modulus of elasticity for steel is 15 times that for concrete, find the stresses in the steel and the concrete. If the stress in the concrete must not exceed 4 MN/m², what area of steel is required in order that the column may support a load of 600 kN?

Solution. Let σ_c and σ_s N/m² be the stresses in the concrete and steel respectively, and E_c and E_s the corresponding values of Young's modulus (also measured in N/m²).

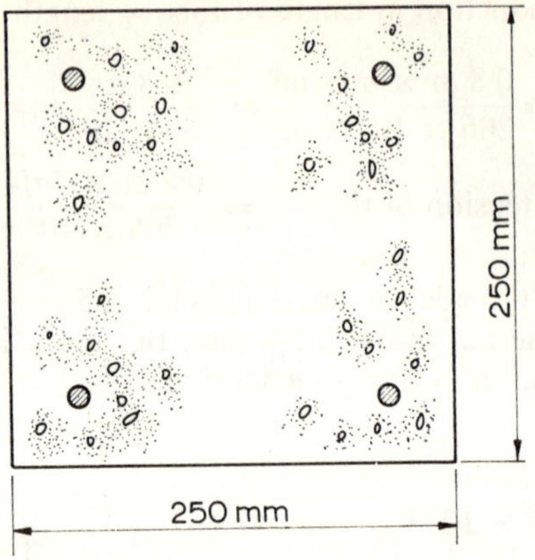

Fig. 1.11

Since the steel and concrete must shorten by the same amount, they suffer equal strains. Also, strain = stress/E and thus

$$\sigma_s/E_s = \sigma_c/E_c \quad \text{or} \quad \sigma_s = (E_s/E_c)\sigma_c = 15\sigma_c \qquad \text{(i)}$$

Area of concrete = 625 − 50 = 575 cm²

= 0·0575 m² (see Fig. 1.11)

Load in concrete + load in steel = total load

i.e.

(σ_c × area of concrete) + (σ_s × area of steel) = total load

or

$$0{\cdot}0575\sigma_c + 0{\cdot}005\sigma_s = 300 \times 10^3 \qquad \text{(ii)}$$

Solving the simultaneous equations (i) and (ii),

$$\sigma_c = 2{\cdot}27 \times 10^6 \quad \text{and} \quad \sigma_s = 34{\cdot}0 \times 10^6$$

Compressive stress in concrete = 2·27 MN/m² (*Ans*)

Compressive stress in steel = 34·0 MN/m² (*Ans*)

For the 600 kN load, let A_s m² be the required area of steel, then ($0{\cdot}0625 - A_s$) m² is the corresponding area of concrete.

Also, if 4 MN/m² is the stress in the concrete, then, by (i), the stress in the steel is 15 × 4 MN/m², i.e. 60 MN/m².

Again, load in concrete + load in steel = total load

i.e.

$$4 \times 10^6(0{\cdot}0625 - A_s) + 60 \times 10^6 A_s = 600 \times 10^3 \text{ from which}$$

$$56\,A_s = 0{\cdot}35 \quad \text{or} \quad A_s = 0{\cdot}00625 \text{ m}^2$$

For the 600 kN load,

$$\text{Area of steel required} = 6{\cdot}25 \times 10^{-3}\ \text{m}^2$$
$$= 62{\cdot}5\ \text{cm}^2 \qquad (Ans)$$

1.11. What is a "coefficient of linear expansion"? A railway is laid so that there is no stress in the rails at 20°C. Calculate the stress in the rails at −6°C if all contraction is prevented.

Take $E = 206\ \text{GN/m}^2$ and $\alpha = 12 \times 10^{-6}/°\text{C}$.

If, however, there is 6 mm allowance for contraction per rail, what is the stress at −6°C? The rails are 27 m long.

Solution. If a bar is heated it will expand, provided that it is free to do so. Similarly it will contract if cooled.

The *coefficient of linear expansion* is defined as the change in length per unit length per degree change in temperature (Celsius or Fahrenheit). It is denoted by α, and thus the expansion (or contraction) of a bar, length l, due to a rise (or fall) in temperature of $t°$ is

$$\alpha l t$$

If the ends of the bar are restrained so that this change in length is prevented, then a stress is induced in the bar. If the ends are rigidly fixed this stress corresponds to a change in length of the bar equal and opposite to that due to the temperature change, i.e.

$$\text{Stress} = \text{strain} \times E = \frac{\text{change in length}}{\text{original length}} \times E$$
$$= (\alpha l t / l) \times E = \alpha t E$$

For a fall in temperature this stress will be tensile, and for a rise in temperature, compressive.

Using the values in the question, the tensile stress in the rails, if all contraction is prevented, is

$$\alpha t E = 12 \times 10^{-6} \times [20 - (-6)] \times (206 \times 10^9\ \text{N/m}^2)$$
$$= 64{\cdot}3\ \text{MN/m}^2 \qquad (Ans)$$

The "free" contraction of one rail is

$$\alpha l t = 12 \times 10^{-6} \times 27\ \text{m} \times 26 = 8{\cdot}42 \times 10^{-3}\ \text{m}$$
$$= 8{\cdot}42\ \text{mm}$$

If there is an allowance of 6 mm then the stress must correspond to an extension of (8·42 − 6) mm, i.e. 2·42 mm. The stress in this case equals

$$\text{Strain} \times E = \frac{\text{extension}}{\text{original length}} \times E$$

$$= \left(\frac{2{\cdot}42 \times 10^{-3}\ \text{m}}{27\ \text{m}}\right) \times (206 \times 10^9\ \text{N/m}^2)$$

$$= 18{\cdot}48\ \text{MN/m}^2 \qquad (Ans)$$

1.12. A 12 mm diameter steel rod passes centrally through a copper tube 48 mm external and 36 mm internal diameter and 2·5 m long. The tube is closed at each end by 25 mm thick steel plates which are secured by nuts. The nuts are tightened until the copper tube is reduced in length to 2·499 5 m and the whole assembly is then raised in temperature by 60°C. Calculate the stresses in the copper and steel before and after the rise in temperature, assuming the thickness of the plates to remain unchanged.

E (steel) = 200 GN/m², *E* (copper) = 100 GN/m². Coefficients of expansion per °C for steel 12×10^{-6}, for copper $17{\cdot}5 \times 10^{-6}$.

(*U.L.*)

Solution. As in Example 1.9, since there are no external forces,

$$\text{Stress in rod} = \text{stress in tube} \times \frac{\text{area of tube}}{\text{area of rod}}$$

$$= \text{stress in tube} \times \frac{\frac{1}{4}\pi(48^2 - 36^2)\ \text{mm}^2}{\frac{1}{4}\pi \times 12^2\ \text{mm}^2}$$

$$= 7 \times \text{stress in tube} \qquad \text{(i)}$$

Before the temperature rise

$$\text{Compressive stress in copper} = \text{strain} \times E(\text{copper})$$

$$= \frac{\text{contraction}}{\text{original length}} \times E(\text{copper})$$

$$= \frac{(2{\cdot}5 - 2{\cdot}499\,5)\ \text{m}}{2{\cdot}5\ \text{m}} \times (100 \times 10^9\ \text{N/m}^2)$$

$$= 20\ \text{MN/m}^2 \qquad (Ans)$$

From (i), tensile stress in steel rod = 7 × 20 = 140 MN/m² (*Ans*)

Let σ N/m² be increase in compressive stress in copper due to temperature rise. Then, by (i), 7σ N/m² is the corresponding increase in tensile stress in the steel.

The "free" expansion of the copper

$$= \alpha lt = 17{\cdot}5 \times 10^{-6} \times 2{\cdot}5 \text{ m} \times 60$$
$$= 2{\cdot}63 \times 10^{-3} \text{ m}$$

and "free" expansion of the steel (which has a total length of 2·55 m)

$$= 12 \times 10^{-6} \times 2{\cdot}55 \text{ m} \times 60$$
$$= 1{\cdot}84 \times 10^{-3} \text{ m}$$

Due to the stresses induced by the temperature change,

$$\text{Contraction of copper} = \frac{l \times \text{stress}}{E}$$
$$= \frac{2{\cdot}5 \text{ m} \times \sigma \text{ N/m}^2}{100 \times 10^9 \text{ N/m}^2}$$
$$= \frac{2{\cdot}5\sigma}{100 \times 10^9} \text{ m}$$
$$\text{Extension of steel} = \frac{2{\cdot}55 \text{ m} \times 7\sigma \text{ N/m}^2}{200 \times 10^9 \text{ N/m}^2}$$
$$= \frac{17{\cdot}85\sigma}{200 \times 10^9} \text{ m}$$

Since the plates are rigid, the total change in length of the copper due to the temperature rise is the same as that for the steel. Hence ("free" expansion of copper) − (contraction of copper due to the stress σ) = ("free" expansion of steel) + (extension in steel due to the stress 7σ).

Substituting the values obtained,

$$(2{\cdot}63 \times 10^{-3} \text{ m}) - \left(\frac{2{\cdot}5\sigma}{100 \times 10^9} \text{ m}\right) = (1{\cdot}84 \times 10^{-3} \text{ m}) + \left(\frac{17{\cdot}85\sigma}{200 \times 10^9} \text{ m}\right)$$

or

$$2{\cdot}63 - 1{\cdot}84 = \left(\frac{17{\cdot}85\sigma}{200 \times 10^6}\right) + \left(\frac{2{\cdot}5\sigma}{100 \times 10^6}\right)$$

from which

$$\sigma = 6{\cdot}92 \times 10^6 \text{ N/m}^2$$

After the temperature rise, therefore,

$$\text{Total stress in copper} = (20 + 6{\cdot}92)\ \text{MN/m}^2$$
$$= 26{\cdot}92\ \text{MN/m}^2\ \text{(compressive)} \qquad (Ans)$$

$$\text{Total stress in steel} = 7 \times 26{\cdot}92\ \text{MN/m}^2$$
$$= 188{\cdot}4\ \text{MN/m}^2\ \text{(tensile)} \qquad (Ans)$$

1.13. A round bar, length L, tapers uniformly from radius r_1 at one end to radius r_2 at the other. Show that the extension produced by a tensile axial load P is

$$PL/\pi E r_1 r_2$$

If $r_2 = 2r_1$, compare this extension with that of a uniform cylindrical bar having a radius equal to the mean radius of the tapered bar.

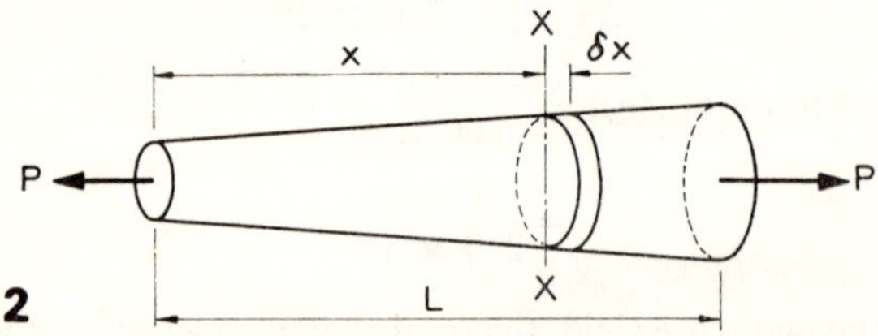

Fig. 1.12

Solution. Let r_1 be the radius at the smaller end. Then, at a cross-section XX, distance x from the smaller end (see Fig. 1.12), the radius is

$$r_1 + \frac{x}{L}(r_2 - r_1)$$

which may be written $r_1(1 + kx)$, where $k = (r_2 - r_1)/r_1 L$

$$\text{Stress at section XX} = \text{load/area}$$

$$= \frac{P}{\pi r_1^2(1 + kx)^2}$$

$$\text{Strain at this section} = \text{stress}/E$$

$$= \frac{P}{E\pi r_1^2(1 + kx)^2}$$

Thus, for a small length δx of the bar at this section the extension is $\delta x \times$ strain or

$$\frac{P}{E\pi r_1^2(1 + kx)^2}\,\delta x$$

By integration from $x = 0$ to $x = L$ we have, for the complete bar,

$$\begin{aligned}
\text{Extension} &= \int_0^L \frac{P}{E\pi r_1^2(1 + kx)^2}\,dx \\
&= \frac{P}{E\pi r_1^2}\int_0^L (1 + kx)^{-2}\,dx \\
&= \frac{P}{E\pi r_1^2}\left[\frac{(1 + kx)^{-1}}{-k}\right]_0^L \\
&= \frac{P}{E\pi r_1^2}\left[\frac{(1 + kL)^{-1}}{-k} - \frac{1}{(-k)}\right] \\
&= \frac{P}{E\pi r_1^2 k}\left(1 - \frac{1}{1 + kL}\right) \\
&= \frac{PL}{E\pi r_1^2(1 + kL)}
\end{aligned}$$

But $k = (r_2 - r_1)/r_1L$ and thus $1 + kL = r_2/r_1$ so that

$$\text{Extension} = \frac{PL}{E\pi r_1^2 \times (r_2/r_1)} = \frac{PL}{\pi E r_1 r_2} \text{ (as required)} \qquad (\textit{Ans})$$

If $r_2 = 2r_1$, this extension is $PL/2\pi E r_1^2$ and the radius of the uniform bar is

$$\tfrac{1}{2}(r_2 + r_1) = 3r_1/2$$

Thus

$$\begin{aligned}
\text{Extension of uniform bar} &= L \times \text{strain} = L \times \frac{\text{stress}}{E} \\
&= \frac{L}{E} \times \frac{P}{\text{area}} \\
&= \frac{L}{E} \times \frac{P}{\pi(\frac{3}{2}r_1)^2} \\
&= \frac{4PL}{9\pi E r_1^2}
\end{aligned}$$

Therefore, the ratio

$$\frac{\text{Extension of uniform bar}}{\text{Extension of tapered bar}} = \frac{4PL}{9\pi E r_1^2} \bigg/ \frac{PL}{2\pi E r_1^2}$$

$$= \frac{4 \times 2}{9} = \frac{8}{9} \qquad (Ans)$$

PROBLEMS

1. A turnbuckle connects two 30 mm diameter rods as shown. Calculate the tensile stress on the section AA (Fig. 1.13) and in the rods (neglecting the effects of the threads).
Answer. Stress on AA = 31·9 MN/m^2; stress in rods = 56·6 MN/m^2.

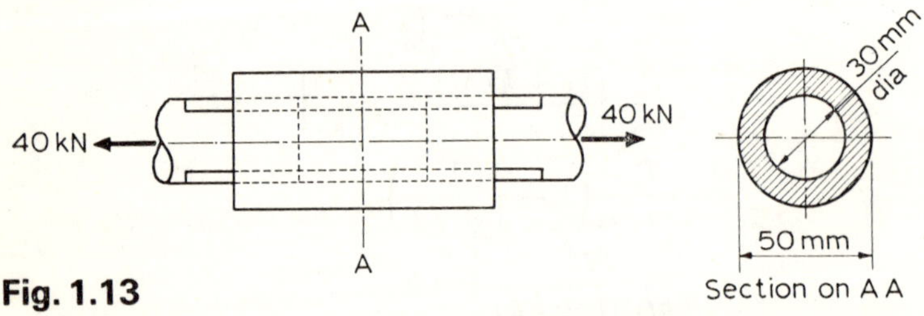

Fig. 1.13

2. The link shown in Fig. 1.14 is loaded in tension by a 25 mm diameter pin which passes through the hole. Calculate the tensile stresses on the sections XX and YY. If the pin is in double shear, what is the shearing stress in it?

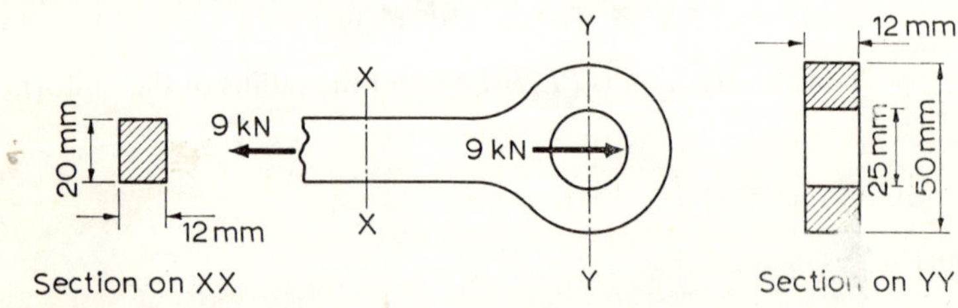

Fig. 1.14

Answer. Stress on XX = 37·5 MN/m^2; stress on YY = 30 MN/m^2; shear stress in pin = 9·17 MN/m^2.

3. A square rigid horizontal plate hangs from 4 vertical steel rods each 6 mm diameter and 6 m long, one at each corner. The plate carries a load, symmetrically placed, of 10 kN (including its own weight). Calculate the stress and extension of each rod. The rods are all rigidly fixed at their upper ends at exactly the same level. (E = 207 GN/m^2)
Answer. Stress in rods = 88·5 MN/m^2; extension of each rod = 2·56 mm.

4. If, in the previous question, two of the rods, diagonally opposite each other, are initially 1·5 mm shorter than the other two, calculate the stresses in the rods.

Hint. First calculate the loads necessary to extend the shorter rods to the same length as the others. The remaining load is shared equally.
Answer. Stress in shorter rods = 114·3 MN/m²; stress in longer rods = 62·5 MN/m².

5. Two pieces of wood are joined end to end by a single close-fitting bolt, 25 mm diameter. Using the dimensions in Fig. 1.15, calculate the (average) shearing stresses in the wood and the bolt when the joint transmits a load of 14 kN. (Use the method of Example 1.6.)

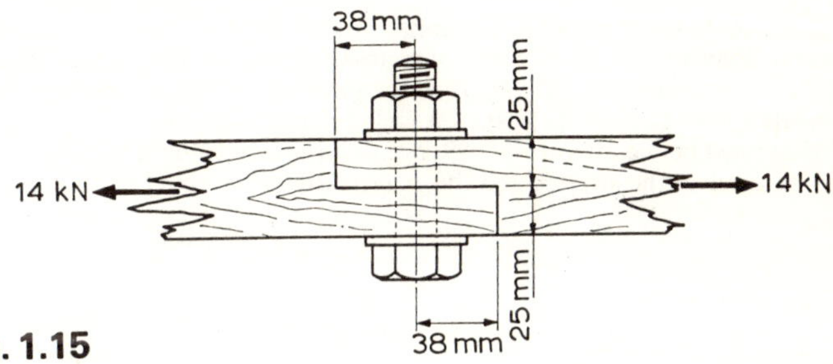

Fig. 1.15

Answer. Stress in wood = 7·37 MN/m²; stress in bolt = 28·52 MN/m².

6. A compound tube is made by shrinking one thin tube on to another. A_1 and A_2 are the cross-sectional areas of the inner and outer tubes, and E_1 and E_2 are the corresponding values of Young's modulus. Show that for a given tensile load, the extension of the compound tube is equal to that of a single tube of the same length and (total) cross-sectional area, but having a Young's modulus of

$$(E_1A_1 + E_2A_2)/(A_1 + A_2).$$

A tube of this kind, 300 mm long, is formed by shrinking an aluminium alloy tube on to a steel tube, the final dimensions being—Internal diameter 47 mm, common diameter 50 mm, external diameter 60 mm.

Calculate the stress in the inner and outer tubes and the extension when the compound tube is subjected to an axial load of 180 kN. E for steel = 207 GN/m², E for alloy = 69 GN/m².

Hint. Use the principles of Example 1.10. In the first part of the question show that the strain in each case is load/$(E_1A_1 + E_2A_2)$.
Answer. Stress in outer tube = 116 MN/m²; stress in inner tube = 348 MN/m²; extension = 0·505 mm.

7. Two steel rods, each 50 mm diameter, are joined end to end by means of a turnbuckle (Fig. 1.16). The other end of each rod is rigidly fixed and there is

Fig. 1.16

initially a small tension in the rods. If the effective length of each rod is 4 m, calculate the increase in this tension when the turnbuckle is tightened one-quarter of a turn.

The thread on each rod has a pitch of 5 mm and the extension of the turn-buckle may be neglected. $E = 206\ GN/m^2$.

If $\alpha = 11{\cdot}5 \times 10^{-6}/°C$, what rise in temperature would nullify this increase?

Answer. Increase in tension = 126·4 kN. This would be nullified by a rise of 27·2°C.

8. Solve Problem 7 if one rod is 50 mm diameter and has a pitch of 5 mm, but the other is 76 mm diameter and has a pitch of 6 mm.

Answer. Increase in tension = 193 kN. This would be nullified by a rise of 29·9°C.

9. A short, hollow, cast-iron column, 250 mm external diameter and 200 mm internal diameter, is filled with concrete as shown in Fig. 1.17. The column carries a total load of 310 kN. If E for cast-iron is six times E for concrete, calculate the stresses in the cast-iron and the concrete.

What must be the internal diameter of the cast-iron column if a load of 400 kN is to be carried, the stresses and the external diameter being unchanged?

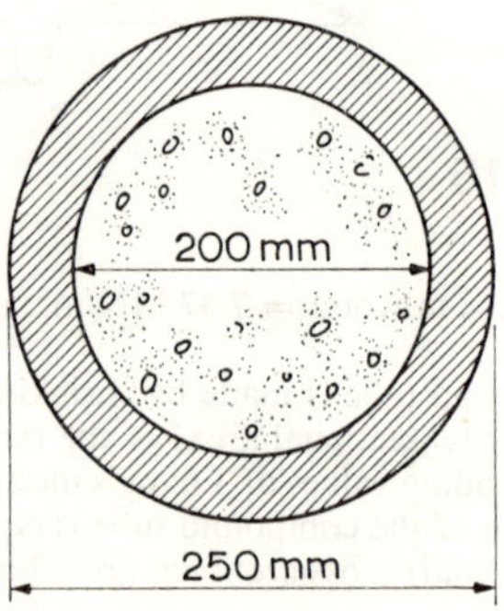

Fig. 1.17

Answer. Stress in cast iron = 13·54 MN/m^2; stress in concrete = 2·26 MN/m^2. For the 400 kN load, the internal diameter of the cast-iron must be 173 mm.

10. Two copper plates, each 25 mm thick, are fastened together by a steel bolt whose thread has a pitch of 2 mm. Calculate the increase in stress in the bolt if the nut is turned an additional 10°, assuming that the bearing surface of the head and nut is sufficient to make the compressive strain in the copper negligible. Take E for steel as 206 GN/m^2.

What rise in temperature would produce the same increase in stress (again neglecting the compressive strain in the copper)?

α (steel) = $11{\cdot}5 \times 10^{-6}/°C$; α (copper) = $17 \times 10^{-6}/°C$.

Answer. Increase in stress = 228·9 MN/m^2. The same increase would be caused by a rise of 202°C.

11. A 25 mm diameter rod passes centrally through a tube 65 mm external diameter, 40 mm internal diameter, and 1·25 m long. The tube is closed by rigid washers of negligible thickness and nuts threaded on the rod. If the rod and tube are of steel ($E = 206\ GN/m^2$), find the stresses in each when the nuts are tightened until the tube is reduced in length by 0·125 mm.

Answer. Stress in tube = 20·6 MN/m² (compressive); stress in rod = 86·5 MN/m² (tensile).

12. Find the increase in the stresses of the tube and rod of Problem 11 if one nut is tightened by $\frac{1}{8}$ of a turn relative to the other, the thread having a pitch of 3 mm.
Answer. Increase in stress in the tube = 47·5 MN/m²; increase in stress in the rod = 199·6 MN/m².

13. Solve problem 12 if the rod is of steel but the tube is of copper.
E (steel) = 206 GN/m², E (copper) = 103 GN/m².
Answer. Increase in stress in tube = 39·9 MN/m²; increase in stress in rod = 167·4 MN/m².

14. A flat bar of steel 26 mm wide and 6 mm thick, is placed between two aluminium alloy bars, each 26 mm wide and 10 mm thick, to form a composite bar 26 mm square, as shown in Fig. 1.18. The three bars are fastened together at their

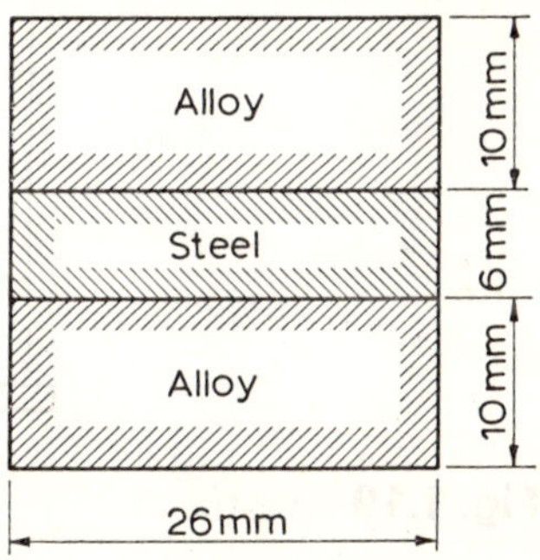

Fig. 1.18

ends when the temperature of each is 10°C. Find the stress in each when the temperature of the whole assembly is raised to 50°C. If at the new temperature a tensile load of 20 kN is applied to the composite bar, what are the final stresses in the steel and the alloy?

E (steel) = 206 GN/m²
E (alloy) = 69 GN/m²
α (steel) = $11{\cdot}5 \times 10^{-6}$/°C
α (alloy) = 23×10^{-6}/°C

Answer. Before the load is applied, stress in the steel = 50 MN/m² (tensile) and stress in the alloy = 15 MN/m² (compressive).

After the load is applied, stress in the steel = 110·6 MN/m² (tensile) and stress in the alloy = 5·3 MN/m² (tensile).

15. If, in Problem 14 the tensile load of 20 kN continues to act while the assembly is heated further, find the temperature at which there will be no stress in the alloy.
Answer. There is no stress in the alloy at 64·1°C.

16. A compound bar 1·5 m long is made up of two pieces of metal; one of steel and the other of copper. The cross-sectional area of the steel is 45 cm² while that of the copper is 32 cm². Both pieces are 1·5 m long and are rigidly connected together at both ends. The temperature of the bar is now raised 250°C. The bar is restrained against bending.

Calculate the stresses in the two materials.

Coefficient of linear expansion of steel = 0·000012 per °C.

Coefficient of linear expansion of copper = 0·0000175 per °C

E for steel = 206 GN/m^2

E for copper = 107 GN/m^2 (*I.Struct.E*)

Answer. Stress in steel = 76·4 MN/m^2 (tensile); stress in copper = 107·4 MN/m^2 (compressive).

17. Two vertical wires are suspended at a distance apart of 500 mm as shown in Fig. 1.19. Their upper ends are firmly secured and their lower ends support

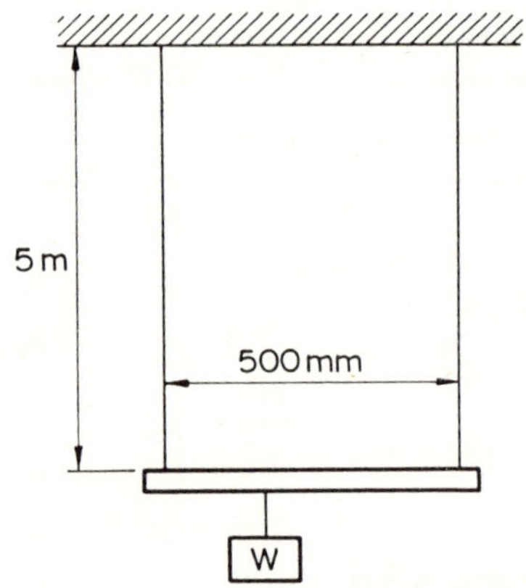

Fig. 1.19

a rigid horizontal bar which carries a load W. The left-hand wire has a diameter of 2 mm and is made of copper and the right-hand wire has a diameter of 1 mm and is made of steel. Both wires initially are exactly 5 m long.

(*a*) Determine the position of the line of action of W, if due to W, both wires extend by the same amount.

(*b*) Determine the load, stress and extension for each wire for $W = 200$ N.

Neglect the weight of the bar and take E for steel = 207 GN/m^2, and E for copper = 107 GN/m^2. (*U.L.*)

Answer. (*a*) 163 mm from the left-hand wire.

(*b*) for copper, load = 134·8 N, stress = 42·8 MN/m^2; for steel, load = 65·2 N, stress = 83 MN/m^2; extension = 2 mm for both.

18. A steel rod 32 mm diameter is fixed concentrically in a brass tube which has outside and inside diameters of 48 mm and 36 mm respectively. Both the rod and the tube are 400 mm long and their ends are level. The compound rod is held between two stops which are exactly 400 mm apart and the temperature of the bar is then raised 80°C.

(*a*) Find the stresses in the rod and tube if the distance between the stops (i) remains constant, (ii) is increased by 0·25 mm.

(*b*) Find the increase in the distance between the stops if the force exerted between them is 80 kN.

Moduli of elasticity: steel, 207 GN/m^2; brass, 92·5 GN/m^2. Coefficients of expansion: steel 11·5 × 10^{-6} per °C; brass, 19 × 10^{-6} per °C. (*U.L.*)
Answer. (*a*) (i) stress in rod = 190·4 MN/m^2; stress in tube = 140·6 MN/m^2; (ii) stress in rod = 61·1 MN/m^2; stress in tube = 82·8 MN/m^2. (*b*) 0·308 mm.

19. (*a*) A bar of steel of rectangular section 30 mm × 6 mm is subjected to an axial pull of 36 kN when it is found that the extension measured on a length of 200 mm is 0·194 mm.

(*b*) The steel bar is then placed between two aluminium alloy bars each of rectangular section 30 mm × 8 mm to form a composite bar of rectangular section 30 mm wide and 22 mm thick. An axial pull of 72 kN is applied to the composite bar when it is found that the extension measured on a length of 200 mm is 0·205 mm.

Determine for (*a*) the value of E for steel and for (*b*) the stresses in the steel and the alloy and hence the value of E for the alloy. (*U.L.*)
Answer. (*a*) 206 GN/m^2. (*b*) 211 MN/m^2 and 70·7 MN/m^2; 69 GN/m^2.

20. A gradually applied load $W = 5$ kN is suspended by ropes as shown in Fig. 1.20a and b. In both cases the ropes have a cross-sectional area of 8 cm^2 and the value of E is 0·98 GN/m^2.

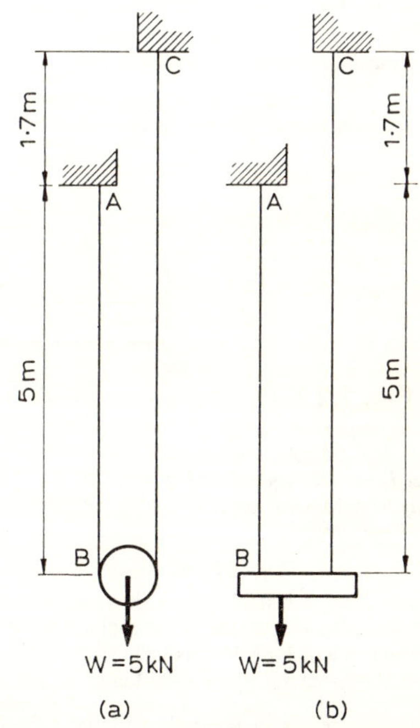

Fig. 1.20

In (*a*) the rope ABC is continuous and W is suspended from a small frictionless pulley. In (*b*) AB and CB are separate ropes joined to a block from which W is suspended in such a way that both ropes stretch by the same amount.

Find, for both (*a*) and (*b*), the stresses in the ropes and the downward movement of the pulley and of the block due to the gradual application of the load. (*U.L.*)

Answer. (*a*) 3·125 MN/m²; 18·65 mm. (*b*) 3·58 MN/m² and 2·67 MN/m²; 18·26 mm.

21. A compound bar consists of two bars of copper, each of rectangular section 30 mm × 5 mm and a steel bar of rectangular section 30 mm × 3 mm. The three bars are securely fixed together with the steel bar in the middle thus forming a compound bar of rectangular section 30 mm wide and 13 mm thick. The temperature of the compound bar is then raised through 140°C and at this temperature the bar is subjected to a steady axial pull.

Find (*a*) the stresses in the steel and copper bars due to the increase in temperature and (*b*) the total pull in the bar if the final stress in the steel is 150 MN/m².

For steel $E = 207$ GN/m²; $\alpha = 11{\cdot}5 \times 10^{-6}$/°C.

For copper $E = 107$ GN/m²; $\alpha = 17{\cdot}5 \times 10^{-6}$/°C. (*U.L.*)

Answer. (*a*) Stress in steel = 109·9 MN/m² (tensile); stress in copper = 33·1 MN/m² (compressive) (*b*) 9·82 kN.

22. A steel bar 330 mm long and 20 mm diameter, is turned down at one end to 16 mm diameter for a length of 100 mm and then placed between two stops as shown in Fig. 1.21. Determine the stresses in the two parts of the rod for the following separate changes of condition.

(*a*) The distance between the stops is reduced by 0·1 mm.
(*b*) The temperature of the rod is increased by 50 deg. C.

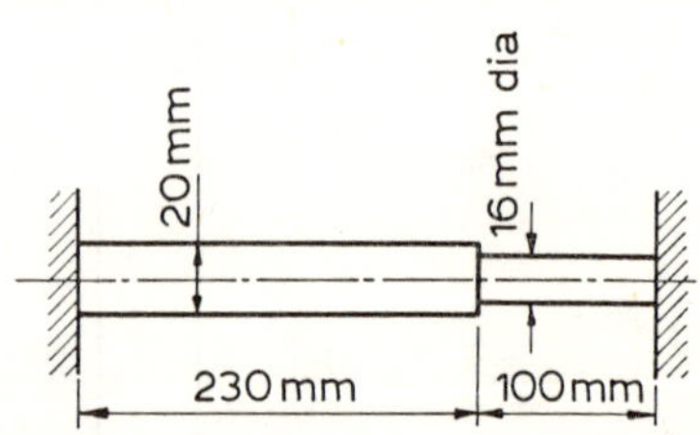

Fig. 1.21

For steel $E = 207$ GN/m² and $\alpha = 12 \times 10^{-6}$/°C. (*U.L.*)

Answer. (*a*) 53·6 MN/m² and 83·7 MN/m² compression; (*b*) 106·2 MN/m² and 166 MN/m² compression.

23. A 27 mm diameter steel bolt with a pitch of 3 mm is placed centrally in a brass tube and also suspended from a rigid body as shown in Fig. 1.22. There is a nut at the top which initially is just tightened, and, at the bottom, the bolt has a head suitable for carrying an axial load.

(*a*) Find the stresses in the bolt and the tube due to tightening the nut by turning it further through 24 degrees.

(*b*) After tightening as in (*a*) a load is then applied which is just sufficient to make the stress in the bolt 140 MN/m² tension; find the magnitude of the load and also the final stress in the tube.

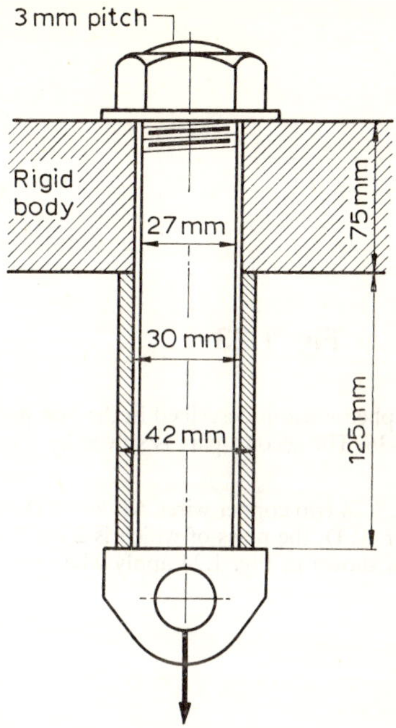

Fig. 1.22

State whether stresses are tensile or compressive, assume no change in the thickness of the rigid body and take E for steel as 206 GN/m^2 and for brass 89 GN/m^2. (*U.L.*)
Answer. (*a*) Stress in bolt = 69·8 MN/m^2 (tensile); stress in tube = 58·9 MN/m^2 (compressive). (*b*) Magnitude of load = 60·8 kN; final stress in tube = 28·6 MN/m^2 (compressive).

24. A bar length L, of conical section, hangs vertically with its apex at the lower end. If the material of the bar weighs w per unit volume show that the stress on a cross-section distance x from this end (due to the bar's own weight) is $wx/3$.

Show also that the extension of the bar due to its own weight is $wL^2/6E$. What is the corresponding result for a uniform bar?
Answer. $wL^2/2E$.

25. A cylindrical tensile specimen, made from a highly ductile material, is subjected to an axial force F that causes slight localized plastic flow. The force is then removed and the deformed zone has the shape of two truncated cones of small taper as shown in Fig. 1.23. (see note on p. 28.)

The specimen is again subjected to an axial tensile force. Assuming that the elongation is confined to the previously deformed zone and that the axial stress is uniform across transverse planes, show that, for forces $<F$, the elongation of the specimen is given by

$$8Fl/\pi EDd$$

where E is Young's modulus. (*U.L.*)

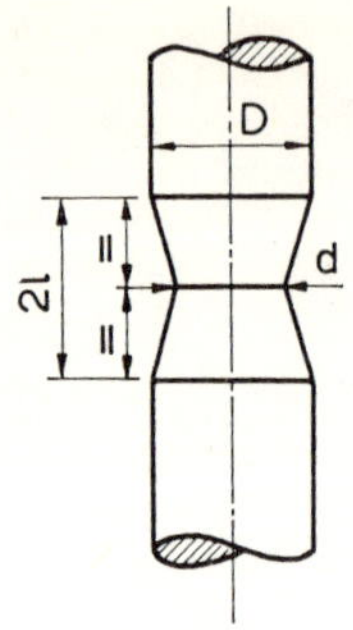

Fig. 1.23

Note. The phenomenon described in the first part of the question is discussed on pages 430–36. The second part is solved by the method used in Example 1.13.

26. In Fig. 1.24 two copper wires AB and CD support a rigid, uniform, horizontal member BFD, the mass of which is 2 kg. The wire FH is made of steel. The dimensions shown in Fig. 1.24 apply when the member BFD is in position and

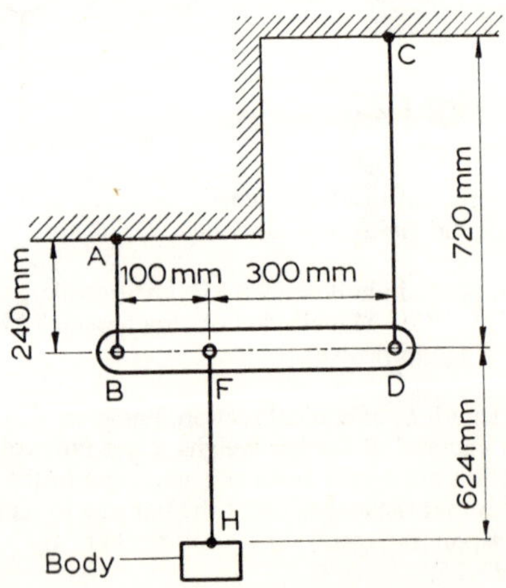

Fig. 1.24

just before a body is hung on the end H of the steel wire. The wire AB has a cross-sectional area of 1·962 mm² and both the wires CD and FH have cross-sectional areas of 0·981 mm². The maximum allowable stresses in the copper and steel are 60 MN/m² and 150 MN/m² respectively and the values of Young's moduli of elasticity are 120 GN/m² and 208 GN/m² respectively. The body is carefully hung on the lower end of the steel wire. Find the maximum mass of the body so that the allowable stresses will not be exceeded and find the downward displacement of this body. The masses of the wires may be neglected. (Assume 1 kgf = 9·81 N.) (*U.L.*)

Answer. $14\frac{2}{3}$ kg (stress in AB is the limiting factor); 0·60 mm.

27. A uniform non-deformable body of mass m is hung from three rods as shown in Fig. 1.25. The system is symmetrical. Each rod has a cross-sectional area 9·81 mm², and all the rods are the same length when unstressed. The inner rod is made of material A and the outer rods of material B. Material A obeys Hooke's law and has Young's modulus 200 GN/m², whereas, for the range of stress and strain in this problem, the relation between the stress σ_B and the strain ϵ_B for material B is given by

$$\sigma_B = (50\,\epsilon_B - 3000\,\epsilon_B^2)\ \mathrm{GN/m^2}$$

Determine the strain and the stresses in the rods when $m = 450$ kg.

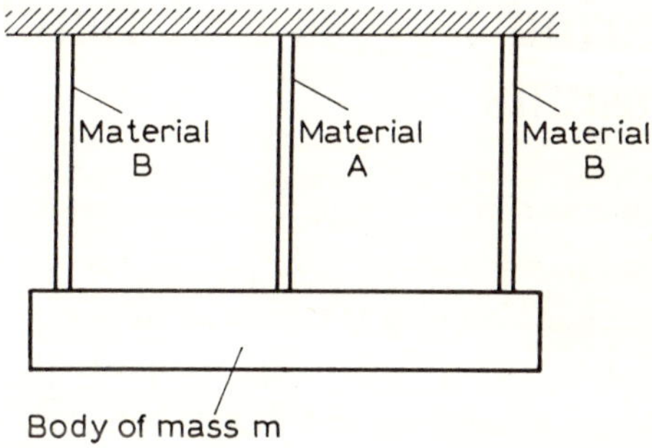

Fig. 1.25

Assume 1 kgf = 9·81 N. (*U.L.*)

Hint. The strains in A and B are equal. Hence $\epsilon_B = \epsilon_A = \sigma_A/E_A$. Substitute this result in the expression for σ_B and then use the relationship that the weight of the body equals the sum of the forces in A and B.

Answer. Strain = 0·001 55 in all rods; stresses are 310 MN/m² (in A) and 70·3 MN/m² (in B).

Chapter 2

Shearing Forces and Bending Moments

The sign convention used in this book is as follows:

"Left up" shearing forces are positive,
"Sagging" bending moments are positive.

In each of the standard cases in Table 1, L is the total length of the beam or cantilever and W is the total load. (For the "uniformly distributed load" cases $W = wL$, where w is the load per unit length.) The scales are approximately the same for all four cases.

At any section of a beam or cantilever, distance x from some datum,

$$\frac{dM}{dx} = F \quad \text{and} \quad \frac{dF}{dx} = -w$$

where w = intensity of loading (i.e. the load per unit length),
F = shearing force,
M = bending moment.

General rules for shearing force and bending moment are:

(*a*) For any part of the span carrying point loads only, the S.F. diagram is a series of horizontal "steps" and the B.M. diagram is a series of (sloping) straight lines.
(*b*) For any part of the span carrying a uniformly distributed load only, the S.F. diagram is a sloping straight line and the B.M. diagram is a parabola.
(*c*) At a point where the S.F. diagram passes through zero (i.e. where the S.F. changes sign) the B.M. is a maximum or a minimum.
(*d*) Over any part of the span where the S.F. is zero, the B.M. has a constant value.

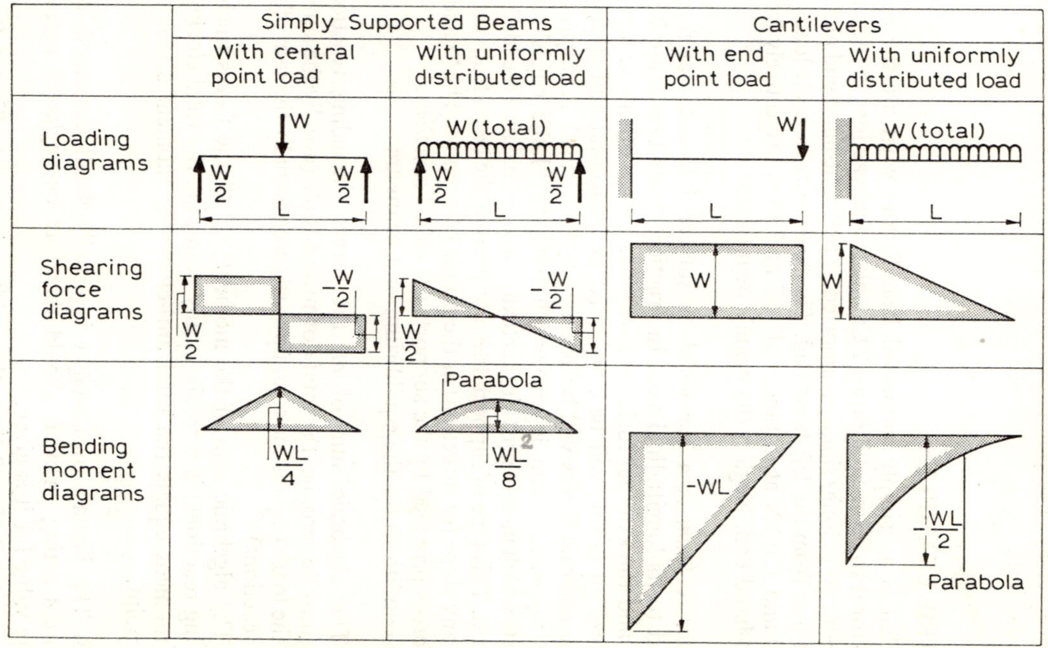

Simply Supported Beams
Cantilevers
With central point load
With uniformly distributed load
With end point load
With uniformly distributed load
Loading diagrams
Shearing force diagrams
Bending moment diagrams
W
W/2
L
W(total)
-W/2
W
-WL
WL/4
Parabola
WL/8
-WL/2
Parabola

Table 1

(*e*) At any point where the B.M. diagram passes through zero (i.e. where the B.M. changes sign) the curvature of the beam changes from concave upwards to concave downwards (or vice versa). The point is called a point of contra-flexure (or inflexion).

WORKED EXAMPLES

2.1. What is a "simply supported beam", and what are the conditions for it to be in equilibrium? Calculate the reactions in the case of a beam, 20 m long, simply supported at the right-hand end and at a point 6 m from the left-hand end. The beam carries point loads of 80, 150 and 120 kN at distances of 1, 7 and 13 m respectively from the left-hand end. Neglect the beam's own weight.

Solution. A *beam* is a bar which is subjected to external forces inclined to its longitudinal axis. In practice, many beams are horizontal and the external forces are weights. Unless otherwise stated, therefore, all beams will be considered as horizontal and straight, and the external forces will be assumed to be vertical. A *simply* (or *freely*) *supported* beam is one which rests on "knife edges", i.e. supports which merely provide (vertical) forces to balance the load on the beam. This type of support does not prevent the beam from rotating to any slope (or direction) at the points of support. The "balancing" forces provided by the supports are called *reactions*.

The conditions for the equilibrium of a beam are

(*a*) The algebraic sum of *all* the forces (including reactions) must be zero, i.e. the total downward force (usually the sum of the loads) equals the total upward force (usually the sum of the reactions).

(*b*) The algebraic sum of the moments of all the forces (including reactions) is zero about any point, i.e. the total clockwise moment equals the total anti-clockwise moment about any point.

N.B. The beam's own weight is usually small compared with the forces acting on it. It should therefore be neglected unless a definite value for it is given.

Let A and B (Fig. 2.1) be the points of support, R_A and R_B the reactions at these points. The simplest procedure in all cases is to take moments about one of the points of support. If this is done, the reaction at that point will not be present in the equation since it has no moment about the point through which it acts.

Taking moments about A,

$$\text{Total anti-clockwise moment} = (80\text{ kN} \times 5\text{ m}) + (R_B \times 14\text{ m}) = (400 + 14R_B)\text{ kN-m}$$

$$\text{Total clockwise moment} = (150\text{ kN} \times 1\text{ m}) + (120\text{ kN} \times 7\text{m}) = 990\text{ kN-m}$$

Since the resultant moment is zero, these two totals are equal, and

$$400 + 14R_B = 990 \qquad 14R_B = 590$$

$$R_B = 590/14 = 42\tfrac{1}{7}\text{ kN} \qquad (Ans)$$

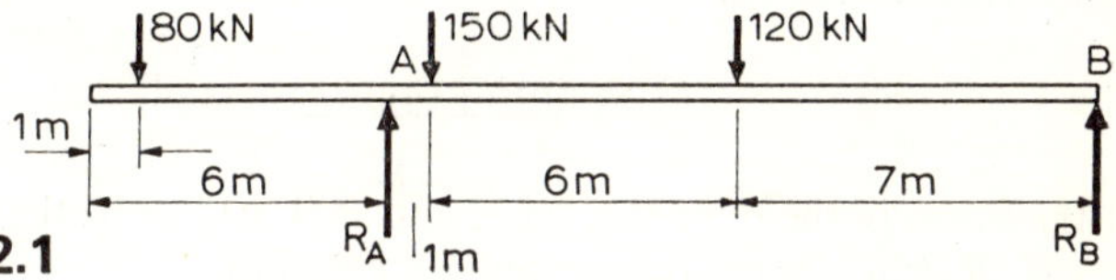

Fig. 2.1

Similarly, taking moments about B,

Total clockwise moment = total anti-clockwise moment

or

$$(R_A \times 14\text{ m}) = (80\text{ kN} \times 19\text{ m}) + (150\text{ kN} \times 13\text{ m}) + (120\text{ kN} \times 7\text{ m})$$

$$14R_A = 1520 + 1950 + 840 = 4310$$

$$R_A = 4310/14 = 307\tfrac{6}{7}\text{ kN} \qquad (Ans)$$

As a check,

$$\text{Sum of the reactions} = 42\tfrac{1}{7} + 307\tfrac{6}{7} = 350\text{ kN}$$

and

$$\text{Total load} = 80 + 150 + 120 = 350\text{ kN}$$

N.B. When the first reaction has been obtained, the second can be found by subtracting the known reaction from the total load. Although there is slightly more work in the method shown here, it does include a useful check.

2.2. Distinguish between *concentrated* loads and *uniformly distributed* loads. A beam, 10 m long, is simply and symmetrically supported over a span of 6 m. It carries a concentrated load of 30 kN at the right-hand end and a uniformly distributed load of 15 kN/m run between the left-hand end and the mid-point of the beam. If the beam weighs 12 kN/m, calculate the reactions.

Solution. A *concentrated* (or *point*) *load* is one which is applied to the beam through a "knife edge", i.e. it is not spread over any measurable part of the span. A *uniformly distributed load* is one which is spread evenly over the whole or a part of the beam, so that the load carried by any portion of the beam under that load is proportional to the length of that portion. A common example of the latter type of load is the weight of the beam itself. The amount of the uniformly distributed load is given in N/m (or kN/m) and it is usually denoted by the symbol w. On diagrams, it is shown thus

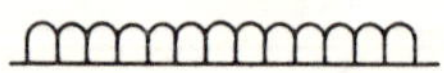

For the beam in the question, let A and B (Fig. 2.2) be the points of support, R_A and R_B the corresponding reactions.

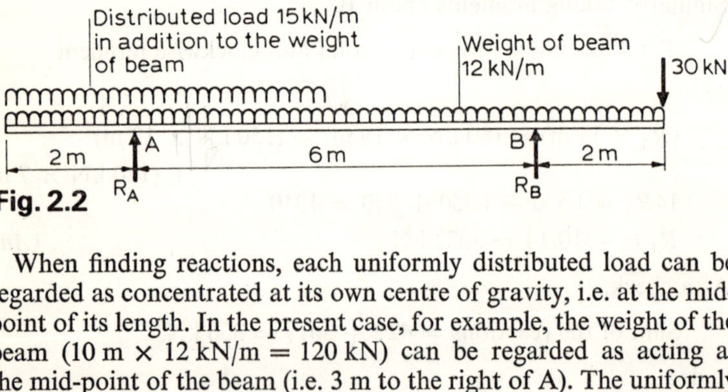

Fig. 2.2

When finding reactions, each uniformly distributed load can be regarded as concentrated at its own centre of gravity, i.e. at the mid-point of its length. In the present case, for example, the weight of the beam (10 m × 12 kN/m = 120 kN) can be regarded as acting at the mid-point of the beam (i.e. 3 m to the right of A). The uniformly distributed load (15 kN/m for 5 m = 75 kN) can be regarded as acting at a point $2\frac{1}{2}$ m from the left-hand end (i.e. $\frac{1}{2}$ m to the right of A).

Taking moments about A,

Total anti-clockwise moment = total clockwise moment

or

$$R_B \times 6 \text{ m} = (75 \text{ kN} \times \tfrac{1}{2} \text{ m}) + (120 \text{ kN} \times 3 \text{ m}) + (30 \text{ kN} \times 8 \text{ m})$$

$$6R_B = 37\tfrac{1}{2} + 360 + 240 = 637\tfrac{1}{2}$$

$$R_B = 637\tfrac{1}{2}/6 = 106\tfrac{1}{4} \text{ kN} \qquad (Ans)$$

Taking moments about B.

Total clockwise moment = total anti-clockwise moment

or

$$(R_A \times 6\text{ m}) + (30\text{ kN} \times 2\text{ m}) = (120\text{ kN} \times 3\text{ m}) + (75\text{ kN} \times 5\tfrac{1}{2}\text{ m})$$

$$6R_A = 360 + 412\tfrac{1}{2} - 60 = 712\tfrac{1}{2}$$

$$R_A = 712\tfrac{1}{2}/6 = 118\tfrac{3}{4}\text{ kN} \qquad (Ans)$$

As a check, $R_A + R_B = 118\frac{3}{4} + 106\frac{1}{4} = 225$ kN which equals the total load.

2.3. Explain the terms "shearing force" and "bending moment". A beam ABCDE 6 m long is simply supported at B and D (Fig. 2.5). It carries point loads of 70 kN and 20 kN at C and E respectively. AB = $\frac{1}{2}$ m, BC = 1 m, CD = 2 m and DE = $2\frac{1}{2}$ m. Calculate the shearing force and bending moment at the following points:

(i) P, which is 1 m from A,
(ii) Q, which is $2\frac{1}{4}$ m from A, and
(iii) R, which is $4\frac{1}{2}$ m from A.

Solution. Consider a simply supported beam, as shown in Fig. 2.3 carrying several point loads. The algebraic sum of all the forces (including reactions) is zero and the algebraic sum of the moments of

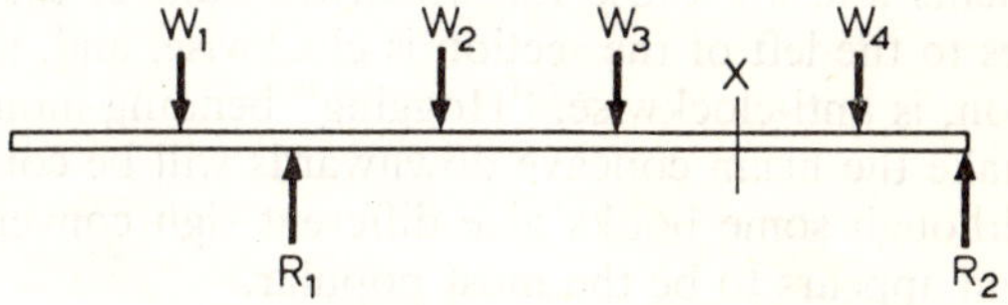

Fig. 2.3

all the forces about any point X is zero. If, however, only the forces to one side of the section at X are considered, these algebraic sums will not, in general, be zero.

The *shearing force* at a section of a beam, such as X, is the algebraic sum of the forces on either side of that section. Since the sum of all the forces on the *whole* beam must be zero, then the sum of those to the left of X must be numerically equal, but opposite in direction, to the sum of those to the right of X. The abbreviation for shearing force is S.F. and it is denoted by the symbol F.

The *bending moment* at a section, such as X, is the algebraic sum of the moments of the forces on either side of the section. Since the sum of the moments of all the forces on the *whole* beam (about X) must be zero, then the sum of those to the left of X must be numerically equal, but opposite in direction, to the sum of those to the right of X. The abbreviation for bending moment is B.M. and it is denoted by the symbol M.

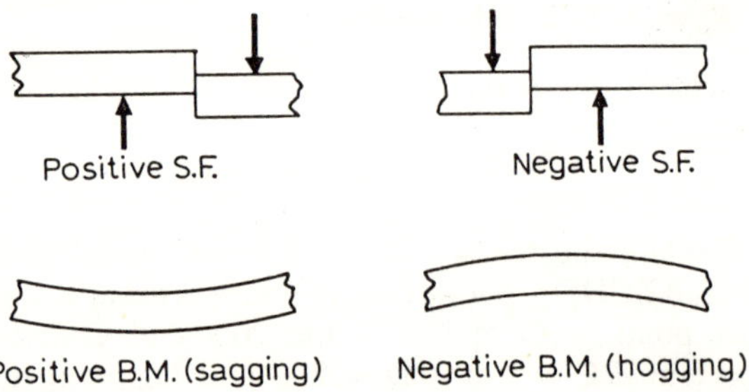

Fig. 2.4

Sign Convention (see Fig. 2.4). In this book shearing forces which tend to make that part of the beam on the left-hand side of the section move up and that part on the right move down will be considered positive. (The rule is, therefore, "left-up, right-down is positive; right-up, left-down is negative".)

Bending moments which tend to make the beam concave upwards will be considered positive. These are called "sagging" bending moments and are those for which the sum of the moments of the forces to the left of the section is clockwise, and, to the right of the section, is anti-clockwise. "Hogging" bending moments which tend to make the beam concave downwards will be considered negative.

Although some books give different sign conventions, the above system appears to be the most popular.

Figure 2.5 illustrates the beam described in the question and the reader should verify (by the method of Example 2.1) that the reactions R_B and R_D are 30 kN and 60 kN respectively.

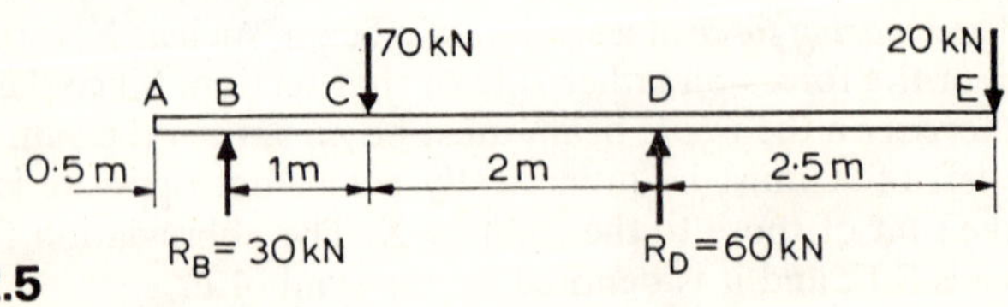

Fig. 2.5

(i) Consider the portion, AP, of the beam, as shown in Fig. 2.6. There is only one force (R_B) acting on this part of the beam. Hence the shearing force at P is equal to R_B and it is "left-up", i.e. positive. Therefore,

S.F. at P = +30 kN (*Ans*)

The moment of this force about P is (30 kN × $\frac{1}{2}$ m), i.e. 15 kN-m. It is clockwise to the left of P (or "sagging"). Therefore,

B.M. at P = +15 kN-m (*Ans*)

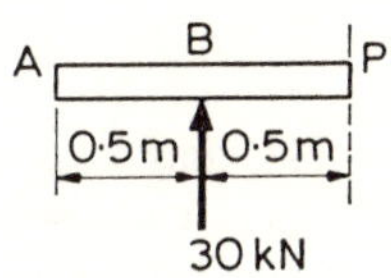

Fig. 2.6

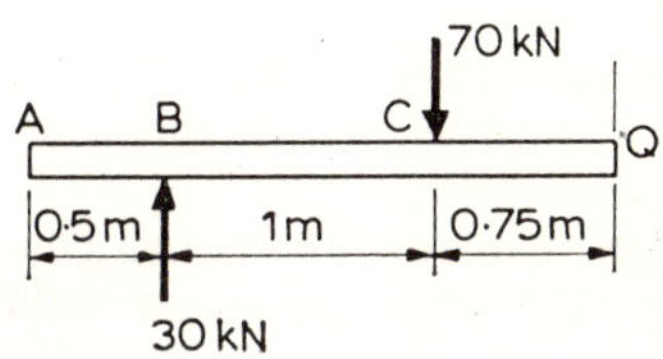

Fig. 2.7

(ii) Consider the portion AQ of the beam (Fig. 2.7). There are two forces to the left of Q (R_B and the 70 kN load). The algebraic sum of these forces is 40 kN (downwards to the left of Q). Since "left-down" is negative

S.F. at Q = −40 kN (*Ans*)

The moment of R_B about Q is (30 kN × $1\frac{3}{4}$ m), i.e. $52\frac{1}{2}$ kN-m clockwise. The moment of the load at C about Q is (75 kN × $\frac{3}{4}$ m) i.e. $52\frac{1}{2}$ kN-m anti-clockwise. The algebraic sum of these two moments is zero. Therefore,

B.M. at Q = 0 (*Ans*)

(iii) For the point R it is easier to deal with the part of the beam to the right of R (Fig. 2.8) since this involves one force only.

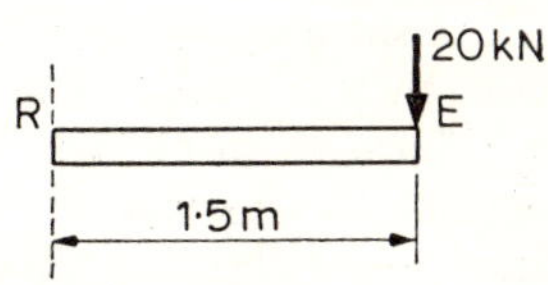

Fig. 2.8

The shear force is equal to the force at E and this is "right-down". Therefore,

S.F. at R = +20 kN (*Ans*)

The moment of this force about R is (20 kN × $1\frac{1}{2}$ m) = 30 kN-m, and since it tends to produce "hogging" at R,

B.M. at R = −30 kN-m (*Ans*)

The reader should now verify that the same results are obtained if the portions PE, QE and AR are considered instead of AP, AQ and RE respectively. A good method in all cases is to cover the loading diagram with a piece of card or paper up to the point under consideration, and then to include all visible forces.

2.4. What are shearing force and bending moment diagrams? Draw these diagrams for the beam of Example 2.3.

Solution. Shearing force and bending moment diagrams are diagrams which show the values of these quantities at each point along the beam. They are drawn immediately underneath the loading diagram to the same horizontal scale. In each case, positive quantities are plotted upwards and negative quantities downwards.

The vertical scales used should be stated and, in addition, the principal values should be shown on the diagrams.

The beam of Example 2.3 can be considered in four sections (Fig. 2.9).

From A *to* B. Since there is no force to the left of any point between A and B then there is no shearing force or bending moment at any point in this range.

From B *to* C. There is one force (R_B) acting to the left of any point in this range. The shearing force for any point between B and C (such as P in Example 2.3) is +30 kN. The bending moment for any such point equals (R_B × its distance from B). At B the bending moment is zero and it will obviously increase uniformly (or linearly) until, at C, it equals 30 kN × 1 m (sagging), i.e. +30 kN-m.

From C *to* D. There are two forces acting to the left of any point between C and D (such as Q in Example 2.3). These two forces together give a shearing force of −40 kN at any such point. If the bending moment is calculated for a series of points between C and D then the values, when plotted, lie on a straight line. (This linear variation applies to all cases of point loading.) At D, the bending moment equals −50 kN-m, but this value is more easily obtained by considering the portion DE.

From D *to* E. There is only one force (20 kN at E) to the *right* of any point between D and E (such as R in Example 2.3). The shearing force at any such point is +20 kN and the bending moment equals 20 kN × (the distance of the point from E). The bending moment diagram is again a straight line, from zero at E to (20 kN × 2½ m) hogging, i.e. −50 kN-m, at D.

Fig. 2.9 shows the resulting diagrams.

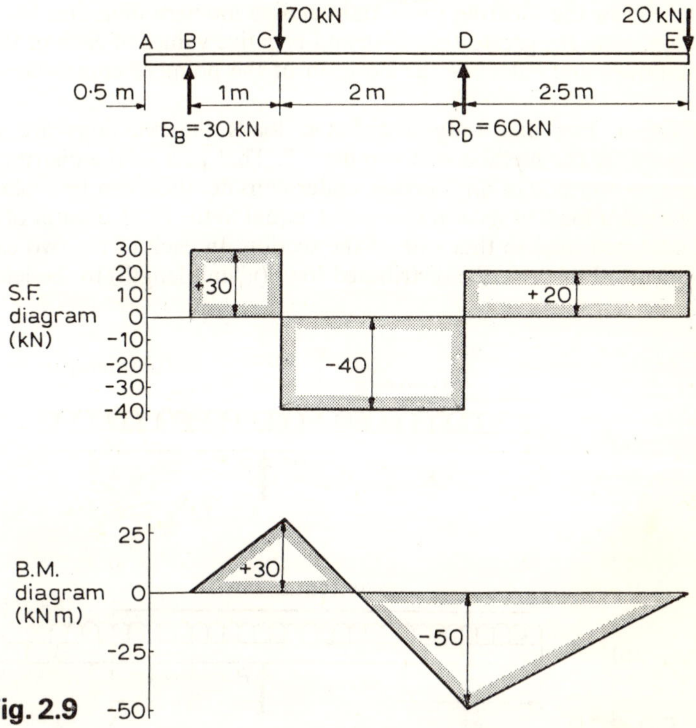

Fig. 2.9

The following method will give the shearing force diagram for all cases of point loading.

Start at the left-hand end of the beam. If there is no load or reaction there, move along the beam to the first force. Draw a vertical line in the direction of the arrow of this force of a length representing the value of the load. Draw a horizontal line to the next force and repeat the procedure. The final force at the right-hand end of the beam should make the shearing force diagram return to the base-line.

The bending moment diagram for such loading is obtained by calculating and plotting the values at each point of loading and joining the points obtained by straight lines.

2.5. Explain how the shearing force and bending moment can be found at any section of a beam carrying a uniformly distributed load. A beam, 10 m long, is simply supported at the left-hand end and at its mid-point. It carries a uniformly distributed load of 48 kN/m between the supports and a point load of 50 kN at the right-hand end. Draw the shearing force and bending moment diagrams for the beam, state the greatest positive and negative values of each of these quantities and calculate the position of the point of contra-flexure.

Solution. For uniformly distributed loads the reactions are calculated by the method of Example 2.2. That part of the distributed load to one side of any section under consideration can be replaced by a point load, at its *own* mid-point, equal to the total amount of the distributed load to that side of the section. In each of the two cases shown in Fig. 2.10, the distributed load (w/unit length) to the left of

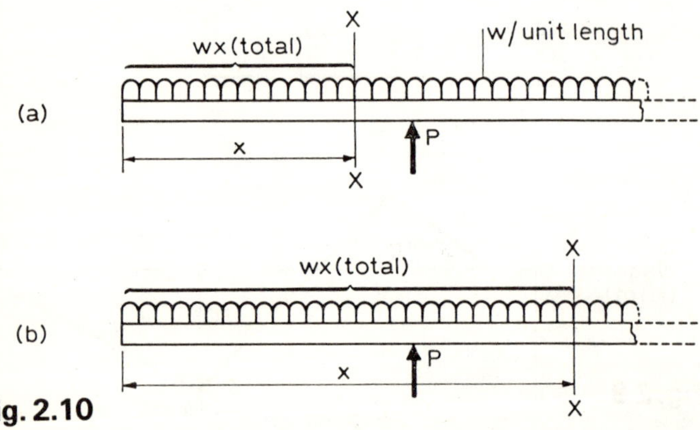

Fig. 2.10

XX can be replaced by a point load wx acting at a point $x/2$ from XX. If diagrams are to be drawn it is best to form equations for the shearing force and bending moment at any such section, in terms of x, the distance along the beam. These equations must be modified each time a point load is passed, or whenever the distributed load ceases or changes its intensity. For example, equations formed for sections to the left of P must not be used for sections to the right of P unless extra terms are included to allow for the reaction at P.

Let the beam in the Example be ABC as shown in Fig. 2.11. Then, by the method of Example 2.2, the reactions at A and B are 70 kN and 220 kN respectively.

The beam must now be considered in two parts:

Between A *and* B. There are two forces acting to the left of any section XX between A and B, distance x m from A as shown. These are the reaction at A (70 kN) and the part of the distributed load between A and the section XX. The latter load is 48 kN/m × x m, i.e. $48x$ kN and its mid-point is $x/2$ m from XX. The shear force at XX is the algebraic sum of 70 kN (left-up) and $48x$ kN (left-down).

Between A *and* B, therefore,

$$F = (70 - 48x) \text{ kN} \qquad \text{(i)}$$

The bending moment at XX is (70 kN × x m) clockwise to the left and ($48x$ kN × $\frac{1}{2}x$ m) anti-clockwise to the left.

Between A and B, therefore,

$$M = (70x - 24x^2) \text{ kN-m} \qquad \text{(ii)}$$

Using equations (i) and (ii) the following values are obtained:

x(m)	0	1	2	3	4	5
F(kN)	70	22	−26	−74	−122	−170
M(kN-m)	0	46	44	−6	−104	−250

Between B *and* C. This part of the beam is under point loading only. By the method of Example 2.4, the shearing force at any section is +50 kN and the bending moment varies linearly from −250 kN-m at B to zero at C.

The diagrams can now be drawn (Fig. 2.11). The shearing force varies linearly between A and B and is constant between B and C. The bending moment diagram is a curve between A and B, and a straight line between B and C.

The greatest positive shearing force is (+) 70 kN at A (*Ans*)
The greatest negative shearing force is (−) 170 kN at B (*Ans*)
The greatest negative bending moment is (−) 250 kN-m at B (*Ans*)

The greatest positive bending moment occurs at a point approximately 1 m from A. To find its exact position, differentiate equation (ii) with respect to x, since this equation applies to this region.

$$dM/dx = 70 - 48x$$

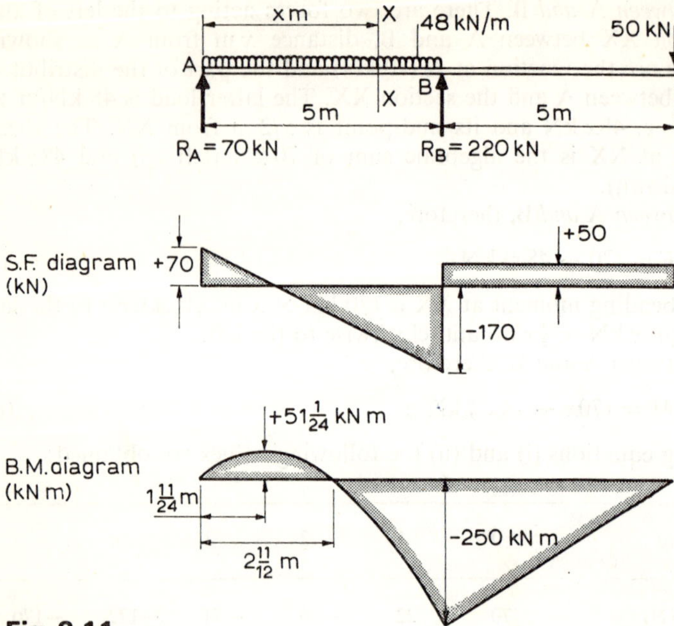

Fig. 2.11

dM/dx is zero when $x = 70/48$. M, therefore, has a maximum or minimum value $1\frac{11}{24}$ m from A. From the diagram it is clearly a maximum and substituting $x = 1\frac{11}{24}$ in equation (ii),

$$\text{Maximum positive bending moment} = (70 \times 1\tfrac{11}{24}) - 24(1\tfrac{11}{24})^2$$

$$= (+)\ 51\tfrac{1}{24}\ \text{kN-m} \quad (Ans)$$

A *point of contra-flexure* (or *inflexion*) is a point where the bending moment passes through zero. It is thus a point where the beam's curvature changes from concave upwards to concave downwards or vice versa. There is clearly one such point between A and B. From equation (ii) we have, for $M = 0$

$$0 = (70x - 24x^2) \quad \text{or} \quad x(70 - 24x) = 0$$

There are two solutions to this equation, $x = 0$ and $x = 70/24$. The first corresponds to the point A at which the bending moment is zero but does not change sign. The second solution gives the required point of contra-flexure.

Therefore, there is a point of contra-flexure at a point $2\frac{11}{12}$ m from A. *(Ans)*

2.6. A horizontal beam AD, 10 m long, carries a uniformly distributed load of 144 N per metre run, together with a concentrated load of 360 N at the left-hand end A. The beam is supported at a point B which is 1 m from A and at C which is in the right-hand half of the beam and x m from the end D.

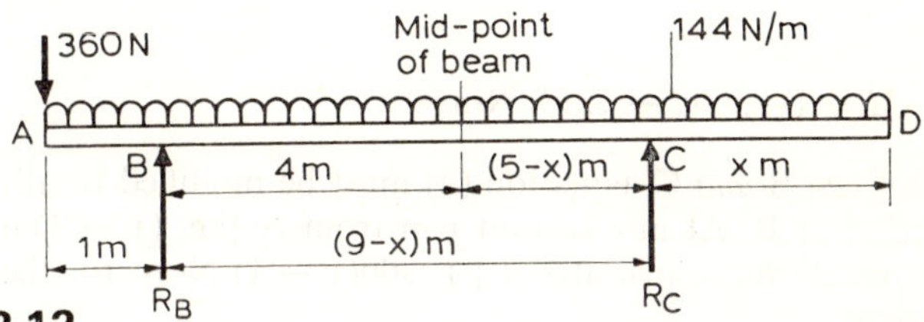

Fig. 2.12

Determine the value of x if the mid-point of the beam is a point of inflexion (or contra-flexure) and for this arrangement plot *on squared paper* the bending moment diagram, indicating the principal numerical values.

Locate any other points of inflexion. *(U.L.)*

Solution. Taking moments about B (Fig. 2.12) to find R_C, (anti-clockwise = clockwise),

$$(360 \text{ N} \times 1 \text{ m}) + R_C(9 - x) \text{ N-m} = (144 \text{ N/m} \times 10 \text{ m} \times 4 \text{ m})$$

$$360 + R_C(9 - x) = 5760$$

$$R_C = \frac{5760 - 360}{9 - x} = \frac{5400}{9 - x} \qquad \text{(i)}$$

If the mid-point of the beam is a point of contra-flexure, then the algebraic sum of the moments of all the forces on the right-hand half of the beam about this point must be zero.

Hence

$$R_C(5 - x) \text{ N-m} = (144 \text{ N/m} \times 5 \text{ m} \times 2\tfrac{1}{2} \text{ m})$$

$$\frac{5400}{(9 - x)} \times (5 - x) = 1800$$

$$5400\,(5 - x) = 1800\,(9 - x)$$

from which

$$x = 3$$

Hence the point C must be 3 m from D. *(Ans)*

Substituting $x = 3$ in equation (i), we have $R_C = 900$ N. The reader should verify that $R_B = 900$ N also.

Between A *and* B, at any section y m from A, the bending moment is

$$-(360 \text{ N} \times y \text{ m}) - (144 \text{ N/m} \times y \text{ m} \times \tfrac{1}{2}y \text{ m})$$

or

$$M = (-360y - 72y^2) \text{ N-m} \qquad \text{(ii)}$$

Between B *and* C, equation (ii) must be modified to allow for the reaction at B. At any section y m from A [i.e. $(y - 1)$ m from B] this reaction contributes (+) $900(y - 1)$ N-m to the bending moment.

Hence

$$\begin{aligned} M &= -360y - 72y^2 + 900(y - 1) \\ &= (-900 + 540y - 72y^2) \text{ N-m} \qquad \text{(iii)} \end{aligned}$$

Between C *and* D, equation (iii) must be modified to allow for the reaction at C. At a section y m from A, this reaction will contribute (+) $900(y - 7)$ N-m to the bending moment.

Hence

$$\begin{aligned} M &= -900 + 540y - 72y^2 + 900(y - 7) \\ &= (-7200 + 1440y - 72y^2) \text{ N-m} \qquad \text{(iv)} \end{aligned}$$

Using each equation in its appropriate range and substituting convenient values of y, we obtain the following table, from which Fig. 2.13 is drawn.

	Equation (ii)				Equation (iii)		
y (m)	0	$\frac{1}{3}$	$\frac{2}{3}$	1 (at B)	2	3	4
M (N-m)	0	−128	−272	−432	−108	72	108

				Equation (iv)		
y (m)	5	6	7 (at C)	8	9	10
M (N-m)	0	−252	−648	−288	−72	0

Alternatively, the last four values can be obtained by taking moments of forces to the right of each section. Since the principal values are required, the maximum sagging (positive) value must be obtained. This occurs between B and C so that, differentiating equation (iii) with respect to y,

$$dM/dy = 540 - 144y$$

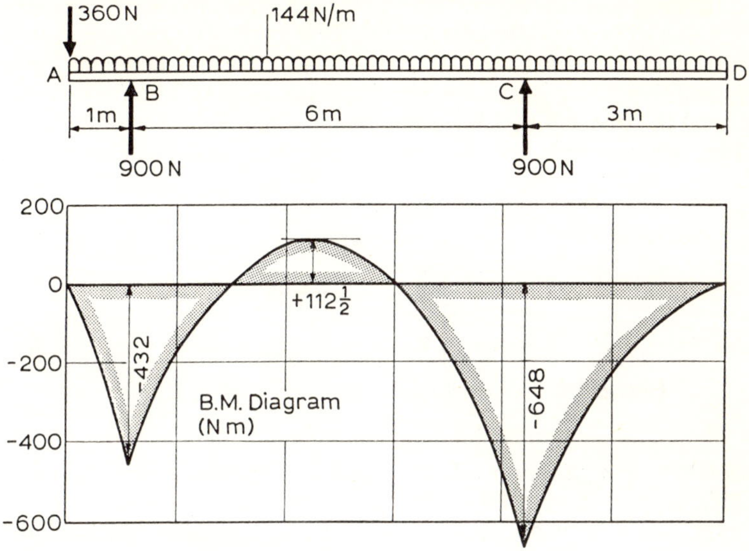

Fig. 2.13

For maximum bending moment, therefore

$$540 - 144y = 0 \qquad y = 3\tfrac{3}{4}$$

Substituting this value in equation (iii), the maximum sagging bending moment is

$$M = -900 + (540 \times 3\tfrac{3}{4}) - [72 \times (3\tfrac{3}{4})^2] = +112\tfrac{1}{2} \text{ N-m}$$

There are clearly two points of inflexion (Fig. 2.13) both between B and C.

Putting $M = 0$ in equation (iii),

$$0 = -900 + 540y - 72y^2$$

or

$$2y^2 - 15y + 25 = 0$$
$$(2y - 5)(y - 5) = 0$$

from which

$$y = 2\tfrac{1}{2} \qquad y = 5$$

Hence, there is another point of inflexion $2\frac{1}{2}$ m from A. (*Ans*)

2.7. What is a "cantilever" and how is its equilibrium maintained?

ABCDE is a cantilever 9 m long. It is built-in at the end A and carries concentrated loads of 2 and 3 kN at B and D respectively. In addition, there is a uniformly distributed load of 2 kN/m between C and the free end E. AB = 2 m, BC = 1 m, CD = 4 m and DE = 2 m. Draw to scale the bending moment diagram for this cantilever.

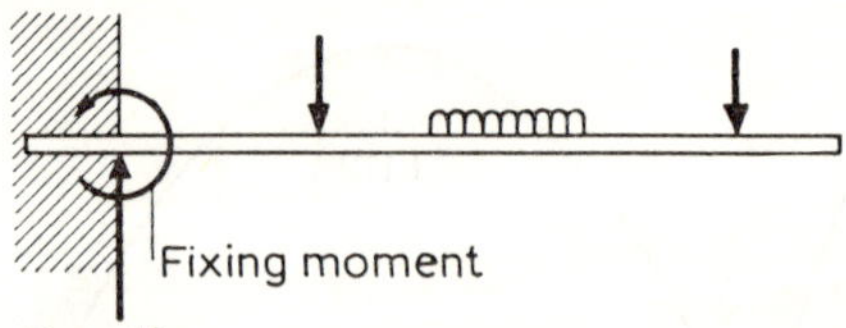

Fig. 2.14

Solution. A cantilever is a beam which is supported at one end only. This support cannot be of the "knife-edge" type or else the cantilever would turn about it. In addition to a vertical reaction, the support must provide a moment equal to the sum of the moments of all the external forces about the supported end.

A common type of cantilever is one which is built into a wall, as shown in Fig. 2.14. The end of a cantilever which is built-in is called the "fixed" end, and the moment supplied by the support is called the "fixing moment."

The fixing moment is clearly equal to the bending moment at the fixed end. Similarly the reaction is (numerically) equal to the shearing force at the same point. The shearing force and bending moment at any section of a cantilever can best be found by considering the forces on the "free" end side of the section (however close to the fixed end this may be). If this is done the reaction and the fixing moment do not enter the calculations.

In the example given, general equations should be formed to give the bending moment between C and E. Starting from E (the free end) we have (Fig. 2.15):

Between E *and* D. At any section XX, distance x m from E, as shown, the load between the section and E is 2 kN/m × x m, and its mid-point is $x/2$ m from XX. Hence there is a (hogging) bending moment at the section of 2 kN/m × x m × $x/2$ m. Therefore,

$$M = -x^2 \text{ kN-m} \qquad \text{(i)}$$

Between D *and* C. Equation (i) must be modified to include the point load at D. At any point x m from E, [i.e. $(x - 2)$ m from D]

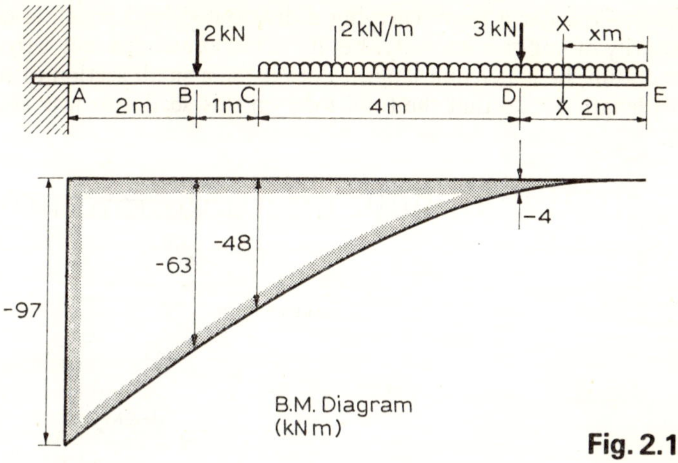

Fig. 2.15

this will contribute a "hogging" moment of 3 kN × $(x - 2)$ m. The equation therefore becomes

$$M = -x^2 - 3(x - 2) = (-x^2 - 3x + 6) \text{ kN-m} \qquad \text{(ii)}$$

Between C *and* B, *and* B *and* A, the bending moment varies linearly and it is necessary to calculate the values at B and A only. For both points the whole distributed load (12 kN) can be considered as acting at its own mid-point (3 m from E).

At B,

$$M = -(3 \text{ kN} \times 5 \text{ m}) - (12 \text{ kN} \times 4 \text{ m})$$
$$= -63 \text{ kN-m}$$

At A,

$$M = -(3 \text{ kN} \times 7 \text{ m}) - (12 \text{ kN} \times 6 \text{ m}) - (2 \text{ kN} \times 2 \text{ m})$$
$$= -97 \text{ kN-m}$$

Using these two values and plotting suitable points from equations (i) and (ii) the required diagram is obtained.

(The reader should now investigate the four standard cases given in Table 1 at the beginning of this chapter. The values of the maximum shearing force and bending moment for these cases are often required.)

2.8. Obtain the mathematical relationships between load, shearing force, and bending moment. What general conclusions can be drawn from these relations?

Solution. Consider a beam carrying a distributed load which is not necessarily of uniform intensity. Let its value, at any distance x from some datum, be w per unit length. Let AB be a small section of the beam, length δx, so that the total load on this section is $w\delta x$ (Fig. 2.16).

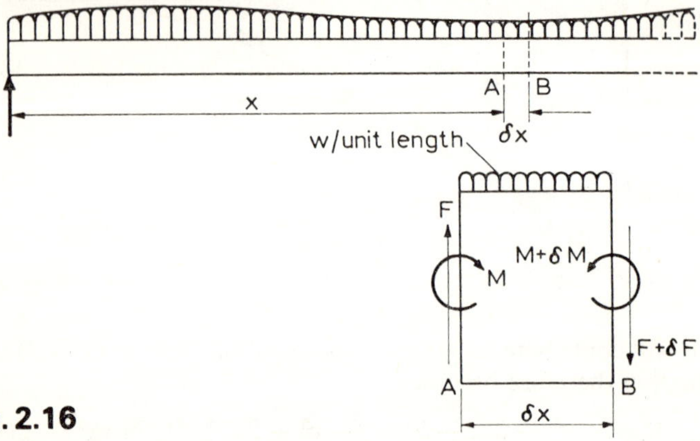

Fig. 2.16

F and M are the shearing force and bending moment acting at the end A of the element; $(F + \delta F)$ and $(M + \delta M)$ are the corresponding values at the end B. [The directions of these forces and moments are in agreement with the sign convention adopted in Example 2.3.]

For the section AB to be in equilibrium, the upward force must equal the downward forces, i.e.

$$F = F + \delta F + w\delta x \qquad \text{from which} \qquad \delta F/\delta x = -w$$

and in the limiting case $dF/dx = -w$ (i)

Also, the algebraic sum of all the moments about any point must be zero. The load on the section $w\delta x$ can be regarded as acting at its mid-point and, therefore, taking moments about A (clockwise = anti-clockwise),

$$M + \delta M = M + w\delta x \, . \tfrac{1}{2}\delta x + (F + \delta F)\delta x$$

or

$$\delta M = w \, . \tfrac{1}{2}(\delta x)^2 + F\delta x + \delta F\delta x$$

Since δx and δF are both small, then $(\delta x)^2$ and $\delta F\delta x$ are negligible compared with the other terms. Therefore,

$$\delta M = F\delta x \qquad \text{from which} \qquad \delta M/\delta x = F$$

and in the limiting case $dM/dx = F$ (ii)

N.B. In books using different sign conventions expressions such as $dF/dx = w$ and $dM/dx = -F$ may be found. The same *numerical* results will be obtained whatever convention is used, provided that it is applied consistently.

The following general conclusions can be drawn from the relations (i) and (ii):

(*a*) If the load per unit length (w) can be expressed as a simple function of x, then expressions giving the shearing force and bending moment at any point can be obtained by integration. (In some complicated cases graphical integration may be necessary.) These expressions will involve constants of integration which can be evaluated by using the known conditions at definite points along the beam.

(*b*) If the load is uniformly distributed then w is a constant.

By integration of (i), $F = -wx + A$
and from (ii), $M = -\frac{1}{2}wx^2 + Ax + B$
where A and B are constants in both cases.

These last two expressions show that, for a uniformly distributed load (alone), the shearing force varies linearly and the bending moment variation is parabolic. (Examples 2.5 and 2.6 provide instances of this. The reader should note the discontinuities at the points where the concentrated loads act and where the uniformly distributed load ends.)

(*c*) For a part of a beam or cantilever which carries no load ($w = 0$), we have

$$F = A \quad \text{and} \quad M = Ax + B$$

where A and B are constants.

This means that over any such part the shearing force is constant and the bending moment varies linearly (see Example 2.4).

(*d*) Over a part of the span where the shearing force is zero ($F = 0$) the bending moment is constant. This is a direct result of the relation (ii).

(*e*) At a point where the shearing force passes through zero, $dM/dx = 0$ which is the condition for M to be a (mathematical) maximum or minimum. At a section where a point load is applied to the beam, the slope of the bending moment diagram changes abruptly and the greatest bending moment may occur at such a section (see Examples 2.5 and 2.6). Thus the greatest value of the bending moment occurs at one of the sections where the shearing force passes through zero, either as a continuous function or with a discontinuity.

2.9. A beam 6 m long is simply supported at its ends. It carries a distributed load which increases uniformly from zero value at the left-hand end to 30 kN/m at mid-span and is then uniform at 30kN/m on the right-hand half of the beam as indicated in Fig. 2.17.

Determine the reactions at the supports and the position and magnitude of the maximum bending moment.

Draw to scale the shearing force and bending moment diagrams for the beam. Scales: space 1 m to 20 mm; shearing force 20 kN to 10 mm; bending moment 20 kN-m to 10 mm. *(U.L.)*

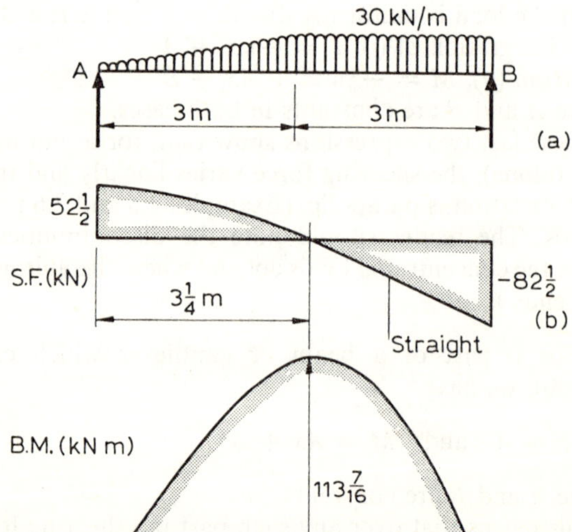

Fig. 2.17

Solution. First method. Let AB, Fig. 2.17a, be the given beam. To determine the reactions consider the load in two parts, the triangular portion of average intensity $(\frac{1}{2} \times 30)$ kN/m with centre of gravity 2 m from A and the rectangular portion of uniform intensity 30 kN/m with centre of gravity $4\frac{1}{2}$ m from A.

Thus, moments about A give

$$6R_B = (\tfrac{1}{2} \times 30 \times 3 \times 2) + (30 \times 3 \times 4\tfrac{1}{2})$$

$$R_B = 82\tfrac{1}{2} \text{ kN} \qquad (Ans)$$

Similarly

$$R_A = 52\tfrac{1}{2} \text{ kN} \qquad (Ans)$$

For the left-hand half of the beam at a section x m from A, the intensity of loading w kN/m is, by proportion

$$w = \tfrac{1}{3}x \times 30$$

Thus the total load to the left of this section is $\tfrac{1}{2}x(\tfrac{1}{3}x \times 30)$ kN and its centre of gravity is $\tfrac{1}{3}x$ m from the section. With these results the shearing force F kN and bending moment M kN-m are

$$F = R_A - \tfrac{1}{2}x(\tfrac{1}{3}x \times 30) = 52\tfrac{1}{2} - 5x^2 \qquad \text{(i)}$$

and

$$M = R_A x - \tfrac{1}{2}x(\tfrac{1}{3}x \times 30) \times \tfrac{1}{3}x$$
$$= 52\tfrac{1}{2}x - 1\tfrac{2}{3}x^3 \qquad \text{(ii)}$$

For a section in the right-hand half of the beam the whole of the triangular load and part of the rectangular load are included. Thus

$$F = R_A - \tfrac{1}{2} \times 3 \times 30 - 30(x - 3)$$
$$= 97\tfrac{1}{2} - 30x \qquad \text{(iii)}$$

and

$$M = R_A x - \tfrac{1}{2} \times 3 \times 30 \times (x - 2) - 30 \times (x - 3) \times \tfrac{1}{2}(x - 3)$$
$$= 97\tfrac{1}{2}x - 15x^2 - 45 \qquad \text{(iv)}$$

Using equations (i) and (ii) from $x = 0$ to $x = 3$ and (iii) and (iv) from $x = 3$ to $x = 6$, we have

x	0	1	2	3	4	5	6
F	$52\tfrac{1}{2}$	$47\tfrac{1}{2}$	$32\tfrac{1}{2}$	$7\tfrac{1}{2}$	$-22\tfrac{1}{2}$	$-52\tfrac{1}{2}$	$-82\tfrac{1}{2}$
M	0	$50\tfrac{5}{6}$	$91\tfrac{2}{3}$	$112\tfrac{1}{2}$	105	$67\tfrac{1}{2}$	0

These figures lead to the diagrams shown in Fig. 2.17b and c.

Second method. Working in kN and m units and using the general relationships established in the solution of 2.8, we have, where $0 \leqslant x \leqslant 3$.

$$w = \tfrac{1}{3}x \times 30 = 10x$$
$$dF/dx = -w = -10x$$
$$dM/dx = F = -5x^2 + C \qquad \text{(v)}$$
$$M = -(5/3)x^3 + Cx + D \qquad \text{(vi)}$$

where C and D are integration constants. When $x = 0$, $M = 0$, and therefore $D = 0$.

With $3 \leqslant x \leqslant 6$

$$dF/dx = -w = -30 \text{ (constant)}$$
$$dM/dx = F = -30x + E \qquad \text{(vii)}$$
$$M = -15x^2 + Ex + G \qquad \text{(viii)}$$

where E and G are further integration constants.

For $x = 3$ equations (v) and (vii) must give the same value of F. This leads to

$$E = C + 45$$

Again for $x = 3$, (vi) and (viii) must give the same value of M. From this condition,

$$G = -45$$

Finally, at $x = 6$, $M = 0$ and thus $C = 52\tfrac{1}{2}$.

With these values of the constants, equations (v)–(viii) are identical to (i)–(iv).

The reactions are equal to the numerical values of F at the ends of the beam. Thus

$$R_A = F_{(x=0)} = C = 52\tfrac{1}{2} \text{ (kN)} \qquad (Ans)$$

and

$$R_B = -F_{(x=6)} = E - (30 \times 6) = 82\tfrac{1}{2} \text{ (kN)} \qquad (Ans)$$

The position of the maximum bending moment corresponds to the section having zero shearing force. It occurs in the right-hand half of the beam and putting $F = 0$ in (iii) we have

$$x = 3\tfrac{1}{4}$$

With this result, (iv) gives

$$M_{max} = (97\tfrac{1}{2} \times 3\tfrac{1}{4}) - [15 \times (3\tfrac{1}{4})^2] - 45$$
$$= 113\tfrac{7}{16} \text{ kN-m} \qquad (Ans)$$

2.10. A horizontal beam 10 m long is freely supported at its ends and carries a vertical load of 1000 N, 3 m from the left-hand end. At a section 6 m from the left-hand end a clockwise couple of 2000 N-m is exerted, the axis of the couple being horizontal and perpendicular

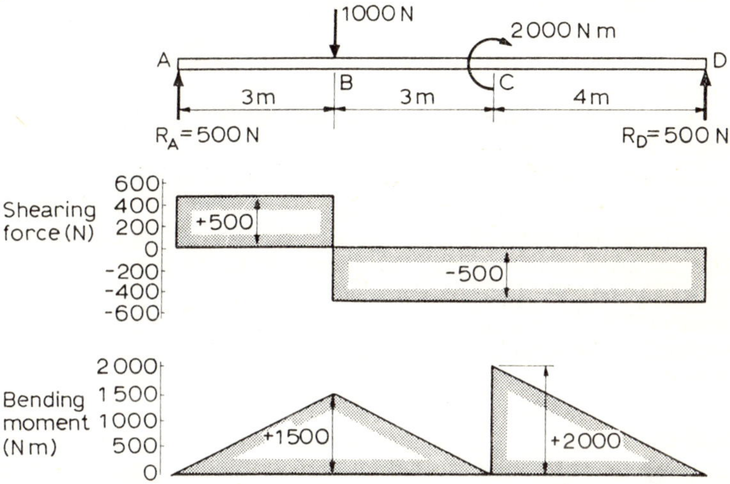

Fig. 2.18

to the longitudinal axis of the beam. Draw to stated scales the bending moment and shearing force diagrams and mark on them the principal dimensions. (*I.Mech.E.*)

Solution. Let ABCD be the beam, as shown in Fig. 2.18. The applied couple must be included in moment equations for all sections of the beam which include C.

Taking moments about A to find R_D (anti-clockwise = clockwise),

$$(R_D \times 10) \text{ N-m} = (1000 \text{ N} \times 3 \text{ m}) + (2000 \text{ N-m})$$

Therefore,

$$10R_D = 3000 + 2000 \qquad R_D = 500 \text{ N}$$

Similarly taking moments about D,

$$R_A = 500 \text{ N}$$

Between A *and* B. At any point there is only one force to the left of the section. It is R_A(=500 N) which is "left-up". Hence the shearing force between A and B is (+) 500 N (constant) and the bending moment varies linearly from zero at A to (500 N × 3 m) sagging, i.e. +1500 N-m at B.

Between B *and* C. At any point there are two forces to the left of the section (R_A and the load at B). The shearing force is the algebraic sum of 500 N (left-up) and 1000 N (left-down), i.e. (−) 500 N (constant).

The bending moment varies linearly from (+) 1500 N-m at B to [(500 N × 6 m) − (1000 N × 3 m)], i.e. zero, at C.

Between C *and* D. The shearing force calculation is identical with that between B and C. There is a sudden increment of (+) 2000 N-m in the bending moment as C is passed due to the applied couple. This means that immediately to the left of C, M is zero, and immediately to the right of C, $M = (+)$ 2000 N-m. The bending moment then decreases linearly to zero at D. (This last stage can also be obtained by working from D to C and considering forces *to the right*.)

N.B. On the diagrams vertical scales have been drawn, but on scale drawings the values represented by 1 cm should be stated.

2.11. State the principle of "superposition". A beam 12 m long, is simply and symmetrically supported over a span of 8 m. It carries concentrated loads of 2 kN and 5 kN at the left-hand and right-hand ends respectively as well as a uniformly distributed load of $1\frac{1}{4}$ kN/m between the supports. Draw the shearing force and bending moment diagrams for this loading, state the principal values of each, and calculate the positions of the points of contra-flexure.

Solution. If a beam or cantilever is subjected to two or more systems of loading, then the resulting shearing force or bending moment at any section is the algebraic sum of the values at that section due to each system acting separately. This principle can also be applied to reactions, bending stresses, and deflections, examples of which will be considered later. It breaks down, however, in the case of laterally-loaded struts (chapter 8), where the axial and lateral loads must be considered simultaneously.

In the beam of the question, the two concentrated loads are considered as one system and the distributed load as another. Let the beam be ABCD, as shown in Fig. 2.19.

For the concentrated loads only:

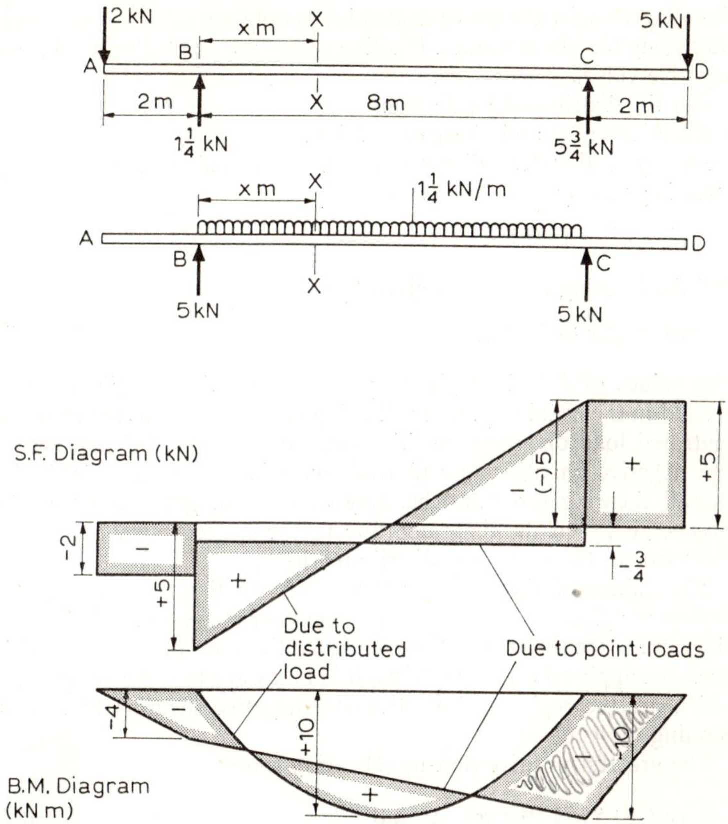

Fig. 2.19

Reactions at B and C are $1\frac{1}{4}$ and $5\frac{3}{4}$ kN respectively.

S.F. between A and B = −2 kN

S.F. between B and C = $-2 + 1\frac{1}{4} = -\frac{3}{4}$ kN

S.F. between C and D = $-2 + 1\frac{1}{4} + 5\frac{3}{4} = +5$ kN

B.M. at A = 0

B.M. at B = −(2 kN × 2 m) = −4 kN-m

B.M. at C = (5 kN × 2 m) = −10 kN-m

B.M. at D = 0

Also at any section XX, distance x m from B (between B and C),

$$M = (-4 - \tfrac{3}{4}x) \text{ kN-m} \qquad \text{(i)}$$

The diagrams for the concentrated loads only are drawn in the usual way using the above values. Positive values are plotted upwards and negative values downwards.

For the distributed load only:

Reactions at B and C are each 5 kN.

At any section XX, distance x m from B as shown, the shearing force is given by

$$F = (5 - 1\tfrac{1}{4}x) \text{ kN} \qquad \text{(ii)}$$

and the bending moment is given by

$$M = (5x - \tfrac{5}{8}x^2) \text{ kN-m} \qquad \text{(iii)}$$

The values of F and M given by expressions (ii) and (iii) are not plotted in the usual way. Instead, using the same base as for the concentrated load diagrams, positive quantities are plotted *downwards* and negative quantities are plotted *upwards*. As a result of this the overall shearing force and bending moment diagrams are the *differences* between the diagrams due to the concentrated loads and those due to the uniformly distributed load.

The values of the shearing force and bending moment at any section are given by the vertical intercepts of the areas shaded in the diagram. Since the rule of "positive quantities are plotted upwards" does not apply to these diagrams, it is necessary to indicate by signs which are the regions of positive and negative shearing force and bending moment.

The greatest positive (left-up) shearing force

= (+) 5 kN [between C and D]. (*Ans*)

The greatest negative (right-up) shearing force

= (−) $5\tfrac{3}{4}$ kN [immediately to the left of C]. (*Ans*)

The greatest negative (hogging) bending moment

= (−) 10 kN-m [at C]. (*Ans*)

The greatest positive (sagging) bending moment occurs between B and C at the point where the overall shearing force is zero. At any section XX, distance x m from B, the shearing force due to the concentrated loads is $-\tfrac{3}{4}$ kN (constant) and due to the uniformly distributed load $F = (5 - 1\tfrac{1}{4}x)$ kN (equation ii).

Hence for maximum B.M. (zero S.F.)

$$5 - 1\tfrac{1}{4}x - \tfrac{3}{4} = 0 \qquad x = 3\tfrac{2}{5} \text{ (m)}$$

Substituting this value in expressions (i) and (iii), the maximum (positive) bending moment is

$$M = [-4 - (\tfrac{3}{4} \times 3\tfrac{2}{5})] + [(5 \times 3\tfrac{2}{5}) - \tfrac{5}{8}(3\tfrac{2}{5})^2]$$
$$= (+)\, 3\tfrac{9}{40} \text{ kN-m} \qquad (Ans)$$

The points of contra-flexure are clearly between B and C, to which region equations (i) and (iii) apply.

Equating the total B.M. to zero we have, from (i) and (iii),

$$(-4 - \tfrac{3}{4}x) + (5x - \tfrac{5}{8}x^2) = 0$$

or

$$-\tfrac{5}{8}x^2 + 4\tfrac{1}{4}x - 4 = 0$$

from which

$$x = 1{\cdot}13 \text{ or } 5{\cdot}67 \text{ (approx.)}$$

Hence there are two points of contra-flexure distances 1·13 m and 5·67 m to the right of B. (*Ans*)

PROBLEMS

1. ABCDE is a simply supported beam 10 m long. AB = $1\frac{1}{2}$ m, BC = 4 m, CD = $3\frac{1}{2}$ m and DE = 1 m. Calculate the reactions when

(i) The beam is supported at A and E, and carries concentrated loads of 40, 70, and 20 kN at B, C, and D respectively.

(ii) The beam is supported at A and D, and carries concentrated loads of 70, 50, and 140 kN at B, C, and E respectively.

(iii) The beam is supported at A, C, and E, and carries concentrated loads of 120 and 150 kN at B and D respectively, the reaction at C being 180 kN.

(iv) The beam is supported at B and D, and there is a uniformly distributed load of 10 kN per metre run between A and D.

(v) The beam is supported at A and C, there is a uniformly distributed load of 40 kN/m between B and C, and there are concentrated loads of 40 kN (each) at D and E.

(vi) The beam is supported at B and D, there being concentrated loads of 20 kN and 10 kN at A and E respectively, together with a clockwise couple of 70 kN-m applied at C (A being treated as the left-hand end).

Answers. The reactions (in kN) in the six cases are

(i) $R_A = 67\frac{1}{2}$, $R_E = 62\frac{1}{2}$ (ii) $R_A = 62\frac{2}{9}$, $R_D = 197\frac{7}{9}$
(iii) $R_A = 36$, $R_E = 54$ (iv) $R_B = 54$, $R_D = 36$
(v) $R_A = 0$, $R_C = 240$ (vi) $R_B = 13\frac{1}{3}$, $R_D = 16\frac{2}{3}$

2. Draw, to scale, the shearing force and bending moment diagrams for the cases (i)–(vi) of Problem 1.

Answers. All shearing forces are in kN, all bending moments in kN-m. In cases (i)–(iii) the S.F. diagrams consist of horizontal "steps" and the B.M. diagrams of (sloping) straight lines.

(i) S.F.: A to B, $+67\frac{1}{2}$; B to C, $+27\frac{1}{2}$; C to D, $-42\frac{1}{2}$; D to E, $-62\frac{1}{2}$.
B.M.: At A, 0; at B, $+101\frac{1}{4}$; at C, $+211\frac{1}{4}$; at D, $+62\frac{1}{2}$; at E, 0.

(ii) S.F.: A to B, $+62\frac{2}{9}$; B to C, $-7\frac{7}{9}$; C to D, $-57\frac{7}{9}$; D to E, +140.
B.M.: At A, 0; at B, $+93\frac{1}{3}$; at C, $+62\frac{2}{9}$; at D, −140; at E, 0.

(iii) S.F.: A to B, +36; B to C, −84; C to D, +96; D to E, −54.
B.M.: At A, 0; at B, +54; at C, −282; at D, +54; at E, 0.

(iv) S.F.: Straight line from zero at A to −15 at B. Straight line from 39 at B to −36 at D, passing through zero at a point $3\frac{9}{10}$ m to the right of B.
B.M.: Parabola from zero at A to $-11\frac{1}{4}$ at B; parabola from $-11\frac{1}{4}$ at B to zero at D, passing through zero at a point (of contra-flexure) $\frac{3}{10}$ m to the right of B and a maximum of $+64\frac{4}{5}$ at a point $3\frac{9}{10}$ m to the right of B.

(v) S.F.: A to B, zero; straight line from zero at B to −160 at C; C to D, +80; D to E, +40.
B.M.: A to B, zero; parabola from zero at B to −320 at C; straight line from −320 at C to −40 at D; straight line from −40 at D to zero at E.

(vi) S.F.: A to B, −20; B to D, $-6\frac{2}{3}$; D to E +10.
B.M.: Straight line from zero at A to −30 at B; straight line from −30 at B to $-56\frac{2}{3}$ at C; straight line from $+13\frac{1}{3}$ at C to −10 at D, passing through zero at a point (of contra-flexure) 2 m to the right of C; straight line from −10 at D to zero at E.

3. A beam ABCD, 10 m long, is simply supported at A and at C, 8 m from A. There is a uniformly distributed load of 1 kN per metre run between the supports and concentrated loads of 50 kN at B, 2 m from A and 20 kN at D. Sketch the shearing force and bending moment diagrams approximately to scale, marking thereon all the principal values; and state the position and magnitude of the maximum bending moment. (*I.Mech.E.*)

Answer. S.F. (kN): Straight line from $+152\frac{1}{2}$ at A to $+92\frac{1}{2}$ at B; straight line from $+42\frac{1}{2}$ at B to $-137\frac{1}{2}$ at C; C to D, +20 (constant).
B.M. (kN-m): Parabola from zero at A to +245 at B; parabola from +245 at B to −40 at C, passing through zero at a point (of contra-flexure) 7·7 m from A; straight line from −40 at C to zero at D.

The maximum bending moment is (+) 345 kN-m and occurs at a point 3·42 m from A.

4. A beam with overhanging ends is loaded as shown in Fig. 2.20. What must be the magnitude of W in order that a point of contra-flexure may occur at a section 1 m to the left of the right-hand support? Also determine (*a*) the maximum

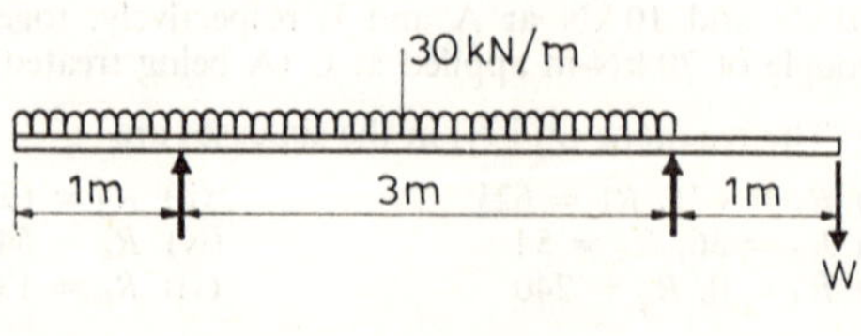

Fig. 2.20

bending moment on that part of the beam between the supports, and (*b*) the position of the other point of contra-flexure. Give a dimensioned sketch of the bending moment diagram. (*I.Mech.E.*)

Answer. $W = 37{\cdot}5$ kN. (*a*) Maximum bending moment between the supports $=(+)$ 8·44 kN-m, at a point 2·25 m from the left-hand end.

(*b*) The other point of contra-flexure is 1·5 m from the left hand end.

B.M. (kN-m): Parabola from zero at the left-hand end to −15 at left-hand support; parabola from −15 at the left-hand support through the points of contra-flexure and the maximum to −37·5 at the right-hand support; straight line from −37·5 at the right-hand support to zero at the right-hand end.

5. A beam with overhanging ends rests freely on two supports A and B, and is loaded as shown in Fig. 2.21. What must be the intensity of loading, in kN per metre run on the beam between C and B if the shearing force is to be zero at a cross-section 1·5 m to the left of support B?

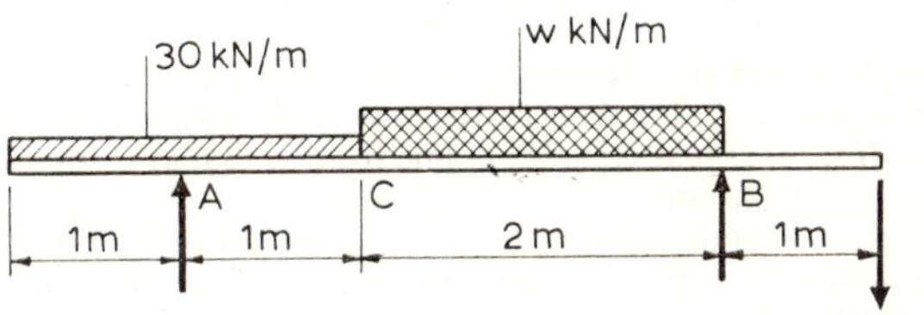

Fig. 2.21

Draw to scale the shearing force and bending moment diagrams and *calculate* the position of the points of contra-flexure. (*I.Mech.E.*)

Answer. $w = 20$ (kN/m) to give zero shearing force as required.

S.F. (kN): Straight line from zero at left-hand end to −30 at A; straight line from +40 at A to +10 at C; straight line from +10 at C to −30 at B; between B and right-hand end, +10 (constant).

B.M. (kN-m): Parabola from zero at left-hand end to −15 at A; parabola from −15 at A to +10 at C; parabola from +10 at C to −10 at B with a maximum of +12·5 at a point 1·5 m to the left of B; straight line from −10 at B to zero at the right-hand end.

The points of contra-flexure are 0·45 m to the right of A and 0·38 m (approx.) to the left of B.

6. Figure 2.22 illustrates a beam under lateral loading. Construct shear and bending moment diagrams, taking care to specify the scales used. (*R.Ae.S.*)

Answer. Reactions (in kN) are $R_A = 12$, $R_B = 10$.

S.F. (kN): Straight line from −5 at left-hand end to −7 at A; straight line from +5 at A to 0 at 2 kN load; from 2 kN load to 3 kN load, +2 (constant); straight line from −1 at 3 kN load to −3·5 at B; straight line from +6·5 at B to +4 at right-hand end.

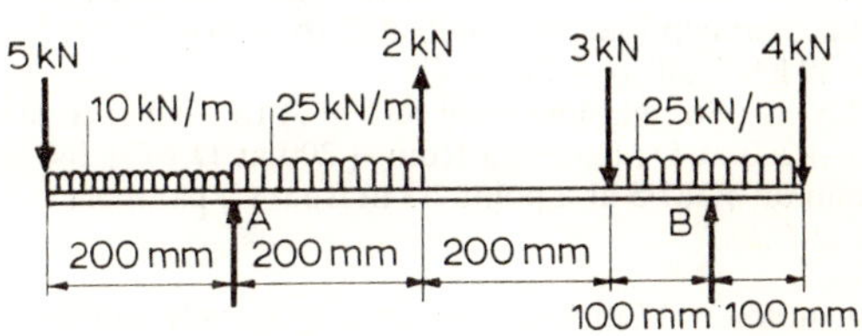

Fig. 2.22

B.M. (kN-m): Parabola from zero at left-hand end to −1·2 at A; parabola from −1·2 at A to −0·7 at 2 kN load; straight line from −0·7 at 2 kN load to −0·3 at 3 kN load; parabola from −0·3 at 3 kN load to −0·525 at B; parabola from −0·525 at B to zero at right-hand end.

7. Define the bending moment and shearing force at a section of a loaded beam and prove that the slope of the bending moment diagram at any section equals the shearing force at that section. A beam AE is 7 m long and is simply supported at one end A and at D, 2·5 m from the other end E. There is a uniformly distributed load of 30 kN/metre run across AD, together with point loads of 60 kN at B, 1·5 m from A; 30 kN at C, 3 m from A, and 20 kN at the end E.

Sketch the bending moment and shearing force diagrams, giving the important numerical values.

Calculate the maximum bending moment and the point at which it occurs. (*I.Struct.E.*)

Answer. Reactions (in kN) are $R_A = 106{\cdot}4$, $R_D = 138{\cdot}6$.
S.F. (kN): Straight line from +106·4 at A to +61·4 at B; straight line from +1·4 at B to −43·6 at C; straight line from −73·6 at C to −118·6 at D; from D to E, +20 (constant).
B.M. (kN-m): Parabola from zero at A to +125·8 at B; parabola from +125·8 at B to +94 at C; parabola from +94 at C to −50 at D; straight line from −50 at D to zero at E.

Maximum B.M. is 125·9 kN-m (approx.) and occurs at a point 1·547 m from A.

8. A horizontal beam 9 m long carries a uniformly distributed load, including its own weight, of 20 kN per metre run and in addition a concentrated load of 22·5 kN at the left-hand end. The beam is supported at two points 6 m apart, so chosen that each support carries half the total load.

Plot *on squared paper* the bending moment and shearing force diagrams, indicating clearly the principal values on each. (*U.L.*)

Answer. The left-hand support must be 1 m from the left-hand end.
S.F. (kN): Straight line from −22·5 at left-hand end to −42·5 at left-hand support; straight line from +58·75 at left-hand support to −61·25 at right hand support; straight line from +40 at right-hand support to zero at right hand end.
B.M. (kN-m): Parabola from zero at left-hand end to −32·5 at left-hand support; parabola from −32·5 at left-hand support to −40 at right-hand support with a maximum of +53·8 at a point 3·94 m from the left-hand end and points of contra-flexure at 1·62 and 6·26 m from the left-hand end.

9. A beam ABC is simply supported at B and C and AB is a cantilevered portion.

AB = 2 m; BC = 6 m. The loading consists of 20 kN concentrated at A, 30 kN concentrated at D, 4 m from C, and 40 kN concentrated at 2 m from C. In addition, the beam carries a uniformly distributed load of 60 kN/m over the length DC. Draw dimensioned sketches of the shearing force and bending moment diagrams. (*U.L.*)

Answer. S.F. (kN): Between A and B, −20 (constant); between B and D, +120 (constant); straight line from +90 at D to −30 at 40 kN load; straight line from −70 at 40 kN load to −190 at C.
B.M. (kN-m): Straight line from zero at A to −40 at B; straight line from −40 at B to +200 at D; parabola from +200 at D to +260 at 40 kN load with a maximum of +267·5 at a point 2·5 m from C; parabola from 260 at 40 kN load to zero at C.

10. A cantilever, 12 m long, carries a uniformly distributed load of 20 kN/m throughout. Draw to scale the bending moment diagram for this loading. Use the method of superposition to obtain the bending moment diagrams when

(i) there is also a concentrated load of 70 kN at a point 8 m from the fixed end;
(ii) there is no concentrated load but the cantilever has a "knife-edge" support at the "free" end which carries $\frac{3}{8}$ of the distributed load. For this case calculate the position of the point of contra-flexure and the maximum hogging and sagging bending moments.

Answer. (i) Maximum bending moment, -2000 kN-m (at fixed end). (ii) The point of contra-flexure is 3 m from the fixed end.

Maximum sagging B.M. = (+) 202·5 kN-m at a point $7\frac{1}{2}$ m from the fixed end.
Maximum hogging B.M. = (−) 360 kN-m at the fixed end.

11. A beam, 10 m long, rests on three supports, one at each end and one at its mid-point. It carries concentrated loads of 90 kN at points 3 m from each end. Use the method of superposition to obtain the bending moment diagrams when the central reaction equals

(i) one-half of the total load;
(ii) two-thirds of the total load.

In each case, state the maximum hogging and sagging bending moment values and locate any points of contra-flexure.

Answer.

(i) Maximum sagging B.M. = (+) 135 kN-m (at each 90 kN load). The B.M. diagram is entirely positive (sagging) and there are no points of contra-flexure.
(ii) Maximum sagging B.M. = (+) 90 kN-m (at each 90 kN load).

Maximum hogging B.M. = (−) 30 kN-m (at central support). There are points of contra-flexure 4·5 m from each end.

12. A beam AB, 6 m long, is simply supported at its ends and carries a distributed load which varies linearly from 10 kN/m at A to 70 kN/m at B.

Calculate

(i) the reactions at A and B,
(ii) the total load carried by the beam,
(iii) the bending moment at mid-span,
(iv) the position and magnitude of the maximum bending moment.

Hint. At any distance x m from A, the loading is

$$w = (\tfrac{1}{2} + \tfrac{1}{4}x) \text{ kN/m}$$

Integrate twice to obtain the bending moment, the constants of integration being evaluated from the conditions that the bending moment is zero at A and B.

Answer.

(i) $R_A = 90$ kN; $R_B = 150$ kN.
(ii) Total load = 240 kN.
(iii) The bending moment at mid-span = 180 kN-m.
(iv) The maximum bending moment is 182·8 kN-m and occurs at a point 3·36 m from A.

13. A beam, 5 m long, is simply supported at its ends and carries a uniformly distributed load of 15 kN/m throughout, together with couples applied to the two ends of the beam at the supports, 24 kN-m anti-clockwise at the left-hand end, and 36 kN-m clockwise at the right-hand end.

Calculate

(i) the bending moment at mid-span,
(ii) the reactions at the supports,
(iii) the position and magnitude of the maximum sagging bending moment,
(iv) the positions of the points of contra-flexure.

Answer.

(i) +16·88 kN-m.
(ii) 35·1 and 39·9 kN at the left- and right-hand ends respectively.
(iii) 17·1 kN-m, at a point 2·34 m from the left-hand end.
(iv) 0·834 and 3·85 m from the left-hand end.

14. A horizontal beam is simply supported at the ends and carries a uniformly distributed load of 32 kN/m between the supports placed 9 m apart. Counter-clockwise moments of 120 and 100 kN-m respectively are applied to the two ends of the beam at the supports. Draw, approximately to scale, the bending moment diagram for the beam and find

(i) the reactions at the supports,
(ii) the position and magnitude of the greatest bending moment. *(U.L.)*

Answer. (i) 168·4 and 119·6 kN; (ii) 323 kN-m at 5·26 m from the end with the 120 kN-m moment.

15. A horizontal beam, simply supported on a span of 10 m, carries a total load of 20 kN; the load distribution varies parabolically from zero at each end to a maximum at mid-span. Calculate the values of the bending moment at intervals of 1 m and plot, on *squared paper*, the bending moment diagram for this loading. State the values of

(*a*) the maximum bending moment,
(*b*) the shearing force at the quarter-span points. *(U.L.)*

Hint. For parabolic distribution $w = ax^2 + bx + c$ but $c = 0$ since $w = 0$ when $x = 0$. Prove that, for a maximum at mid-span $b = -aL$ (L = span). Hence show that the total load is $-aL^3/6$ and that $a = -0{\cdot}12$ (kN and m units). Then proceed as in Problem 12.

Answer. (*a*) 31·25 kN-m; (*b*) 6·875 kN.

16. A beam, 8 m long is simply supported at two points 5 m apart, one at 2 m from the left-hand end and the other at 1 m from the right-hand end. The beam carries a uniformly distributed load of 15 kN/m on the whole length.

(*a*) Calculate the shearing forces and bending moments at the supports.
(*b*) Determine where the sagging bending moment is greatest and the value of this moment.
(*c*) Calculate the positions of the points of contra-flexure.

Make well-proportioned freehand sketches of the shearing force and bending moment diagrams for the beam and show on them all the calculated values. *(U.L.)*

Answer. (*a*) At left-hand support, S.F. = −30 kN on left and 42 kN on right. B.M. = −30 kN-m. At right-hand support, S.F. = −33 kN on left and 15 kN on right. B.M. = −7·5 kN-m.

(*b*) 4·8 m from left-hand end; 28·8 kN-m.
(*c*) 2·84 and 6·76 m from left-hand end.

17. A beam AD, 8 m long, is supported at B and C where AB = 2 m, BC = 5 m and CD = 1 m, AB and CD being overhanging ends. It carries loads of 10 kN at A, 20 kN at D and 20 kN at E, the mid-point of BC, and in addition a uniformly distributed load of 10 kN/m from B to C.

Calculate the bending moments at B, E and C, and, *on squared paper*, sketch neatly, with dimensions, the bending moment and shearing force diagrams. (*U.L.*)

Answer. B.M. at B, −20 kN-m; at E, 36·25 kN-m; at C, −20 kN-m. S.F. diagram (kN): From A to B, −10 constant; straight line from +35 at B to +10 at E; straight line from −10 at E to −35 at C; from C to D, +20 constant.
B.M. diagram (kN-m): Straight line from 0 at A to −20 at B; parabola from −20 at B to +36·25 at E; parabola from +36·25 at E to −20 at C; straight line from −20 at C to 0 at D.

18. A beam 10 m long is simply supported at the left-hand end and at a point 2 m from the right-hand end; it carries a uniformly distributed load of 24 kN/m extending from the left-hand end to the mid-point of the beam, and a point load of 40 kN at the right-hand end.

Draw *on squared paper* the shearing force and bending moment diagrams for the beam. Determine where the greatest bending moment occurs and give its value. Find also the positions of the points of contra-flexure. Use the following scales

Space, 1:100; shearing force, 10 mm = 20 kN;
bending moment, 10 mm = 20 kN-m.

Show on your diagrams the calculated values and any other values obtained for plotting them. (*U.L.*)

Answer. Maximum B.M. = 109·5 kN-m at 3·02 m from left-hand end. Point of contra-flexure 6·31 m from left-hand end.
S F. diagram (kN): Straight line from +72·5 at left-hand end to −47·5 at 5 m from left-hand end. From this point to right-hand support, −47·5 (constant). From right-hand support to right-hand end, +40 (constant).
B.M. diagram (kN-m): Parabola from 0 at left-hand end to +62·5 at a point 5 m from left-hand end with the maximum given above. Straight line from +62·5 at this point to −80 at right-hand support. Straight line from −80 at right-hand support to 0 at right-hand end.

19. Fig. 2.23 shows a beam 8·5 m long which rests on supports 6 m apart.

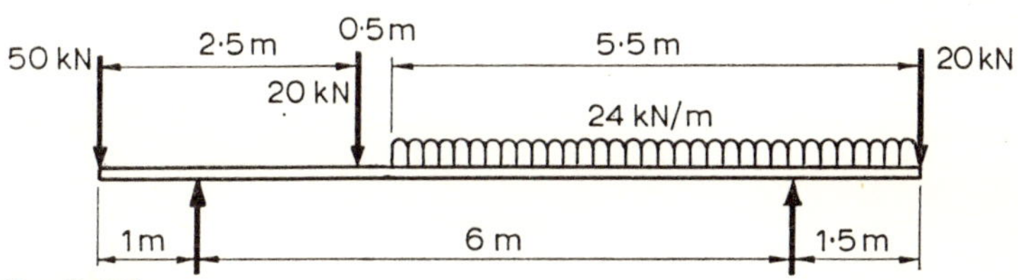

Fig. 2.23

Determine the reactions and draw to scale the shearing force and bending moment diagrams for the given loading. Scales—Space 1:100; S.F. 10 m = 20 kN; B.M. 10 mm = 10 kN-m.

State the maximum positive and negative values of the shearing force and bending moment and the positions of the points of contra-flexure. (*U.L.*)

Answer. Left-hand reaction =95·8 kN; right-hand reaction = 126·2 kN. If x m is distance from left-hand end then

S.F. diagram (kN): From $x = 0$ to $x = 1$, -50 (constant); from $x = 1$ to $x = 2{\cdot}5$, $+46$ (constant); from $x = 2{\cdot}5$ to $x = 3$, $+26$ (constant); straight line from $+26$ at $x = 3$ to -70 at $x = 7$; straight line from $+56$ at $x = 7$ to $+20$ at $x = 8{\cdot}5$.

B.M. diagram (kN-m): Straight line from 0 at $x = 0$ to -50 at $x = 1$; straight line from -50 at $x = 1$ to $+19$ at $x = 2{\cdot}5$; straight line from $+19$ at $x = 2{\cdot}5$ to $+31{\cdot}7$ at $x = 3$; parabola from $+31{\cdot}7$ at $x = 3$ to -57 at $x = 7$ passing through a maximum value of $+46{\cdot}1$ at $x = 4{\cdot}08$; parabola from -57 at $x = 7$ to 0 at $x = 8{\cdot}5$.

Maximum positive S.F. = 56, maximum negative S.F. = (−) 70.

Maximum positive B.M. = 46·1, maximum negative B.M. = (−) 57.

20. A beam having a total length of 10 m rests on supports 7·5 m apart; one support is 1·5 m from the left-hand end and the other support is 1 m from the right-hand end. The beam carries a uniformly distributed load of 20 kN/m run extending for a length of 6 m measured from the left-hand end of the beam and a point load of 12 kN at the right-hand end of the beam. Draw to scale the shearing force and bending moment diagrams due to the given loading. Show on the diagrams all the maximum positive and negative values and the positions of the points of contra-flexure.

Scales: space 1:100; shearing force 10 mm = 20 kN; bending moment 10 mm = 20 kN-m. (*U.L.*)

Answer. If x m is the distance from left-hand end, then

S.F. diagram (kN): Straight line from 0 at $x = 0$ to -30 at $x = 1{\cdot}5$; straight line from $+64{\cdot}4$ at $x = 1{\cdot}5$ to $-25{\cdot}6$ at $x = 6$; from $x = 6$ to $x = 9$, $-25{\cdot}6$ (constant); from $x = 9$ to $x = 10$, $+12$ (constant).

B.M. diagram (kN-m): Parabola from 0 at $x = 0$ to $-22{\cdot}5$ at $x = 1{\cdot}5$; parabola from $-22{\cdot}5$ at $x = 1{\cdot}5$ to $+64{\cdot}8$ at $x = 6$, passing through a maximum of 81·1 at $x = 4{\cdot}72$; straight line from 64·8 at $x = 6$ to -12 at $x = 9$; straight line from -12 at $x = 9$ to 0 at $x = 10$.

There are two points of contra-flexure: at $x = 1{\cdot}87$ and $x = 8{\cdot}53$.

Chapter 3

Longitudinal Stresses in Beams

The bending equation for an initially straight unstressed beam is

$$\sigma/y = M/I = E/R$$

where

σ = stress in N/m^2 at a distance y m from the neutral axis,
M = the bending moment (or moment of resistance) in N-m,
I = the second moment of area (or moment of inertia) of the section about the neutral axis in m^4,
E = Young's modulus of elasticity in N/m^2,
R = radius of curvature of the neutral surface in metres.

Also

$$M = Z\sigma_{max}$$

where σ_{max} = maximum stress in N/m^2 occurring at the section, and Z = section modulus in m^3 and is defined as I/y_{max} where y_{max} is the distance from the neutral axis to the most strained fibre.

Note. To avoid very large or very small numbers I is often given in cm^4, E in GN/m^2, σ in MN/m^2 and Z in cm^3. When substituting in formulae the units of these quantities should be converted to those given above.

Combined bending and direct stress: the total longitudinal stress at any point is the algebraic sum of the direct stress and the bending stress at that point.

WORKED EXAMPLES

3.1. What is a "neutral axis"? Derive an expression for the stress σ at any distance y from the neutral axis in terms of Young's modulus

E for an initially straight (and unstressed) beam which is bent until the radius at the neutral axis is R.

Calculate the maximum stress in a piece of rectangular steel strip, 25 mm wide and 3 mm thick, when it is bent round a drum, 2·4 m diameter. $E = 206\ \text{GN/m}^2$.

Solution. If a beam, simply supported at its ends, carries concentrated or distributed loads (acting downwards) then the upper layers (or fibres) of the beam are compressed and the lower layers are extended. Similarly, in the case of a loaded cantilever, the upper layers are in tension and the lower layers are in compression. In either case there is some intermediate surface, called the *neutral surface*, which is neither in tension nor compression. The line of intersection of the neutral surface and any cross-section is called the *neutral axis* of that cross-section. It is usually shown thus:

n —‑———‑———‑——— *a*

For symmetrical cross-sections in which bending takes place about an axis of symmetry, the neutral axis is that axis of symmetry.

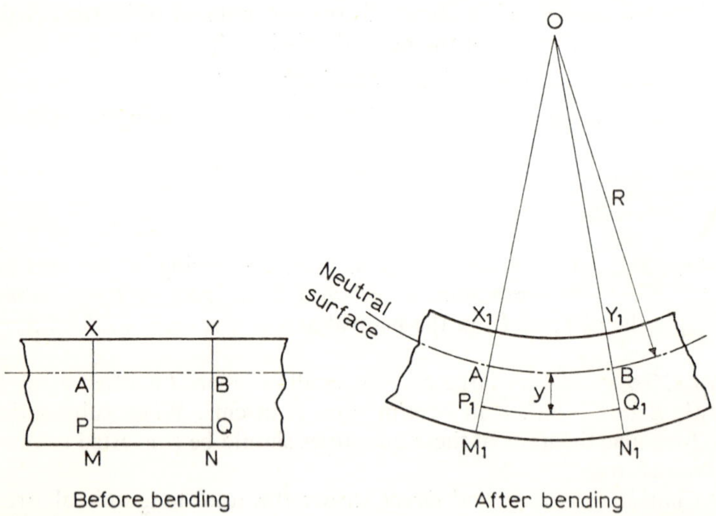

Fig. 3.1

Suppose (Fig. 3.1) XAPM and YBQN represent two cross-sections of a beam, originally parallel. AB is a "layer" in the neutral surface, XY and MN are the top and bottom layers of the beam, while PQ is another horizontal layer in the beam, distance y below the neutral surface.

When the beam is bent, as shown, the two cross-sections tilt, and, assuming that they remain plane sections, they will have a line of intersection at O. The layer XY will be compressed and the layer MN will be stretched. The layer PQ will also be in tension while AB, which is in the neutral surface, is unstressed, and thus retains its original length.

If the angle between the planes after bending is θ radians and R is the radius of the neutral surface, then $AB = R \times \theta$. Since, from the first diagram PQ = AB, we have

$$PQ = R \times \theta \qquad \text{(i)}$$

After bending, the new length of PQ is

$$P_1Q_1 = (R + y)\theta \qquad \text{(ii)}$$

Using (i) and (ii), the stress at a distance y below the neutral axis is

$$\begin{aligned}\sigma &= E \times (\text{strain of the layer PQ})\\ &= E \times \frac{\text{extension of PQ}}{\text{original length of PQ}}\\ &= E\left[\frac{P_1Q_1 - PQ}{PQ}\right]\\ &= E\left[\frac{(R+y)\theta - R\theta}{R\theta}\right]\\ &= E(y/R)\end{aligned}$$

This is usually written

$$\sigma/y = E/R \qquad \textit{(Ans)}$$

The layers above the neutral axis are in compression and the same expression will give the compressive stress at any distance y above the neutral axis. For hogging type of bending the layers above the neutral axis are in tension, and those below are in compression.

Since E/R is constant across any one section of a beam (of one material) the stress σ is proportional to the distance y from the neutral axis. The greatest stresses will occur at the top and bottom edges of the section.

The steel strip given in the question is of symmetrical cross-section and the neutral axis is the axis of symmetry (see Fig. 3.2).

The radius to which the neutral surface is bent is

$$R = 1{\cdot}2\ \text{m} + 1{\cdot}5\ \text{mm} = 1{\cdot}202\ \text{m (approx)}$$

The maximum value of y is 1·5 mm (to the inside or outside edges of the strip).

The maximum stress is therefore

$$\sigma_{max} = \frac{E}{R} y_{max} = \frac{206 \times 10^9 \text{ N/m}^2}{1{\cdot}202 \text{ m}} \times (1{\cdot}5 \times 10^{-3} \text{ m})$$

$$= 257 \text{ MN/m}^2 \qquad (Ans)$$

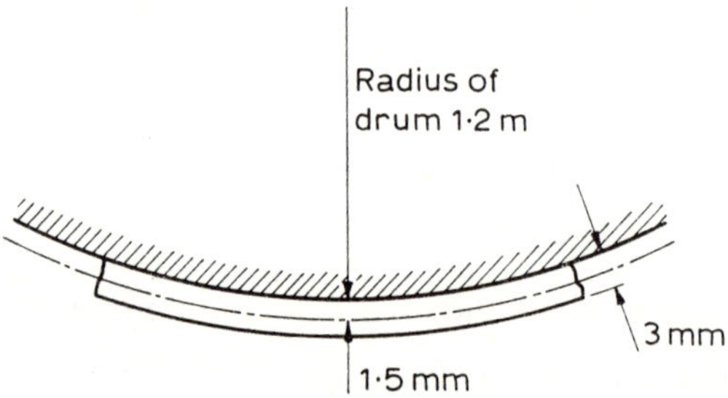

Fig. 3.2

This is the value of the compressive stress at the inside edge of the strip and also of the tensile stress at the outside edge. (It should be noted that sufficient accuracy is obtained if the radius of the neutral surface is taken to be equal to the radius of the drum.)

3.2. What is a "moment of resistance"? An I-section beam has flanges each 150 mm by 13 mm and an overall depth of 250 mm. Calculate the moment of resistance at a section where the flange stress is 77 MN/m². Neglect the effects of the web and assume that the stress in each flange is uniform. (There is no axial force in the beam which is in equilibrium under lateral loads.)

Solution. For a beam to be in equilibrium the sum of the moments of the internal forces acting at any section must equal the bending moment at that section. The total internal moment is called the *moment of resistance*, and it is usually calculated for a prescribed maximum stress.

Suppose, in the example given, that the upper flange is in compression and the lower flange is in tension, as shown in Fig. 3.3. The compressive *force* in the upper flange is

$$P = \text{stress} \times \text{flange area}$$
$$= (77 \times 10^6) \text{ N/m}^2 \times (150 \times 13 \times 10^{-6}) \text{ m}^2$$
$$= 150 \text{ kN}$$

The tensile force in the lower flange is numerically the same but opposite in direction, since the resultant force in the longitudinal direction is zero.

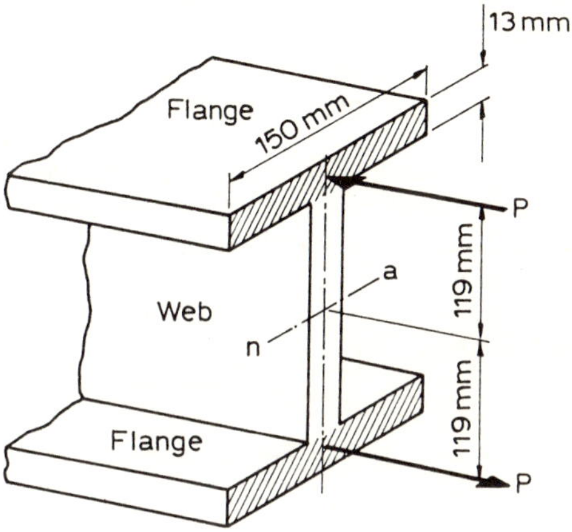

Fig. 3.3

These two forces can be regarded as acting at the mid-points of the flanges, as shown in Fig. 3.3, and they provide a couple, which is the moment of resistance, of

$$M = 2 \times 150 \times 10^3 \text{ N} \times 0{\cdot}119 \text{ m} = 35{\cdot}7 \text{ kN-m} \qquad (Ans)$$

The above method is applicable only to cross-sections in which nearly all the area is situated at an approximately constant distance from the neutral axis. The result obtained here should be compared with that found by the more exact method in Example 3.4.

3.3. Derive the bending equation $\sigma/y = M/I = E/R$, stating the assumptions involved.

Calculate the moment of resistance of the steel strip of Example 3.1 when it is bent round the drum.

Solution. The theory of simple bending depends upon the following assumptions:

(*a*) The material of the beam is uniform throughout.
(*b*) Each cross-section of the beam is symmetrical about the plane of bending. This means that for a horizontal beam carrying vertical loads the section must be symmetrical about a vertical axis.

(*c*) The beam is initially straight and unstressed.
(*d*) The loads are applied to the beam in the plane of bending.
(*e*) Plane cross-sections of the beam before bending remain plane after bending.
(*f*) Hooke's Law applies to each longitudinal layer of the beam and Young's modulus of elasticity has the same value in compression as in tension.
(*g*) The resultant force perpendicular to any cross-section is zero.

From the relationship $\sigma/y = E/R$, derived in Example 3.1, the stress at any point, distance y from the neutral axis, is $\sigma = (E/R)y$. Hence the force on a small strip, area δA, parallel to the neutral axis, as shown in Fig. 3.4, is

$$\sigma\delta A = (E/R)y\delta A \qquad \text{(i)}$$

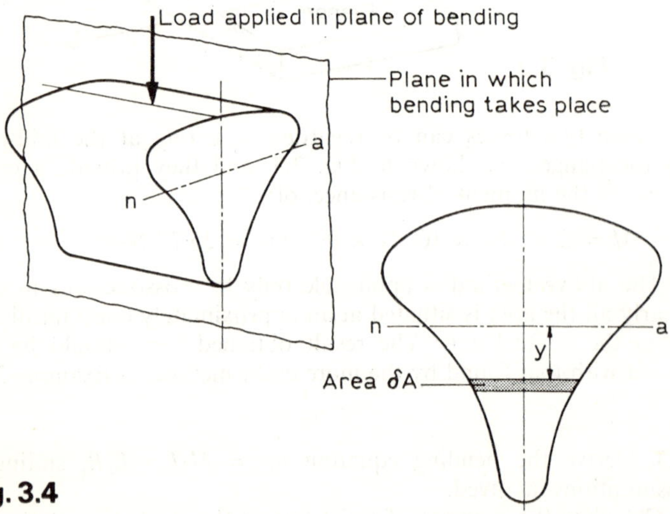

Fig. 3.4

The moment of this force about the neutral axis is

$$\sigma\delta Ay = (E/R)y^2\delta A \qquad \text{(ii)}$$

The total force on the cross-section is the sum of all terms such as (i). This can be expressed as $\Sigma(E/R)y\delta A$. In the notation of the calculus, the total force perpendicular to the section is

$$\int(E/R)ydA = (E/R)\int ydA \qquad \text{(iii)}$$

the integral being taken over the whole cross-section. E/R is taken outside of the integral because it has a constant value for any one section.

From assumption (g) the total force given by (iii) is zero. The integral $\int y dA$ is called the *first moment of area* for the section (see Appendix 1). Since E/R is not zero, this integral must be zero and the first moment of area can be zero only about an axis passing through the centroid. Hence the *neutral axis passes through the centroid of the cross-section.*

The moment of resistance is the sum of all expressions such as (ii), i.e. $\Sigma(E/R)y^2\delta A$, or, in the calculus notation,

$$M = \int(E/R)y^2dA = (E/R)\int y^2dA \qquad \text{(iv)}$$

The last integral, taken over the whole cross-section, is called the *second moment* (*or moment of inertia*) *of the area.* It is denoted by I and must be calculated about the neutral axis. In Appendix 1 values of I for common geometrical shapes are given together with the useful Theorem of Parallel Axes.

Using the symbol I, expression (iv) can be written

$$M = (E/R)I$$

or, in its usual form,

$$M/I = E/R$$

Combining this last result with equation (iii) of Example 3.1, we obtain the bending equation

$$\sigma/y = M/I = E/R \qquad \text{(v)}$$

The units of each quantity in this important relation are given at the beginning of this chapter.

In many practical cases, I is constant along the beam (uniform section) but the bending moment M varies. In these cases the value of R varies from one section to another, and R is no longer the radius of a circular arc but is called the radius of curvature. This matter is considered in more detail in Chapter 5.

The steel strip of Example 3.1 is of rectangular cross-section. Using the appropriate formula for I (see Appendix 1) we have

$$I = bd^3/12 = 0{\cdot}025 \text{ m} \times (0{\cdot}003 \text{ m})^3/12 = 56{\cdot}3 \times 10^{-12} \text{ m}^4$$

Therefore, the moment of resistance is

$$M = \frac{E}{R}I = \frac{206 \times 10^9 \text{ N/m}^2}{1{\cdot}202 \text{ m}} \times (56{\cdot}3 \times 10^{-12} \text{ m}^4) = 9{\cdot}65 \text{ N-m}$$

(*Ans*)

3.4. Calculate the moment of resistance for the I-section of Example 3.2 using the bending equation of Example 3.3 and a maximum stress of 77 MN/m². The web thickness is 13 mm.

Also find its value when the same section is used so that the neutral axis is parallel to the web.

Solution. The value of I for the section can be found using the rectangle formula. The section can be treated as a rectangle 150 mm wide and 250 mm deep, with two rectangular "cut-outs" each 68·5 mm wide and 224 mm deep (Fig. 3.5).

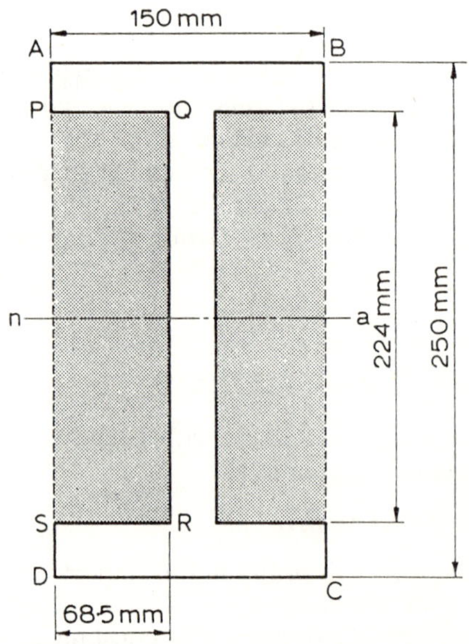

Fig. 3.5

Each second moment of area (or moment of inertia) must however, be taken about the neutral axis. Using the lettering of Fig. 3.5.

I about the n.a.

= (I for rectangle ABCD about n.a.) − (2 × I for rectangle PQRS about n.a.)

$$= \frac{0{\cdot}15 \text{ m} \times (0{\cdot}25 \text{ m})^3}{12} - 2 \times \frac{0{\cdot}685 \text{ m} \times (0{\cdot}224 \text{ m})^3}{12}$$

$= 67{\cdot}1 \times 10^{-6} \text{ m}^4$

The maximum value of y is 125 mm (from the n.a. to the top or bottom edges).

The moment of resistance is, therefore,

$$M = \frac{\sigma}{y} I$$
$$= \frac{(77 \times 10^6 \text{ N/m}^2) \times (67{\cdot}1 \times 10^{-6} \text{ m}^4)}{125 \times 10^{-3} \text{ m}}$$
$$= 41{\cdot}3 \text{ kN-m} \qquad (Ans)$$

(This result is about 16 per cent higher than that obtained by the approximate method of Example 3.2).

If the section is used so that the neutral axis is parallel to the web (Fig. 3.6) the value of I is best found by adding the values for the two flanges (each 13 mm wide and 150 mm deep) to that of the web (224 mm wide and 13 mm deep). If this is done, the rectangle formula $bd^3/12$ is applicable to each.

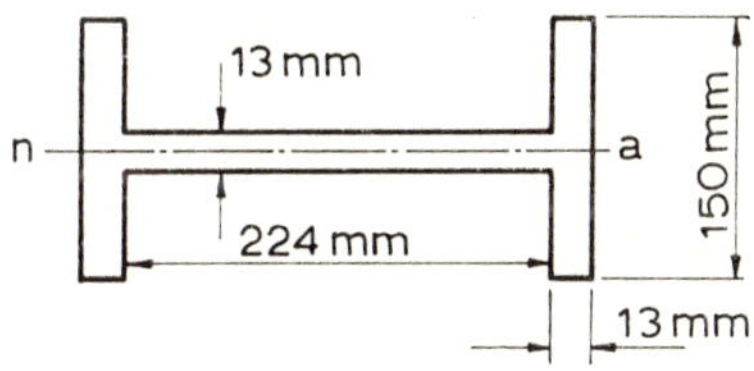

Fig. 3.6

I about the n.a.

$$= (2 \times I \text{ for one flange about n.a.}) + (I \text{ for web about n.a.})$$
$$= \frac{2 \times 0{\cdot}013 \text{ m} \times (0{\cdot}15 \text{ m})^3}{12} + \frac{0{\cdot}224 \text{ m} \times (0{\cdot}013 \text{ m})^3}{12}$$
$$= 7{\cdot}36 \times 10^{-6} \text{ m}^4$$

The maximum value of y is now 75 mm and thus the moment of resistance is

$$M = \frac{\sigma}{y} I = \frac{(77 \times 10^6 \text{ N/m}^2) \times (7{\cdot}36 \times 10^{-6} \text{ m}^4)}{75 \times 10^{-3} \text{ m}}$$
$$= 7{\cdot}56 \text{ kN-m} \qquad (Ans)$$

The last result is much smaller than that obtained with the neutral axis parallel to the flanges. In practice I and T sections are used extensively for beams. They are designed to have a large value of I about one axis and should be positioned so that bending takes place about this axis.

3.5. Calculate the maximum stress in a round steel bar, 200 mm diameter and 15 m long, due to its own weight when it is simply supported at its ends. Density of steel, 7·75 Mg/m³.

Solution

Weight of steel per unit volume

$= \text{density} \times \text{gravitational acceleration}$

$= (7{\cdot}75 \times 10^3\ \text{kg/m}^3) \times (9{\cdot}81\ \text{m/sec}^2)$

$= 76{\cdot}0\ \text{kN/m}^3$

$$\text{Volume of the bar} = \tfrac{1}{4}\pi d^2 \times L = \tfrac{1}{4}\pi(0{\cdot}2\ \text{m})^2 \times (15\ \text{m}) = 0{\cdot}15\pi\ \text{m}^3$$

$$\text{Weight of the whole bar} = \text{volume} \times \text{weight per unit volume} = (0{\cdot}15\pi\ \text{m}^3) \times 76\ \text{kN/m}^3 = 11{\cdot}4\pi\ \text{kN}$$

The weight of the bar can be regarded as a uniformly distributed load, and the maximum bending moment (which occurs at mid-span) is

$$M = WL/8 = (11{\cdot}4\pi\ \text{kN} \times 15\ \text{m})/8 = 21{\cdot}4\pi\ \text{kN-m}$$

The neutral axis of the cross-section is the horizontal diameter and, using the appropriate formula (see Appendix 1),

$$I = \pi d^4/64 = \pi(0{\cdot}2\ \text{m})^4/64 = 10^{-4}\pi/4\ \text{m}^4$$

The maximum value of y equals the radius (0·1 m), and, from the bending equation, the maximum stress is

$$\sigma = \frac{M}{I}y = \frac{(21{\cdot}4\pi\ \text{kN-m}) \times 0{\cdot}1\ \text{m}}{10^{-4}\pi/4\ \text{m}^4} = 85{\cdot}6\ \text{MN/m}^2 \qquad (Ans)$$

This stress will be compressive at the top, and tensile at the bottom, of the mid-span section of the bar.

3.6. A cast-iron bracket subject to bending has a cross-section of I-form with unequal flanges. The total depth of the section is 280 mm and the metal is 40 mm thick throughout. The top flange is 200 mm wide and the bottom is 120 mm. Calculate the position of the neutral axis and the moment of inertia of the section about this axis and determine the maximum bending moment that should be imposed on this section if the tensile stress in the top flange is not to exceed 16 MN/m^2. What is then the value of the compressive stress in the bottom flange? *(I.Mech.E.)*

Solution. The section consists of three rectangles, the upper flange, the web, and the lower flange, as shown in Fig. 3.7. Suppose the centroid of the whole section is at a distance $\bar{x}$ mm from the top edge AB. It is convenient to work initially in mm.

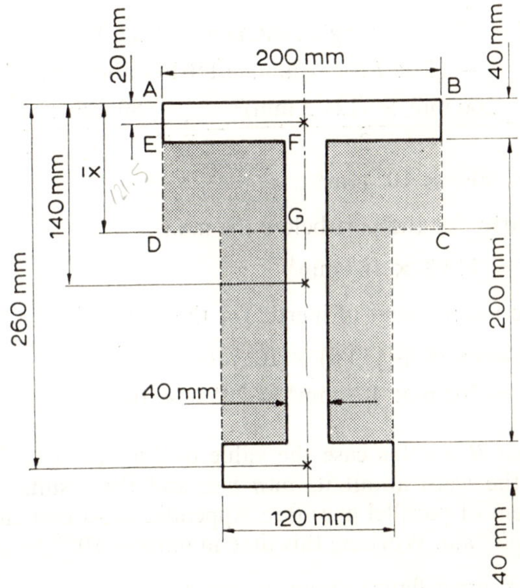

Fig. 3.7

Then, taking first moments of area about the top edge,

$\bar{x} \times$ (total area of section)

= sum of 1st moments of the three separate rectangles about AB

= (area of upper flange × 20 mm) + (area of web × 140 mm) + (area of lower flange × 260 mm)

= (200 mm × 40 mm × 20 mm) + (40 mm × 200 mm × 140 mm) + (120 mm × 40 mm × 260 mm)

$= 2{\cdot}528 \times 10^6\ \text{mm}^3$

and

$$\bar{x} = \frac{2{\cdot}528 \times 10^6\ \text{mm}^3}{(200\ \text{mm} \times 40\ \text{mm}) + (40\ \text{mm} \times 200\ \text{mm}) + (120\ \text{mm} \times 40\ \text{mm})}$$

$= 121{\cdot}5$ mm

Thus, the neutral axis is 121·5 mm below the top edge of the section. (*Ans*)

The value of I can be obtained in two ways.

Method A. Consider rectangles and rectangular "cut-outs" each having one end coincident with the neutral axis as shown in Fig. 3.7. (In each case the expression $bd^3/3$ is applicable). For that part of the section above the n.a.

$$I = (I \text{ for rectangle ABCD about n.a.})$$
$$- (2 \times I \text{ for rectangle DEFG about n.a.})$$
$$= \frac{200\text{ mm} \times (121{\cdot}5\text{ mm})^3}{3} - 2 \times \frac{80\text{ mm} \times (81{\cdot}5\text{ mm})^3}{3}$$
$$= 90{\cdot}8 \times 10^6\text{ mm}^4$$

Similarly, for the part below the n.a.,

$$I = 114{\cdot}8 \times 10^6\text{ mm}^4$$

Hence the moment of inertia for the section is

$$I = (90{\cdot}8 + 114{\cdot}8) \times 10^6\text{ mm}^4$$
$$= 205{\cdot}6 \times 10^6\text{ mm}^4 = 205{\cdot}6 \times 10^{-6}\text{ m}^4$$ (*Ans*)

Method B. In this case the value of I is calculated for each flange (and the web) about its *own n.a.* and the results adjusted by the theorem of parallel axes (see Appendix 1) so that they apply to the "*overall*" *n.a.* Working this time in units of $10^{-6} \times \text{m}^4$,

		10^{-6} m^4
I for upper flange about its own n.a.		
$= 0{\cdot}2\text{ m} \times (0{\cdot}04\text{ m})^3/12$	=	1·07
I for upper flange due to "transferring" to overall n.a.		
$= (0{\cdot}2\text{ m} \times 0{\cdot}04\text{ m}) \times (0{\cdot}1215\text{ m} - 0{\cdot}02\text{ m})^2$	=	82·46
I for web about its own n.a.		
$= 0{\cdot}04\text{ m} \times (0{\cdot}2\text{ m})^3/12$	=	26·67
I for web due to "transferring" to overall n.a.		
$= (0{\cdot}04\text{ m} \times 0{\cdot}2\text{ m}) \times (0{\cdot}14\text{ m} - 0{\cdot}1215\text{ m})^2$	=	2·72
I for lower flange about its own n.a.		
$= 0{\cdot}12\text{ m} \times (0{\cdot}04\text{ m})^3/12$	=	0·64
I for lower flange due to "transferring" to overall n.a.		
$= (0{\cdot}12\text{ m} \times 0{\cdot}04\text{ m}) \times (0{\cdot}26\text{ m} - 0{\cdot}1215\text{ m})^2$	=	92·03
	Total I =	205·60

The distance from the n.a. to the edge with the greatest tensile stress is 121·5 mm. Hence the maximum bending moment which can be applied for the given stress, is

$$M = \frac{\sigma}{y} I = \frac{(16 \times 10^6 \text{ N/m}^2) \times (205{\cdot}6 \times 10^{-6} \text{ m}^4)}{(121{\cdot}5 \times 10^{-3} \text{ m})}$$

$$= 27{\cdot}08 \times 10^3 \text{ N-m} \qquad (Ans)$$

The distance from the n.a. to the bottom edge of the section is 158·5 mm and, since the stress at any point is proportional to the distance of that point from the n.a., then

$$\frac{\text{stress at bottom edge}}{158{\cdot}5 \text{ mm}} = \frac{\text{stress at top edge}}{121{\cdot}5 \text{ mm}}$$

Hence, the maximum compressive stress in the bottom flange is

$$\sigma = \frac{158{\cdot}5 \text{ mm}}{121{\cdot}5 \text{ mm}} \times 16 \text{ MN/m}^2 = 20{\cdot}9 \text{ MN/m}^2 \qquad (Ans)$$

3.7. What is a "section modulus"? Calculate the two moduli for a T-section flange, 125 mm by 25 mm, web thickness 25 mm and overall depth 150 mm.

A beam of this section, 6 mm long, is simply supported (with the flange uppermost) at the left-hand end and at a point 1·5 m from the right-hand end. It carries a uniformly distributed load of 8 kN/m over its entire length. Calculate the maximum hogging and sagging bending moments for the beam and hence find the greatest tensile and compressive stresses occurring in the beam.

Solution. At any section of a beam the moment of resistance (or bending moment) M is related to the maximum stress σ_{max} by the bending equation,

$$M = \left(\frac{I}{y_{max}}\right) \sigma_{max}$$

where y_{max} is the distance from the neutral axis to the edge of the section. The quantity I/y_{max} is called the *section modulus* and it is usually denoted by Z. It is a geometrical property of the section and has the units m³. For symmetrical sections the neutral axis is midway between the edges and the section modulus has one value only. For sections which are not symmetrical about the neutral axis the values of y_{max} to the top and bottom edges are not equal. In these cases two moduli are required so that the moment can be related to the stresses at the top and bottom edges.

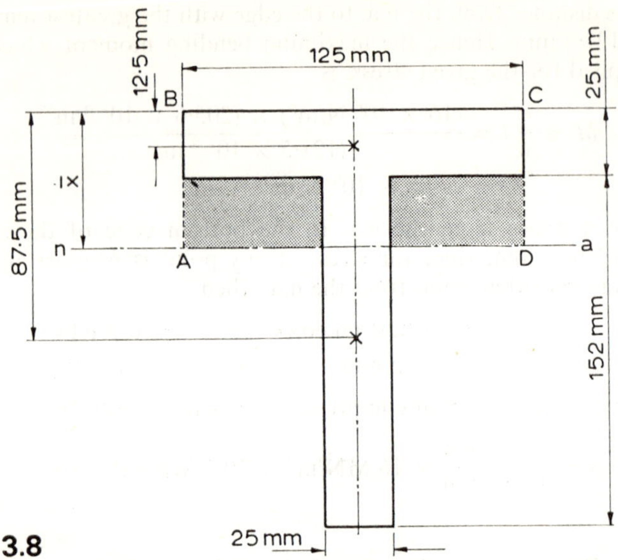

Fig. 3.8

The T-section of the question is shown in Fig. 3.8. If $\bar{x}$ is the distance of the n.a. from the top edge then, taking first moments of area about the top edge, and working in mm units,

$$\bar{x} \times (\text{total area}) = (\text{area of flange} \times 12{\cdot}5\text{ mm}) + (\text{area of web} \times 87{\cdot}5\text{ mm})$$

or

$$\bar{x} = \frac{(125\text{ mm} \times 25\text{ mm} \times 12{\cdot}5\text{ mm}) + (25\text{ mm} \times 125\text{ mm} \times 87{\cdot}5\text{ mm})}{(125\text{ mm} \times 25\text{ mm}) + (25\text{ mm} \times 125\text{ mm})}$$

$$= 50\text{ mm}$$

As in Example No. 3.6, *I* can be evaluated in two ways.

Method A. Considering rectangles and rectangular cut-outs each having one edge coincident with the neutral axis,

$$I = (I\text{ for rectangle ABCD about n.a.}) - (I\text{ for two shaded areas about n.a.}) + (I\text{ for area below n.a. about n.a.})$$

$$= \frac{125\text{ mm} \times (50\text{ mm})^3}{3} - 2 \times \frac{50\text{ mm} \times (25\text{ mm})^3}{3} + \frac{25\text{ mm} \times (100\text{ mm})^3}{3}$$

$$= \frac{100 \times 25^4}{3}\text{ mm}^4$$

Method B mm^4

I for flange about its own n.a.

$$= 125\text{ mm} \times (25\text{ mm})^3/12 \qquad = \tfrac{5}{12} \times 25^4$$

I for flange due to "transferring" to overall n.a.

$$= (125\text{ mm} \times 25\text{ mm}) \times (37{\cdot}5\text{ mm})^2 \qquad = \tfrac{45}{4} \times 25^4$$

I for web about its own n.a.

$$= 25\text{ mm} \times (125\text{ mm})^3/12 \qquad = \tfrac{125}{12} \times 25^4$$

I for web due to "transferring" to overall n.a.

$$= (25\text{ mm} \times 125\text{ mm}) \times (37{\cdot}5\text{ mm})^2 \qquad = \tfrac{45}{4} \times 25^4$$

$$\text{Total } I = \tfrac{100}{3} \times 25^4$$

The section moduli are, for the top edge,

$$Z = \frac{I}{y_{max}} = \frac{100 \times 25^4/3\text{ mm}^4}{50\text{ mm}}$$
$$= 2 \times 25^4/3\text{ mm}^3 \qquad (Ans)$$

for the bottom edge,

$$Z = \frac{100 \times 25^4/3\text{ mm}^4}{100\text{ mm}} = 25^4/3\text{ mm}^3 \qquad (Ans)$$

Let A and B be the points at which the beam is supported, as shown in Fig. 3.9. The reactions, found by taking moments about B and A respectively, are

$$R_A = 16\text{ kN} \quad \text{and} \quad R_B = 32\text{ kN}$$

The maximum hogging bending moment clearly occurs at B and is

$$M = (-)\,1{\cdot}5\text{ m} \times 8\text{ kN/m} \times 0{\cdot}75\text{ m} = (-)\,9\text{ kN-m} \qquad (Ans)$$

The maximum sagging bending moment occurs between A and B at the section where the shearing force is zero. The position of this section is such that the distributed load to its left equals

R_A i.e. it is 16 kN $\div$ 8 kN/m = 2 m to the right of A.

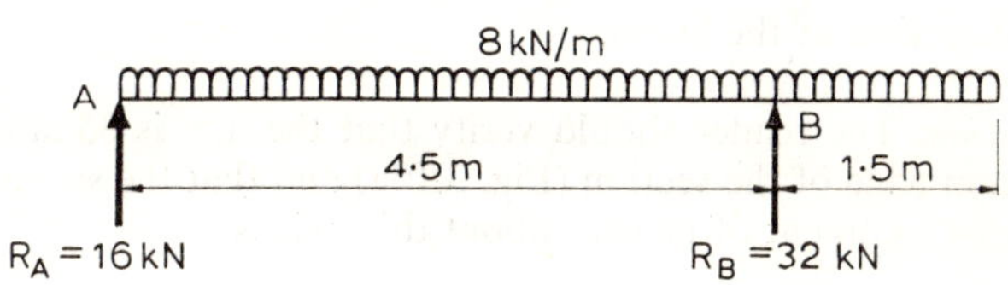

Fig. 3.9

The maximum sagging bending moment is, therefore,

$$M = (16 \text{ kN} \times 2 \text{ m}) - (2 \text{ m} \times 8 \text{ kN/m} \times 1 \text{ m}) = 16 \text{ kN-m}$$

(Ans)

Since the beam is of uniform section, the maximum stresses occur at the edges of the sections of greatest bending moments.

For the section of maximum sagging bending moment, the compressive stress at the top edge of the section is

$$\sigma = M/Z = \frac{16 \text{ kN-m}}{\frac{2}{3} \times 25^4 \text{ mm}^3} = 61{\cdot}4 \text{ MN/m}^2$$

and the tensile stress at the bottom of the section is

$$\sigma = \frac{16 \text{ kN-m}}{\frac{1}{3} \times 25^4 \text{ mm}^3} = 123 \text{ MN/m}^2$$

Similarly at **B**, where the bending moment has its greatest hogging value, the tensile stress at the top edge is

$$\sigma = \frac{9 \text{ kN-m}}{\frac{2}{3} \times 25^4 \text{ mm}^3} = 34{\cdot}6 \text{ MN/m}^2$$

and the compressive stress at the bottom of the section is

$$\sigma = \frac{9 \text{ kN-m}}{\frac{1}{3} \times 25^4 \text{ mm}^3} = 69{\cdot}1 \text{ MN/m}^2$$

The maximum tensile stress occurring in the beam is 123 MN/m^2 and the maximum compressive stress occurring in the beam is 69·1 MN/m^2. *(Ans)*

The reader should notice that these two stresses do not occur at the same section.

3.8. A horizontal beam, of the section shown in Fig. 3.10a is 4 m long and is simply supported at its ends. Calculate the maximum uniformly distributed load it can carry if the tensile and compressive stresses must not exceed 32 and 56 MN/m^2 respectively.

Draw a diagram showing the variation of stress over the mid-span section of the beam.

Solution. The reader should verify that the n.a. is 35 mm from the bottom edge of the section (Fig. 3.10a) and that the second moment of area (moment of inertia) about this axis is

$$I = 581 \text{ cm}^4 \quad \text{or} \quad 581 \times 10^{-8} \text{ m}^4$$

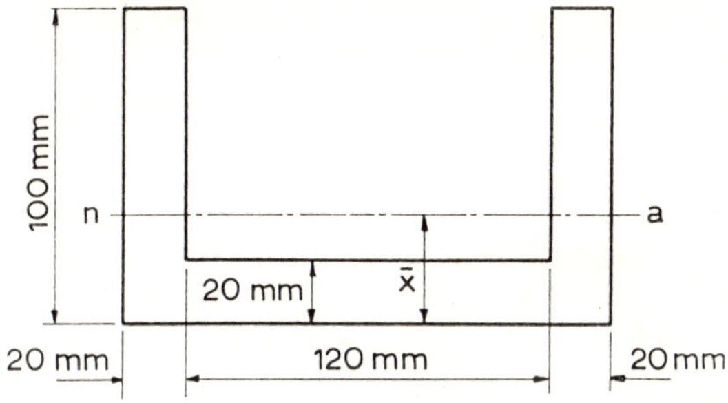

(a)

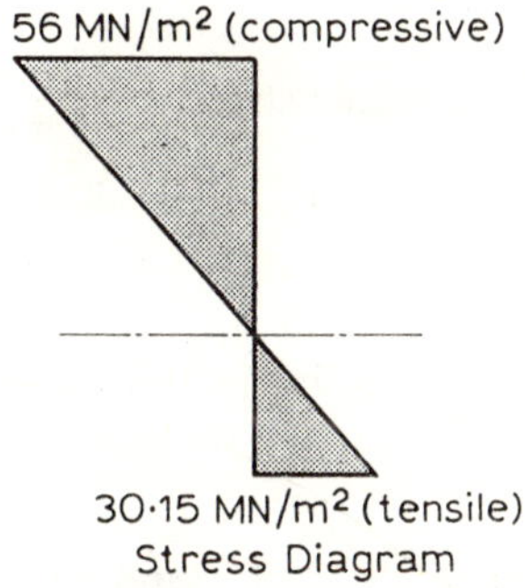

(b)

Fig. 3.10

Since the beam is everywhere "sagging" the maximum tensile stress occurs at the bottom edge for which y (the distance from the n.a.) is 35 mm. Hence, for a maximum tensile stress of 32 MN/m² the moment of resistance is

$$M = \frac{I}{y}\sigma = \frac{581 \times 10^{-8}\ \text{m}^4}{0{\cdot}035\ \text{m}} \times (32 \times 10^6\ \text{N/m}^2)$$

$$= 5{\cdot}31\ \text{kN-m}$$

Similarly, for a maximum compressive stress of 56 MN/m² (at the top of the section where $y = 65$ mm),

$$M = \frac{581 \times 10^{-8}\ \text{m}^4}{0{\cdot}065\ \text{m}} \times (56 \times 10^6\ \text{N/m}^2) = 5{\cdot}01\ \text{kN-m}$$

If both conditions are to be satisfied, the bending moment must not exceed 5·01 kN-m. The maximum bending moment occurs at mid-span and, if W = total load, it is given by $M = WL/8$. Rearranging and substituting known values,

$$W = 8M/L = (8 \times 5{\cdot}01 \text{ kN-m})/4 \text{ m} = 10{\cdot}0 \text{ kN (total)}$$
$$\text{or } 2{\cdot}5 \text{ kN/m} \qquad (Ans)$$

At the mid-span section, the maximum compressive stress is the prescribed 56 MN/m² but the maximum tensile stress is less than the permissible 32 MN/m². Since the stress at any point is proportional to the distance of that point from the neutral axis, then

$$\frac{\text{maximum tensile stress}}{35 \text{ mm}} = \frac{\text{maximum compressive stress}}{65 \text{ mm}}$$

or

$$\text{maximum tensile stress} = \frac{35 \text{ mm}}{65 \text{ mm}} \times 56 \text{ MN/m}^2$$
$$= 30{\cdot}15 \text{ MN/m}^2$$

The diagram showing the variation of stress is drawn alongside the view of the section (Fig. 3.10b). The stress varies linearly from 56 MN/m² compressive at the top edge to 30·15 MN/m² tensile at the bottom edge passing through zero at the neutral axis.

3.9. A vertical flagstaff standing 10 m above the ground is of square section throughout, the dimensions being 120 mm by 120 mm at the ground tapering uniformly to 60 mm by 60 mm at the top. A horizontal pull of 300 N is applied at the top, the direction of loading being along a diagonal of the section. Calculate the maximum stress due to bending. *(U.L.)*

Solution. The second moment of area for a square, side d, about a diagonal is $d^4/12$. (This result follows from the fact that the maximum and minimum values of I for a rectangle are $bd^3/12$ and $db^3/12$. Thus, putting $b = d$, $I = d^4/12$ for a square about any central axis.) The maximum value of y is the semi-diagonal, i.e. $\sqrt{2}d/2 = d/\sqrt{2}$. Thus, the section modulus is

$$Z = \frac{I}{y_{max}} = \frac{d^4/12}{d/\sqrt{2}} = \frac{d^3}{6\sqrt{2}}$$

At any section XX of the given flagstaff, distance x m from the top as shown in Fig. 3.11, the side of the square of the section is

$$d = 60 \text{ mm} + \left(\frac{x \text{ m}}{10 \text{ m}} \times 60 \text{ mm}\right) = 60\left(1 + \frac{x}{10}\right) \text{ mm}$$

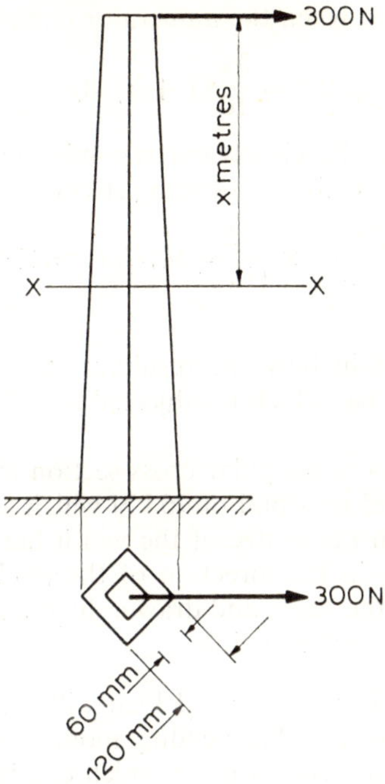

Fig. 3.11

Hence the modulus of the section at XX is

$$Z = \frac{[60(1 + x/10)\ \text{mm}]^3}{6\sqrt{2}}$$
$$= 18\sqrt{2}(1 + x/10)^3 \times 10^{-6}\ \text{m}^3$$

Also the bending moment at XX is

$$M = 300\ \text{N} \times x\ \text{m} = 300\ x\ \text{N-m}$$

Thus, the maximum stress at this section is

$$\sigma = M/Z = \frac{300x\ \text{N-m}}{18\sqrt{2}\,(1 + x/10)^3 \times 10^{-6}\ \text{m}^3}$$
$$= \frac{25\sqrt{2}}{3} \times \frac{x}{(1 + x/10)^3}\ \text{MN/m}^2 \qquad \text{(i)}$$

The section having the greatest bending stress is found by differentiating this expression for σ with respect to x

$$\frac{d\sigma}{dx} = \frac{25\sqrt{2}}{3}\left[\frac{(1 + x/10)^3 - x\,.\,3\,(1 + x/10)^2\,.\,\frac{1}{10}}{(1 + x/10)^6}\right]$$

This is zero when the numerator is zero, i.e. when

$$(1 + x/10)^3 = \frac{3x}{10}(1 + x/10)^2 \qquad x = 5 \text{ m}$$

This value of x clearly corresponds to a maximum and, substituting in (i), the maximum bending stress is

$$\sigma = \frac{25\sqrt{2}}{3} \times \frac{5}{(1{\cdot}5)^3} \text{ MN/m}^2 = 17{\cdot}5 \text{ MN/m}^2 \qquad (Ans)$$

3.10. Explain how the resultant stress may be calculated for any point in a bar which is subjected to a direct axial load and a bending moment.

A bar of rectangular cross-section 60 mm wide and 40 mm thick is subjected to a pull of 96 kN which acts parallel to the axis of the bar and in the centre of the width but at a distance of 5 mm from the centre in the direction of the thickness. Calculate the extreme values of the stress and draw a diagram showing the variation of the stress across the section.

Solution. The direct axial load will cause a stress which is uniform over the section. The bending moment will cause an additional stress at any point, the value of which can be obtained from the bending equation. The resultant stress at any point is the algebraic sum of the direct and bending stresses at that point.

Hence, if σ_d is the uniform direct stress and σ_b is the bending stress at a distance y from the neutral axis, then the resultant stress at the same point is $\sigma_d + \sigma_b$. The resultant stress diagram is a straight line as shown in Fig. 3.12. It can be obtained by displacing the bending

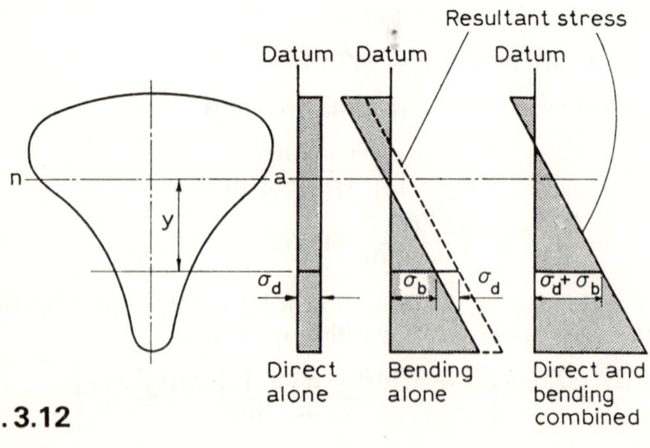

Fig. 3.12

stress line, parallel to itself, a horizontal distance equivalent to the direct stress (to the right in the diagrams of Fig. 3.12). Alternatively, the datum can be displaced in the opposite direction by the same amount.

The neutral surface (at which the stress is zero) does not pass through the centroid of the section. The direct stress displaces it to a position where the direct and bending stresses are equal and opposite. It is clear that if the direct stress is numerically greater than the maximum bending stress of opposite type, then the stress over the section will be wholly tensile or compressive (according to the nature of the direct load). In this case there is no neutral surface in the bar.

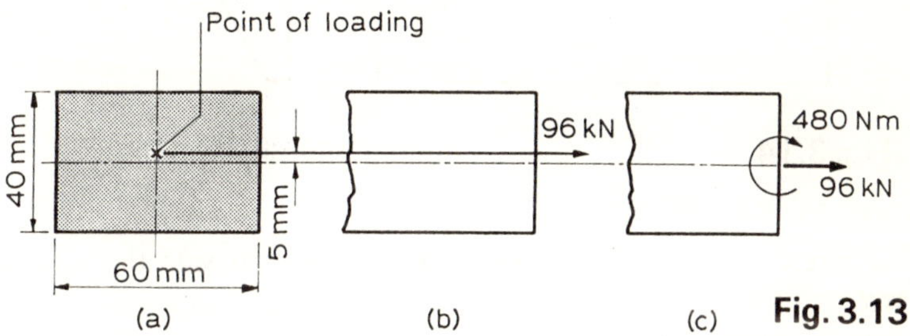

Fig. 3.13

The eccentric load of the question (Fig. 3.13b) can be replaced by a direct axial load together with a bending moment (Fig. 3.13c) equal to the load multiplied by the eccentricity, i.e. 96 kN × 5 mm = 480 N-m.

$$\text{Direct stress} = \frac{\text{load}}{\text{cross-sectional area}}$$

$$= \frac{96\ \text{kN}}{0{\cdot}06\ \text{m} \times 0{\cdot}04\ \text{m}}$$

$$= 40\ \text{MN/m}^2\ \text{(tensile)}$$

The maximum tensile and compressive bending stresses occur at the top and bottom edges of the section. Since the section is symmetrical, they are each

$$\sigma = \frac{M}{I}y = \frac{480\ \text{N-m}}{[0{\cdot}06\ \text{m} \times (0{\cdot}04\ \text{m})^3/12]} \times 0{\cdot}02\ \text{m}$$

$$= 30\ \text{MN/m}^2$$

At the top of the section, the direct stress is 40 MN/m² (tensile) and the bending stress is 30 MN/m² (tensile). Hence the resultant stress is 70 MN/m² (tensile). *(Ans)*

At the bottom of the section the direct stress is 40 MN/m^2 (tensile) and the bending stress is 30 MN/m^2 (compressive). Hence the resultant stress is 10 MN/m^2 (tensile). (*Ans*)

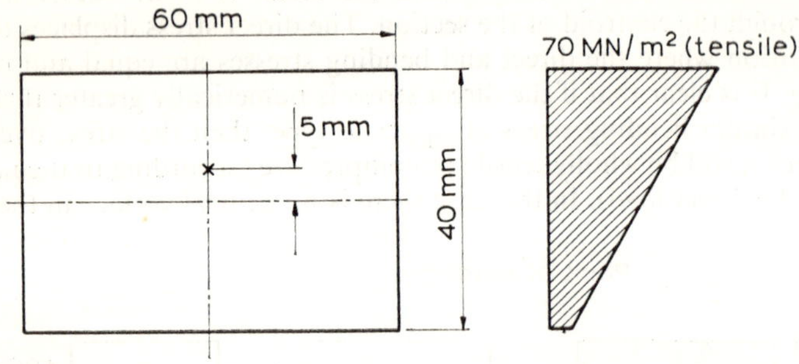

Fig. 3.14

The stress variation is shown in Fig. 3.14 alongside the section view.

3.11. Explain the term "reverse stress" and prove the "middle-third" rule relating to it.

A bar of rectangular section, 40 mm wide and 20 mm thick, is subjected to an axial compressive load of 60 kN. By how much can the width of the section be reduced by removing material from one edge only, if there is to be no tensile stress in the bar and the axis of the load is unchanged? For this condition, calculate the maximum compressive stress in the bar.

Solution. For all eccentric loads, the bending effect alone produces stresses which are tensile on one side of the neutral axis and compressive on the other. If the eccentricity of the load is small, then the bending stress will not outweigh the direct stress at any point. If the eccentricity is sufficiently great, however, the bending stress at one edge of the section will outweigh the direct stress. This means that a compressive stress may be obtained at some points with a tensile load, and a tensile stress with a compressive load. This condition is known as *reverse stress*.

Suppose a bar of rectangular section (Fig. 3.15) is subjected to a tensile load P which acts at a point on the vertical centre-line but with an eccentricity e to the horizontal centre-line.

The bending moment it produces is Pe and the maximum bending stresses are each

$$\sigma = \frac{M}{I}y = \frac{Pe}{(bd^3/12)} \times \tfrac{1}{2}d = 6Pe/bd^2$$

(tensile at the top of the section and compressive at the bottom edge).

The direct stress is P/bd (tensile).

There will be reverse (compressive) stress at the bottom edge if the bending stress there (which is compressive) outweighs the direct stress.

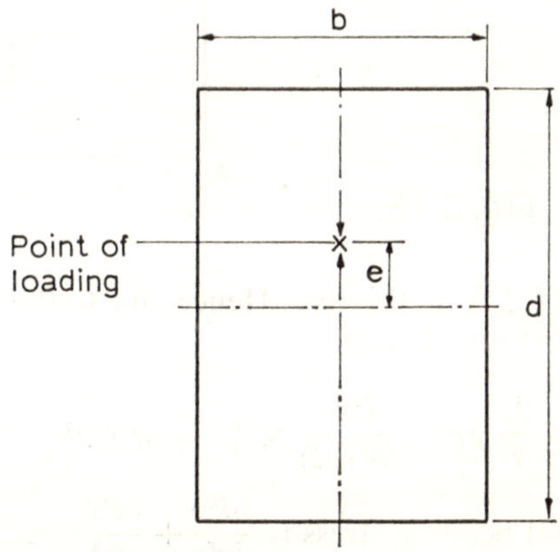

Fig. 3.15

Hence, for no reverse stress,

$6Pe/bd^2$ must be less than P/bd

or, on simplification, e must be less than $d/6$. A similar condition applies if the load is compressive or if it acts below the neutral axis. Altogether, therefore, for there to be no reverse stress, *the load must act within the "middle third" of the depth.*

If the load is eccentric to the vertical centre-line only, the middle-third rule is again applicable, but if the load is eccentric to both centre-lines then bending about each must be considered. Suppose the load is applied at a point distances x and y from the centre-lines, as shown in Fig. 3.16. The corner A has the greatest reverse bending stress since it is on the opposite side of both centre-lines to the load. Due to the eccentricity y, the bending stress at A is $6Py/bd^2$ (as in the previous case).

Due to the eccentricity x, there is a bending moment of Px about the vertical centre-line. I about this axis is $db^3/12$ and the point A is a

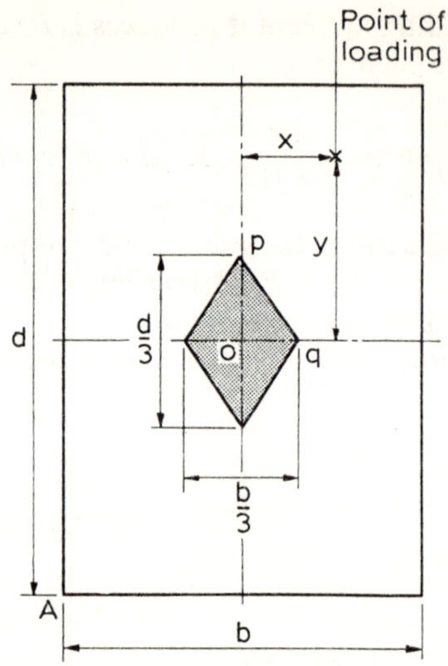

Fig. 3.16

distance $b/2$ from this axis. Hence, due to the eccentricity x, there is a bending stress at A of

$$\sigma = \frac{M}{I}y = \frac{Px}{(db^3/12)} \times \tfrac{1}{2}b = 6Px/db^2$$

$$\text{Total bending stress is } \frac{6Py}{bd^2} + \frac{6Px}{db^2}$$

For no reverse stress at A, this must be less than the direct stress, i.e.

$$\left(\frac{6Py}{bd^2} + \frac{6Px}{db^2}\right) \text{ must be less than } \frac{P}{bd}$$

or

$$\left(\frac{y}{d} + \frac{x}{b}\right) \text{ must be less than } \tfrac{1}{6} \qquad (Ans)$$

When $y = 0$ or $x = 0$ this condition becomes the middle-third rule. If the load is eccentric to both axes, then it must act within the triangle *opq* for there to be no reverse stress at A (assuming that it acts in the quadrant opposite A). A similar condition applies when the load acts in the other quadrants and, if there is to be no reverse stress at any point in the section, then the load must act within the shaded diamond-shaped area (Fig. 3.16). This is called the *core* or *kernel* of the section.

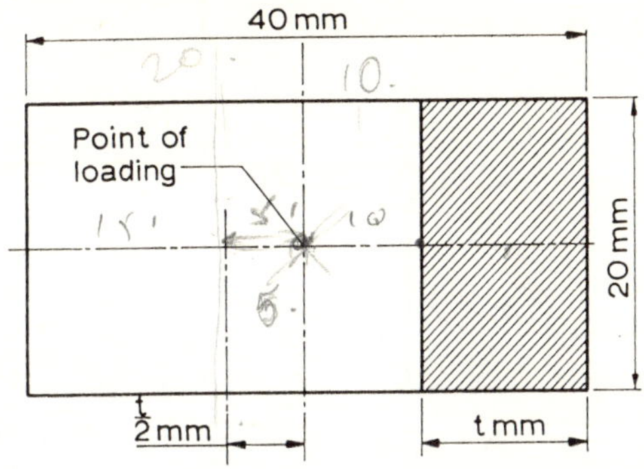

Fig. 3.17

Suppose, for the given bar, a thickness of t mm is removed from the width, as shown in Fig. 3.17. The load is now eccentric to the vertical centre-line of the remaining section by an amount $t/2$ mm.

In the limiting case of zero resultant stress at the left-hand edge, the eccentricity will be one-sixth of the new width of the section (by the middle-third rule) i.e.

$$t/2 = \tfrac{1}{6} \times (40 - t) = (20/3) - (t/6)$$

$$t = 10 \text{ mm} \qquad (Ans)$$

The direct stress is, therefore,

$$\frac{\text{load}}{\text{area}} = \frac{60 \text{ kN}}{(0{\cdot}04 \text{ m} - 0{\cdot}01 \text{ m}) \times 0{\cdot}02 \text{ m}}$$

$$= 100 \text{ MN/m}^2 \text{ (compressive)}$$

Since the stress is zero at the left-hand edge, the bending stress there must be 100 MN/m² (tensile). The bending stress at the right-hand edge is numerically the same but of opposite sign, 100 MN/m² (compressive).

The maximum compressive stress in the bar occurs at the right-hand edge and is (100 + 100) MN/m² = 200 MN/m². (*Ans*)

3.12. A short vertical column has a cross-section as shown in Fig. 3.18. If the working stress (tensile and compressive) is not to exceed 92 MN/m², find the greatest vertical load, its line of action passing through P, which can be carried by the column. Also find the position of the neutral axis in terms of the intercepts on the AB and CD axes.

Properties of each 300 mm × 150 mm I-section:

$I_{AB} = 15630$ cm⁴ $\qquad I_{EF} = 1176$ cm⁴ $\qquad$ Area = 103 cm²

(*I.Mech.E.*)

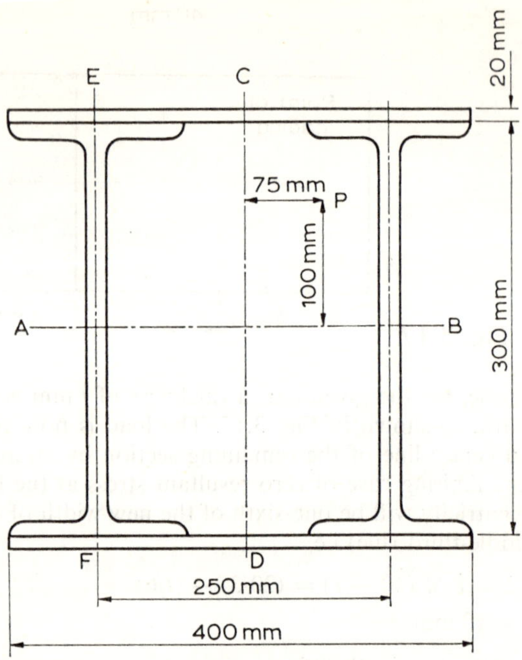

Fig. 3.18

Solution. The values of I about AB and CD are required and the theorem of parallel axes must be used so that the value for each item can be adjusted to the proper axis (see Appendix 1). It is convenient to use cm^4 for the units of I. This avoids the very large or very small numbers which occur if mm^4 or m^4 are used.

I about AB

	cm^4
For the two I-sections $= 2 \times 15630\ cm^4$	$= 31260$
For the two plates:	
for each about its own n.a.	
$= \dfrac{40\ cm \times (2\ cm)^3}{12} = 26{\cdot}7\ cm^4$	
for transferring each to AB	
$= 40\ cm \times 2\ cm \times (16{\cdot}25\ cm)^2 = 21120\ cm^4$	
Total for each about AB $= 21147\ cm^4$	
Total for both about AB $= 2 \times 21147$	$= 42294$
Total	$= 73554$

I about CD

cm^4

For the two plates about CD

$$= 2 \times \frac{(2\ \text{cm}) \times (40\ \text{cm})^3}{12} \qquad = 21\,333$$

For the two I-sections:
for each about its own n.a. = 1176 cm^4
for transferring each to CD

$$= 103\ \text{cm}^2 \times (13\ \text{cm})^2 = 17\,402\ \text{cm}^4$$

Total for each about CD = 18578 cm^4
Total for both about CD = 2 × 18578 = 37156

Total = 58489

Total area of the section

$$= (\text{area of two plates}) + (\text{area of two I-sections})$$
$$= 2 \times (40 \times 2\ \text{cm}) + 2 \times 103\ \text{cm}^2 = 366\ \text{cm}^2$$

Suppose the required load is W kN. The direct stress due to this load is

$$W\ \text{kN}/366\ \text{cm}^2 = 27{\cdot}32\ W\ \text{kN/m}^2 \qquad \text{(i)}$$

The bending moment about AB due to this load is

$$M = W\ \text{kN} \times 100\ \text{mm} = 0{\cdot}1\,W\ \text{kN-m}$$

The maximum stress due to this bending moment is

$$\sigma = \frac{M}{I}\,y_{max} = \frac{0{\cdot}1\,W\ \text{kN-m}}{73\,554 \times 10^{-8}\ \text{m}^4} \times 0{\cdot}1625\ \text{m}$$
$$= 22{\cdot}09\,W\ \text{kN/m}^2 \qquad \text{(ii)}$$

Similarly the bending moment about CD is $0{\cdot}075W$ kN-m and the corresponding maximum stress is

$$\sigma = \frac{0{\cdot}075\,W\ \text{kN-m}}{58\,489 \times 10^{-8}\ \text{m}^4} \times 0{\cdot}2\ \text{m} = 25{\cdot}64\,W\ \text{kN-m}^2 \qquad \text{(iii)}$$

The three stresses (i), (ii) and (iii) are additive at the top right-hand corner of the section (regarding upright as shown in the diagram). Since the maximum permissible stress is 92 MN/m^2 we have

$$27{\cdot}32W + 22{\cdot}09W + 25{\cdot}64W = 92 \times 10^3$$

from which

$$W = 1226\ \text{kN} = 1{\cdot}226\ \text{MN} \qquad (\textit{Ans})$$

The stress is zero at the point on the CD axis where the stress due to the bending moment about AB is equal and opposite to the direct stress. Suppose this occurs at a distance y mm below AB (it is only below AB that the bending stress is opposite in sign to the direct stress). Then

$$\text{Bending stress} = \frac{M}{I}y = \frac{0{\cdot}1W\text{ kN-m}}{73554 \times 10^{-8}\text{ m}^4} \times (y \times 10^{-3}\text{ m})$$

and if this equals the direct stress, we have

$$Wy/7{\cdot}355 = 27{\cdot}32W \qquad y = 201\text{ (mm)} \qquad (Ans)$$

Similarly the stress is zero at a point on the AB axis, distance x mm to the left of CD where

$$0{\cdot}75Wx/5{\cdot}849 = 27{\cdot}32W \qquad x = 213\text{ (mm)} \qquad (Ans)$$

These two distances are the intercepts of the neutral axis on the AB and CD axes.

3.13. A flitched beam is made up of two timber joists each 100 mm wide by 240 mm deep, with a 20 mm steel plate 150 mm deep placed symmetrically between them and firmly attached to both. The plate is recessed into grooves cut in the inner faces of the joists so that the overall dimensions of the built-up section may be taken as 200 mm by 240 mm.

Calculate the moment of resistance of the combined section when the maximum bending stress in the timber is 8·4 MN/m². What is then the maximum stress in the steel?

Take E for steel as 206×10^9 N/m² and for timber as 12×10^9 N/m². (*U.L.*)

Solution. A flitched beam is a composite beam in which two timber joists and a steel plate are rigidly bolted together (Fig. 3.19).

Since the steel and timber must bend to the same radius of curvature and, also, from the bending equation,

$$R = EI/M$$

then EI/M must have the same value for the steel and the timber. If suffixes s and t denote the steel and timber respectively,

$$\frac{E_s I_s}{M_s} = \frac{E_t I_t}{M_t} \qquad \text{(i)}$$

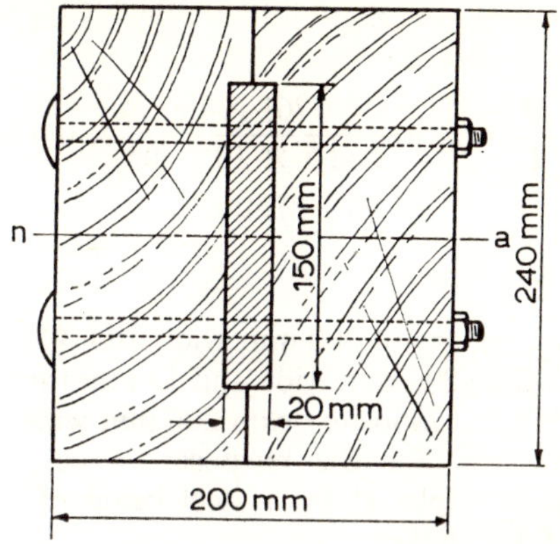

Fig. 3.19

It is again convenient to calculate the I values in cm units and, with the dimensions given in the question,

$$I_s = \frac{2\ \text{cm} \times (15\ \text{cm})^3}{12} = 562{\cdot}5\ \text{cm}^4$$

and

$$I_t = \frac{20\ \text{cm} \times (24\ \text{cm})^3}{12} - \frac{2\ \text{cm} \times (15\ \text{cm})^3}{12} = 22498\ \text{cm}^4$$

For the timber $y_{max} = 120$ mm and, for the given stress,

$$M_t = \frac{I_t}{y_{max}} \times \sigma$$

$$= \frac{(22498 \times 10^{-8}\ \text{m}^4) \times (8{\cdot}4 \times 10^6\ \text{N/m}^2)}{120 \times 10^{-3}\ \text{m}}$$

$$= 15750\ \text{N-m}$$

Thus from (i),

$$M_s = M_t\left(\frac{E_s I_s}{E_t I_t}\right)$$

$$= 15750\ \text{N-m} \times \frac{(206 \times 10^9\ \text{N/m}^2) \times 562{\cdot}5\ \text{cm}^4}{(12 \times 10^9\ \text{N/m}^2) \times 22498\ \text{cm}^4}$$

$$= 6761\ \text{N-m}$$

Total moment of resistance is

$$M_t + M_s = (15750 + 6761)\ \text{N-m} = 22{\cdot}5\ \text{kN-m} \qquad (Ans)$$

For the steel $y_{max} = 75$ mm and hence the maximum stress in the steel is

$$\sigma = \frac{M_s}{I_s} y_{max} = \frac{6761 \text{ N-m}}{562{\cdot}5 \times 10^{-8} \text{ m}^4} \times 0{\cdot}075 \text{ m}$$
$$= 90{\cdot}2 \text{ MN/m}^2$$

3.14. A timber beam 200 mm wide and 300 mm deep is reinforced by a steel plate 200 mm wide and 13 mm thick, bolted to its bottom edge, giving a composite beam 200 mm wide and 313 mm deep.

Explain how, for the purposes of calculation, the composite beam can be replaced by a steel beam of ⊥ section. E for steel $= 20 \times E$ for timber.

Calculate the maximum stresses in the steel and timber when the composite beam carries a uniformly distributed load of 15 kN/m and is simply supported over a span of 6 m.

Solution. The neutral axis of a composite beam does not in general pass through the centroid of the section. Since E for steel is much greater than E for timber then, for a given strain, the stress in the steel is much greater than that in the timber. Assuming that plane sections remain plane (for the composite beam) the longitudinal strain of a layer AB in the timber (Fig. 3.20) is proportional to its

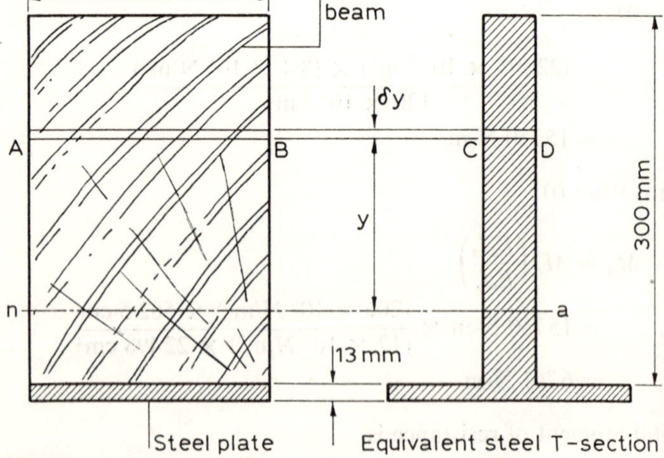

Fig. 3.20

distance y from the neutral axis. The corresponding stress acts on an area which, in the present example, is 200 mm wide. For the same radius of curvature, the stress in a steel strip CD, at the same distance y from the neutral axis, is much greater. In order that the steel strip CD shall carry the same longitudinal *force* as the timber strip AB, of the same thickness δy, it must be much narrower. In the present case, where E for steel $= 20 \times E$ for timber, the width of the equivalent steel section (where it replaces the timber) must be 1/20th of the width of the timber. The equivalent steel section has the same overall height and neutral axis position as the composite beam. For the same radius of curvature it has the same moment of resistance. The stress at any point in the steel is found in the usual way. The stress in the timber at any point is found by multiplying the stress at the corresponding point of the equivalent steel section by the ratio E timber/E steel.

In the example given, the timber beam, 200 mm wide, is replaced by a steel web whose width is

$$(E_t/E_s) \times 200 \text{ mm} = \tfrac{1}{20} \times 200 \text{ mm} = 10 \text{ mm}$$

the suffixes t and s referring to the timber and steel respectively.

The equivalent steel section is shown in Fig. 3.20b, and the reader should verify that its neutral axis is 87 mm from the bottom edge. Also using any of the methods given previously, the second moment of area about the neutral axis is

$$I = 5638 \text{ cm}^4$$

The maximum bending moment on the beam is

$$M = \frac{WL}{8} = \frac{15 \text{ kN/m} \times 6 \text{ m} \times 6 \text{ m}}{8} = 67{\cdot}5 \text{ kN-m}$$

The maximum stress in the steel (of the composite beam) is equal to the stress at the bottom edge of the equivalent section and is

$$\sigma = \frac{My}{I} = \frac{67{\cdot}5 \text{ kN-m} \times (87 \times 10^{-3} \text{ m})}{5638 \times 10^{-8} \text{ m}^4}$$

$$= 104{\cdot}2 \text{ MN/m}^2 \text{ (tensile)} \qquad (Ans)$$

The stress at the top edge of the equivalent section is

$$\sigma = \frac{My}{I} = \frac{67{\cdot}5 \text{ kN-m} \times (226 \times 10^{-3} \text{ m})}{5638 \times 10^{-8} \text{ m}^4}$$

$$= 270{\cdot}6 \text{ MN/m}^2 \text{ (compressive)}$$

The corresponding (maximum) stress in the timber is E_t/E_s times the amount, i.e.

$$\tfrac{1}{20} \times 270\cdot6 \text{ MN/m}^2 = 13\cdot5 \text{ MN/m}^2 \text{ (compressive)} \qquad (Ans)$$

3.15. Two rectangular bars, one of brass and the other of steel, each 36 mm by 10 mm are placed together to form a beam 36 mm wide and 20 mm deep, on two supports 0·8 m apart. The brass bar is on top of the steel.

Determine the maximum central load which can be applied to the beam, if the bars are (*a*) separate and can bend independently, (*b*) firmly secured to each other throughout their length.

For brass $E = 86 \times 10^9$ N/m², for steel $E = 206 \times 10^9$ N/m².

Maximum allowable stress in brass is 70 MN/m² and in steel 105 MN/m². (*U.L.*)

Solution. In case (*a*) each bar bends about its own neutral axis, but the radius of curvature at any section is (for practical purposes) the same for both.

$$Z \text{ for each bar (separately)} = 3\cdot6 \text{ cm} \times (1 \text{ cm})^2/6 = 0\cdot6 \text{ cm}^3$$

Let suffixes *b* and *s* refer to the brass and steel respectively. If *R* has the same value for both bars, then

$$\frac{E_b I_b}{M_b} = \frac{E_s I_s}{M_s} \ (=R)$$

or

$$\frac{M_s}{M_b} = \frac{E_s I_s}{E_b I_b} = \frac{206 \times 10^9 \text{ N/m}^2}{86 \times 10^9 \text{ N/m}^2} = 2\cdot4 \text{ (since } I_s = I_b)$$

For the given stress in the brass,

$$M_b = Z \times 70 \text{ MN/m}^2 = (0\cdot6 \times 10^{-6} \text{ m}^3) \times (70 \times 10^6 \text{ N/m}^2) = 42 \text{ N-m}$$

and

$$M_s = 2\cdot4\, M_b = 2\cdot4 \times 42 \text{ N-m} = 100\cdot8 \text{ N-m}$$

For the given stress in the steel,

$$M_s = Z \times 105 \text{ MN/m}^2 = (0\cdot6 \times 10^{-6} \text{ m}^3) \times (105 \times 10^6 \text{ N/m}^2) = 63 \text{ N-m}$$

and

$$M_b = M_s/2{\cdot}4 = 63 \text{ N-m}/2{\cdot}4 = 26{\cdot}25 \text{ N-m}$$

The stress in the steel is therefore the limiting factor and the moment of resistance for the composite beam is

$$M = M_s + M_b = 63 \text{ N-m} + 26{\cdot}25 \text{ N-m} = 89{\cdot}25 \text{ N-m}$$

This equals the maximum bending moment ($WL/4$ for a simply supported beam with a central point load), and thus the maximum central load is

$$W = \frac{4}{L} \times 89{\cdot}25 \text{ N-m} = \frac{4}{0{\cdot}8 \text{ m}} \times 89{\cdot}25 \text{ N-m} = 446 \text{ N}$$

For case (b) an "equivalent steel section" is used. The brass bar is replaced by steel whose width is

$$\frac{E_b}{E_s} \times \text{width of brass} = \frac{86 \text{ GN/m}^2}{206 \text{ GN/m}^2} \times 36 \text{ mm}$$
$$= 15 \text{ mm}$$

The equivalent section is shown in Fig. 3.21. The neutral axis is 7·95 mm from the bottom edge of the section and the second moment of area of the equivalent section is $I = 1{\cdot}484 \text{ cm}^4$.

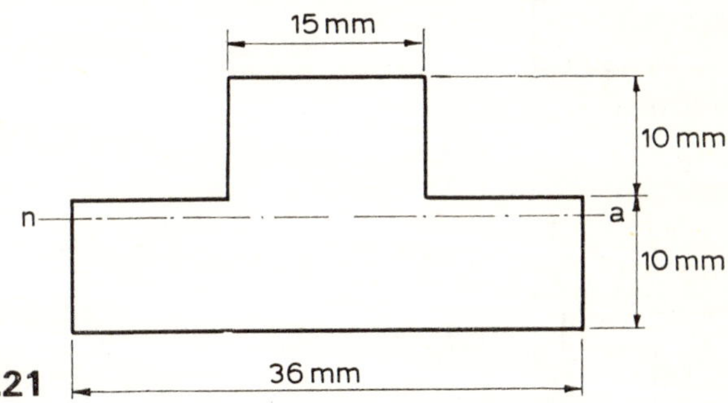

Fig. 3.21

The maximum stress in the steel bar (105 MN/m²) corresponds to the stress at the bottom edge of the equivalent section ($y =$ 7·95 mm). For this stress, the moment of resistance is

$$M = \frac{\sigma}{y} I = \frac{105 \text{ MN/m}^2}{7{\cdot}95 \times 10^{-3} \text{ m}} \times (1{\cdot}484 \times 10^{-8} \text{ m}^4) = 196 \text{ N-m}$$

The maximum stress in the brass bar (75 MN/m²) corresponds to a stress of $(E_s/E_b) \times 75 \text{ MN/m}^2 = 180 \text{ MN/m}^2$ at the top edge of the

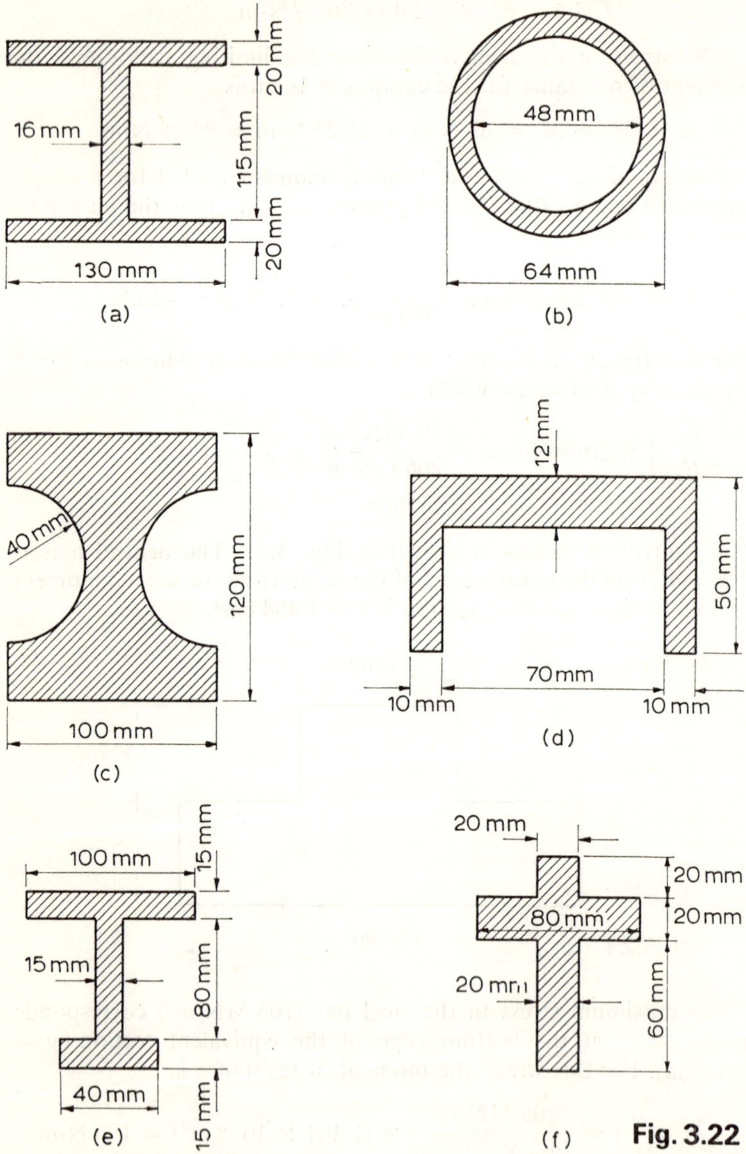
20 mm
16 mm
115 mm
20 mm
130 mm
(a)
48 mm
64 mm
(b)
40 mm
120 mm
100 mm
(c)
12 mm
50 mm
70 mm
10 mm
10 mm
(d)
100 mm
15 mm
15 mm
80 mm
40 mm
15 mm
(e)
20 mm
20 mm
20 mm
80 mm
20 mm
60 mm
(f)

Fig. 3.22

equivalent section ($y = 12{\cdot}05$ mm). For this stress the moment of resistance is

$$M = \frac{\sigma}{y} I = \frac{180 \times 10^6 \text{ N/m}^2}{12{\cdot}05 \times 10^{-3} \text{ m}} \times (1{\cdot}484 \times 10^{-8} \text{ m}^4)$$
$$= 221{\cdot}8 \text{ N-m}$$

In order that both conditions may be satisfied, the maximum bending moment is limited to 196 N-m. Hence, in a manner similar to case (*a*), the maximum central load is

$$W = \frac{4}{L} \times 196 \text{ N-m} = \frac{4}{0{\cdot}8 \text{ m}} \times 196 \text{ N-m} = 980 \text{ N} \qquad (Ans)$$

PROBLEMS

1. The sections shown in Fig. 3.22 are to be used (upright as shown) for horizontal beams. Calculate the position of the neutral axis in each case and the second moment of area (or moment of inertia) about this axis.
Answer.

(*a*) Horizontal centre line; 2589 cm^4.
(*b*) Horizontal centre-line; 56·3 cm^4.
(*c*) Horizontal centre-line; 1239 cm^4.
(*d*) 16·3 mm from top edge; 38·3 cm^4.
(*e*) 42·0 mm from top edge; 486 cm^4.
(*f*) 42·5 mm from top edge; 200·7 cm^4.

2. In the following problems the cross-sections referred to are those shown in Fig. 3.22.

(i) Use the approximate method of Example 3.2 to calculate the moment of resistance for the section (*a*) the flange stress being 70 MN/m². What is the flange stress for a moment of 10 kN-m?

(ii) A tube with the section (*b*) is bent round a circular former 15 m radius. What is the maximum stress in the tube if $E = 206$ GN/m²?

(iii) A cast-iron beam with the section (*c*) is to carry a uniformly distributed load of 1·5 kN/m in addition to its own weight. What is the maximum span over which the beam may be simply supported (at its ends) if the stress is limited to 15 MN/m²? Density of cast iron, 7·14 Mg/m³.

(iv) A cantilever, 2 m long, has the section (*d*) and carries a uniformly distributed load of 300 N/m throughout, together with a concentrated load of 220 N at a point 0·6 m from the free end. Calculate the maximum tensile and compressive stresses due to bending.

(v) A beam 6 m long is simply and symmetrically supported on a span of 4 m and has the section shown at (*e*). If the tensile and compressive bending stresses must not exceed 30 and 22 MN/m² respectively, what is the maximum load that can be carried by the beam, assuming that it is uniformly distributed over its whole length?

(vi) The section shown at (*f*) is used for a cantilever 3 m long. The cantilever carries a uniformly distributed load of 24 kN total and receives additional support from a prop at the free end, which carries 3/8 of the total load. Calculate the maximum tensile and compressive stresses in the cantilever.

Answer.

(i) 24·6 kN-m; 28·5 MN/m^2.
(ii) 439 MN/m^2.
(iii) 3·53 m.
(iv) 38·7 MN/m^2 (tensile) and 79·8 MN/m^2 (compressive).
(v) 2·86 kN (total).
(vi) 190·5 MN/m^2 (tensile) and 257·8 MN/m^2 (compressive).

3. A cantilever, 4 m long, carries a uniformly distributed load of 800 N total and is so proportioned that a maximum bending stress of 4 MN/m^2 is reached at each section. If the cross-section is rectangular throughout, calculate its dimensions at points 0, 1, 2 and 3 m from the fixed end when

(*a*) the width is 100 mm throughout,
(*b*) the depth is 150 mm throughout,
(*c*) the depth is twice the width throughout.

Answer.

(*a*) Depths are 155, 116, 77·5 and 38·7 mm respectively.
(*b*) Widths are 107, 60, 26·7 and 6·7 mm respectively.
(*c*) Widths are 84·4, 69·6, 53·1, and 33·5 mm respectively and depths are 168·8, 139·2, 106·2 and 67·0 mm respectively.

4. A beam is made up of four 300 mm by 15 mm plates, riveted to four 50 mm by 50 mm angles as shown in Fig. 3.23a. Calculate the second moments of area about the principal axes AA and BB, neglecting the effects of rivets.

Properties of one 50 mm by 50 mm angle (Fig. 3.23b)

cross-sectional area = 5·69 cm^2 $\qquad I_{xx} = I_{yy} = 12{\cdot}8$ cm^4

Distance of centroid from side = 14·5 mm.

When used with the axis AA horizontal, the maximum stress in the beam, under a certain loading, was 108 MN/m^2. Calculate the corresponding stress when the axis BB is horizontal, the loading being unchanged.

Answer. $I_{AA} = 33\,300$ cm^4; $I_{BB} = 17\,500$ cm^4. With BB horizontal, the maximum stress is 187 MN/m^2.

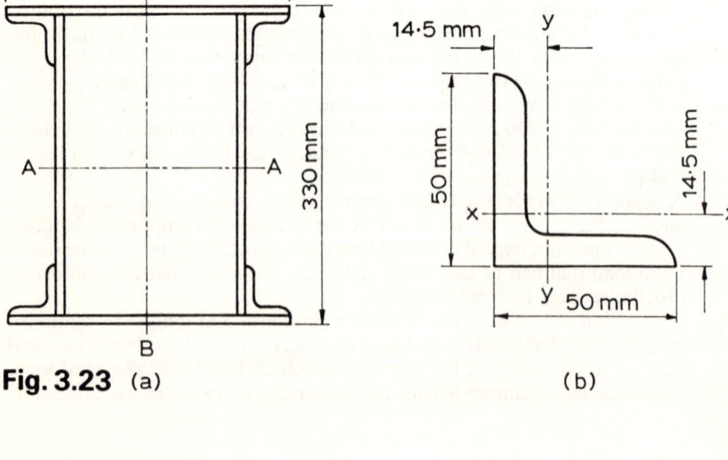

Fig. 3.23 (a) (b)

5. Explain what is meant by the terms "stress" and "strain", giving a relationship between the two, and specifying their respective dimensions.

A bar of diameter 25 mm has an axial hole 19 mm bored out concentric with the outer circumference. The bar is submitted to the following loads simultaneously:

(*a*) An axial tension of 13 kN.
(*b*) A bending moment of 100 N-m.

Determine the maximum and minimum direct stresses on the section of the bar. (*R.Ae.S.*)

Answer. Extreme stresses are 160·6 MN/m² (tensile) and 35·2 MN/m² (compressive).

6. A short hollow, cylindrical column carries a compressive load of 400 kN. Find the maximum permissible eccentricity of this load if (*a*) the tensile stress in the column must not exceed 15 MN/m² and (*b*) the compressive stress must not exceed 75 MN/m². The diameters are 150 mm internal and 200 mm external.

For this eccentricity, draw a diagram showing the distribution of stress across a section of the column.

Answer. The maximum eccentricity for both conditions is 59·2 mm; (*a*) is the limiting case. Extreme stresses are 15 MN/m² (tensile) and 72 MN/m² (compressive).

7. Figure 3.24 shows a clamp which is tightened up to a load of 400 N. Calculate the stress at the top and bottom edges of the section XX and draw a diagram showing the variation of stress across this section.

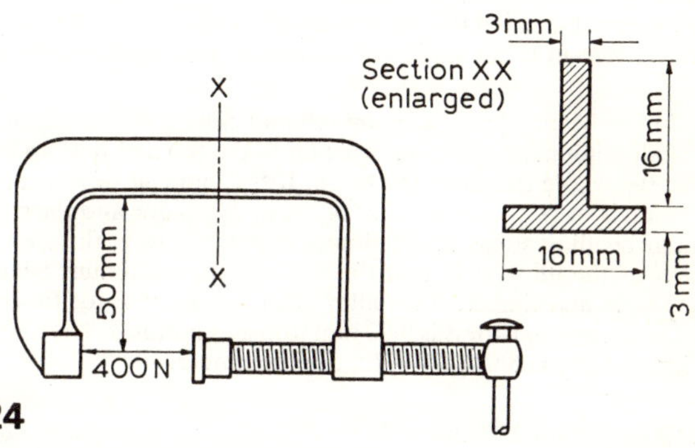

Fig. 3.24

Answer. 84·8 MN/m² (compressive) at top edge; 47·8 MN/m² (tensile) at bottom edge.

8. A composite beam consists of a timber joist 240 mm deep × 150 mm wide with a steel plate 150 mm × 10 mm bolted on each side, the steel plates being symmetrical about the axis of bending. If the stresses in the timber and steel are not to exceed 7 MN/m² and 140 MN/m² respectively, find the maximum bending moment the beam will carry and the maximum stresses in the two materials when carrying this moment.

Compare the value of this moment with that for the timber joist alone.

E (steel) = 206 GN/m². *E* (timber) = 10 GN/m² (*I.Struct.E.*)

Answer. Maximum bending moment is 16·84 kN-m, the stresses in the timber and steel being 7 MN/m² and 90 MN/m² respectively. This moment is 1·67 times that for the timber joist alone.

9. Figure 3.25 shows the cross-section of a compound bar which has been formed by brazing a steel strip 40 mm by 10 mm to a brass strip 40 mm by 20 mm. The compound bar is bent to a circular arc with neutral axis parallel to AB. If E for brass is 82 GN/m² and E for steel is 206 GN/m², find (*a*) the position of the

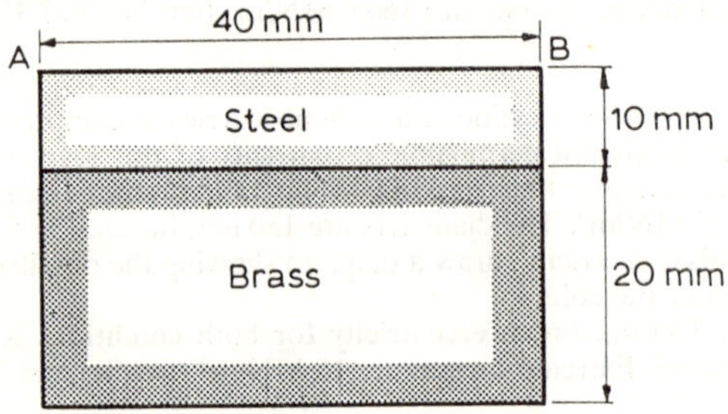

Fig. 3.25

neutral axis, and (*b*) the ratio of the maximum stress in the steel to that in the brass. Also find the radius of curvature of the neutral surface when the maximum stress in the steel becomes 70 MN/m². (*I.Mech.E.*)
Answer. Neutral axis is 11·65 mm from AB. The ratio of the maximum stress in the steel to that in the brass is 1·60. Required radius of curvature is 34·3 m.

10. A timber beam 80 mm wide by 160 mm deep is to be reinforced by bonding strips of aluminium alloy 80 mm wide on to the top and bottom faces of the timber over the whole length of the beam. If the moment of resistance of the composite beam is to be four times that of the timber alone and the same value of the maximum bending stress in the timber is reached in both cases, determine the thickness of the alloy strip and the ratio of the maximum bending stresses in the alloy strip and timber. E for alloy strip = 7·15 × E for timber. (*U.L.*)
Answer. Required thickness is 9·92 mm (top and bottom). The maximum bending stress in the alloy is 8·035 times that in the timber.

11. A timber beam 100 mm wide by 200 mm deep is to be reinforced with two steel plates 16 mm thick. Compare the moments of resistance for the same value of the maximum bending stress in the timber when the plates are: (i) 100 mm wide and fixed to the top and bottom surfaces of the beam; or (ii) 200 mm deep and fixed to the vertical sides of the beam. (E for steel = 20 × E for timber.) (*U.L.*)
Answer. The moment of resistance in case (i) is 1·416 times that in case (ii).

12. A short hollow pier, 2 m square outside and 1·2 m square inside, supports a vertical point load of 100 kN located on a diagonal and 1·2 m from the vertical axis of the pier. Neglecting the self-weight of the pier, calculate the normal stresses at the four outside corners on a horizontal section of the pier. (*U.L.*)
Answer. (All kN/m²) 185 compressive; 39 compressive (twice); 107 tensile.

13. A cast-iron beam has a section as shown in Fig. 3.26 being symmetrical about the axis YY. Determine the position of the neutral axis of the section XX, and the second moment of area about XX.

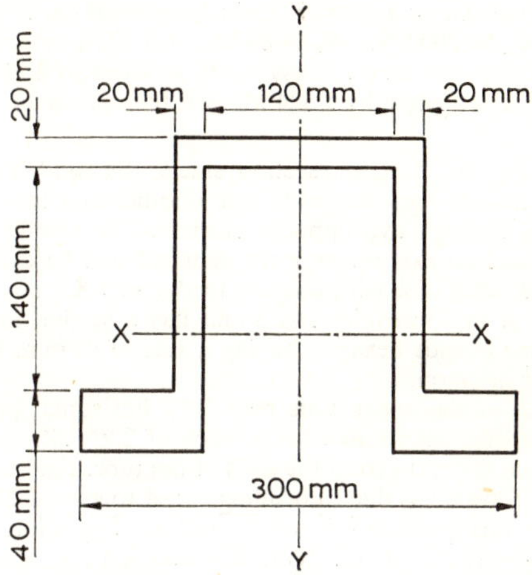

Fig. 3.26

If, when subjected to a bending moment in the plane YY, the tensile stress at the bottom edge is 24 MN/m^2, find

(*a*) the value of the bending moment, expressing it in kN-m, and
(*b*) the stress at the top edge. (*U.L.*)

Answer. 85·5 mm from bottom edge: 7940 cm^4. (*a*) 22·3 kN-m, (*b*) 32·1 MN/m^2 (compressive).

14. Figure 3.27 shows the section of a short column on which there is an axial load W_1 at O and a load W_2 at P in a direction parallel to that of W_1. The stress at the edge BC is 15·6 MN/m^2 compression and that at DE is 84 MN/m^2 compression. Determine the magnitude of W_1 and W_2.

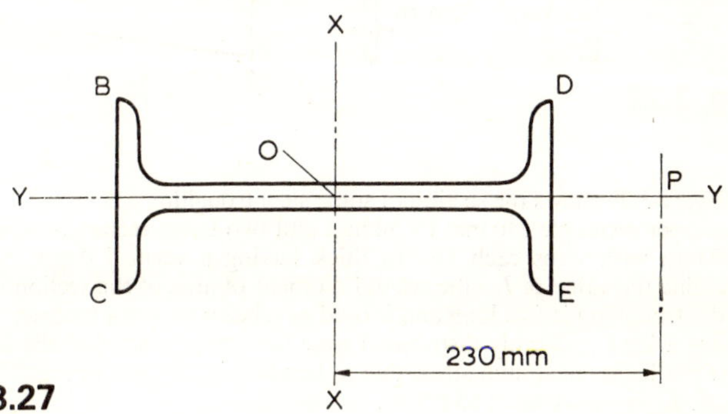

Fig. 3.27

What would be the changes in the given stresses if

(*a*) W_1 were increased by 10 kN?
(*b*) W_2 were increased by 10 kN?

Section constants: area = 57 cm^2; I_{xx} = 8 608 cm^4. (*U.L.*)
Answer. W_1 = 200 kN. W_2 = 84 kN. (*a*) Both stresses increased by 1·754 MN/m^2 (compressive). (*b*) Stress on BC reduced by 2·316 MN/m^2 (compressive) stress on DE increased by 5·824 MN/m^2 (compressive).

15. For a regular hexagonal section, side *b*, XX and YY are two axes in the plane of the section at right angles to one another and intersecting at the centroid; XX passes through two opposite corners. The second moment of area of the section about an axis through the centroid can be expressed in the form kb^4. Obtain the value of *k* with relation to the axis XX and also to YY.

A tube of uniform thickness 3 mm has a section in the shape of a regular hexagon the outside hexagon having a side of 40 mm. Find the value of I_{xx} for the hollow hexagon.

A length of the above tube rests in a horizontal position on supports 2 m apart. The tube carries two loads, each of 2 kN situated between the supports and at equal distances from the ends of the tube. Find the least distance between the loads if the stress due to bending is not to exceed 100 MN/m^2. (*U.L.*)

Note. In this question *k* is not the radius of gyration.
Answer. $k = (5\sqrt{3})/16$ for both XX and YY; I_{xx} = 42·2 cm^4; least distance = 0·783 m.

16. Figure 3.28 shows the dimensions of a welded steel bracket which acts as a cantilever. Determine the greatest stress due to bending produced at the fixed end of the bracket by the load shown. (*U.L.*)
Answer. 128 MN/m^2 (compressive).

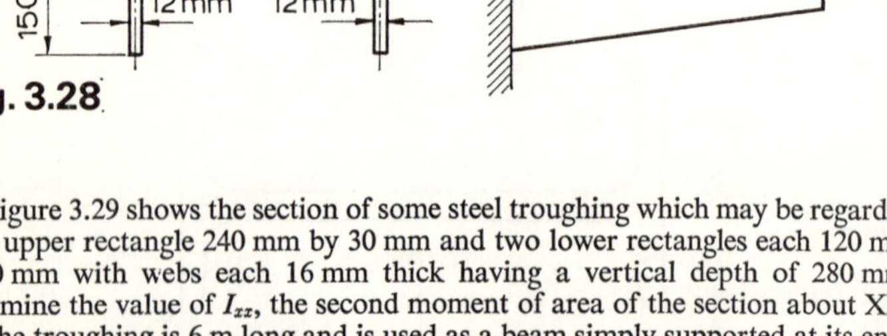

Fig. 3.28

17. Figure 3.29 shows the section of some steel troughing which may be regarded as an upper rectangle 240 mm by 30 mm and two lower rectangles each 120 mm by 30 mm with webs each 16 mm thick having a vertical depth of 280 mm. Determine the value of I_{xx}, the second moment of area of the section about XX.

If the troughing is 6 m long and is used as a beam simply supported at its ends to carry a load uniformly distributed over the whole span, find the load in kN per m run if the maximum stress due to bending is 120 MN/m^2. (*U.L.*)
Answer. I_{xx} = 41 470 cm^4; 65 kN/m.

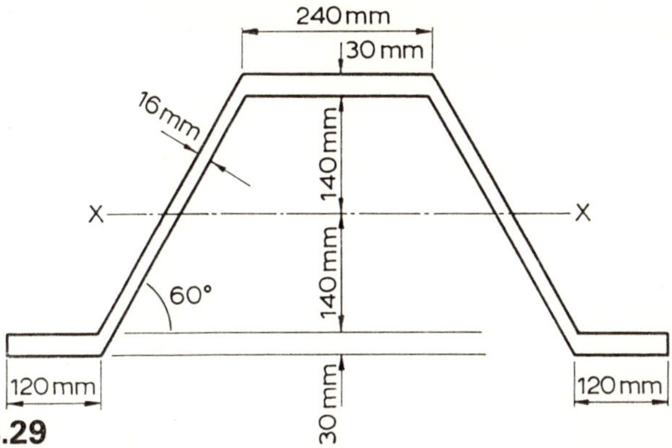

Fig. 3.29

18. A short vertical pillar has a section of uniform thickness in the form of a hollow square of side 160 mm externally and 100 mm internally. A vertical load of 210 kN is applied at a distance of 75 mm from the central axis of the pillar and on one of the diagonals of the square. What is the maximum stress on the cross-section?

Comment on the determination of the second moment of area in special cases where the section is a regular polygon. (*U.L.*)

Answer. 52·0 MN/m²; in the case of a regular polygon the second moment of area has the same value for all axes through the centroid.

19. A tension-member has a T-section symmetrical about the vertical centre-line and of the following dimensions: flange 100 mm wide by 16 mm deep, web 80 mm deep by 16 mm wide, thus making the total depth 96 mm.

The member transmits a longitudinal pull P which acts on the section at a point on the centre-line and 45 mm up from the bottom edge of the web

Find (*a*) the magnitude of P if the greatest tensile stress on the section is 120 MN/m², and (*b*) the minimum stress on the section when P is being transmitted.

Make a diagram showing the variation in stress across the section. (*U.L.*)

Answer. (*a*) 124·8 kN; (*b*) 9·68 MN/m² (tensile).

20. A 375 mm × 150 mm I-section column is loaded along an axis which is parallel to the axis of the column. The load acts at a point P which is not on either of the principal axes XX and YY. The longitudinal stresses are measured at three points A, B and C as shown in Fig. 3.30, A and B being 120 mm apart. The stresses are as follows: at A, 84·4 MN/m² compression; at B, 49·9 MN/m² compression; at C, 21·6 MN/m² compression. Determine

(*a*) The magnitude of the load and the position of the point P relative to the axes XX and YY, and

(*b*) The position of the points on XX and YY where the stress would be zero. For a 375 mm × 150 mm I-section, $I_{xx} = 20450$ cm⁴, $I_{yy} = 825$ cm⁴, cross-sectional area 85·4 cm².

Make a sketch of the section showing on it A, B and C and the answers obtained. (*U.L.*)

Answer. (*a*) 379 kN (compressive) at co-ordinates $x = -6{\cdot}48$ mm and $y = 65{\cdot}5$ mm. (*b*) On XX at $x = +149$ mm; on YY at $y = -365$ mm.

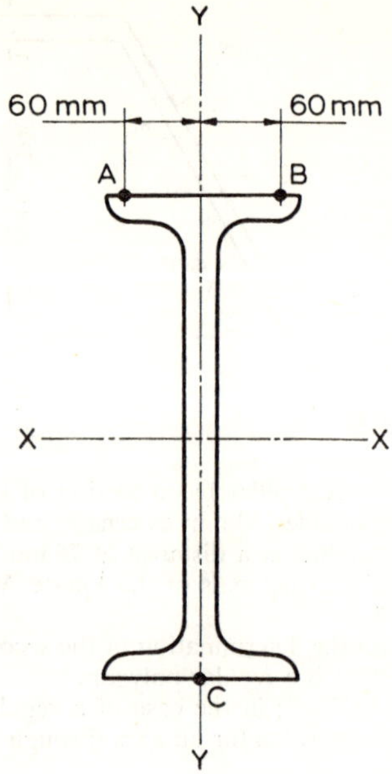

Fig. 3.30

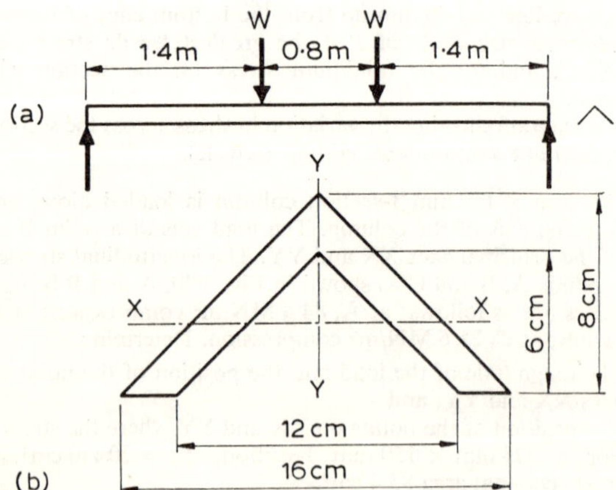

Fig. 3.31

21. A brass strip 50 mm by 13 mm in section is riveted to a steel strip 64 mm × 10 mm in section to form a compound beam of total depth 23 mm, the brass strip being on top and the beam section being symmetrical about the vertical axis. The beam is simply supported on a span of 1·2 m and carries a load of 1·6 kN at mid-span.

(*a*) Determine the maximum stresses in each of the materials due to bending;
(*b*) make a diagram showing the distribution of bending stress over the depth of the beam;
(*c*) determine the maximum deflection (see Chapter 5).

Take E for steel = 206 GN/m²; E for brass = 103 GN/m². (*U.L.*)
Answer. (*a*) In steel 110·9 MN/m² (tensile); in brass, 90·1 MN/m² (compressive); (*c*) 7·3 mm.

22. A simply supported beam 3·6 m long has a section symmetrical about the YY axis as shown in Fig. 3·31b; the beam carries two loads of equal magnitude as shown in Fig. 3.31a.

The maximum compressive stress must not exceed 108 MN/m² and the maximum tensile stress must not exceed 124 MN/m². Determine the maximum possible value for W to comply with these conditions. State clearly the values of the maximum tensile and compressive stresses and where they occur. (*U.L.*)
Answer. W = 2·05 kN; 108 MN/m² (compressive) and 85 MN/m² (tensile).

23. A cast-iron column of hollow square section and uniform wall thickness as shown in Fig. 3.32 is subjected to a load which is perpendicular to the section and which acts at a point on the XX axis.

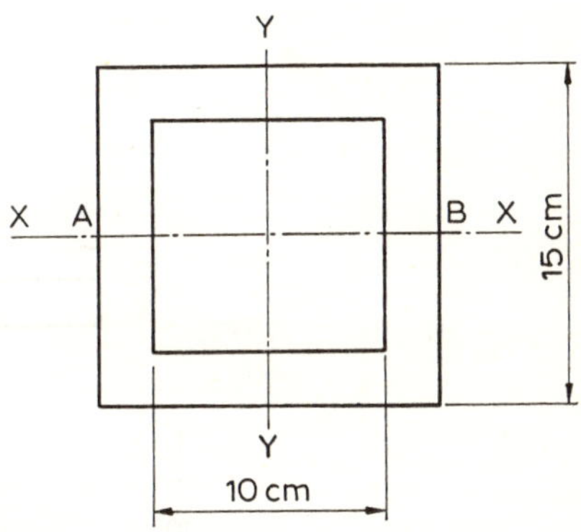

Fig. 3.32

Find the magnitude of the load and the position of its point of application if the stress at A is 140 MN/m² compression and the stress at B is 30 MN/m² tension. (*U.L.*)
Answer. 688 kN compression; 19·2 mm from A.

24. A compression member has a channel cross-section of dimensions shown in Fig. 3.33. The load carried is 250 kN and it acts longitudinally at a point P on the XX axis and 5 cm from the back of the channel. Determine the greatest and least stresses acting at the section and make a diagram to show clearly the distribution of stress across the section. (*U.L.*)
Answer. 116 MN/m² and 8 MN/m² (both compressive).

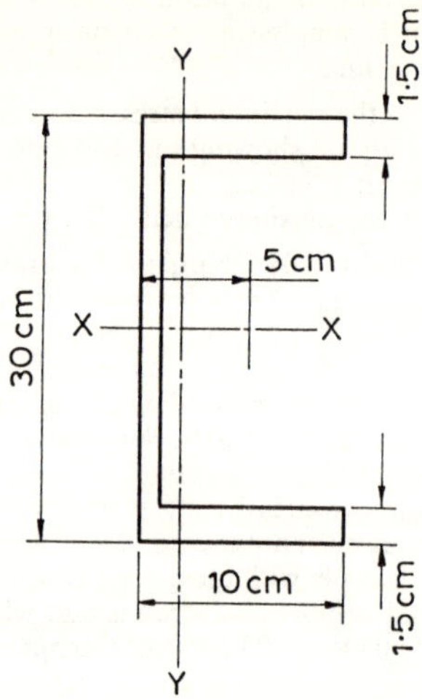

Fig. 3.33

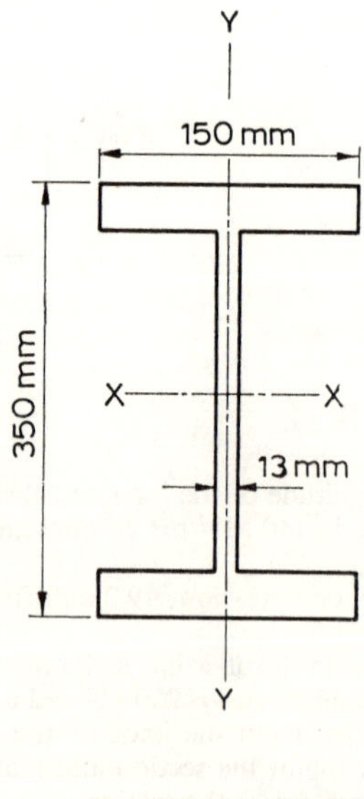

Fig. 3.34

25. The steel beam of Fig. 3.34 is of I-section and has flanges 150 mm wide. It is 350 mm deep overall and its web is 13 mm thick. Find the flange thickness required to make the value of $I_{xx} = 25000\ cm^4$; find also the value of I_{yy}. The beam is required to carry a uniformly distributed load of 41 kN/m on a simply supported span of l m; find the value of l for a maximum bending stress of 140 MN/m^2. (*U.L.*)
Answer. 28·7 mm; 1615 cm^4; $l = 6{\cdot}25$ m.

26. (*a*) Define the term "core of the section" and discuss its importance.
(*b*) A rectangular section of size $b \times h$ is subjected to a direct compressive load. Show that when the eccentricity of loading is $b/6$ or $h/6$ along the relevant axis of symmetry, the stress at the extreme outer fibres on the other side of the axis is zero. Give a sketch of the core of this section explaining how it is arrived at.
(*c*) A 300 mm × 200 mm rolled steel I-section column has a cross-sectional area 123 cm^2 and the values of its second moments of area are $I_{xx} = 20300\ cm^4$ and $I_{yy} = 2710\ cm^4$. Make a sketch of the section showing also with dimensions the core of the section. (*U.L.*)
Answer. (*c*) The core is a diamond whose corners are ±22·0 mm on the XX axis and ±110·0 mm on the YY axis from the centroid.

27. A floor is carried on 5 m long simply supported beams placed at 1 m centres. The cross-section of each beam is shown in Fig. 3.35, the 50 mm wide flange being uppermost. The web thickness is 25 mm.

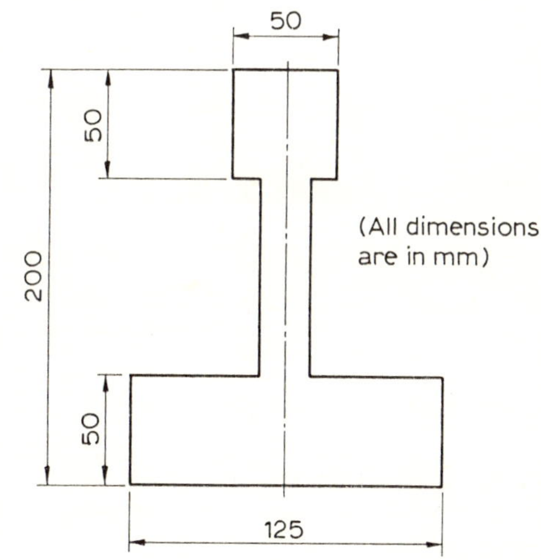

Fig. 3.35

The safe stress for the material is 75 MN/m^2 in tension and 150 MN/m^2 in compression. Determine the safe weight the beams will carry per square metre of floor area, and for this condition calculate the maximum tensile and compressive stresses in the beams. Ignore the weight of the beams. (*U.L.*)
Hint. Assume that each beam supports a floor area 5 m × 1 m.
Answer. 11·75 kN; 75 MN/m^2 (tensile) and 125 MN/m^2 (compressive).

28. With the usual assumptions used in the simple theory of bending show that

$$\sigma/y = E/R$$

where σ is the stress at distance y from the neutral axis, E is the modulus of elasticity and R is the radius of curvature. What modification to this equation would be necessary if the modulus of elasticity in tension were different from that in compression? It may be assumed that plane sections before bending remain plane after bending.

A length of straight wire is of diameter 6 mm. The wire is made of material for which the elastic limit is 140 MN/m² and the modulus of elasticity is 56 GN/m². Find the least radius to which the axis of the wire may be bent by pure couples such that there will be no permanent deformation, and determine the magnitude of the couples. (*U.L.*)

Answer. If the moduli of elasticity in tension and compression are different the neutral axis does not pass through the centroid of the cross-section. However, the equation given in the question still holds provided the appropriate value of E is used in calculating σ. $R = 1{\cdot}2$ m; $M = 2{\cdot}97$ N-m.

29. A beam has a hollow cellular cross-section as shown in Fig. 3.36. The beam spans 2 m and is simply supported at its ends. Given that the stress due to

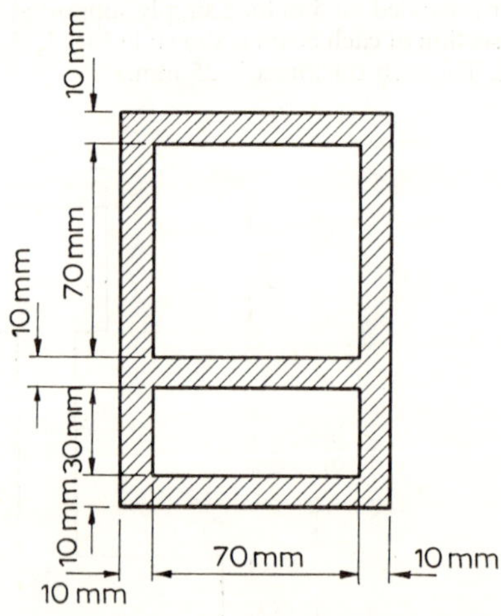

Fig. 3.36

bending is not to exceed 120 MN/m² find the maximum force uniformly distributed over the whole span which may act vertically downwards on the beam. (*U.L.*)

Answer. 63·2 kN.

Chapter 4

The Torsion of Circular Shafts

The torsion equation for a circular shaft (solid or hollow) is

$$\tau/r = T/J = G\theta/L$$

where τ = shear stress in N/m^2 at a radius r m from the axis,

T = the twisting moment (or torque) in N-m,

J = the polar second moment of area (or polar moment of inertia) in m^4,

G = modulus of rigidity (shear or torsion modulus) in N/m^2

θ = angle of twist, in radians, in a length of L m.

N.B. τ is often given in MN/m^2, J in cm^4, and G in GN/m^2.

For a solid shaft, diameter d,

$$J = \pi d^4/32$$

and for a hollow shaft, external diameter D and internal diameter d,

$$J = \frac{\pi}{32}(D^4 - d^4)$$

The power P, in watts (i.e. N-m/s) transmitted by a shaft is given by

$$P = 2\pi NT/60$$

where N = number of revolutions per minute (rev/min).

WORKED EXAMPLES

4.1. Figure 4.1 shows a horizontal shaft carrying three pulleys and supported in bearings at A and B. All the pulley belts can be considered as vertical. Calculate the unknown tension F, the reactions R_1 and R_2, and draw to scale the shearing force, bending moment, and torque diagrams for the shaft, inserting the principal values on each.

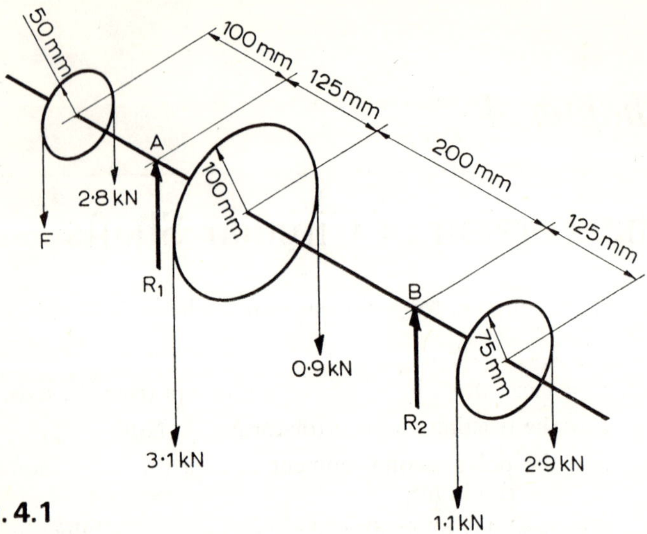
50 mm
100 mm
125 mm
200 mm
125 mm
A
100 mm
2·8 kN
F
R1
B
75 mm
0·9 kN
R2
3·1 kN
2·9 kN
1·1 kN

Fig. 4.1

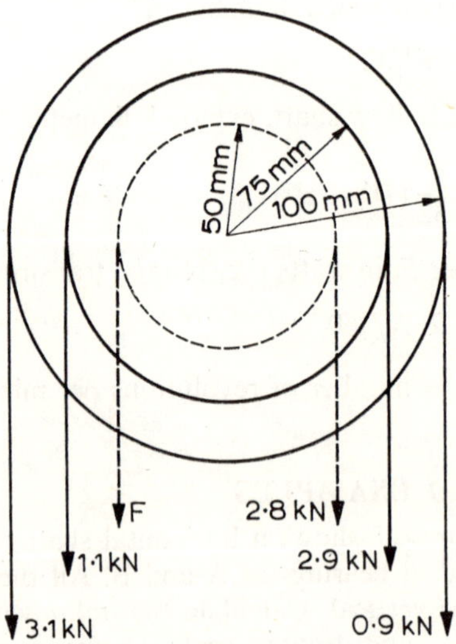
50 mm
75 mm
100 mm
F
2·8 kN
1·1 kN
2·9 kN
3·1 kN
0·9 kN

Fig. 4.2

Solution. Viewing the shaft from the right-hand end (Fig. 4.2) and taking moments about the axis of the shaft, for equilibrium, total anti-clockwise moment = total clockwise moment or, (working in kN and mm)

$$(3{\cdot}1 \text{ kN} \times 100 \text{ mm}) + (1{\cdot}1 \text{ kN} \times 75 \text{ mm}) + (F \times 50 \text{ mm})$$
$$= (2{\cdot}8 \text{ kN} \times 50 \text{ mm}) + (2{\cdot}9 \text{ kN} \times 75 \text{ mm}) + (0{\cdot}9 \text{ kN} \times 100 \text{ mm})$$

from which

$$F = 1{\cdot}1 \text{ kN} \qquad (Ans)$$

The shaft is now considered as a simply supported beam (Fig. 4.3) carrying a point load at each pulley equal to the sum of the two tensions for that pulley belt.

Using the side view of the shaft (Fig. 4.3) the reactions R_1 and R_2 are found by taking moments about the points of support.

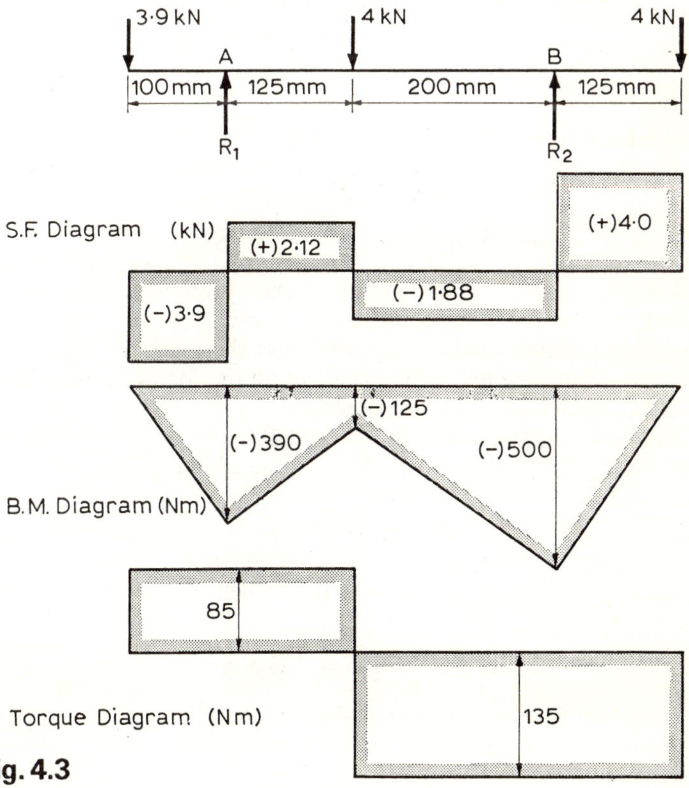

Fig. 4.3

Moments about A, (anti-clockwise = clockwise), using kN and mm,

$$(3{\cdot}9\text{ kN} \times 100\text{ mm}) + (R_2 \times 325\text{ mm})$$
$$= (4\text{ kN} \times 125\text{ mm}) + (4\text{ kN} \times 450\text{ mm})$$

from which

$$R_2 = 5{\cdot}88\text{ kN} \qquad (Ans)$$

Similarly, taking moments about B,

$$R_1 = 6{\cdot}02\text{ kN} \qquad (Ans)$$

S.F. between 50 mm pulley and A = (−) 3·9 kN

S.F. between A and 100 mm pulley = (−) 3·9 kN + 6·02 kN
= 2·12 kN

S.F. between 100 mm pulley and B = (−) 3·9 kN + 6·02 kN − 4 kN
= (−) 1·88 kN

S.F. between B and 75 mm pulley = +4 kN (considering forces to the right)

B.M. at A = −(3·9 kN × 100 mm) = −390 N-m

B.M. at 200 mm pulley = −(3·9 kN × 225 mm) + (6·02 kN × 125 mm)
= −125 N-m

B.M. at B = −(4 kN × 125 mm) = −500 N-m

A torque diagram is similar to a bending moment diagram and shows the torque being transmitted at each section of the shaft. The torque transmitted at any section is the algebraic sum of the moments about the axis of the shaft of all the forces on one side (it does not matter which) of that section.

The torque between the 50 mm and 100 mm pulleys is found by considering the tensions of the belt on the 50 mm pulley, i.e.

$$(2{\cdot}8\text{ kN} - 1{\cdot}1\text{ kN}) \times 50\text{ mm} = 85\text{ N-m}.$$

Similarly the torque between the 100 mm and 75 mm pulleys is found by considering the tensions of the belt on the 75 mm pulley, i.e.

$$(2{\cdot}9\text{ kN} - 1{\cdot}1\text{ kN}) \times 75\text{ mm} = 135\text{ N-m}$$

These two torques act in opposite senses and are plotted upwards and downwards respectively. It is not necessary however, to specify a sign convention for torque.

4.2. What is the torque transmitted by a thin tube 50 mm mean diameter and 3 mm wall thickness, if the shear stress is 69 MN/m^2 and is assumed to be uniform throughout the wall of the tube?

Solution. When a tube is transmitting torque there is a tendency at any section for shearing to take place. This is resisted by internal shearing forces (see Fig. 4.4). The total area resisting shear is the cross-sectional area of the tube. In the present case,

$$\begin{aligned}\text{Area resisting shear} &= \text{mean circumference} \times \text{thickness}\\ &= \pi \times 50\text{ mm} \times 3\text{ mm}\\ &= 150\,\pi\text{ mm}^2\end{aligned}$$

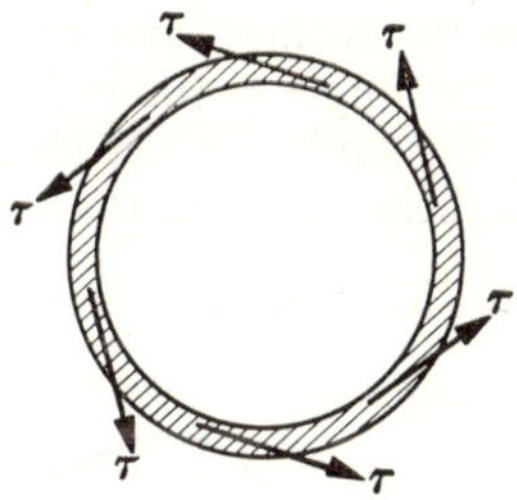

Fig. 4.4

If the shear stress is assumed uniform over this area, then the total tangential shearing force equals

$$\begin{aligned}\text{Shear stress} \times \text{area} &= (69 \times 10^6\text{ N/m}^2) \times (150\pi \times 10^{-6}\text{ m}^2)\\ &= 10\,350\pi\text{ N}\end{aligned}$$

This force is acting at a radius of 25 mm and hence the torque transmitted is

$$\begin{aligned}T = \text{force} \times \text{radius} &= 10\,350\pi\text{ N} \times (25 \times 10^{-3}\text{ m})\\ &= 813\text{ N-m} \qquad (Ans)\end{aligned}$$

4.3. Derive an expression giving the shear stress τ at a radius r from the axis of a circular shaft subjected to a twisting moment (or torque) T, which produces an angle of twist θ in a length L. (The modulus of rigidity for the material of the shaft is G.)

What is the angle of twist, in degrees, in a 3 m length of a hollow shaft, 150 mm external and 90 mm internal diameter, when it is subjected to a twisting moment which produces a maximum shear stress of 70 MN/m^2? $G = 77$ GN/m^2.

Find also the shear stress at the inside edge of the shaft and draw a diagram showing the variation of the shear stress through the wall of the shaft.

Solution. The theory of torsion depends on certain assumptions analogous to those made in the theory of simple bending. They are

(*a*) The material of the shaft is uniform throughout.
(*b*) The twist is uniform along the shaft.
(*c*) The cross-sections remain plane.
(*d*) All radii remain straight.

Consider a shaft length L rigidly fixed at one end A, as shown in Fig. 4.5. (The same results will hold if the shaft is rotating and transmitting power.) If a torque is applied at the free end, then a line AB on the surface, originally parallel to the axis, becomes a helix AB′. The angle γ (in radians) is the shearing strain at all points along the surface of the shaft.

Hence the shear stress (at the surface) is

$$\tau = G \times \text{shear strain} = G\gamma \qquad \text{(i)}$$

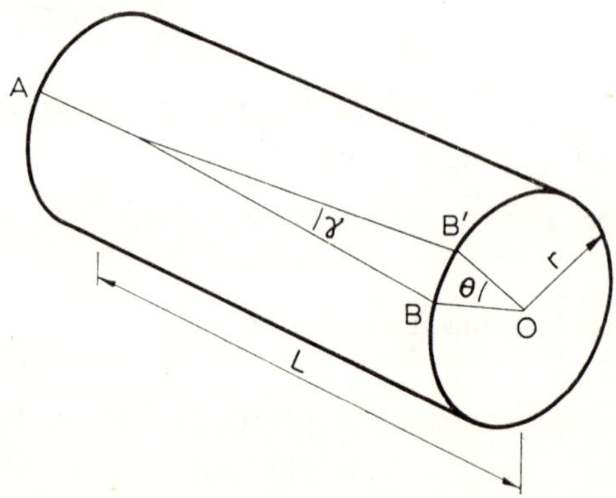

Fig. 4.5

The displacement of B can be expressed in terms of γ or θ. Since γ is a small angle, then BB′ $= L\gamma$. Also if θ is the angle of twist, in radians, then BB′ $= r\theta$.

From these two relationships,

$$L\gamma = r\theta \quad \text{or} \quad \gamma = r\theta/L$$

Substituting for γ in (i) we obtain

$$\tau = Gr\theta/L$$

or, as it is usually written

$$\tau/r = G\theta/L \qquad (Ans.)$$

This expression holds, by similar reasoning, at any other radius in the shaft, and hence the shear stress at any point in a shaft is proportional to its distance from the axis, the maximum occurring at the outside edge.

For the shaft in the question we have (rearranging the expression obtained)

$$\theta = \tau L/rG = \frac{70\ \text{MN/m}^2 \times 3\ \text{m}}{75\ \text{mm} \times 77\ \text{GN/m}^2}$$
$$= 0{\cdot}036 \text{ radians or (approx) } 2{\cdot}08^\circ \qquad (Ans)$$

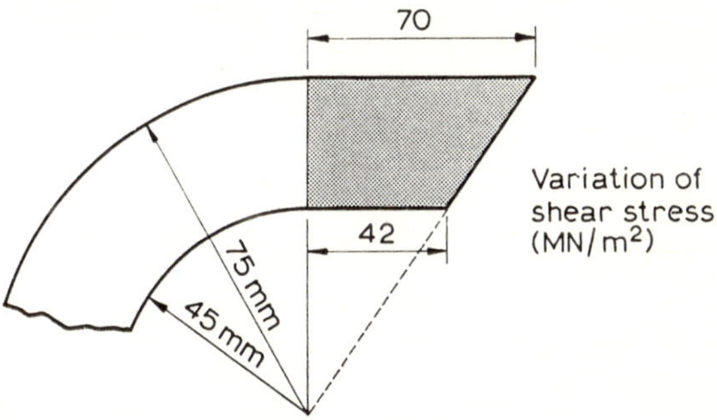

Fig. 4.6

Since the shear stress is proportional to the radius then

Stress at inside edge/45 mm = stress at outside edge/75 mm

or

$$\text{Stress at inside edge} = (45\ \text{mm}/75\ \text{mm}) \times 70\ \text{MN/m}^2$$
$$= 42\ \text{MN/m}^2 \qquad (Ans)$$

The stress variation is linear (Fig. 4.6); the diagram can be obtained by drawing a straight line from zero at the axis to 70 MN/m^2 at the outside edge, though it is applicable between r = 45 mm and r = 75 mm only.

4.4. Derive the complete torsion equation

$$\tau/r = T/J = G\theta/L$$

Use it to calculate the torque transmitted by the tube of Example 4.2 when the maximum shear stress is 69 MN/m².

Solution. From the expression obtained in Example 4.3, the shear stress at radius r is

$$\tau = rG\theta/L$$

Consider a hollow shaft, internal and external diameters d and D, Fig. 4.7. The shaft can be considered as made up of concentric tubes such as the one shown, radius r, thickness δr.

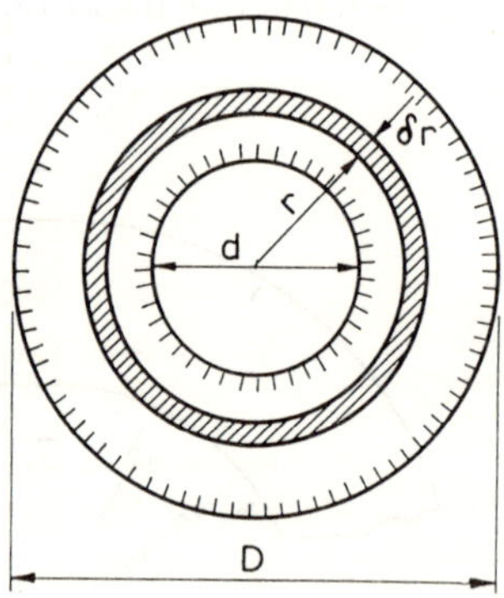

Fig. 4.7

The contribution of this tube to the total twisting moment transmitted by the shaft is (as in the case of the thin tube of Example 4.2)

$$\begin{aligned}&\text{Shear stress} \times \text{area} \times \text{radius}\\ &= \tau \times \text{circumference} \times \text{thickness} \times \text{radius}\\ &= \tau \times 2\pi r \times \delta r \times r\\ &= 2\pi\tau r^2\delta r\\ &= 2\pi(rG\theta/L)r^2\delta r\\ &= \frac{G\theta}{L}2\pi r^3\delta r\end{aligned}$$

The total twisting moment is the sum of all such moments, i.e.

$$\sum\frac{G\theta}{L}2\pi r^3\delta r \qquad \text{from } r = d/2 \text{ to } r = D/2$$

In the calculus notation, the total twisting moment is

$$T = \int_{d/2}^{D/2} \frac{G\theta}{L} 2\pi r^3 dr = \frac{G\theta}{L} \int_{d/2}^{D/2} 2\pi r^3 dr$$

The integral in the last expression is called the *polar second moment of area* or *polar moment of inertia* and it is denoted by *J*.

By direct integration

$$J = \left[\frac{\pi r^4}{2}\right]_{d/2}^{D/2} = \frac{\pi}{2}\left[\left(\frac{D}{2}\right)^4 - \left(\frac{d}{2}\right)^4\right]$$

$$= \frac{\pi}{32}(D^4 - d^4)$$

For a solid shaft, diameter d, the limits of integration are 0 and $d/2$, and

$$J = \left[\frac{\pi r^4}{2}\right]_0^{d/2} = \pi d^4/32$$

In both cases we have

$$T = (G\theta/L)\,J \quad \text{or} \quad T/J = G\theta/L$$

From the expression obtained in Example 4.3,

$$\tau/r = G\theta/L$$

and altogether, the torsion equation is

$$\tau/r = T/J = G\theta/L \qquad (Ans)$$

The units of each quantity in the equation are given in the formulae at the beginning of this chapter.

For the tube of Example 4.2,

$$J = \frac{1}{32}\pi[(5{\cdot}3\ \text{cm})^4 - (4{\cdot}7\ \text{cm})^4] = 29{\cdot}5\ \text{cm}^4$$

From the torsion equation

$$T = \frac{\tau}{r}J = \frac{69 \times 10^6\ \text{N/m}^2}{26{\cdot}5 \times 10^{-3}\ \text{m}} \times (29{\cdot}5 \times 10^{-8}\ \text{m}^4)$$

$$= 768\ \text{N-m} \qquad (Ans)$$

It is worth noting that if the *mean* stress had been taken as 69 MN/m^2, the torque would have been 814 N-m which is very close to the value obtained in Example 4.2.

4.5. Derive an expression for the power transmitted by a shaft at N revolutions per minute in terms of the mean twisting moment, T N-m.

Calculate the power which can be transmitted at 50 rev/min by a hollow circular shaft, 250 mm external diameter and 150 mm internal diameter, if the maximum twisting moment is 40 per cent greater than the mean and the shear stress must not exceed 69 MN/m^2.

Find also the maximum twist in a length of 4·6 m, taking $G = 79$ GN/m^2.

Solution. The work done by a torque in turning a shaft is equal to the average torque × angle in radians through which it turns.

Hence the work done in one revolution (2π radians) is

$$T \text{ N-m} \times 2\pi = 2\pi T \text{ N-m}$$

If the shaft is rotating at N revolutions per minute, the work done per second is

$$2\pi T \text{ N-m} \times \frac{N}{60} = \frac{2\pi NT}{60} \text{ N-m}$$

Thus the power, which is the rate of doing work, is

$$\text{Power} = 2\pi NT/60 \text{ watts}$$

where watt is the name given to the unit of power, i.e.

$$1 \text{ watt} = 1 \text{ N-m/s}$$

In many installations the torque fluctuates and the *mean* torque must be used in calculating the power, but the *maximum* value in calculating the maximum stress.

For the shaft in question,

$$J = \frac{\pi}{32}(D^4 - d^4) = \frac{\pi}{32}[(25 \text{ cm})^4 - (15 \text{ cm})^4] = 33\,370 \text{ cm}^4$$

From the equation $\tau/r = T/J$ the maximum torque is

$$T_{max} = \frac{\tau J}{r_{max}} = \frac{(69 \times 10^6 \text{ N/m}^2) \times (33\,370 \times 10^{-8} \text{ m}^4)}{125 \times 10^{-3} \text{ m}}$$
$$= 1{\cdot}84 \times 10^5 \text{ N-m}$$

The mean torque is

$$T_{mean} = \frac{100}{140} \times (1{\cdot}84 \times 10^5 \text{ N-m})$$
$$= 1{\cdot}32 \times 10^5 \text{ N-m}$$

Hence the shaft can transmit a power of

$$\frac{2\pi NT}{60} = \frac{2\pi \times 50 \times (1{\cdot}32 \times 10^5)}{60} \text{ watt} = 689 \text{ kW} \qquad (Ans)$$

From the equation $\tau/r = G\theta/L$ the maximum twist in a length of 4·6 m is

$$\theta = \frac{\tau L}{rG} = \frac{(69 \times 10^6 \text{ N/m}^2) \times 4{\cdot}6 \text{ m}}{(125 \times 10^{-3} \text{ m}) \times (79 \times 10^9 \text{ N/m}^2)}$$
$$= 0{\cdot}0322 \text{ rad} = 1{\cdot}84°$$

4.6. A solid shaft is to transmit 370 kW at 120 rev/min. If the shear stress of the material must not exceed 92 MN/m², find the diameter required.

What percentage saving in weight would be obtained if this shaft were replaced by a hollow one whose internal diameter equals 0·6 × external diameter, the length, material, and maximum shear stress remaining unchanged?

Solution. The torque is assumed constant since there is no information to the contrary. Rearranging the power expression, the torque is

$$T = \frac{60 \times \text{power}}{2\pi N} = \frac{60 \times (370 \times 10^3)}{2\pi \times 120}$$
$$= 29{\cdot}4 \text{ kN-m}$$

For a solid shaft, diameter d, $J = \pi d^4/32$ and the maximum stress occurs where $r = d/2$. Substituting in the torsion equation $\tau/r = T/J$, we have

$$\frac{92 \times 10^6 \text{ N/m}^2}{d/2} = \frac{29{\cdot}4 \times 10^3 \text{ N-m}}{\pi d^4/32}$$

from which

$$d^3 = 1{\cdot}63 \times 10^{-3} \text{ m}^3 \qquad d = 0{\cdot}1177 \text{ m}$$

The required diameter is 117·7 mm. (*Ans*)

Let D be the external diameter of the hollow shaft. Then 0·6 D is its internal diameter.

If the two shafts are to transmit the same torque at the same maximum shear stress then, since

$$T/\tau_{max} = J/r_{max}$$

the value of J/r_{max} must be the same for both.

For the solid shaft

$$\frac{J}{r_{max}} = \frac{\pi d^4/32}{d/2} = \frac{\pi d^3}{16} = \frac{\pi}{16} \times (1{\cdot}63 \times 10^{-3}\ \text{m}^3)$$

and for the hollow shaft

$$\frac{J}{r_{max}} = \frac{\frac{1}{32}\pi[D^4 - (0{\cdot}6D)^4]}{(D/2)}$$
$$= \tfrac{1}{16}\pi D^3[1 - (0{\cdot}6)^4]\ \text{m}^3$$

Equating the two values of J/r_{max},

$$D^3[1 - (0{\cdot}6)^4] = 1{\cdot}63 \times 10^{-3}$$
$$D^3 = 1{\cdot}874 \times 10^{-3} \qquad D = 0{\cdot}1233\ \text{m} \quad (\text{or } 123{\cdot}3\ \text{mm})$$

Since the two shafts are of the same material and length, the saving in weight is proportional to the reduction in cross-sectional area.

Hence, percentage saving in weight

$$= \frac{\text{area of solid shaft} - \text{area of hollow shaft}}{\text{area of solid shaft}} \times 100$$
$$= \left[1 - \frac{\text{area of hollow shaft}}{\text{area of solid shaft}}\right] \times 100$$
$$= \left\{1 - \frac{\frac{1}{4}\pi[(123{\cdot}3)^2 - (0{\cdot}6 \times 123{\cdot}3)^2]}{\frac{1}{4}\pi \times (117{\cdot}7)^2}\right\} \times 100$$
$$= 29{\cdot}8$$

A saving in weight of 29·8 per cent would be obtained. *(Ans)*

4.7. Show that for a given maximum shear stress the minimum diameter of solid circular shaft required to transmit P watts at N revolutions per minute can be written as

$$d = \text{constant} \times \sqrt[3]{(P/N)}$$

What value of the maximum shear stress has been used if the constant equals 8, d being in millimetres?

Solution. Rearranging the power formula, the torque to be transmitted is

$$T = 60P/2\pi N\ \text{N-m}$$

Assuming that this torque is constant, we have, from the torsion equation,

$$J/r_{max} = T/\tau_{max} = 60P/2\pi\tau_{max}N$$

Since, however, $J = \pi d^4/32$ and $r_{max} = d/2$, this gives

$$J/r_{max} = \pi d^3/16 = 60P/2\pi\tau_{max}N$$

or

$$d^3 = \frac{16 \times 60 \times P}{2 \times \pi^2 \tau_{max} N}$$

and

$$d = \sqrt[3]{\left[\left(\frac{16 \times 60}{2\pi^2\tau_{max}}\right)\frac{P}{N}\right]} \text{ metres}$$

or

$$d = 1000\sqrt[3]{\left[\left(\frac{16 \times 60}{2\pi^2\tau_{max}}\right)\frac{P}{N}\right]} \text{ millimetres}$$

For a given value of τ_{max} the expression in brackets is a constant and hence

$$d = \text{constant} \times \sqrt[3]{(P/N)}$$

If the constant is 8, then

$$8 = 1000 \times \sqrt[3]{\frac{16 \times 60}{2\pi^2\tau_{max}}}$$

from which

$$\tau_{max} = \frac{16 \times 60 \times 10^9}{2\pi^2 \times 512} = 95 \text{ MN/m}^2 \qquad (Ans)$$

4.8. A hollow marine propeller shaft turning at 110 rev/min is required to propel a vessel at 25 knots for the expenditure of 6·3 MW shaft power, the efficiency of the propeller being 68 per cent. The diameter ratio of the shaft is to be 2/3 and the direct stress due to the thrust is not to exceed 8 MN/m². Calculate

(*a*) the shaft diameters,
(*b*) the maximum shearing stress due to the torque.

Take 1 knot = 0·52 m/s. (*U.L.*)

Solution. Since the propeller is 68 per cent efficient then its thrust, F newtons, does work corresponding to $0{\cdot}68 \times 6{\cdot}3$ MW. Hence, thrust × distance moved per second = work done per second or

$$F\,\text{N} \times (25 \times 0{\cdot}52\ \text{m/s}) = 0{\cdot}68 \times 6{\cdot}3 \times 10^6\ \text{N-m/s}$$

from which

$$F = \frac{0{\cdot}68 \times 6{\cdot}3 \times 10^6}{25 \times 0{\cdot}52} = 330\ \text{kN} \qquad (Ans)$$

If D is the external diameter, then $\frac{2}{3}D$ is the internal diameter and, since area = load/stress, we have

$$\tfrac{1}{4}\pi[D^2 - (\tfrac{2}{3}D)^2] = \frac{330 \times 10^3\ \text{N}}{8 \times 10^6\ \text{N/m}^2}$$

from which

$$D^2 = \frac{4 \times 330 \times 10^3}{\frac{5}{9}\pi \times 8 \times 10^6}$$

and

$$D = 0{\cdot}3075\ \text{m} \quad \text{or} \quad 307{\cdot}5\ \text{mm}$$

The shaft diameters are 307·5 mm external and $307{\cdot}5 \times \frac{2}{3}$ = 205 mm internal. (*Ans*)

$$J \text{ for the shaft} = \pi\tfrac{1}{32}(307{\cdot}5^4 - 205^4) \times 10^{-12}$$
$$= 75{\cdot}13 \times 10^{-5}\ \text{m}^4$$

and

$$\text{Torque } T = \frac{60 \times P}{2\pi N} = \frac{60 \times 6{\cdot}3 \times 10^6}{2\pi \times 110}$$
$$= 547\ \text{kN-m}$$

From the torsion equation, the maximum shear stress is

$$\tau = \frac{Tr_{max}}{J} = \frac{(547 \times 10^3\ \text{N-m}) \times (0{\cdot}1538\ \text{m})}{75{\cdot}13 \times 10^{-5}\ \text{m}^4}$$
$$= 112\ \text{MN/m}^2 \qquad (Ans)$$

4.9. Explain the terms "torsional rigidity" and "polar modulus of section".

A hollow steel shaft, 200 mm internal and 300 mm external diameter, is to be replaced by a solid alloy shaft. If the polar modulus has the same value for both, calculate the diameter of the latter and the ratio of the torsional rigidities. G for steel $= 2{\cdot}4 \times G$ for alloy.

If, alternatively, the torsional rigidity has the same value for both, calculate the ratio of the polar moduli.

Solution. From the torsion equation, the angle of twist is

$$\theta = TL/GJ$$

Hence, for a given torque and length of shaft, the twist is inversely proportional to the product GJ. This product is called the *torsional rigidity* for the shaft.

If the torsion equation is rearranged to express the torque in terms of the maximum shear stress, we have

$$T = \frac{J}{r_{max}} \tau_{max}$$

The ratio J/r_{max} is called the *polar modulus* of the section and is usually denoted by Z_p. It is a geometrical property of the section, having the units m^3, and is analogous to the section modulus used in bending theory. (For a solid circular shaft its value is $Z_p = \pi d^3/16$).

For the steel shaft in the question,

$$J = \frac{1}{32}\pi\left[(0{\cdot}3)^4 - (0{\cdot}2)^4\right] = \left(\frac{65\pi}{32} \times 10^{-4}\,\text{m}^4\right)$$

If the polar modulus has the same value for both shafts, then

$$\frac{J \text{ for steel}}{r_{max} \text{ for steel}} = \frac{J \text{ for alloy}}{r_{max} \text{ for alloy}} \qquad \text{(i)}$$

or

$$\frac{(65\pi \times 10^{-4}/32)\text{m}^4}{0{\cdot}15} = \frac{\pi d^4/32}{d/2}$$

where $d =$ diameter of alloy shaft in metres.

Hence

$$d^3 = (21{\cdot}67 \times 10^{-3})\ \text{m}^3$$
$$d = 0{\cdot}279\ \text{m} \quad \text{or} \quad 279\ \text{mm}$$

Ratio of torsional rigidities

$$= \frac{G \text{ for steel} \times J \text{ for steel}}{G \text{ for alloy} \times J \text{ for alloy}}$$

$$= \frac{G \text{ for steel}}{G \text{ for alloy}} \times \frac{r_{max} \text{ for steel}}{r_{max} \text{ for alloy}} \text{ [from (i)]}$$

$$= 2{\cdot}4 \times \frac{300/2 \text{ mm}}{279/2 \text{ mm}}$$

$$= 2{\cdot}58$$

i.e. the torsional rigidity of the steel shaft is 2·58 times that of the alloy shaft. *(Ans)*

If the torsional rigidity has the same value for both, then

$$G \text{ for steel} \times J \text{ for steel} = G \text{ for alloy} \times J \text{ for alloy} \qquad \text{(ii)}$$

or

$$J \text{ for alloy} = \frac{G \text{ for steel}}{G \text{ for alloy}} \times J \text{ for steel}$$

$$\pi d^4/32 = 2{\cdot}4 \times (65\pi \times 10^{-4}/32)$$

from which

$$d^4 = (156 \times 10^{-4}) \text{ m}^4$$

$$d = 0{\cdot}353 \text{ m} \quad \text{or} \quad 353 \text{ mm}$$ *(Ans)*

Ratio of polar moduli

$$= \frac{J \text{ for steel}}{r_{max} \text{ for steel}} \times \frac{r_{max} \text{ for alloy}}{J \text{ for alloy}}$$

$$= \frac{G \text{ for alloy}}{G \text{ for steel}} \times \frac{r_{max} \text{ for alloy}}{r_{max} \text{ for steel}} \text{ [using (ii)]}$$

$$= \frac{1}{2{\cdot}4} \times \frac{353/2 \text{ mm}}{300/2 \text{ mm}} = 0{\cdot}491$$

i.e. the polar modulus of the steel shaft is 0·491 times that of the alloy shaft. *(Ans)*

4.10. Two shafts are connected end to end by means of a coupling in which there are 12 bolts, the pitch circle diameter being 250 mm. The maximum shear stress is limited to 55 MN/m² in the shafts and 21 MN/m² in the bolts. If one shaft is solid, 50 mm in diameter, and the other is hollow, 100 mm external diameter, calculate the internal diameter of the latter and the bolt diameter so that both shafts and the coupling are all equally strong.

Solution. The torque transmitted by the whole system can be calculated from the conditions for the solid shaft.

$$T = \frac{\tau_{max}}{r_{max}} \times J = \frac{55 \times 10^6 \text{ N/m}^2}{25 \times 10^{-3} \text{ m}} \times \frac{\pi}{32}(50 \times 10^{-3} \text{ m}^4)$$
$$= 430\pi \text{ N-m}$$

For the hollow shaft $r_{max} = 100$ mm, and if d mm is the internal diameter, $J = \frac{\pi}{32}(100^4 - d^4) \times 10^{-12}$ m^4.

From the torsion equation $J = Tr_{max}/\tau_{max}$, and, on substitution,

$$\frac{\pi}{32}(100^4 - \text{d}^4) \times 10^{-12}\, m^4 = \frac{(430\pi \text{ N-m}) \times (50 \times 10^{-3} \text{ m})}{55 \times 10^6 \text{ N/m}^2}$$

or

$$10^8 - d^4 = 1250 \times 10^4$$
$$d^4 = 8750 \times 10^4 \qquad d = 96{\cdot}7 \text{ (mm)}$$

The internal diameter of the hollow shaft is 96·7 mm. (*Ans*)

The coupling must transmit the same torque, and, since there are 12 bolts, the torque transmitted by each bolt is

$430\pi/12$ N-m

Also since the force F in each bolt acts at a radius of 125 mm as shown in Fig. 4.8 then

$$F = \frac{430\pi/12 \text{ N-m}}{0{\cdot}125 \text{ m}} = 286{\cdot}7\pi \text{ N}$$

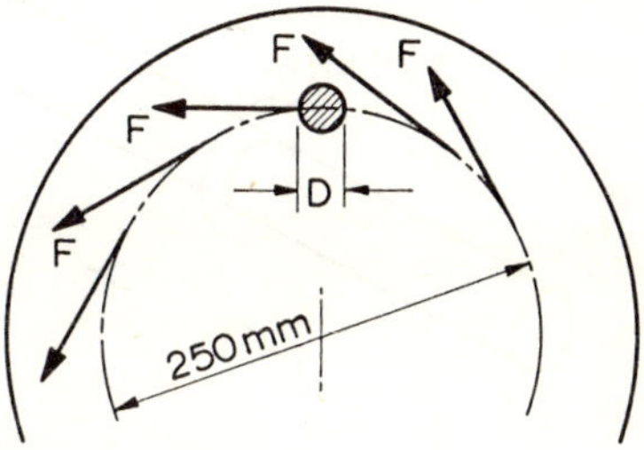

Fig. 4.8

The bolts are in single shear and therefore, assuming uniform shear stress throughout,

$$\text{Area of each bolt} = \frac{\text{force}}{\text{stress}} = \frac{286{\cdot}7\pi \text{ N}}{21 \times 10^6 \text{ N/m}^2}$$
$$= 13{\cdot}66\pi \times 10^{-6} \text{ m}^2$$

The diameter required is

$$D = \sqrt{\left(\frac{4}{\pi} \times \text{area}\right)} = \sqrt{\left(\frac{4}{\pi} \times 13{\cdot}66\pi \times 10^{-6} \text{ m}^2\right)}$$
$$= 0{\cdot}0074 \text{ m} \quad \text{or} \quad 7{\cdot}4 \text{ mm} \qquad (Ans)$$

4.11. A solid steel shaft 6 m long is securely fixed at each end. A torque of 1·25 kN-m is applied to the shaft at a section 2·4 m from one end. What are the "fixing" torques set up at the ends of the shaft?

If the diameter of the shaft is 40 mm, what are the maximum shear stresses in the two portions? Calculate also the angle of twist for the section where the torque is applied. $G = 82$ GN/m².

Solution. Let the shaft be AB as shown in Fig. 4.9. The "fixing" couples T_A and T_B will oppose the applied couple of 1·25 kN-m.

For equilibrium the fixing couples together equal the applied couple or

$$T_A + T_B = 1{\cdot}25 \text{ kN-m} \qquad \text{(i)}$$

Also, since the ends are securely fixed, the total angle of twist in the left-hand portion of the shaft must equal that in the right-hand portion. From the torsion equation we get

$$\theta = TL/JG$$

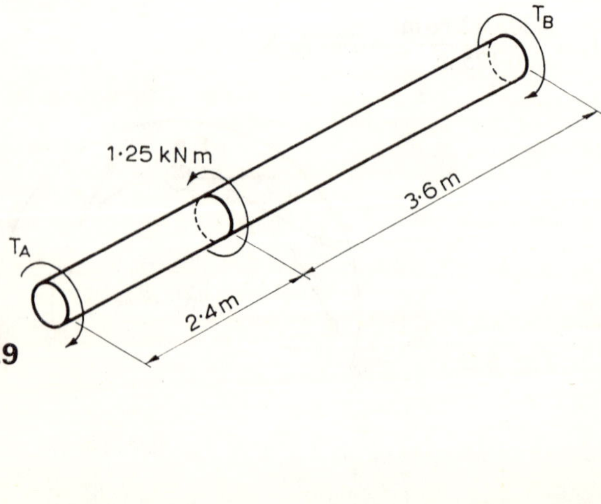

Fig. 4.9

and since J and G have the same values for both portions of the shaft, then the product TL has the same value for both portions, i.e.

$$T_A \times 2{\cdot}4\text{ m} = T_B \times 3{\cdot}6\text{ m}$$

or

$$T_A = 1\tfrac{1}{2}T_B \qquad \text{(ii)}$$

Solving the simultaneous equations (i) and (ii), we obtain

$$T_A = 0{\cdot}75\text{ kN-m} \quad \text{and} \quad T_B = 0{\cdot}5\text{ kN-m}$$

For the left-hand portion of the shaft the maximum shear stress is

$$\tau_{max} = \frac{T_A r_{max}}{J} = \frac{750\text{ N-m} \times (20 \times 10^{-3}\text{ m})}{(\pi/32) \times (40 \times 10^{-3}\text{ m})^4}$$
$$= 59{\cdot}7\text{ MN/m}^2 \qquad (Ans)$$

Similarly for the right-hand portion

$$\tau_{max} = 39{\cdot}8\text{ MN/m}^2$$

The angle of twist can be found by considering either portion of the shaft. Taking the left-hand part,

$$\theta = \frac{\text{TL}}{\text{JG}} = \frac{750\text{ N-m} \times 2{\cdot}4\text{ m}}{(\pi/32) \times (40 \times 10^{-3}\text{ m})^4 \times (82 \times 10^9\text{ N/m}^2)}$$
$$= 0{\cdot}087\text{ rad or } 5^\circ\text{ approx} \qquad (Ans)$$

PROBLEMS

1. Figure 4.10 shows a shaft carrying three gear wheels and (simply) supported in bearings at A and B. The forces can be considered as acting vertically. Calculate the unknown force P, the reactions at A and B, and draw to scale the shearing force, bending moment and torque diagrams for the shaft.
Answer. $P = 800$ N; $R_A = 820$ N and $R_B = 1780$ N.

S.F. (N): Between A and 150 mm wheel, +820; between 150 mm and 200 mm wheels −1180; between 200 mm wheel and B, −380; between B and 100 mm wheel, +1140. S.F. is constant in each range.

B.M. (N-m): At A, zero; at 150 mm wheel, +82; at 200 mm wheel, −94; at B, −140. Linear variation in each range.

Torque (N-m): Between 150 mm and 200 mm wheels, 300; between 200 mm and 100 mm wheels, 140.

2. A line shaft ABCD is driven at 20 rad/s through a pulley at C. Power is led off through pulleys at A, B and D, the amounts being 5, 11 and 8 kW respectively. Calculate the torque (in N-m) transmitted by each portion of the shaft.
Answer. For AB, 250 N-m; BC, 800 N-m and CD, 400 N-m.

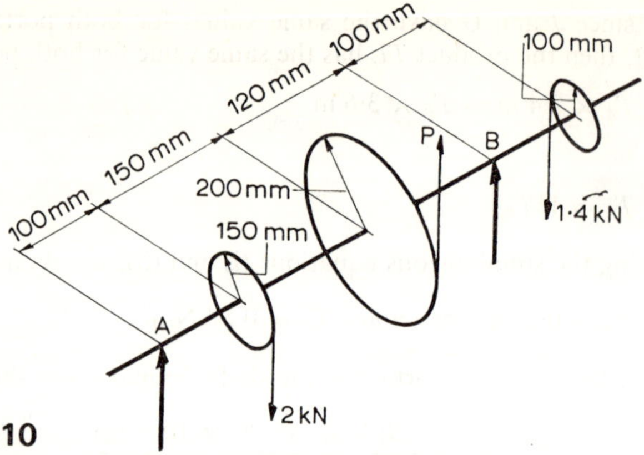

Fig. 4.10

3. Use the method of Example 4.2 to find the torque that can be transmitted by a thin tube, 80 mm mean diameter and wall thickness 5 mm, the shear stress being 60 MN/m².

What diameter is necessary for a torque of 4·5 kN-m, the wall thickness and shear stress being unchanged?

Answer. 3·02 kN-m; for a torque of 4·5 kN-m a mean diameter of 97·7 mm is required.

4. Select diameters (to the nearest mm) of round steel bar for the following conditions, taking $G = 82$ GN/m².

(*a*) To transmit a twisting moment of 900 N-m with a maximum shear stress of 40 MN/m².
(*b*) To transmit 4 kN-m with a twist of 1° in a length of 10 m.
(*c*) To twist through 10° in a length of 2 m with a maximum shear stress of 50 MN/m².
(*d*) To transmit 30 kW at 1000 rev/min with a maximum shear stress of 20 MN/m².
(*e*) To twist through 5° in a length of 3 m when transmitting 60 kW at 25 rad/s.

Answer. (*a*) 49 mm. (*b*) 130 mm. (*c*) 14 mm. (*d*) 42 mm. (*e*) 75 mm.

5. Solve the following problems in connexion with hollow shafts.

(*a*) If the internal and external diameters are 150 mm and 250 mm respectively, what is the maximum shear stress under a torque of 200 kN-m?
(*b*) What is the maximum permissible length for a torque of 3·8 kN-m if the twist must not exceed 2°, the diameters being 80 mm and 100 mm? $G =$ 80 GN/m².
(*c*) Calculate the internal diameter of a shaft to transmit 350 kW at 35 rad/s, if the maximum shear stress is to be 50 MN/m² and the maximum twisting moment is 40 per cent greater than the mean. The external diameter is 150 mm.
(*d*) Find the diameters if the twist is 3° in a length of 8 m under a torque of 1 kN-m. The diameter ratio is 5/8 and $G = 80$ GN/m².
(*e*) Find the diameters if, for a maximum shear stress of 75 MN/m², the twist is 4° in 6 m and the torque is 2 kN-m. $G = 82$ GN/m².

Answer. (*a*) 74·9 MN/m^2. (*b*) 4·26 m. (*c*) 131 mm. (*d*) 43·3 mm and 69·3 mm. (*e*) 155·8 mm and 157·2 mm.

6. The following problems relate to a solid shaft, 100 mm diameter, and a hollow shaft, of the same length and material. The diameter ratio of the latter is 3/4.

(*a*) If the weights of the two shafts are the same, find the diameters of the hollow shaft. What is the ratio of the permissible torques for the two shafts, the maximum shear stress being the same for both?

(*b*) If the torsional rigidity for the hollow shaft is to be 1·5 times that for the solid shaft, calculate the diameters of the hollow shaft.

(*c*) Find the diameters of the hollow shaft, so that the two shafts transmit the same power at the same speed and same maximum shear stress.

(*d*) What power can be transmitted by the solid shaft at a speed of 800 rev/min if the maximum twisting moment is 50 per cent greater than the mean and the shear stress is limited to 15 MN/m^2? At what speed must the hollow shaft run to transmit the same power at the same maximum stress (with the same torque fluctuation), its outside diameter being 125 mm?

(*e*) If the solid shaft has a total twist of 2° when transmitting 300 kW find the twist in the hollow shaft when it transmits 375 kW at the same speed and maximum stress.

Answer. (D and d refer to the external and internal diameters respectively.)
(*a*) $D = 151{\cdot}2$ mm, $d = 113{\cdot}4$ mm; the hollow shaft can transmit a torque of 2·36 times that of the solid. (*b*) $D = 121{\cdot}7$ mm, $d = 91{\cdot}3$ mm. (*c*) $D = 113{\cdot}5$ mm, $d = 85$ mm. (*d*) 164·5 kW, 599 rev/min. (*e*) 1·64°.

7. The following problems on couplings should be solved by the method used in Example 4.10.

(*a*) Calculate the torque which can be transmitted by a coupling with six bolts, each 20 mm diameter, the pitch circle diameter being 150 mm. Shear stress in bolts, 28 MN/m^2.

(*b*) Calculate the necessary bolt diameter (to the nearest mm) in the case of a coupling transmitting 750 kW at 20 rad/s. There are four bolts, the pitch circle diameter being 300 mm and the shear stress is limited to 15 MN/m^2.

(*c*) A coupling is to have six bolts, the diameter of the pitch circle being twelve times that of the bolts. Calculate these diameters (to the nearest mm) if the coupling is to transmit 500 hp at 150 rev/min. The maximum twisting moment is 30 per cent greater than the mean and the shear stress in the bolts must not exceed 20 MN/m^2.

Answer. (*a*) 3·96 kN-m. (*b*) 73 mm. (*c*) Bolt diameter 38 mm., P.C.D., 456 mm.

8. Explain what is meant by elastic shear strain and by the modulus of rigidity. What torque could be transmitted by a solid round shaft 75 mm diameter if the maximum shear stress intensity is 69 MN/m^2? Through what angle, in degrees, would this shaft twist in a length of 3 m if the modulus of rigidity is 82 GN/m^2? (*I.Mech.E.*)

Answer. 5·72 kN-m and 3·86°.

9. Find the external diameter of a hollow steel shaft to transmit 3 MW at 20 rad/s if the internal diameter is 0·75 of the external and the maximum shear stress due to torsion is not to exceed 56 MN/m^2. If the modulus of rigidity of the steel is

82 GN/m^2, find the elastic twist of the shaft in a length of 4 m when stressed to the maximum allowed intensity. (*I.Mech.E.*)
Answer. 271 mm and 1·155°.

10. Define shear strain and modulus of rigidity.

A hollow shaft 150 mm external and 75 mm internal diameters, is transmitting 750 kW at 15 rad/s. A torsion meter records an angle of twist of 2·3° in a length of 3 m. Determine the shear strain in the outer fibres of the shaft and the modulus of rigidity for the material. (*I.Mech.E.*)
Answer. Shear strain in outer fibres = 0·00100. τ_{max} = 80·5 MN/m^2; G = 80·2 GN/m^2.

11. A solid circular shaft of diameter d and length L is subjected to a twisting moment T. If θ is the total angle of twist in this length, establish from first principles, expressions giving the relationship between (*a*) angle of twist and maximum shearing stress, τ_s, in the fibres of the shaft, and (*b*) twisting moment and maximum shearing stress.

A solid circular shaft is to transmit 20 kW at 15 rad/s. If the maximum shearing stress is not to exceed 37 MN/m^2 find the minimum diameter of the shaft.

Find also, the angle of twist per metre length of shaft for this diameter. (Modulus of rigidity, $G = 82$ GN/m^2.) (*I.Struct.E.*)
Answer. (*a*) $\theta = 2\tau_s L/Gd$; (*b*) $T = (\pi/16)d^3\tau_s$

For the given shaft d = 56·8 mm and θ = 0·91° per metre length.

12. Derive an expression for the angle of twist of a tube of length L, internal diameter D_1, external diameter D_2, shear modulus G, under a torque T.

A tube of length 600 mm has identical rigid levers at each end (of length 150 mm) which are disposed similarly. The levers are loaded so as to produce a torque of 340 N-m. If the outer diameter of the tube is 25 mm, the inner is 12·5 mm and the shear modulus is 82 GN/m^2, determine the relative movement at the ends of the levers. (*R.Ae.S.*)

Answer. $\theta = \dfrac{32TL}{\pi G(D_2^4 - D_1^4)}$; 10·4 mm.

13. A hollow shaft, of diameter ratio 3/5, is required to transmit 600 kW at 12 rad/s, the maximum torque being 12 per cent greater than the mean. The shearing stress is not to exceed 60 MN/m^2 and the twist in a length of 3 m is not to exceed one degree. Calculate the minimum external diameter of the shaft satisfying these conditions. Take $G = 82$ GN/m^2. (*U.L.*)
Answer. 192·5 mm. (The twist is the limiting factor; the shear stress condition is satisfied by an external diameter of 176·1 mm.)

14. A solid alloy shaft of 50 mm diameter is to be coupled in series with a hollow steel shaft of the same external diameter. Find the internal diameter of the steel shaft if the angle of twist per unit length is to be 75 per cent of that of the alloy shaft.

Determine the speed at which the shafts are to be driven to transmit 20 kW if the limits of shearing stress are to be 55 MN/m^2 and 78 MN/m^2 in the alloy and steel respectively.

Modulus of rigidity for steel = 2·2 × modulus of rigidity for alloy. (*U.L.*)
Answer. 39·6 mm and 17·24 rad/s. (The stress in the steel is the limiting condition for the speed.)

15. In the case of a solid round shaft of diameter d, an axial torque T produces a maximum shearing stress τ. Derive the expression for T in terms of d and τ.

A solid shaft is required to transmit 450 kilowatt when running at 239 rev/min. If the maximum allowable shearing stress is 54 MN/m^2 determine

(*a*) the diameter of the shaft,
(*b*) the angle of twist in degrees per metre of length.
$G = 81$ GN/m^2. (*U.L.*)

Answer. $T = \pi d^3\tau/16$; (*a*) 119 mm; (*b*) 0·642°/m.

16. A steel shaft ABCD having a total length of 1·5 m is made of three lengths AB, BC, CD each 500 mm long. AB and BC are solid, having diameters of 60 mm and 80 mm respectively, and CD is hollow having outside and inside diameters of 80 and 50 mm respectively. When an axial torque of 2 kN-m is transmitted from one end of the shaft to the other the total angle of twist from A to D is 0·94 deg.

Determine

(*a*) the maximum shearing stress in the shaft and state where this occurs,
(*b*) the angle of twist for each of the three lengths AB, BC and CD,
(*c*) the modulus of rigidity of the material. (*U.L.*)

Answer. (*a*) 47·2 MN/m^2 in AB. (*b*) 0·556°; 0·176°; 0·208°. (*c*) 81 GN/m^2.

17. A 32 mm diameter solid shaft forms the transmission in the drill of an oil-well 3350 m deep. The drilling speed is 500 rev/min and the maximum shear stress is limited to 50 MN/m^2. Find

(*a*) the limiting value of the torque transmitted,
(*b*) the angle of twist of one end of the shaft relative to the other, and
(*c*) the power required for the drive.
Take the modulus of rigidity G as 81 GN/m^2. (*U.L.*)

Answer. (*a*) 322 N-m, (*b*) 129·2 rad, (*c*) 16·9 kW.

18. A steel shaft, AB, 50 mm diameter, is screwed into the end of a steel tube BC, 45 mm internal diameter and 57 mm external diameter. The effective lengths of the shaft and tube after assembly are AB = 1 m and BC 0·6 m, the total length AC thus being 1·6 m. The open end C of the tube is firmly held and a torque applied to the end A of the shaft.

If the greatest shearing stress allowed in either the shaft or tube is 46 MN/m^2, determine

(*a*) the maximum torque which may be transmitted,
(*b*) the actual values of the maximum shearing stress in the shaft and tube,
(*c*) the angle of twist of the end A.

For the steel take $G = 82$ GN/m^2. (*U.L.*)

Answer. (*a*) 1·025 kN-m (stress in tube is limiting factor). (*b*) 41·7 MN/m^2 (shaft) and 46 MN/m^2 (tube). (*c*) 1·84 degree.

19. A torsion test on a mild-steel specimen 20 mm diameter showed that the limit of proportionality in shear at the surface was reached when the torque was 250 N-m.

Determine the diameter required, to the nearest mm, for a shaft made of the material used in the test, to transmit 50 hp at 350 rev/min with a maximum shearing stress one-third of that found in the test. Find also the angle of twist of the shaft in degrees per metre length. $G = 81$ GN/m^2. (*U.L.*)

Answer. 46 mm; 1·63°.

20. A solid shaft 100 mm diameter is to be replaced by a hollow shaft having an outside diameter of 120 mm and which will transmit the same torque as the solid shaft with the same maximum shearing stress. Determine the inside diameter of the hollow shaft and find the ratio of the angles of twist, per unit length, of the second shaft to the original.

If the shaft speed is 400 rev/min and the maximum shearing stress is 52 MN/m^2 what is the power transmitted? (*U.L.*)

Answer. 96·7 mm; 5:6; 428 kW.

21. A 6 cm diameter steel shaft, 1·50 m long, is subjected to an axial torque that induces a maximum shear strain of $3{\cdot}5 \times 10^{-4}$. Determine

(*a*) the power that the shaft can transmit at a speed of 3000 rev/min,
(*b*) the relative angular deflection of the ends of the shaft, and
(*c*) the magnitude and direction of the maximum tensile stress in the shaft and where it occurs.

Shear modulus = 86 GN/m^2. (*U.L.*)

Note. The solution of (c) depends on the theory given in Chapter 9.

Answer. (*a*) 401 kW. (*b*) 0·501 degrees. (*c*) 30·1 MN/m^2 at 45° to the axis of the shaft on the surface.

22. A hollow steel shaft, having an internal to external diameter ratio of 0·6 is required to transmit 5·5 MW when rotating at a constant speed of 300 rev/min. Given that the shear stress is not to exceed 55 MN/m^2 find the minimum external diameter of the shaft. What is the minimum shear stress in the shaft?

The shear modulus for the material is 80 GN/m^2. Determine the angle of twist, in degrees, of a length of this shaft equal to twenty times the external diameter, when transmitting the required power. (*U.L.*)

Answer. 265 mm; 0·358 degrees.

23. When a hollow, circular cylinder of outer diameter D and inner diameter d, is twisted by a pure couple T it can be shown that at any radius r within the material of the cylinder, the shear stress τ is directly proportional to r. Hence show that

$$\tau = 2\tau_{max} r/D$$

and

$$T = \pi\tau_{max}(D^4 - d^4)/16D$$

ABCE are four points along the centre line of a shaft. From A to B the shaft is solid and is 60 mm diameter; from B to C the shaft is also solid but is 80 mm diameter and from C to E the shaft is hollow with 80 mm outside diameter and 60 mm inside diameter. Given that the shear stress is not to exceed 60 MN/m^2 find the maximum permissible torque that may act on the shaft. Find the actual maximum shear stress in each section of the shaft when this torque is applied. Neglect stress concentrations at the changes of section. (*U.L.*)

Answer. 2544 N-m; 60 MN/m^2 (in AB), 25·3 MN/m^2 (in BC) and 37·0 MN/m^2 (in CE).

Chapter 5

The Deflection of Beams

Bending to an arc of a circle occurs when M/EI is a constant, and in such cases

for a beam:
Maximum slope $= ML/2EI$
Maximum deflection $= ML^2/8EI$

for a cantilever:
Maximum slope $= ML/EI$
Maximum deflection $= ML^2/2EI$

The product EI is the *flexural rigidity*. A uniform beam is one in which E and I are constants throughout.

The formulae in Table 2 apply only to uniform beams (and cantilevers) and take no account of shearing effects. In each case, L is the total length and W is the total load. For the distributed load cases, $W = wL$, where w is the load per unit length.

The following relationships apply to all uniform beams and cantilevers, the origin of x being the left-hand end.

$$\text{Deflection (positive downwards)} = y$$

$$\text{Slope (positive if down to the right)} \quad i = \frac{dy}{dx}$$

$$\text{Bending moment (positive if "sagging")} \quad M = -EI\frac{d^2y}{dx^2}$$

$$\text{Shearing force (positive if "left-up")} \quad F = \frac{dM}{dx} = -EI\frac{d^3y}{dx^3}$$

$$\text{Load per unit length} \quad w = -\frac{dF}{dx} = EI\frac{d^4y}{dx^4}$$

Mohr's theorems, as applied to uniform beams and cantilevers, are

(1) The change in slope between any two points

$$= \frac{1}{EI} \times \text{area of the B.M. diagram between the points.}$$

(2) The deflection of any point relative to the tangent at a second point

$$= \frac{1}{EI} \times \text{first moment of area of the B.M. diagram between the two points about the first point.}$$

Case	Loading Diagram	Maximum Slope	Maximum Deflection
Cantilever, end point load	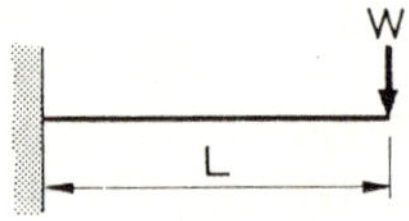	$\frac{WL^2}{2EI}$	$\frac{WL^3}{3EI}$
Cantilever, uniformly distributed load	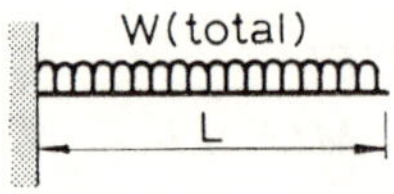	$\frac{WL^2}{6EI}$	$\frac{WL^3}{8EI}$
Simply supported beam, central point load	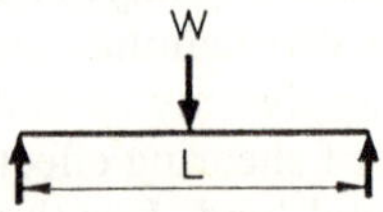	$\frac{WL^2}{16EI}$	$\frac{WL^3}{48EI}$
Simply supported beam, uniformly distributed load	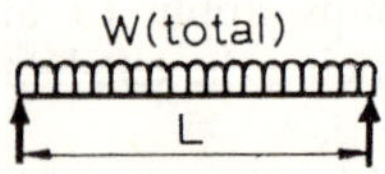	$\frac{WL^2}{24EI}$	$\frac{5WL^3}{384EI}$

Table 2

WORKED EXAMPLES

5.1. Under what conditions will a beam bend in a circular arc? Derive expressions for the maximum slope and deflection of a uniform beam, length L, when couples, each M, are applied to the ends. The couple at one end is clockwise and that at the other end is anti-clockwise.

A brass bar, 1·2 m long, is simply and symmetrically supported on a span of 1 m. It is of rectangular section, 25 mm wide and 10 mm deep, and carries concentrated loads at the ends of 120 N (each). Calculate the (upward) deflection at mid-span due to these loads, taking $E = 96\ \text{GN/m}^2$.

What is the maximum stress in the bar?

Solution. From the bending equation, we obtain

$$1/R = M/EI$$

Assuming that E is constant throughout, the value of R is inversely proportional to M/I. If this latter ratio is constant along the beam, then R will have the same value at all sections. In this case, the beam will bend to an arc of a circle, radius $R = EI/M$. This type of bending is called "circular bending" and occurs if

(*a*) the beam is of constant cross-section (constant I) and the bending moment is also constant throughout; or
(*b*) the beam is so proportioned that the value of I at any section is proportional to the bending moment M at that section.

N.B. In this chapter the deflections due to the bending moments only are considered. The deflections due to the shearing forces are, however, comparatively small in most practical cases. For case (*a*) above, the shearing force is zero (since M is constant and $F = dM/dx$). Consequently, the following theory, which neglects the shearing effects, is "exact" for case (*a*) only.

Let AB be the beam, length L, and suppose the clockwise couple to be applied at the left-hand end (Fig. 5.1a). The bending moment will be sagging and equal to M at every section. Since the beam is uniform, then M/I is constant and the beam bends to a circular arc, radius $R = EI/M$. Let O be the centre for the arc, Q the mid-point of the deflected beam and P the mid-point of the chord AB (Fig. 5.1b).

The maximum slope i, which occurs at each end, is small and can be taken as the angle in radians made by the tangent TB with the chord AB. (In the diagram the slopes and deflections have been exaggerated for clarity.)

By simple geometry, the maximum slope is

$$i = \angle \text{TBP (radians)} = \tfrac{1}{2}\pi - \angle \text{PBO} = \angle \text{POB}$$

$$= \frac{\text{arc QB}}{\text{radius OB}} = \frac{\frac{1}{2}L}{R}$$

$$= \frac{L/2}{EI/M} \text{ (since } R = EI/M\text{)}$$

$$= ML/2EI \qquad (Ans)$$

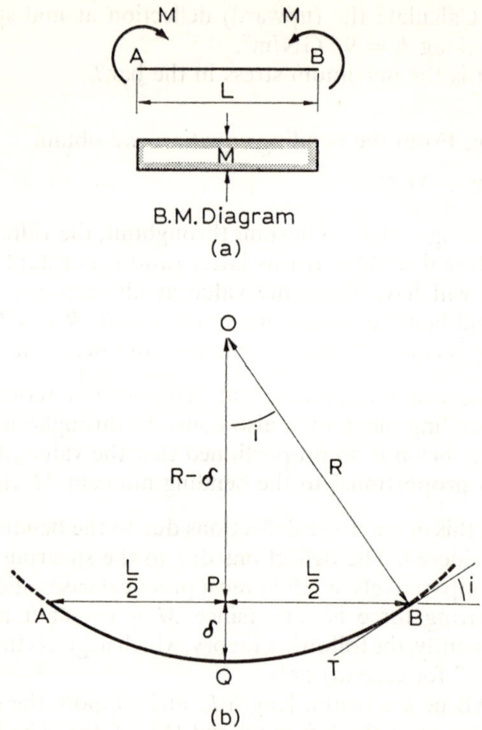

Fig. 5.1

The maximum deflection δ occurs at mid-span, by symmetry. Since it is small, then PB is sensibly equal to $\frac{1}{2}L$.

Also $OB = R$ and $OP = R - \delta$, so that, applying Pythagoras' theorem to the triangle OBP,

$$OB^2 = OP^2 + PB^2$$
$$R^2 = (R - \delta)^2 + (\tfrac{1}{2}L)^2$$
$$= R^2 - 2R\delta + \delta^2 + \tfrac{1}{4}L^2$$

from which

$$2R\delta = \tfrac{1}{4}L^2 + \delta^2$$

Since δ is small, then δ^2 is negligible compared with $L^2/4$ and the maximum deflection is

$$\delta = \frac{L^2}{4(2R)} = \frac{L^2}{8(EI/M)} = \frac{ML^2}{8EI} \qquad (Ans)$$

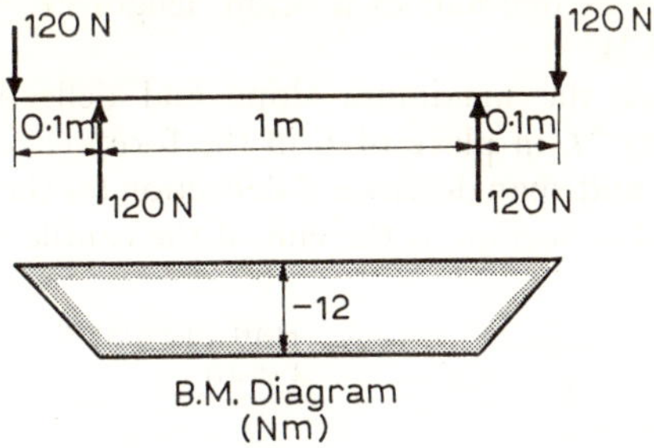

Fig. 5.2

The bending moment diagram for the brass bar in the question is shown in Fig. 5.2. The portion of the bar between the supports is subjected to a bending moment of (—)12 N-m at all sections. This portion, therefore, bends to a circular arc and, taking L as the distance between the supports, the upward deflection at mid-span is

$$\delta = \frac{ML^2}{8EI} = \frac{12 \text{ N-m} \times (1{\cdot}2 \text{ m})^2}{8 \times (96 \times 10^9 \text{ N/m}^2) \times (25 \times 10^3/12 \text{ mm}^4)}$$

$$= 10{\cdot}8 \text{ mm} \qquad (Ans)$$

The maximum stress is $\sigma_{max} = (M/I)\, y_{max}$

$$= \frac{12 \text{ N-m} \times (5 \times 10^{-3} \text{ m})}{(25 \times 10^3/12) \times 10^{-12} \text{ m}^4}$$

$$= 28{\cdot}8 \text{ MN/m}^2 \qquad (Ans)$$

5.2. Use the formulae derived in Example 5.1 to find the slope and deflection at the free end of a cantilever bent to a circular arc.

A timber cantilever, 2 m long, is of rectangular cross-section and has a uniform depth of 200 mm. It carries a uniformly distributed load of 2 kN/m run and the breadth at each point is such that a maximum bending stress of 4 MN/m² is reached at every section. Taking E for timber as 10^{10} N/m², calculate the deflection at the free end.

To what radius does the cantilever bend and what are the required widths at sections 0, $\frac{2}{3}$ and $1\frac{1}{3}$ m from the fixed end?

Solution. There is no slope at the fixed end of a cantilever since it is prevented from rotating at that end. The maximum slope and maximum deflection both occur at the free end. The cantilever, length L, behaves as one-half of a beam, length $2L$, bent to the same radius (Fig. 5.3).

Hence the maximum slope and deflection are found by substituting $2L$ in place of L in the formulae of Example 5.1. (In this case, a mid-span downward deflection for the beam corresponds to an upward deflection at the end of the cantilever.)

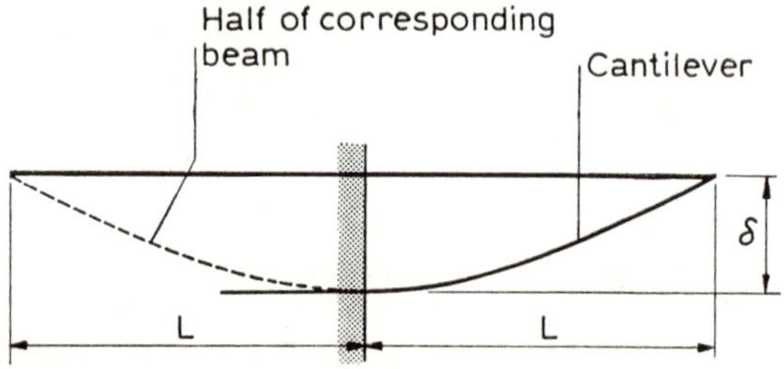

Fig. 5.3

The maximum slope

$$i = M(2L)/2EI = ML/EI$$

The maximum deflection

$$\delta = M(2L)^2/8EI = ML^2/2EI$$

With the numerical values given, we have, from the bending equation,

$$\frac{M}{I} = \frac{\sigma_{max}}{y_{max}} = \frac{4 \times 10^6 \text{ N/m}^2}{0{\cdot}1 \text{ m}} = 40 \times 10^6 \text{ N/m}^3$$

This is constant at all sections and therefore the cantilever bends to an arc of a circle. The deflection at the free end is

$$\delta = \frac{L^2}{2E}\left(\frac{M}{I}\right) = \frac{(2 \text{ m})^2 \times (40 \times 10^6 \text{ N/m}^3)}{2 \times (10^{10} \text{ N/m}^2)} = 8 \text{ mm} \qquad (Ans)$$

The radius to which the cantilever bends is

$$R = \frac{Ey_{max}}{\sigma_{max}} = \frac{(10^{10} \text{ N/m}^2) \times 0{\cdot}1 \text{ m}}{4 \times 10^6 \text{ N/m}^2} = 250 \text{ m} \qquad (Ans)$$

To find the required width, b, we have

$$I/y_{max} = M/\sigma_{max} \quad \text{or} \quad bd^2/6 = M/\sigma_{max}$$

σ_{max} and d are constant and hence

$$b = \frac{6M}{d^2\sigma_{max}} = \frac{(6M)\text{ N-m}}{(0{\cdot}2\text{ m})^2 \times 4\text{ MN/m}^2} = \frac{3M}{80000}\text{ m}$$

At the fixed end, $M = 2\text{ kN/m} \times 2\text{ m} \times 1\text{ m} = 4000\text{ N-m}$ and

$$b = \frac{3 \times 4000}{80000} = 0{\cdot}15\text{ m} = 150\text{ mm} \qquad (\textit{Ans})$$

At $\frac{2}{3}$ m from the fixed end, $M = 2\text{ kN/m} \times 1\frac{1}{3}\text{ m} \times \frac{2}{3}\text{ m}$ and

$$b = \frac{3 \times (2000 \times 1\frac{1}{3} \times \frac{2}{3})}{80000}\text{ m} = 67\text{ mm} \qquad (\textit{Ans})$$

At $1\frac{1}{3}$ m from the fixed end, $M = 2\text{ kN/m} \times \frac{2}{3}\text{ m} \times \frac{1}{3}\text{ m}$ and

$$b = \frac{3 \times (2000 \times \frac{2}{3} \times \frac{1}{3})}{80000}\text{ m} = 16{\cdot}7\text{ mm} \qquad (\textit{Ans})$$

5.3. Derive the differential equation of flexure. A uniform beam 2·5 m long, is simply supported at its ends and is loaded in such a manner that the deflection y m at any distance x m from one of the supports is

$$y = Ax^2 + Bx$$

where A and B are constants. Show that the bending moment is constant along the beam and find its numerical value if the deflection is zero at each end and 25 mm at mid-span, the flexural rigidity EI being 57·5 kN-m².

Solution. Suppose the deflection of a beam is represented by x and y co-ordinates as shown in Fig. 5.4. The x-axis represents the neutral surface of the beam before bending takes place and the deflection curve represents it after bending.

It is usual to take the origin at the left-hand end of the beam, and to consider downward deflections as positive.

Let P and Q be two points on the deflection curve whose co-ordinates are $[x, y]$ and $[x + \delta x, y + \delta y]$ respectively and whose tangents make angles of i and $(i + \delta i)$, in radians, with the x-axis. The normals at P and Q meet at C and, by simple geometry, the angle between them is δi. In all beam problems the angle of slope, i, is very small and it is sufficiently accurate to put

$$i = \tan i = dy/dx \text{ and hence } di/dx = d^2y/dx^2 \qquad \text{(i)}$$

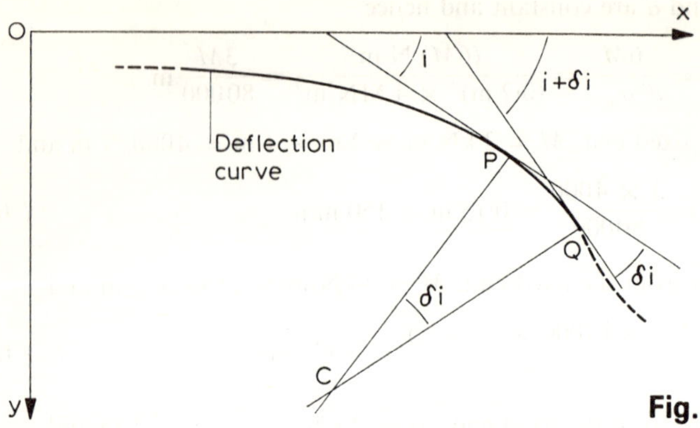

Fig. 5.4

and also

$$\text{arc PQ} = \text{difference in } x \text{ co-ordinates of P and Q}$$
$$= \delta x \qquad \text{(ii)}$$

As Q approaches P, the lengths CP and CQ tend to the same limiting value, denoted by R. Approximately,

$$R = \text{arc PQ}/\delta i$$

or

$$1/R = \delta i \,/\, \text{arc PQ} = \delta i/\delta x \quad \text{(from (ii))}$$

In the limiting case,

$$\frac{1}{R} = \frac{di}{dx} = \frac{d^2y}{dx^2} \qquad \text{(iii)}$$

R is called the *radius of curvature* and $1/R$ is called the *curvature* at the given point on the curve.

Equation (iii) is, for the reasons stated, an approximation. In text-books on mathematics it is shown that an exact expression for the curvature is

$$\frac{1}{R} = \frac{(d^2y/dx^2)}{[1 + (dy/dx)^2]^{3/2}}$$

For beam problems, it is unlikely that dy/dx will exceed 0·1, and equation (iii) will thus be sufficiently accurate for practical purposes.

From the bending equation, the curvature can be expressed as

$$1/R = M/EI \qquad \text{(iv)}$$

Since downward deflections are considered positive, however, then d^2y/dx^2 will be positive for hogging (i.e. negative) bending moments. Eliminating $1/R$ between (iii) and (iv), and adjusting the signs, we obtain

$$d^2y/dx^2 = -M/EI$$

or

$$EI(d^2y/dx^2) = -M \qquad (Ans)$$

This result is called the *differential equation of flexure* and the product EI is the *flexural rigidity*. (The term "stiffness" is sometimes used for flexural rigidity but some authors define the stiffness of a beam as *maximum deflection/span*.)

In the beam of the question,

$$y = Ax^2 + Bx$$

Differentiating,

$$dy/dx = 2Ax + B \quad \text{and} \quad d^2y/dx^2 = 2A$$

Hence

$$M = -EI(d^2y/dx^2) = -2AEI$$

Since A, E and I are constants, the bending moment is constant.

A and B are found from the conditions at mid-span and the right-hand end of the beam.

When $x = 1{\cdot}25$, $y = 0{\cdot}025$ and $0{\cdot}025 = A \,.\, 1{\cdot}25^2 + B \,.\, 1{\cdot}25$ (v)

When $x = 2{\cdot}5$, $y = 0$ and $0 = A \,.\, 2{\cdot}5^2 + B \,.\, 2{\cdot}5$ (vi)

Solving the simultaneous equations (v) and (vi),

$$A = -16 \times 10^{-3} \quad \text{and} \quad B = 40 \times 10^{-3}$$

The constant bending moment is, therefore,

$$M = -2AEI = -2(-16 \times 10^{-3})(57{\cdot}5 \times 10^3)$$
$$= 1840 \text{ N-m} \qquad (Ans)$$

5.4. Derive expressions for the slope and deflection at the free end of a uniform cantilever, length L, which carries a uniformly distributed load w per unit length.

Calculate the deflection at the free end of a steel cantilever, 6 m long, when it carries a uniformly distributed load of 100 kN total. $I = 957 \times 10^{-6}$ m^4 and $E = 206$ GN/m^2.

What couple, applied at the free end of the cantilever, will reduce this deflection by one-half?

Solution. Taking an origin at the left-hand end (Fig. 5.5), the hogging bending moment at any section XX, distance x from the origin, is

$$M = -\text{ (load to the right of XX)} \times \text{(distance of its C.G. from XX)}$$
$$= -w(L - x) \times \tfrac{1}{2}(L - x) = -\tfrac{1}{2}w(L^2 - 2Lx + x^2)$$

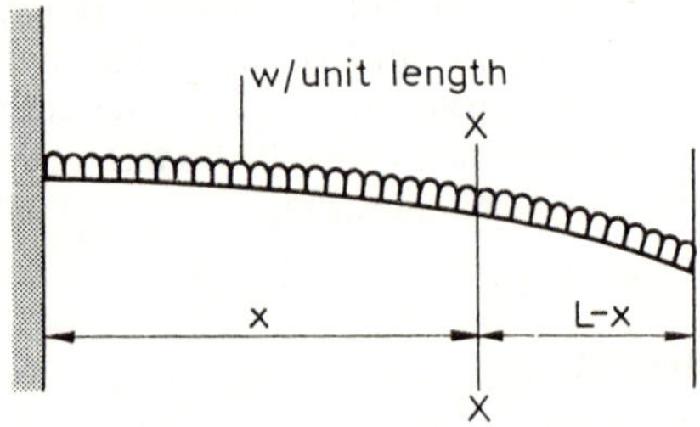

Fig. 5.5

Substituting for M in the differential equation of flexure,

$$EI(d^2y/dx^2) = -M = \tfrac{1}{2}w(L^2 - 2Lx + x^2)$$

Since EI is a constant, by direct integration

$$EI(dy/dx) = \tfrac{1}{2}w(L^2x - Lx^2 + \tfrac{1}{3}x^3) + A \qquad \text{(i)}$$

A is a constant of integration and its value can be found using the condition that the slope is zero at the origin.

Since $dy/dx = 0$ when $x = 0$, then $A = 0$.

Integrating again,

$$EIy = \frac{w}{2}\left(\frac{L^2x^2}{2} - \frac{Lx^3}{3} + \frac{x^4}{12}\right) + B \qquad \text{(ii)}$$

B, the second constant of integration, is also zero since the deflection y is zero at the origin ($x = 0$).

At the free end $x = L$, and substituting this value in (i) and (ii) we have

Slope at the free end,

$$\frac{dy}{dx} = \frac{w}{2EI}(L^3 - L^3 + \tfrac{1}{3}L^3) = \frac{wL^3}{6EI}$$

Deflection at the free end,

$$y = \frac{w}{2EI}\left(\frac{L^4}{2} - \frac{L^4}{3} + \frac{L^4}{12}\right) = \frac{wL^4}{8EI}$$

Writing $W = wL$, the total load, these expressions become $WL^2/6EI$ and $WL^3/8EI$ respectively.

With the numerical values given in the question, the deflection at the free end is

$$\delta = \frac{WL^3}{8EI} = \frac{(100 \times 10^3 \text{ N}) \times (6 \text{ m})^3}{8 \times (206 \times 10^9 \text{ N/m}^2) \times (957 \times 10^{-6} \text{ m}^4)}$$

$$= 13{\cdot}7 \text{ mm} \qquad (Ans)$$

The method of superposition applies to deflections and if the required couple is M, then

Deflection due to $M = \frac{1}{2} \times$ (deflection due to the load)

$$\frac{ML^2}{2EI} = \frac{1}{2} \times \frac{WL^3}{8EI}$$

and

$$M = \frac{WL}{8} = \tfrac{1}{8} \times (100 \times 10^3 \text{ N}) \times 6 \text{ m}$$

$$= 75 \text{ kN-m} \qquad (Ans)$$

5.5. Derive an expression for the maximum deflection of a uniform beam, length L, simply supported at its ends and carrying a concentrated load W at mid-span.

A uniform beam, length l, is simply and symmetrically supported on a span L, L less than l (Fig 5.7). Find the ratio l/L so that, due to a central point load, the upward deflection at each end equals the downward deflection at mid-span.

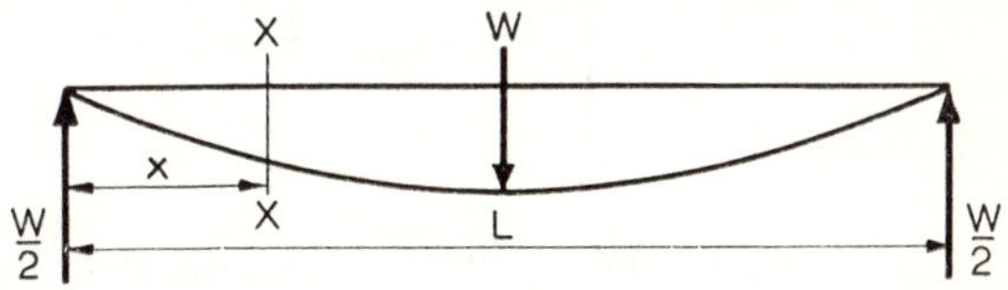

Fig. 5.6

Solution. With an origin at the left-hand end (Fig. 5.6), the (sagging) bending moment at any section XX, distance x from the origin is

$$M = \tfrac{1}{2}Wx \qquad (\text{provided that } x < \tfrac{1}{2}L)$$

It is important to notice that this expression and any integrals of it do not apply to the right-hand half of the beam. Substituting in the differential equation of flexure,

$$EI(d^2y/dx^2) = -M = -\tfrac{1}{2}Wx$$

Integrating,

$$EI(dy/dx) = -\tfrac{1}{4}Wx^2 + A \qquad \text{(i)}$$

The constant of integration can be evaluated using the mid-span conditions.

When $x = \frac{1}{2}L$ (at mid-span) the slope dy/dx is zero, by symmetry. Hence

$$0 = -\tfrac{1}{4}W(\tfrac{1}{2}L)^2 + A \text{ and } A = WL^2/16$$

Substituting for A in (i) and integrating again,

$$EIy = -\frac{Wx^3}{12} + \frac{WL^2x}{16} + B \qquad \text{(ii)}$$

$B = 0$ since the deflection y is zero at the origin ($x = 0$). The maximum deflection occurs at mid-span ($x = \frac{1}{2}L$) and is, from (ii),

$$y_{max} = \frac{1}{EI}\left[\frac{-W(\frac{1}{2}L)^3}{12} + \frac{WL^2(\frac{1}{2}L)}{16}\right]$$

$$= \frac{WL^3}{48EI} \qquad (Ans)$$

For the beam in the question, the overhanging portions deflect upwards, but remain straight since they carry no bending moments

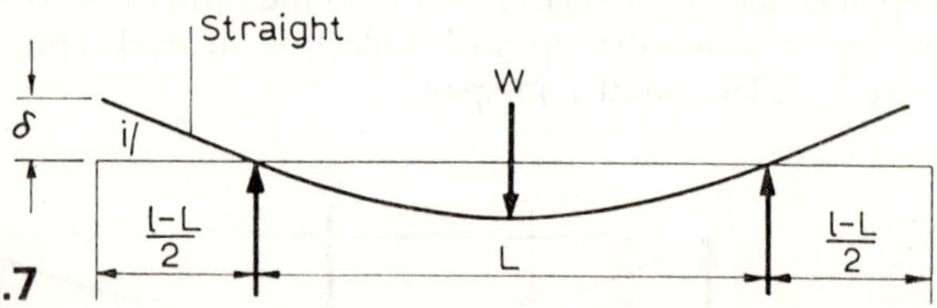

Fig. 5.7

(Fig. 5.7). From the analysis given above, the slope at the left-hand support is, substituting $x = 0$ in (i),

$$i\left(\text{or } \frac{dy}{dx}\right) = \frac{A}{EI} = \frac{WL^2}{16EI}$$

The upward deflection at each end is, since i is a small angle,

$$\delta = i \times (\text{length of the overhanging portion})$$

$$= \frac{WL^2}{16EI} \times \left(\frac{l-L}{2}\right)$$

The upward deflection at each end = downward deflection at mid-span, if

$$\frac{WL^2}{16EI} \times \left(\frac{l-L}{2}\right) = \frac{WL^3}{48EI} \qquad \text{i.e. } l/L = 5/3 \qquad (Ans)$$

5.6. A 600 mm length of flat spring steel, 12 mm by 3 mm, is bent between pegs as shown in Fig. 5.8. The pegs, which are set in a

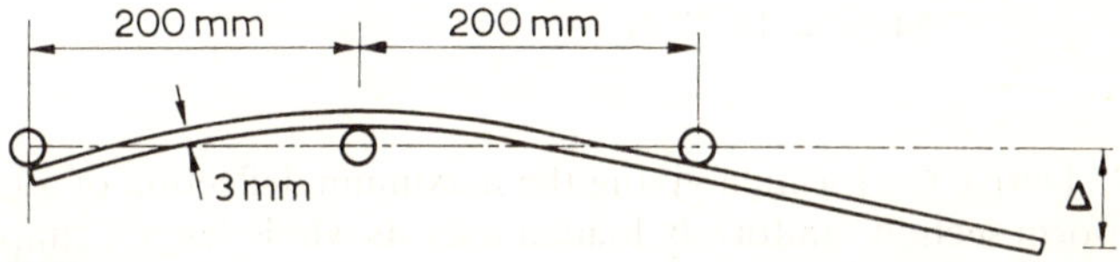

Fig. 5.8

straight line, are 6 mm diameter. Find the deflection Δ at the free end and also the reactions on the pegs. ($E = 206$ GN/m^2).

(*I.Mech.E.*)

Solution. The length of steel can be regarded as a uniform beam simply supported on a span of 400 mm with a concentrated (upward) load at mid-span equal to the reaction on the middle peg. The deflection at mid-span equals

$$\begin{aligned}(\text{Thickness of the steel}) + (\text{Diameter of the peg}) &= 3\,\text{mm} + 6\,\text{mm}\\ &= 9\ \text{mm}\end{aligned}$$

The reaction on the middle peg is

$$W = \text{deflection} \times \frac{48EI}{L^3} = 9\ \text{mm} \times 48EI/L^3 \qquad \text{(i)}$$

The slope at the outer pegs is $WL^2/16EI$ so that the deflection at the free end is, substituting from (i),

$$\begin{aligned}\Delta &= [(\text{length of overhanging portion}) \times (\text{slope at outer pegs})]\\ &\quad + \tfrac{1}{2}[(\text{thickness of spring}) + (\text{diameter of peg})]\\ &= \left[200\ \text{mm} \times W \times \frac{L^2}{16EI}\right] + \tfrac{1}{2}[3\ \text{mm} + 6\ \text{mm}]\\ &= \left[200\ \text{mm} \times \left(9\ \text{mm} \times \frac{48EI}{L^3}\right) \times \frac{L^2}{16EI}\right] + 4{\cdot}5\ \text{mm}\\ &= \frac{5400\ \text{mm}^2}{L\ \text{mm}} + 4{\cdot}5\ \text{mm} = \frac{5400\ \text{mm}^2}{400\ \text{mm}} + 4{\cdot}5\ \text{mm}\\ &= 18\ \text{mm} \qquad (Ans)\end{aligned}$$

Substituting the given numerical values in (i), the reaction on the central peg is

$$W = 9\text{mm} \times \frac{48 \times (206 \times 10^9 \text{ N/m}^2) \times [12 \text{ mm} \times (3 \text{ mm})^3/12]}{(400 \text{ mm})^3}$$

$$= \frac{(9 \times 10^{-3} \text{ m}) \times 48 \times (206 \times 10^9 \text{ N/m}^2) \times (27 \times 10^{-12} \text{ m}^4)}{64 \times 10^{-3} \text{ m}^3}$$

$$= 37{\cdot}54 \text{ N} \qquad (Ans)$$

By symmetry, the reaction on each of the other pegs is

$$\tfrac{1}{2} \times 37{\cdot}54 \text{ N} = 18{\cdot}77 \text{ N} \qquad (Ans)$$

5.7. Derive the formula giving the maximum deflection of a beam of uniform section, uniformly loaded over its whole length, simply supported at its ends.

If such a beam is a symmetrical I-section made of steel, having $E = 200$ GN/m² and in which the maximum stress due to bending is 120 MN/m², show that the deflection Δ may be written $\Delta = KL^2/d$, where L is the span and d the overall depth. Determine the value of the constant K when L is expressed in metres and Δ and d are in millimetres. (*U.L.*)

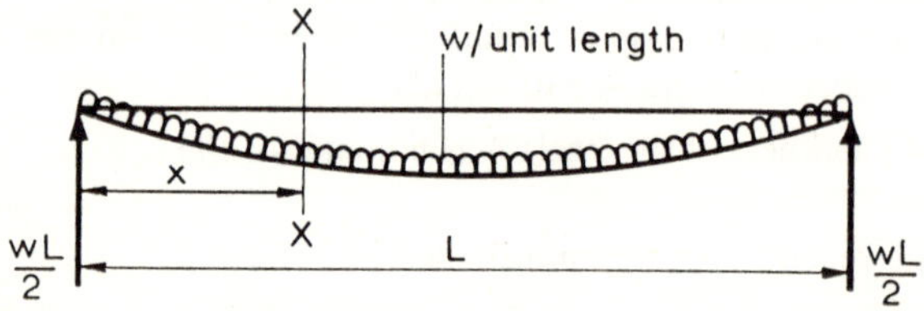

Fig. 5.9

Solution. By symmetry, the reactions are each $\frac{1}{2}wL$ and, taking an origin at the left-hand end (Fig. 5.9), the bending moment at any section XX is

$$M = (\tfrac{1}{2}wL)x - wx(\tfrac{1}{2}x) = \tfrac{1}{2}wLx - \tfrac{1}{2}wx^2$$

Substituting in the differential equation of flexure,

$$EI\frac{d^2y}{dx^2} = -M = -\tfrac{1}{2}wLx + \tfrac{1}{2}wx^2$$

Integrating,

$$EI\frac{dy}{dx} = -\tfrac{1}{4}wLx^2 + \tfrac{1}{6}wx^3 + A \qquad \text{(i)}$$

At mid-span the slope dy/dx is zero. Putting $dy/dx = 0$, when $x = \frac{1}{2}L$, we have

$$0 = -\tfrac{1}{4}wL(\tfrac{1}{2}L)^2 + \tfrac{1}{6}w(\tfrac{1}{2}L)^3 + A$$

and

$$A = wL^3/24$$

Substituting for A in (i) and integrating again,

$$EIy = -\frac{wLx^3}{12} + \frac{wx^4}{24} + \frac{wL^3x}{24} \qquad \text{(ii)}$$

(the constant being zero since $y = 0$ when $x = 0$).

The maximum deflection occurs at mid-span and hence putting $x = \frac{1}{2}L$ in (ii), we obtain

$$y_{max} = \frac{1}{EI}\left[-\frac{wL(\frac{1}{2}L)^3}{12} + \frac{w(\frac{1}{2}L)^4}{24} + \frac{wL^3(\frac{1}{2}L)}{24}\right]$$

$$= \frac{5wL^4}{384EI}$$

If $W = wL$ (the total load) the formula can be written as

$$\Delta = \frac{5WL^3}{384EI}$$

For a simply-supported beam with a uniformly distributed load the maximum bending moment (see Table 1 on page 31) is $WL/8$. From the bending equation

$$M = \frac{\sigma_{max}}{y_{max}} I = \frac{2\sigma_{max}}{d} I$$

where d is the overall depth.

Thus

$$\frac{WL}{8} = \frac{2\sigma_{max}}{d} I \quad \text{or} \quad W = \frac{16I}{L}\frac{\sigma_{max}}{d}$$

Substituting in the deflection formula above, we have

$$\Delta = \left(\frac{16I}{L}\frac{\sigma_{max}}{d}\right)\frac{5L^3}{384EI}$$

$$= \left(\frac{5}{24}\frac{\sigma_{max}}{E}\right)\frac{L^2}{d}$$

For given values o σ_{max} and E, therefore, the deflection $\Delta = KL^2/d$ as required.

If the span = L metres = $1000L$ mm and the depth of the beam is d mm then the deflection Δ (also in mm) is given by

$$\Delta = \left(\frac{5}{24}\frac{\sigma_{max}}{E}\right)\frac{(1000L)^2}{d}$$

and, with the values of σ_{max} and E given in the question,

$$\Delta = \left(\frac{5}{24} \times \frac{120 \times 10^6 \text{ N/m}^2}{200 \times 10^9 \text{ N/m}^2}\right) \times \frac{(1000\,L)^2}{d}$$
$$= 125\,L^2/d$$

The value of K is therefore 125. *(Ans)*

5.8. A cantilever, 3 m long and of symmetrical cross-section 500 mm deep, carries a uniformly distributed load of 32 kN per metre run throughout its length. If $I = 51\,000$ cm^4 and $E = 206$ GN/m^2, calculate the deflection at the free end.

What is the maximum point load which the cantilever can carry at a distance 2·1 m from the fixed end (in addition to the distributed load) if

(*a*) the bending stress must nowhere exceed 140 MN/m^2;
(*b*) the deflection at the free end must not exceed 6 mm?

Solution. For the distributed load alone, the deflection at the free end is

$$\delta = \frac{WL^3}{8EI} = \frac{(32 \times 10^3 \text{ N/m} \times 3 \text{ m}) \times 27 \text{ m}^3}{8 \times (206 \times 10^9 \text{ N/m}^2) \times (5{\cdot}1 \times 10^{-4} \text{ m}^4)}$$
$$= 3{\cdot}084 \text{ mm}$$ *(Ans)*

(*a*) For a maximum bending stress of 140 MN/m^2 the maximum permissible bending moment is

$$M = \frac{\sigma_{max}}{y_{max}} I = \frac{140 \times 10^6 \text{ N/m}^2}{0{\cdot}25 \text{ m}} \times (5{\cdot}1 \times 10^{-4} \text{ m}^4)$$
$$= 286 \text{ kN-m}$$

The maximum bending moment occurs at the fixed end. Due to the distributed load, the B.M. here is

$$32 \text{ kN/m} \times 3 \text{ m} \times 1{\cdot}5 \text{ m} = 144 \text{ kN-m}$$

Hence the B.M. due to the point load must not exceed

$$(286 - 144) \text{ kN-m} = 142 \text{ kN-m}$$

Since the point load is 2·1 m from the fixed end then its maximum value is

$$142\ \text{kN-m}/2{\cdot}1\ \text{m} = 67{\cdot}7\ \text{kN} \qquad (\textit{Ans})$$

(*b*) If the total deflection at the free end is 6 mm then the deflection at the free end due to the point load is (6 − 3·084) mm = 2·916 mm. This deflection can be considered in two parts: δ_1 the deflection under the load; and δ_2 the "extra" deflection due to the fact that the part of the cantilever to the right of the load deflects but remains straight (Fig. 5.10). If W newtons is

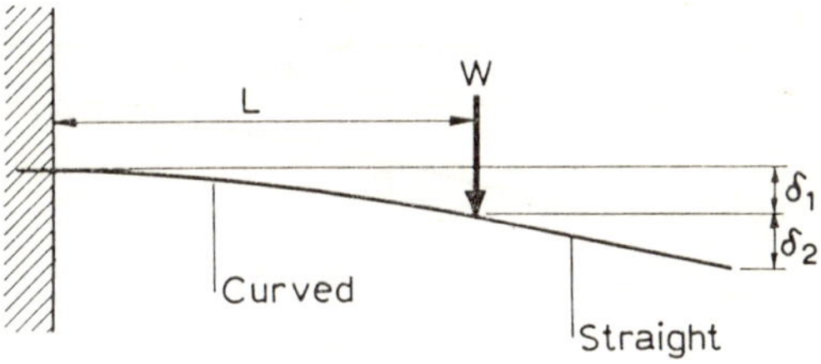

Fig. 5.10

the point load then, taking L as 2·1 m (the distance between the fixed end and the load) we have

$$\delta_1 = \frac{WL^3}{3EI} = \frac{W \times (2{\cdot}1\ \text{m})^3}{3 \times (206 \times 10^9\ \text{N/m}^2) \times (5{\cdot}1 \times 10^{-4}\ \text{m}^4)}$$

$$\delta_2 = (\text{slope at the point of loading}) \times (\text{length of the straight portion})$$

$$= \frac{WL^2}{2EI} \times 0{\cdot}9\ \text{m} = \frac{W \times (2{\cdot}1\ \text{m})^2 \times 0{\cdot}9\ \text{m}}{2 \times (206 \times 10^9\ \text{N/m}^2) \times (5{\cdot}1 \times 10^{-4}\ \text{m}^4)}$$

Equating $(\delta_1 + \delta_2)$ to the permissible end deflection (due to W),

$$\frac{W \times 2{\cdot}1^3}{3 \times 206 \times 5{\cdot}1 \times 10^5}\ \text{m} + \frac{W \times 2{\cdot}1^2 \times 0{\cdot}9}{2 \times 206 \times 5{\cdot}1 \times 10^5}\ \text{m} = 2{\cdot}916 \times 10^{-3}\ \text{m}$$

or

$$\left(\frac{W \times 2{\cdot}1^2}{206 \times 5{\cdot}1 \times 10^5}\right)\left(\frac{2{\cdot}1}{3} + \frac{0{\cdot}9}{2}\right) = 2{\cdot}916 \times 10^{-3}$$

$$W = 60{\cdot}4\ \text{kN} \qquad (\textit{Ans})$$

The maximum load which satisfies both conditions is 60·4 kN.

5.9. Prove Mohr's theorems as applied to uniform beams. A beam, 5 m long, is simply supported at its ends and carries concentrated loads of 20 kN each, at points 1 m from the ends. Calculate the maximum slope and deflection of the beam and the slope and deflection under each load. $EI = 10$ MN-m^2.

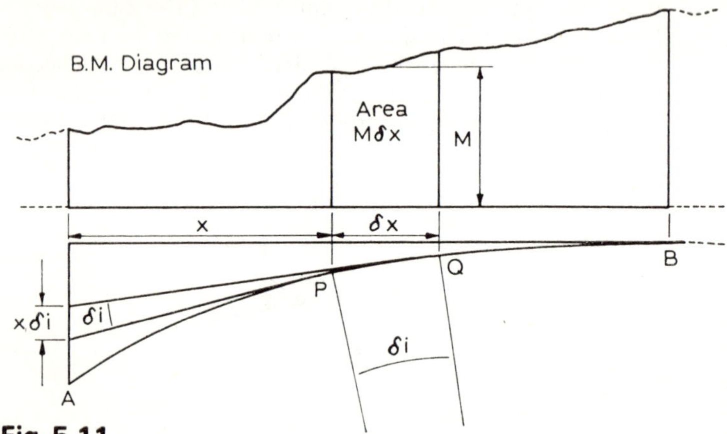

Fig. 5.11

Solution. Let A and B (Fig. 5.11) be two points on the deflection curve of a uniform beam. Suppose B is a point of zero slope and deflection, and P and Q are two points whose horizontal distances from A are x and $x + \delta x$ respectively.

The angle δi between the tangents at P and Q is, from the theory of Example 5.3,

$$\delta i = \frac{\delta x}{R} = \frac{\delta x}{(EI/M)} = \frac{M\delta x}{EI} \qquad \text{(i)}$$

Since B is a point of zero slope (e.g. the fixed end of a cantilever) then the total slope at A is the sum of all terms such as (i). $M\delta x$ is the area under the B.M. curve between P and Q and the total slope at A is

$$i = \frac{1}{EI} \sum_{x=0}^{x=AB} M\delta x = \frac{1}{EI} \times \begin{pmatrix} \text{area of the B.M. diagram} \\ \text{between A and B} \end{pmatrix}$$

If B is not a point of zero slope, we have, *the total change in slope between A and B is the area of the B.M. diagram between these points divided by the flexural rigidity.*

The part of the deflection at A due to the bending of the portion PQ is given by

$$\delta y = x\,\delta i \qquad \text{(Fig. 5.11)}$$

Substituting for δi from (i), we have

$$\delta y = Mx.\delta x/EI \tag{ii}$$

The total deflection at A is the sum of all terms such as (ii). $Mx\delta x$ is the first moment about A of the area under the B.M. curve between P and Q, and hence the total deflection at A is

$$y = \frac{1}{EI}\sum_{x=0}^{x=AB} Mx.\delta x = \frac{1}{EI} \times \begin{pmatrix}\text{the first moment about A of the} \\ \text{area of the B.M. diagram between} \\ \text{A and B}\end{pmatrix}$$

If B is not a point of zero slope and deflection, then *the deflection of* A *relative to the tangent at* B *is the first moment about* A *of the area of the B.M. diagram between* A *and* B.

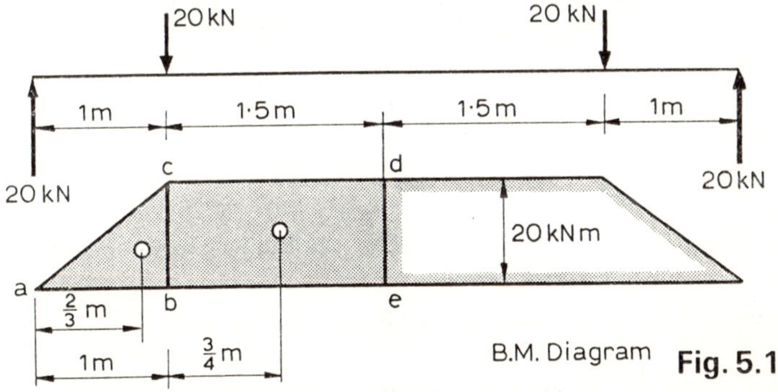

Fig. 5.12

These results are known as Mohr's theorems. Formulae for the areas and centroid position of shapes commonly occurring in B.M. diagrams are given in Appendix 1.

The beam given in the question is symmetrically loaded (Fig. 5.12). Hence the reactions are each 20 kN and the slope at mid-span is zero.

For the first part of the calculation the origin is taken at the left-hand end. The slope at the ends is clearly the maximum and hence

$$i_{max} = \frac{1}{EI} \times \begin{pmatrix}\text{area of the B.M. diagram between the left-hand} \\ \text{end and mid-span}\end{pmatrix}$$

$$= \frac{1}{EI} \times (\text{area abc} + \text{area bcde})$$

$$= \frac{1}{10\ \text{MN-m}^2} \times [\tfrac{1}{2}(1\ \text{m} \times 20\ \text{kN-m}) + (1{\cdot}5\ \text{m} \times 20\ \text{kN-m})]$$

$$= 0{\cdot}004\ \text{rad} \qquad (Ans)$$

Using the same origin, Mohr's theorem gives the "deflection" of the left-hand end relative to the tangent at mid-span, which is numerically equal to the deflection at mid-span. Therefore

$$\delta_{max} = \frac{1}{EI} \times \begin{pmatrix}\text{first moment about the left-hand end of the} \\ \text{B.M. diagram between this end and mid-span}\end{pmatrix}$$

$$= \frac{1}{EI} \times [(\text{area abc} \times \tfrac{2}{3}\text{ m}) + (\text{area bcde} \times 1{\cdot}75\text{ m})]$$

$$= \frac{1}{10\text{ MN-m}^2} \times [\tfrac{1}{2}(1\text{ m} \times 20\text{ kN-m} \times \tfrac{2}{3}\text{ m})$$

$$+ (1{\cdot}5\text{ m} \times 20\text{ kN-m} \times 1{\cdot}75\text{ m})]$$

$$= 5{\cdot}92\text{ mm} \qquad (Ans)$$

To find the slope and deflection under each load, it is convenient to take an origin at the point where the left-hand load acts.

Since the slope at mid-span is zero, the slope at this point is

$$i = \frac{1}{EI} \times (\text{area bcde}) = \frac{1}{10\text{ MN-m}^2} \times (1{\cdot}5\text{ m} \times 20\text{ kN-m})$$

$$= 0{\cdot}003\text{ rad} \qquad (Ans)$$

The deflection at this point relative to the tangent at mid-span is

$$\delta = \frac{1}{EI} \times (\text{area bcde}) \times 0{\cdot}75\text{ m}$$

$$= \frac{1}{10\text{ MN-m}^2} \times (1{\cdot}5\text{ m} \times 20\text{ kN-m} \times 0{\cdot}75\text{ m}) = 2{\cdot}25\text{ mm}$$

The deflection at mid-span is 5·92 mm and thus the actual deflection under the load is (5·92 − 2·25) mm

$$= 3{\cdot}67\text{ mm} \qquad (Ans)$$

[The reader should now verify the formulae for the standard cases of slope and deflection quoted at the beginning of this chapter, using Mohr's theorems.]

5.10. A uniform beam, length $4L$, is simply and symmetrically supported over a span of $2L$. It carries a load of W_1 at each end and a load of W_2 at mid-span. Find the ratio of W_1 to W_2 if the deflection at mid-span is equal to that at each end.

Solution. The central load and the end loads can be considered as separate loading systems. The B.M. diagram due to W_2 alone is

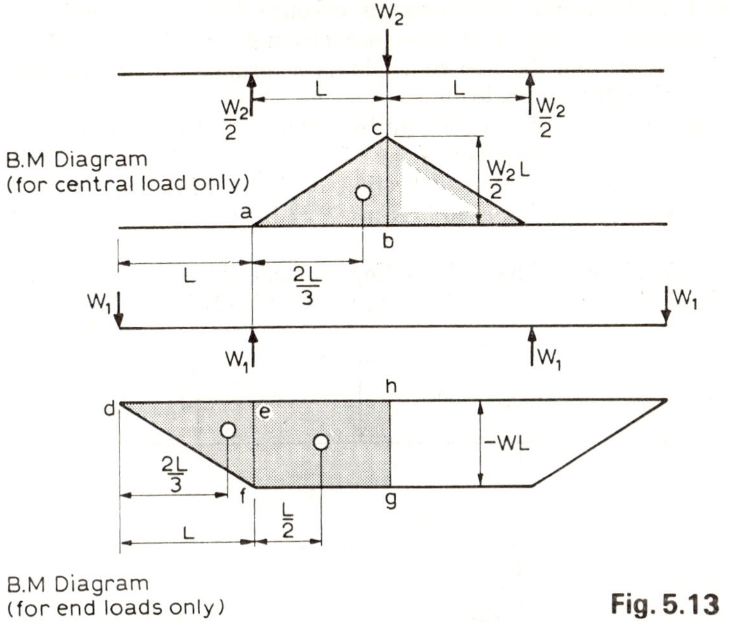

Fig. 5.13

positive and that due to the end loads is negative, as shown in Fig. 5.13. These signs must be observed in applying Mohr's theorem. By symmetry, the slope at mid-span is zero. Hence, the deflection at the left-hand end relative to the tangent at mid-span is

$$\delta = \frac{1}{EI} \times \begin{pmatrix}\text{total first moment about the L.H. end of the B.M.} \\ \text{diagram between this end and mid-span}\end{pmatrix}$$

$$= \frac{1}{EI}\{[(\text{area abc}) \times (L + \tfrac{2}{3}L)] - [(\text{area def}) \times \tfrac{2}{3}L]$$
$$- [(\text{area efgh}) \times (L + \tfrac{1}{2}L)]\}$$

$$= \frac{1}{EI}\{[\tfrac{1}{2} \times L \times \tfrac{1}{2}W_2L \times (L + \tfrac{2}{3}L)]$$
$$- [\tfrac{1}{2} \times L \times W_1L \times \tfrac{2}{3}L] - [L \times W_1L \times (L + \tfrac{1}{2}L)]\}$$

$$= \frac{1}{EI}\left\{\frac{5W_2L^3}{12} - \frac{W_1L^3}{3} - \frac{3W_1L^3}{2}\right\}$$

If the deflection at mid-span is equal to that at each end then δ must be zero. Hence

$$\frac{5W_2L^3}{12} - \frac{W_1L^3}{3} - \frac{3W_1L^3}{2} = 0$$

$$W_1/W_2 = 5/22 \qquad (Ans)$$

5.11. A horizontal cantilever of effective length $3a$, carries two concentrated loads, W at a distance a from the fixed end and W' at a distance a from the free end. Obtain a formula for the maximum deflection due to this loading.

If the cantilever is a 250 mm by 150 mm steel I-beam, 3 m long, having a second moment of area $I = 8\,500\ \text{cm}^4$, determine W and W' to give a maximum deflection of 6 mm when the maximum stress due to bending is $90\ \text{MN/m}^2$. $E = 185\ \text{GN/m}^2$. *(U.L.)*

Solution. The loading and bending moment diagrams are shown in Fig. 5.14. The maximum bending moment is $(2W'a + Wa)$ and the

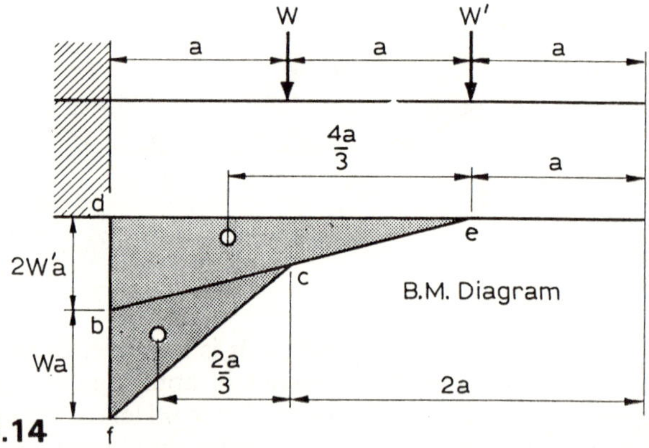

Fig. 5.14

B.M. diagram is the sum of two triangles, fbc and bde, as shown. Since the fixed end has zero slope and deflection, the maximum deflection, which occurs at the free end, is found from Mohr's theorem, and is

$$\delta = \frac{1}{EI} \times \text{(first moment about free end of total B.M. diagram)}$$

$$= \frac{1}{EI} \times \{[(\text{area fbc}) \times (2a + \tfrac{2}{3}a)] + [(\text{area bde}) \times (a + \tfrac{4}{3}a)]\}$$

$$= \frac{1}{EI} \times \{(\tfrac{1}{2} \times Wa \times a \times \tfrac{8}{3}a) + (\tfrac{1}{2} \times 2W'a \times 2a \times \tfrac{7}{3}a)\}$$

$$= \frac{2a^3}{3EI}(2W + 7W')$$

Working in newton and metre units,

$$3a = 3 \qquad \text{and} \qquad a = 1$$

For a maximum deflection of 6 mm,

$$\frac{2a^3}{3EI}(2W + 7W') = 6 \times 10^{-3}$$

$$2W + 7W' = (6 \times 10^{-3}) \times 3EI/2a^3$$
$$= (6 \times 10^{-3}) \times 3 \times (185 \times 10^9) \times (85 \times 10^{-6})/(2 \times 1^3)$$
$$2W + 7W' = 141\,500 \qquad \text{(i)}$$

For the given maximum stress, the maximum bending moment is

$$M = \frac{\sigma_{max}}{y_{max}} I$$

or

$$Wa + 2W'a = \frac{90 \times 10^6}{0{\cdot}125} \times (85 \times 10^{-6})$$

$$W + 2W' = \frac{90 \times 85}{0{\cdot}125 \times 1} = 61\,200 \qquad \text{(ii)}$$

Solving the simultaneous equations (i) and (ii),

$W = 48{\cdot}47$ kN and $W' = 6{\cdot}37$ kN (*Ans*)

5.12. Explain the Macaulay method for obtaining the deflections of any point in a uniform beam which is subjected to point loads only.

A brass bar, 900 mm long, is simply supported at its ends and carries a concentrated load of 90 N at a point 200 mm from the right-hand end. If $I = 0{\cdot}22$ cm^4, calculate the deflection at mid-span. $E = 96$ GN/m^2.

Solution. Suppose that AB (Fig. 5.15) is a uniform beam simply supported at its ends. Consider an origin at the left-hand end.

For any section XX between A and the load W_1 the bending moment is

$$M = R_A x \qquad \text{(i)}$$

For any section between the loads W_1 and W_2, the bending moment is

$$M = R_A x - W_1[x - a] \qquad \text{(ii)}$$

For any section between the load W_2 and B, the bending moment is

$$M = R_A x - W_1[x - a] - W_2[x - b] \qquad \text{(iii)}$$

To avoid using three different equations, the following method is adopted. The bending moment at *any* section in the beam is given by (iii), provided that *any term is omitted when the expression inside its square brackets becomes negative.*

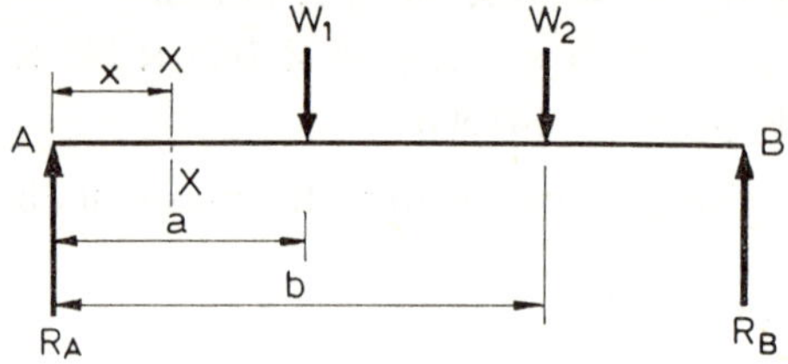

Fig. 5.15

Using (iii), the differential equation of flexure becomes

$$EI\frac{d^2y}{dx^2} = -M = -R_Ax + W_1[x - a] + W_2[x - b] \qquad \text{(iv)}$$

Equation (iv) is integrated to find expressions for the slope and deflection and each square bracket must be integrated as a whole, e.g. from the first integration,

$$EI\frac{dy}{dx} = -R_A\frac{x^2}{2} + W_1\frac{[x - a]^2}{2} + W_2\frac{[x - b]^2}{2} + C \qquad \text{(v)}$$

The constant C has the same value for each portion of the beam since the value of the slope at each load will clearly be the same whether the term allowing for that load is included or not. Similarly,

$$EIy = -R_A\frac{x^3}{6} + W_1\frac{[x - a]^3}{6} + W_2\frac{[x - b]^3}{6} + Cx + D$$

In using the equations for slope and deflection, the rule concerning square brackets must be observed. At any specific point, however, the brackets can be evaluated in the usual way after the appropriate terms are omitted. The constants C and D are usually found by using the conditions at each support.

The above method (Macaulay's) can be extended for any number of loads.

Figure 5.16 shows the brass beam of the question. The reactions,

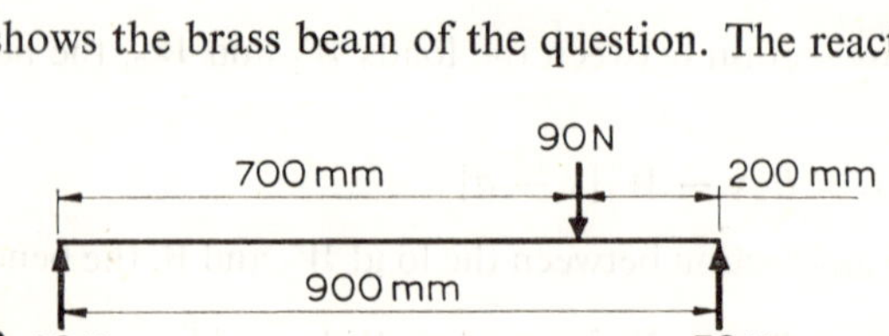

Fig. 5.16

found by taking moments about each end in turn are 20 N and 70 N at the left-hand and right-hand ends respectively.

Working in newton and metre units throughout, the bending moment at any distance x from the left-hand end is

$$M = 20x - 90[x - 0{\cdot}7]$$

the term in square brackets being omitted when $x < 0{\cdot}7$. The differential equation of flexure becomes

$$EI\frac{d^2y}{dx^2} = -M = -20x + 90[x - 0{\cdot}7]$$

Integrating twice, we obtain

$$EIy = -\frac{10}{3}x^3 + 15[x - 0{\cdot}7]^3 + Ax + B \qquad \text{(vi)}$$

At the left-hand end, where $x = 0$, the deflection (y) is zero, the term in square brackets is omitted and hence $B = 0$.

At the right-hand end, where $x = 0{\cdot}9$, the deflection (y) is again zero, the term in square brackets must be included and, substituting in (vi),

$$0 = -\left(\frac{10}{3} \times 0{\cdot}9^3\right) + 15[0{\cdot}9 - 0{\cdot}7]^3 + 0{\cdot}9A$$

from which

$$A = 2{\cdot}567$$

At mid-span the term in square brackets is omitted. Hence substituting $x = 0{\cdot}45$ and $A = 2{\cdot}567$ in (vi) we have

$$y = \frac{1}{EI}\left[-\left(\frac{10}{3} \times 0{\cdot}45^3\right) + (2{\cdot}567 \times 0{\cdot}45)\right]$$

$$= \frac{0{\cdot}851}{EI} = \frac{0{\cdot}851}{(96 \times 10^9) \times (0{\cdot}22 \times 10^{-8})} = 4{\cdot}03 \text{ mm} \qquad (\textit{Ans})$$

5.13. A simply supported beam AB of length L carries a concentrated load W at a distance a from the end A and distance b from B, a being greater than b. Find the magnitude of the maximum deflection and its position, measured from A, and show that for any value of a this position is approximately within $L/13$ of the centre. (*U.L.*)

Solution. Taking moments about A and B (Fig. 5.17) the reactions are

$$R_B = Wa/L \qquad R_A = Wb/L$$

The bending moment at any distance x from A is

$$M = \frac{Wbx}{L} - W[x - a]$$

the term in square brackets being omitted when $x \leqslant a$.

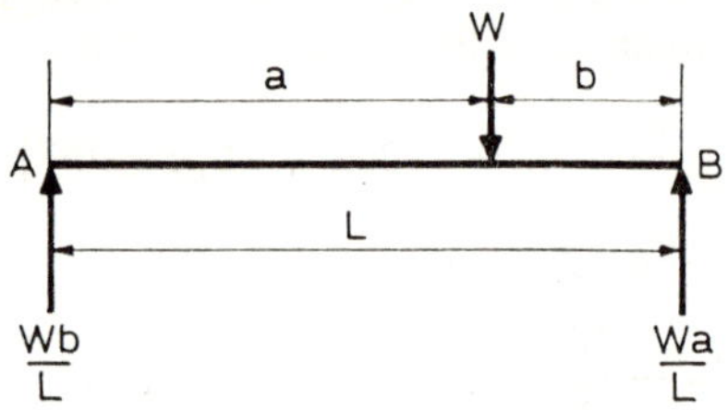

Fig. 5.17

The differential equation of flexure becomes

$$EI\frac{d^2y}{dx^2} = -M = -\frac{Wbx}{L} + W[x - a] \qquad \text{(i)}$$

Integrating (i), the slope is given by

$$EI\frac{dy}{dx} = -\frac{Wbx^2}{2L} + \tfrac{1}{2}W[x - a]^2 + C \qquad \text{(ii)}$$

Integrating again, the deflection is given by

$$EIy = -\frac{Wbx^3}{6L} + \tfrac{1}{6}W[x - a]^3 + Cx + D \qquad \text{(iii)}$$

As usual, $y = 0$ when $x = 0$, and hence $D = 0$.
Also, since $y = 0$ when $x = L$, we have

$$0 = -\frac{WbL^3}{6L} + \tfrac{1}{6}W[L - a]^3 + CL$$

from which

$$C = \tfrac{1}{6}W\left(bL - \frac{(L - a)^3}{L}\right)$$

$$= \tfrac{1}{6}W\left(bL - \frac{b^3}{L}\right) = \frac{Wb}{6L}(L^2 - b^2)$$

(since $b = L - a$).

The maximum deflection occurs where the slope is zero, and this point is clearly between the end A and the load W (i.e. $x < a$).

Hence, substituting for C, omitting the term in square brackets and putting $dy/dx = 0$ we have, from (ii),

$$0 = -\frac{Wbx^2}{2L} = \frac{Wb}{6L}(L^2 - b^2)$$

from which

$$x^2 = \tfrac{1}{3}(L^2 - b^2)$$
$$x = \sqrt{[\tfrac{1}{3}(L^2 - b^2)]} \qquad \text{(iv)}$$

Using this value of x, the maximum deflection is, from (iii) omitting the term in square brackets and substituting for C,

$$y_{max} = \frac{1}{EI}\left\{-\frac{Wb}{6L}\left(\frac{L^2 - b^2}{3}\right)^{3/2} + \frac{Wb}{6L}(L^2 - b^2)\sqrt{\left(\frac{L^2 - b^2}{3}\right)}\right\}$$
$$= \frac{Wb(L^2 - b^2)^{3/2}}{6(\sqrt{3})EIL}(-\tfrac{1}{3} + 1)$$
$$= \frac{Wb(L^2 - b^2)^{3/2}}{9(\sqrt{3})EIL} \qquad (Ans)$$

The position of maximum deflection is given by (iv) and b can take values from 0 to $\frac{1}{2}L$.

The maximum value of x is clearly found by putting $b = 0$ in (iv). This gives

$$x = \sqrt{(\tfrac{1}{3}L^2)} = L/\sqrt{3} = 0{\cdot}5773L \qquad \text{(v)}$$

The minimum value of x occurs when $b = \frac{1}{2}L$ and is clearly

$$x = \tfrac{1}{2}L \qquad \text{(vi)}$$

The point having the maximum deflection is always within the range whose limits are given by (v) and (vi). Hence the maximum deflection is always within

$0{\cdot}0773L$ or approximately $L/13$ of the centre. *(Ans)*

5.14. A horizontal beam of uniform section and length L rests on supports at its ends. It carries a uniform load w per unit length, which extends over a length l from the right-hand support.

Determine the value of l in order that the maximum deflection may occur at the left-hand end of the load, and if the maximum deflection is wL^4/kEI, find the value of k. *(U.L.)*

Solution. Taking moments about the right-hand support (Fig. 5.18), the reaction at the left-hand end is $wl^2/2L$.

The bending moment at any distance x from the left-hand end is

$$M = \frac{wl^2x}{2L} - \tfrac{1}{2}w[x - (L - l)]^2$$

the term containing square brackets being omitted when $x < (L - l)$.

The differential equation of flexure becomes

$$EI\frac{d^2y}{dx^2} = -M = -\frac{wl^2x}{2L} + \tfrac{1}{2}w[x - (L - l)]^2$$

Integrating once, the slope is given by

$$EI\frac{dy}{dx} - = \frac{wl^2x^2}{4L} + \tfrac{1}{6}w[x - (L - l)]^3 + A \qquad \text{(i)}$$

Integrating again, the deflection is given by

$$EIy = -\frac{wl^2x^3}{12L} + \frac{w}{24}[x - (L - l)]^4 + Ax \qquad \text{(ii)}$$

the second constant of integration being zero, since $y = 0$ when $x = 0$.

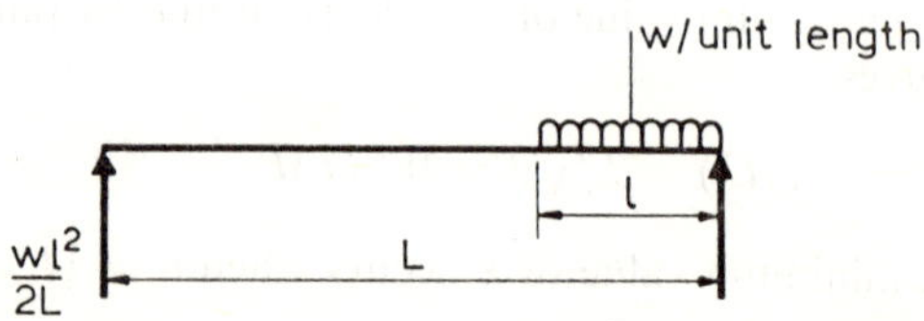

Fig. 5.18

To find A we use the condition that the deflection (y) is zero when $x = L$. Substituting in (ii),

$$0 = -\frac{wl^2L^3}{12L} + \frac{w}{24}[L - (L - l)]^4 + AL$$

from which

$$A = \frac{wl^2L^3}{12L^2} - \frac{w}{24L}(L - L + l)^4$$

$$= \frac{wl^2L}{12} - \frac{wl^4}{24L} \qquad \text{(iii)}$$

Since the maximum deflection occurs at the left-hand end of the load, the slope (dy/dx) is zero when $x = L - l$.

Substituting these values in (i) and omitting the term in square brackets, we obtain

$$0 = -\frac{wl^2(L-l)^2}{4L} + \frac{wl^2L}{12} - \frac{wl^4}{24L}$$

from which

$$\frac{(L^2 - 2Ll + l^2)}{4L} - \frac{L}{12} + \frac{l^2}{24L} = 0$$

$$7l^2 - 12Ll + 4L^2 = 0$$

and

$$l = \left\{\frac{12 \pm \sqrt{[12^2 - (4 \times 7 \times 4)]}}{2 \times 7}\right\} L$$

$$= \left(\frac{6 - \sqrt{8}}{7}\right) L = 0{\cdot}4531L \qquad \text{(iv)}$$

(The positive square root gives a result which is inadmissible since l must be less than L.)

Using (iv) and (iii) the value of A is given by

$$A = \frac{w(0{\cdot}4531L)^2L}{12} - \frac{w(0{\cdot}4531L)^4}{24L}$$

$$= 0{\cdot}01535wL^3$$

The maximum deflection occurs where

$$x = L - l = L - 0{\cdot}4531L = 0{\cdot}5469L$$

Substituting these values of A and x, the maximum deflection is, from (ii),

$$y_{max} = \frac{1}{EI}\left\{-\frac{(0{\cdot}4531L)^2(0{\cdot}5469L)^3w}{12L} + 0{\cdot}01535wL^3 \times 0{\cdot}5469L\right\}$$

$$= \frac{wL^4}{EI}\{-0{\cdot}00280 + 0{\cdot}00840\}$$

$$= 0{\cdot}0056\frac{wL^4}{EI}$$

Hence the constant $k = 1/0{\cdot}0056 = 178{\cdot}6$ (*Ans*)

5.15. A uniform beam, length L, is simply supported at its ends and carries a uniformly distributed load of w per unit length between mid-span and a point $L/4$ from the right-hand end.

Obtain an expression for the deflection at mid-span.

Solution. The Macaulay method can only be applied to distributed loads which extend to the right-hand end. In the present case the given distributed load is extended as shown in Fig. 5.19 and a "compensating" load of $-w$ per unit length is applied between the right-hand end and the point distance $L/4$ from it.

Taking moments about the right-hand end, the reaction at the left-hand end is $3wL/32$. The bending moment at any distance x from the left-hand end is

$$M = \frac{3wLx}{32} - \tfrac{1}{2}w[x - \tfrac{1}{2}L]^2 + \tfrac{1}{2}w[x - \tfrac{3}{4}L]^2$$

the second and third terms being omitted when

$x < \frac{1}{2}L$ and $x < \frac{3}{4}L$ respectively.

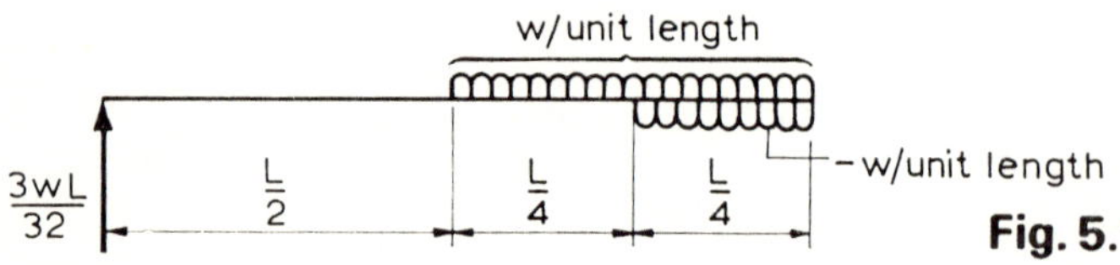

Fig. 5.19

The differential equation of flexure becomes

$$EI\frac{d^2y}{dx^2} = -M = -\frac{3wLx}{32} + \tfrac{1}{2}w[x - \tfrac{1}{2}L]^2 - \tfrac{1}{2}w[x - \tfrac{3}{4}L]^2$$

Integrating twice, we obtain

$$EIy = -\frac{wLx^3}{64} + \frac{w}{24}[x - \tfrac{1}{2}L]^4 - \frac{w}{24}[x - \tfrac{3}{4}L]^4 + Cx + D \quad \text{(i)}$$

C and D are constants of integration which are evaluated using the end conditions.

When $x = 0$, $y = 0$, and hence $D = 0$.

When $x = L$, $y = 0$, and, including the terms in square brackets,

$$0 = -\frac{wL^4}{64} + \frac{w}{24}\left(\frac{L}{2}\right)^4 - \frac{w}{24}\left(\frac{L}{4}\right)^4 + CL$$

from which

$$C = \frac{wL^3}{64} - \frac{wL^3}{(24 \times 16)} + \frac{wL^3}{(24 \times 256)}$$

$$= \frac{27wL^3}{2048}$$

From (i), the deflection at mid-span ($x = L/2$) is, omitting the terms in square brackets,

$$y = \frac{1}{EI}\left\{-\frac{wL(\frac{1}{2}L)^3}{64} + \frac{27wL^3}{2048}(\tfrac{1}{2}L)\right\}$$

$$= \frac{19wL^4}{4\,096EI} \qquad (Ans)$$

5.16. A uniform beam, length L, is simply supported at its ends. The loading consists of a concentrated load W at a distance $L/4$ from the left-hand end and a clockwise couple T applied in the plane of bending at a distance $3L/4$ from the same end.

Obtain an expression for the deflection at mid-span and show that this is zero when $T = 11WL/36$.

Solution. Let AB be the beam (Fig. 5.20). Taking moments about B,

$$R_AL = W \times \tfrac{3}{4}L - T \quad \text{and} \quad R_A = \tfrac{3}{4}W - T/L \qquad \text{(i)}$$

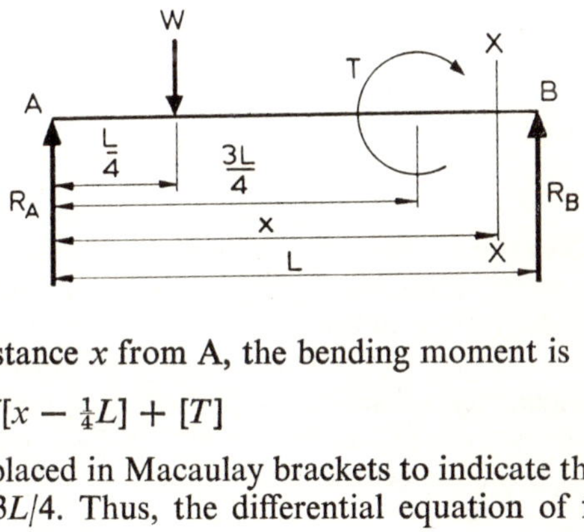

Fig. 5.20

At a section XX, distance x from A, the bending moment is

$$M = R_Ax - W[x - \tfrac{1}{4}L] + [T]$$

and the couple T is placed in Macaulay brackets to indicate that it is omitted when $x < 3L/4$. Thus, the differential equation of flexure becomes

$$EI\frac{d^2y}{dx^2} = -M = -R_Ax + W[x - \tfrac{1}{4}L] - [T]$$

In the subsequent integrations $[T]$ becomes $T[x - \frac{3}{4}L]$ and $\frac{1}{2}T \times [x - \frac{3}{4}L]^2$ to ensure continuity of slope and deflection at $x = 3L/4$. Integrating twice we obtain

$$EIy = -\tfrac{1}{6}R_Ax^3 + \tfrac{1}{6}W[x - \tfrac{1}{4}L]^3 - \tfrac{1}{2}T[x - \tfrac{3}{4}L]^2 + Cx + D$$

where C and D are integration constants. When $x = 0$, $y = 0$, the Macaulay terms are omitted and $D = 0$. When $x = L$, $y = 0$ and

$$0 = -\tfrac{1}{6}R_AL^3 + \tfrac{1}{6}W(\tfrac{3}{4}L)^3 - \tfrac{1}{2}T(\tfrac{1}{4}L)^2 + CL$$

from which

$$C = \tfrac{1}{6}R_A L^2 - \frac{9WL^2}{128} + \frac{TL}{32} \qquad \text{(ii)}$$

At mid-span, $x = L/2$ and the central deflection is given by

$$EIy_c = -\tfrac{1}{6}R_A(\tfrac{1}{2}L)^3 + \tfrac{1}{6}W(\tfrac{1}{4}L)^3 + \tfrac{1}{2}CL$$
$$= -\frac{R_A L^3}{48} + \frac{WL^3}{384} + \frac{CL}{2}$$

Substituting for R_A and C from (i) and (ii) the required expression becomes

$$y_c = \frac{11WL^3 - 36TL^2}{768\,EI}$$

If this is to be zero, $11WL^3 = 36TL^2$ and

$$T = \frac{11WL}{36}$$

as required.

5.17. Give the expressions for bending moment M, shearing force F and intensity of loading w per unit length in terms of the derivatives of y the deflection, in the case of a uniform beam.

A uniform beam of flexural rigidity EI is simply supported at its ends and carries a distributed load whose intensity increases linearly from zero at one end to w_0 per unit length at the other. Obtain an expression for the central deflection.

Solution. In Example 5.3 it was shown that $EI(d^2y/dx^2) = -M$. From Chapter 2 we have the general relationships $F = dM/dx$ and $w = -dF/dx$. Collecting results we obtain for a uniform beam, with the sign conventions previously adopted,

$$M = -EI\frac{d^2y}{dx^2}$$

$$F = \frac{dM}{dx} = \frac{d}{dx}\left(-EI\frac{d^2y}{dx^2}\right) = -EI\frac{d^3y}{dx^3}$$

$$w = -\frac{dF}{dx} = -\frac{d}{dx}\left(-EI\frac{d^3y}{dx^3}\right) = EI\frac{d^4y}{dx^4}$$

The loading for the second part of the question is illustrated in Fig. 5.21. At distance x from the lighter end the intensity of loading is, by proportion

$$w = \frac{x}{L} w_0$$

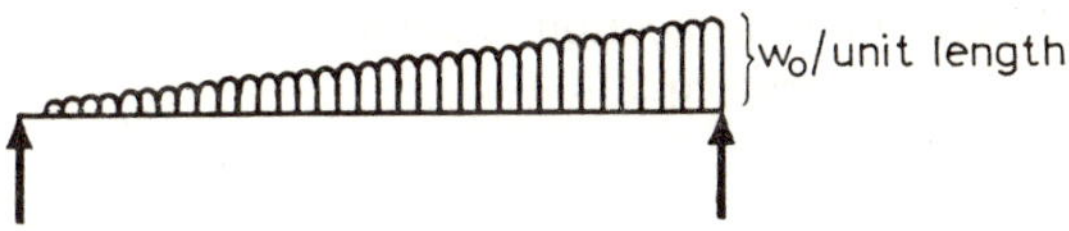

Fig. 5.21

Hence

$$EI \frac{d^4y}{dx^4} = \frac{w_0 x}{L}$$

By successive integrations

$$F = -EI \frac{d^3y}{dx^3} = -\frac{w_0 x^2}{2L} + C$$

and

$$M = -EI \frac{d^2y}{dx^2} = -\frac{w_0 x^3}{6L} + Cx + D$$

where C and D are constants. When $x = 0$, $M = 0$, thus $D = 0$. Also when $x = L$, $M = 0$, thus $C = w_0 L/6$. With these values of C and D, two more integrations give

$$EI \frac{dy}{dx} = \frac{w_0 x^4}{24L} - \frac{w_0 L x^2}{12} + H$$

and

$$EIy = \frac{w_0 x^5}{120L} - \frac{w_0 L x^3}{36} + Hx + K$$

where H and K are further integration constants. When $x = 0$, $y = 0$, thus $K = 0$. When $x = L$, $y = 0$, thus $H = 7w_0 L^3/360$. With this result the central deflection is given by $x = L/2$ and

$$EIy_c = \frac{w_0}{120L}\left(\frac{L}{2}\right)^5 - \frac{w_0 L}{36}\left(\frac{L}{2}\right)^3 + \frac{7w_0 L^3}{360}\left(\frac{L}{2}\right)$$

from which

$$y_c = \frac{5w_0 L^4}{768EI}$$

This result can also be obtained as follows. If a second load whose intensity varies from zero at the right-hand end to w_0 at the left-hand end is added to the first the result is a uniformly distributed load of intensity w_0. At mid-span the deflections due to these two varying loads are the same and each must therefore be equal to one-half of that due to a uniform load w_0, i.e. one-half of $5w_0L^4/384EI$. However, the general expressions obtained in this solution apply to all points. Note that the location of the maximum deflection requires the solution of a quartic equation.

5.18. The plan view of a horizontal cantilevered plate of uniform thickness is an equilateral triangle ABC, having sides of length L. The cantilever is supported from the side AB, where the second moment of area of the vertical section through the cantilever is I. There is a concentrated load of P at the free end C and a uniformly distributed load of Q per unit area on the surface ABC.

Show that the deflection y at C is given by the equation

$$y = \frac{3L^3}{128EI}[QL^2 + 8(\sqrt{3})P]$$

where E is the Modulus of Elasticity. (*U.L.*)

Solution. Figure 5.22 shows the cantilevered plate in side elevation at (*a*) and in plan at (*b*). From the geometry of the equilateral triangle the effective length of the cantilever is $\sqrt{3}L/2$. Since the plate thickness is constant the second moment of area of the cross-section is proportional to the width of the plate. At the fixed end the value of the second moment is I and, hence, at a distance x from the free end the corresponding value is

$$\frac{x}{\sqrt{3}L/2} \times I = \frac{2xI}{\sqrt{3}L}$$

It is convenient to consider the loads separately. Due to the concentrated load P there is a bending moment Px at the section XX and the ratio (bending moment)/(second moment of area) is

$$\frac{Px}{2xI/\sqrt{3}L} = \frac{\sqrt{3}LP}{2I}$$

This is constant and therefore we have circular arc bending as described in Example 5.1. The curvature is given by

$$\frac{1}{R} = \frac{1}{E} \times \left(\frac{\text{bending moment}}{\text{second moment}}\right) = \frac{\sqrt{3}LP}{2EI} \quad \text{(i)}$$

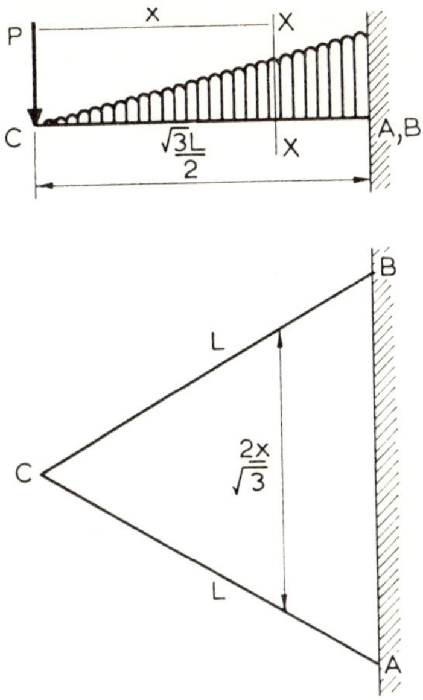

Fig. 5.22

By an analysis similar to that of Example 5.2 the deflection at the free end is, using (i),

$$y = \frac{(\text{span})^2}{2R} = \tfrac{1}{2}\left(\frac{\sqrt{3}L}{2}\right)^2 \times \frac{\sqrt{3}LP}{2EI}$$

$$= \frac{3\sqrt{3}PL^3}{16EI} \qquad \text{(ii)}$$

The deflection due to the distributed load is obtained by integration and it is convenient to take the origin at the free end C. At section XX, distance x from C, the width of the plate is, by proportion, $2x/\sqrt{3}$ and the intensity of the distributed load at this section is therefore $2xQ/\sqrt{3}$ per unit length. Between C and XX the *average* intensity is $xQ/\sqrt{3}$, the total load is $x^2Q/\sqrt{3}$ and its centre of gravity is $x/3$ from XX.

The bending moment at XX is therefore

$$M = -\frac{x^2Q}{\sqrt{3}} \times \frac{x}{3} = -\frac{x^3Q}{3\sqrt{3}}$$

(the negative sign indicating a hogging moment).

Since the second moment of area is not constant the differential equation of flexure is re-arranged as

$$E\frac{d^2y}{dx^2} = -\frac{M}{\text{second moment}}$$

$$= \frac{x^3Q}{3\sqrt{3}} \times \frac{\sqrt{3}L}{2xI} = \frac{LQx^2}{6I}$$

Integrating once, the slope is given by

$$E\frac{dy}{dx} = \frac{LQx^3}{18I} + K$$

where K is a constant. When $x = \sqrt{3}L/2$, $dy/dx = 0$ and thus $K = -3L^4Q/48I$. Using this value and integrating again, the deflection is given by

$$Ey = \frac{LQx^4}{72I} - \frac{\sqrt{3}L^4Qx}{48I} + J$$

where J is another constant.

When $x = 3L/2$, $y = 0$, thus $J = 3L^5Q/128I$

Using this value, and putting $x = 0$, the deflection at the free end is given by

$$Ey = \frac{3L^5Q}{128I} \quad \text{or} \quad y = \frac{3L^5Q}{128EI} \qquad \text{(iii)}$$

Adding (ii) and (iii) the total deflection is

$$y = \frac{3\sqrt{3}PL^3}{16EI} + \frac{3L^5Q}{128EI}$$

$$= \frac{3L^3}{128EI}(QL^2 + 8\sqrt{3}P)$$

as required.

PROBLEMS

1. A steel bar, pin-jointed at its ends, 25 mm wide and 3 mm thick is 1·25 m long. What is the deflection at mid-span when it is subjected to end couples, each 0·85 N-m in magnitude but opposite in sense? Take $E = 206$ GN/m².

What will be the deflection if the magnitude of the couples is increased until a maximum stress of 40 MN/m² is attained? Calculate the radius to which the bar bends in the latter case.

Answer. 14·35 mm; 25·3 mm; 7·73 m.

2. A beam, 6 m long, is simply supported at points 1·5 m from each end. It carries two point loads of 40 kN, one at each end. If $E = 206\ GN/m^2$ and $I = 8\,600\ cm^4$ calculate the deflection at mid-span. What central point load is necessary to reduce this deflection by one-half?
Answer. 3·81 mm; 60 kN.

3. A uniform cantilever, 3 m long, is subjected to a bending couple at a point 2 m from the fixed end. Calculate the ratio of the deflection at the free end to that at the point where the couple is applied.

Estimate the value of E for the material of the cantilever if the deflection at the free end is 9·7 mm when the magnitude of the couple is 60 N-m. $I = 12\ cm^4$.
Answer. Ratio is 2; 206 GN/m^2.

4. A uniform cantilever, 250 mm deep and 3 m long, is subjected to various loadings each of which produces a maximum bending stress of 80 MN/m^2. If the section of the cantilever is symmetrical and $I = 12\,500\ cm^4$ calculate the magnitude of the total load and the deflection at the free end in each of the following cases. (Take $E = 200\ GN/m^2$.)

(i) The load is concentrated at the free end.
(ii) The load is uniformly distributed throughout.
(iii) The load is concentrated at a point 1 m from the free end.
(iv) The load is uniformly distributed over the cantilever between its mid-point and the free end.
(v) Half the load is concentrated at the free end and half at the mid-point of the cantilever.
(vi) One-third of the load is uniformly distributed over the middle third of the cantilever and the remainder is concentrated at the mid-point.

Answer. (i) 26·7 kN and 9·60 mm. (ii) 53·3 kN and 7·20 mm. (iii) 40 kN and 7·47 mm. (iv) 35·6 kN and 8·20 mm. (v) 35·6 kN and 8·40 mm. (vi) 53·3 kN and 6·07mm.

5. A 50 mm diameter steel bar, 6 m long, rests on supports at its ends. Find the deflection at mid-span due to its own weight. The density of steel is 7700 kg/m^3 and $E = 206\ GN/m^2$. (Take the weight of 1 kg as 9·81 N.)
Answer. 39·6 mm.

6. A uniform circular bar, length L, extends an amount x under a tensile pull P. Show that the maximum deflection of the bar when it is used as a beam, simply supported at its ends and carrying a central point load W, is given by

$$\delta = WxL^2/3Pd^2$$

d being the diameter.

What is the ratio δ/x if $L = 100d$ and the maximum bending stress due to W is equal to the tensile stress due to the pull P?
Answer. $\delta/x = 50/3$.

7. When a simply supported beam, 3 m long of negligible weight, carries a load of 200 kN at its mid-point, the maximum deflection is 4 mm. What would be the deflection at the free end of a cantilever of the same material, length and cross-section if it carries a load of 100 kN at a point 1 m from the free end?
Answer. 16·6 mm.

8. A cantilever, 3 m long, and of symmetrical section 250 mm deep, carries a uniformly distributed load of 16 kN per metre run throughout, together with a point load of 40 kN at a section 1·2 m from the fixed end. Find the deflection at the free end and the maximum stress due to bending. $E = 206$ GN/m², $I = 13\,500$ cm⁴.
Answer. 8·52 mm and 111 MN/m².

9. Figure 5.23 shows a horizontal beam freely supported at its ends on two similar cantilevers. If I, the second moment of area of the cantilever section, is twice that for the beam, derive an expression, in terms of W, L, E and I, for the deflection at the centre of the beam. (*I.Mech.E.*)
Answer. $73WL^3/648EI$.

Fig. 5.23

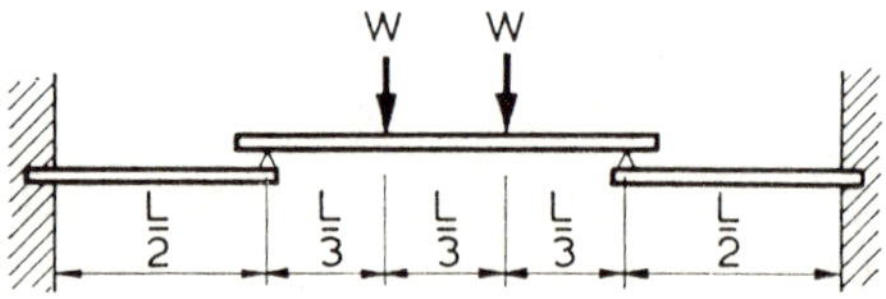

10. A simply supported beam 6 m long has a single point load of 80 kN applied at a point 1·5 m from one end. Calculate the position and value of the maximum deflection and also the slope of the beam at the centre point. ($EI = 28{\cdot}6$ MN-m².) (*I.Struct.E.*)
Answer. 3·35 m from the right end; 8·8 mm; 0·00079 rad.

11. State the conditions necessary that a loaded beam should bend in a circular arc.

A uniform horizontal beam, ABCD 4·8 m long, is supported at the same level at B and C, B being 1·2 m from end A and C being 0·6 m from end D. Loads of 30 kN and 60 kN are suspended from ends A and D respectively.

Calculate

(*a*) The vertical displacement of the mid-point of BC from the level of the supports.
(*b*) The slope of the beam at support B.
(*c*) The deflection of the end A below the level of the supports. ($EI =$ 32 MN-m².) (*I.Struct.E.*)

Answer. (*a*) 1·29 mm. (*b*) 0·001 69 rad. (*c*) 2·54 mm.

12. Horizontal pulls p_1 and p_2 are applied to a vertical pole, 50 mm in diameter, as shown in Fig. 5.24. If the deflection at the top of the pole is to be zero find the ratio of p_1 to p_2. Draw the bending moment diagram. Also, if the bending stress is not to exceed 14 MN/m² find p_1 and p_2. (*I.Mech.E.*)
Answer. $p_1/p_2 = 5/16$; maximum B.M. $= 3{\cdot}6\, p_1$ (N-m); $p_1 = 47{\cdot}7$ N and $p_2 =$ 153 N.

13. A cantilever of length L carries two loads each W, one at the free end and the other at a distance a ($<L$) from the free end. Derive a formula for the maximum deflection. If the cantilever consists of a steel tube of circular cross-section, 100 mm outside diameter and 6 mm thick, $L = 2$ m and $a = 0{\cdot}7$ m, calculate the value of W to give a maximum deflection of 6 mm. Take $E = 206$ GN/m². (*U.L.*)

Answer. $\dfrac{W}{6EI}(4L^3 - 3aL^2 + a^3)$; 610 N.

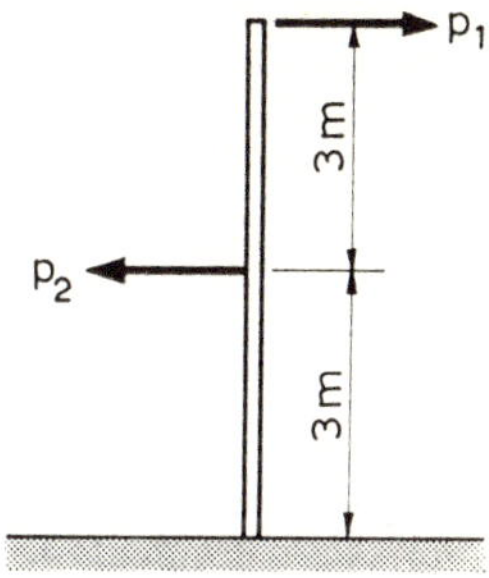

Fig. 5.24

14. A 300 mm × 300 mm timber beam is strengthened by the addition of 300 mm × 6 mm steel plates secured to its top and bottom surfaces. The composite beam is simply supported at its ends and carries a uniformly distributed load of 24 kN per metre run over an effective span of 6 m. Find the maximum bending stresses in the steel and timber at mid-span and the deflection at a quarter-span. E (steel) $= 200 \times 10^9$ N/m^2; E (wood) $= 10 \times 10^9$ N/m^2. *(U.L.)*

Hint. The deflection is the same as that of the "equivalent" steel beam with the same total load (see Example 3.14).

Answer. Stress in steel = 143 MN/m^2 and stress in timber = 6·87 MN/m^2; 12·2 mm.

15. A cantilever of length L and constant flexural rigidity EI carries a uniformly distributed load of intensity w/unit length on the middle half of its length. Determine the ratio of the slopes and the ratio of the deflections at the centre and free end of the cantilever. *(U.L.)*

Answer. Ratio of slopes = 25/26; ratio of deflections = 43/112.

16. A steel beam of uniform section has a length of 7 m and is simply supported at points 5 m apart and 1 m from the ends of the beam. The beam carries three point loads; 20 kN at the left-hand end, 40 kN at the right-hand end and 140 kN at 3 m from the left-hand end. Determine the deflection at each of the points of loading, stating in each case whether the deflection is upwards or downwards. Take $EI = 37$ MN-m^2. *(U.L.)*

Answers. (All in mm) at 20 kN load, 4·24 upwards; at 40 kN load, 2·68 upwards; at 140 kN load, 6·7 downwards.

17. A beam of uniform section and length l is simply supported at the ends and loaded at mid-span with a point load W. Derive the expression for the deflection at mid-span in terms of W, l, E and I.

A beam as above has a span $l = 4{\cdot}2$ m and when $W = 32$ kN the deflection at mid-span is 7·1 mm. If I for the beam is 3 380 cm^4 determine the value of E for the material.

If the load of 32 kN had been uniformly distributed over the whole span, what would be the value of the maximum deflection? *(U.L.)*

Answer. $Wl^3/48EI$; 206 GN/m^2; 4·44 mm.

18. A horizontal cantilever of uniform section and of length L carries a point load W at a distance l from the fixed end. Derive expressions for the slope and deflection at the load and from these or otherwise obtain an expression for deflection at the free end.

A horizontal cantilever of uniform section has an effective length of 2·5 m and carries a load of 50 kN at the free end. If the 50 kN load is replaced by two equal loads, one at the free end and the other at 1·5 m from the fixed end, such that the maximum deflection is the same as in the first case find

(*a*) the magnitude of the equal loads,
(*b*) the maximum bending stress in the second case expressed as a percentage of that in the first case. (*U.L.*)

Answer. $\frac{Wl^2}{2EI}(L - \frac{1}{3}l)$; (*a*) 34·9 kN; (*b*) 111·7 per cent.

19. Figure 5.25 shows beams of uniform section which rest on supports distant l apart. Derive expressions, either analytically or by use of the bending moment diagrams, for the slope at the supports and the deflection at mid-span (i) if the beam carries a uniformly distributed load as in Fig. 5.25a, (ii) if the beam is subjected to equal opposite couples at the ends as in Fig. 5.25b.

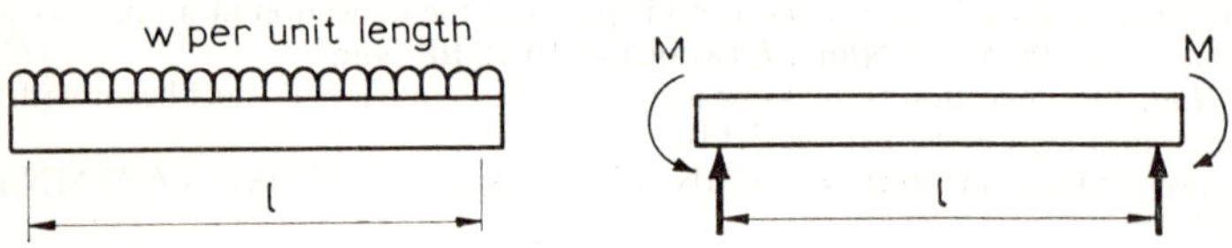

Fig. 5.25

A beam of uniform section rests on supports 6 m apart and carries a uniformly distributed load of 24 kN/m on the whole span including its own weight. Find the magnitude of the equal opposite couples which must be applied to the ends of the beam to make the beam just horizontal at the supports; find also the deflection at mid-span under these conditions. Take $E = 200$ GN/m² and $I = 8100$ cm⁴. (*U.L.*)

Answer. (i) $wl^3/24EI$; $5wl^4/384EI$. (ii) $Ml/2EI$; $Ml^2/8EI$; 72 kN-m; 5 mm.

20. Figure 5.26 shows a cantilever and a simply supported beam, each being of uniform section. The particular values of slope and deflection stated below are to be found by Mohr's method, i.e. using the M/EI diagram.

(*a*) Cantilever, Fig. 5.26(*a*); determine the slope at B and the deflection at B.
(*b*) Beam, Fig. 5.26(*b*); determine the deflection at C and the slopes at A and E. (*U.L.*)

Answer. (*a*) $Wl^2/2EI$; $Wl^3/3EI$. (*b*) $19Wl^3/384EI$; $5Wl^2/32EI$.

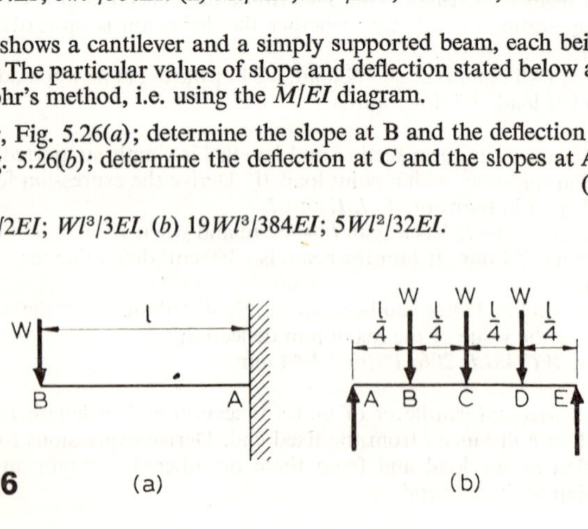

Fig. 5.26

21. A rolled steel beam of I section is 360 mm deep and has flanges 150 mm wide. The beam is simply supported at its ends and is required to carry a uniformly distributed load of 24 kN/m on its whole length. This load includes its own weight.

Find the greatest allowable span to satisfy the following conditions:

(*a*) the bending stress not to be greater than 140 MN/m^2,
(*b*) the maximum deflection not to be greater than 1/400th of the span.

What will be the actual values of the maximum stress and the maximum deflection for the span calculated?

$E = 200$ GN/m^2; second moment of area of the section, $I = 18300$ cm^4. (*U.L.*)

Answer. 6·64 m; (*b*) is the limiting factor; 130 MN/m^2; 16·6 mm.

22. A 13 mm diameter steel rod is bent to form a square with sides $2a = 500$ mm long. The ends meet at the mid-point of one side and are separated by equal opposite forces of 65 N applied in a direction perpendicular to the plane of the

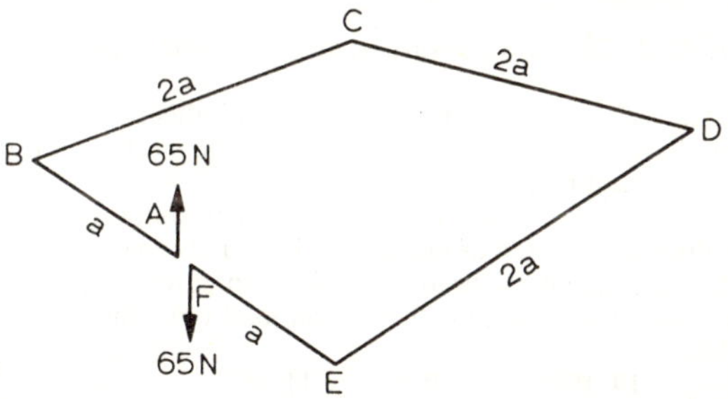

Fig. 5.27

square as shown in perspective in Fig. 5.27. Calculate the amount by which they will be out of alignment. $E = 206$ GN/m^2; $G = 81$ GN/m^2. (*U.L.*)
Answer. 74·8 mm.

23. A 360 mm × 150 mm rolled steel beam of I section 7·5 m long rests on two supports 6 m apart, one of these supports being at the left-hand end of the beam. The beam carries a uniformly distributed load of 24 kN/m on its whole length. Determine

(*a*) the maximum stress in the beam due to bending,
(*b*) the deflection midway between the supports,
(*c*) the deflection at the right-hand end of the beam.

State whether the deflection is upwards or downwards. For this beam $I = 18400$ cm^4; take $E = 200$ GN/m^2. (*U.L.*)
Answer. (*a*) 93 MN/m^2. (*b*) 9·35 mm (downwards). (*c*) 6·2 mm (upwards).

24. A beam of uniform section has a length of 10 m and is simply supported at each end. There are three loads on the beam as follows: 45 kN (2·5 m), 50 kN (4·5 m) and 25 kN (8·5 m), the distances in brackets being the distances of the loads from the left-hand end.

Determine the deflection at mid-span and the slope at each end. The constants for the beam are: $I = 35000$ cm^4; $E = 206$ GN/m^2. (*U.L.*)
Answer. 26·4 mm; 0·00795 rad; (—) 0·00874 rad.

25. A beam ABC of uniform section and 9 m long is simply supported at B and C; AB = $1\frac{1}{2}$ m and BC = $7\frac{1}{2}$ m, whilst the value of EI for the beam is 28 MN-m^2. Find the deflection at the overhanging end A and at D, the mid-point of BC, when the beam carries a uniformly distributed load of 20 kN/m:

(*a*) on the whole length AC;
(*b*) on the $7\frac{1}{2}$ m length between the supports at B and C.

State in each case whether the deflection is up or down. (*U.L.*)
Answer. (*a*) at A, 1·85 mm (upwards); at D, 2·66 mm (downwards); (*b*) at A, 1·88 mm (upwards); at D, 2·94 mm (downwards).

26. A beam ABCDE is of uniform section and the length AE is 9 m; AB = 1 m, BC = $2\frac{1}{2}$ m, CD = 4 m and DE = $1\frac{1}{2}$ m. The beam is simply supported at B and D which are $6\frac{1}{2}$ m apart and the loads carried are a point load of 40 kN at A and a uniformly distributed load of 25 kN/m and $5\frac{1}{2}$ m long extending from C to E.

Determine the deflection at the end A and also at F where AF = 4 m; state whether the deflections are up or down. E = 200 GN/m^2 and I = 6000 cm^4. (*U.L.*)
Answer. At A, 38·1 mm (upwards); at F, 1·85 mm (upwards).

27. State, with explanation, Mohr's second theorem connecting the bending moment diagram of a beam with the slope and deflection at points of the beam. A proof is not required.

A beam of uniform section and length l is simply supported at its ends. It carries three loads, each of magnitude W equally spaced at $l/4$, $l/2$ and $3l/4$ from an end. Determine the deflection at mid-span in terms of W, l, E and I due to (*a*) the three loads together, (*b*) the central load only and (*c*) one only of the non-central loads. (*U.L.*)
Answer. (*a*) $\frac{19}{384}\frac{Wl^3}{EI}$; (*b*) $\frac{1}{48}\frac{Wl^3}{EI}$; (*c*) $\frac{11}{768}\frac{Wl^3}{EI}$.

28. A beam is supported horizontally at two points at a distance L apart. The beam overhangs each support by a distance $L/3$. A uniformly distributed load of total magnitude W is carried by the portion of the beam between the supports and a concentrated load $W/4$ is carried at each free end. If the deflection at the middle of the beam is to be equal to that at the free ends, determine the constant second moment of area of the section of the overhanging portions, if that for the part between the supports is I. (*U.L.*)
Answer. $64I/81$.

29. A horizontal steel beam, simply supported at the ends, carries a load which varies uniformly from 20 kN/m at one end to 70 kN/m at the other end over a span of 5 m. Find the magnitude of the maximum bending moment.

If the depth of the beam is 0·4 m and the maximum bending stress is 100 MN/m^2, find the deflection at mid-span. E = 200 GN/m^2. (*U.L.*)
Answer. 142 kN-m; 0·039 4 m.

Chapter 6

Propped Beams and Cantilevers; Built-in Beams

In each of the cases of uniform built-in beams in Table 3, L is the total length and W is the total load.

For a uniform built-in beam AB, length L, carrying a concentrated load W at distances a and b from A and B respectively, the fixing moments are

$$M_A = \frac{Wab^2}{L^2} \quad \text{and} \quad M_B = \frac{Wa^2b}{L^2}$$

and the reactions are

$$R_A = \frac{Wb^2}{L^3}(L + 2a) \text{ and } R_B = \frac{Wa^2}{L^3}(L + 2b)$$

WORKED EXAMPLES

6.1. A cantilever of constant cross-section carries a uniformly distributed load throughout. What (upward) force must be applied at the free end to reduce the deflection there to zero?

A uniform cantilever, 6 m long, is propped at the free end to the level of the fixed end. Calculate the reaction on the prop when the cantilever carries a load of 120 kN uniformly distributed over its whole length. Draw, to scale, the shearing force and bending moment diagrams and state the principal values on each.

Solution. Let W be the total distributed load and P the required force at the free end.

The deflection at the free end due to W alone is, by the formula of Chapter 5,

$$\delta_1 = WL^3/8EI \text{ (downwards)}$$

Case	Load concentrated at mid-span	Load uniformly distributed throughout
Loading Diagram	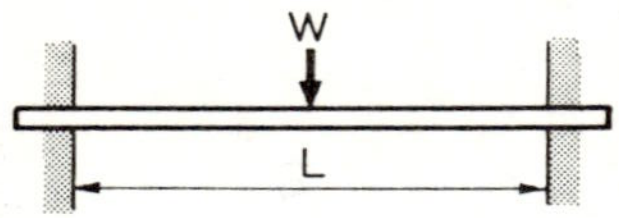	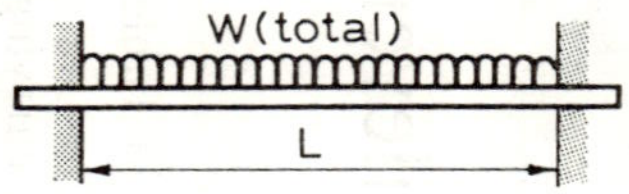
S.F. Diagram	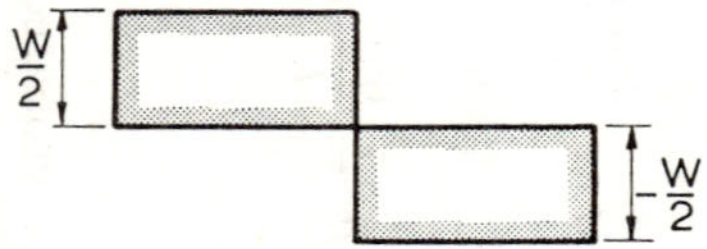	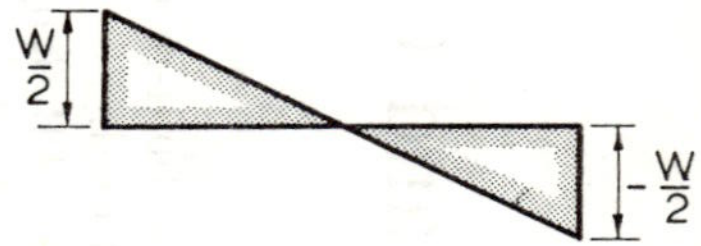
B.M. Diagram	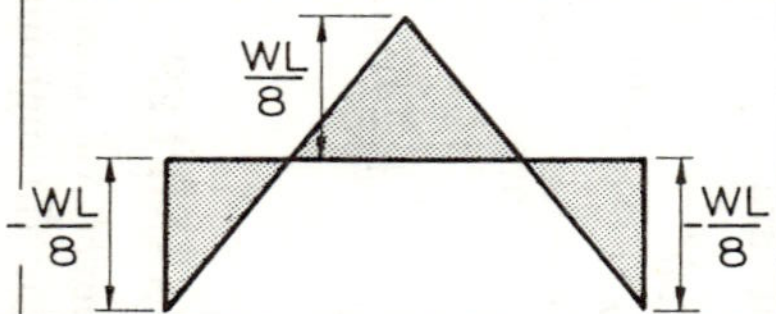	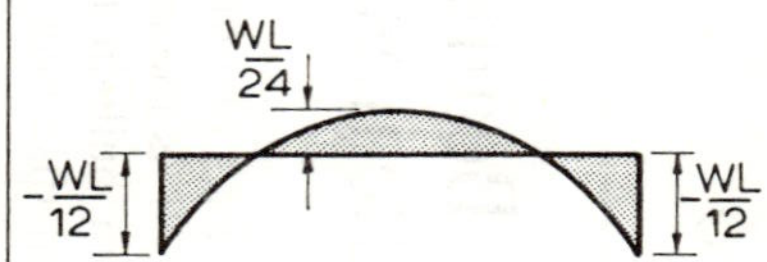
Maximum Deflection	$\frac{WL^3}{192EI}$	$\frac{WL^3}{384EI}$

Table 3

Similarly, the deflection due to P, using the formula for an end concentrated load, is

$$\delta_2 = PL^3/3EI \text{ (upwards)}$$

In each case, L, E, and I have their usual meanings.

If the final deflection at the free end is zero, then $\delta_2 = \delta_1$, or

$$PL^3/3EI = WL^3/8EI$$

and

$$P = \tfrac{3}{8}W \qquad \text{(i)}$$

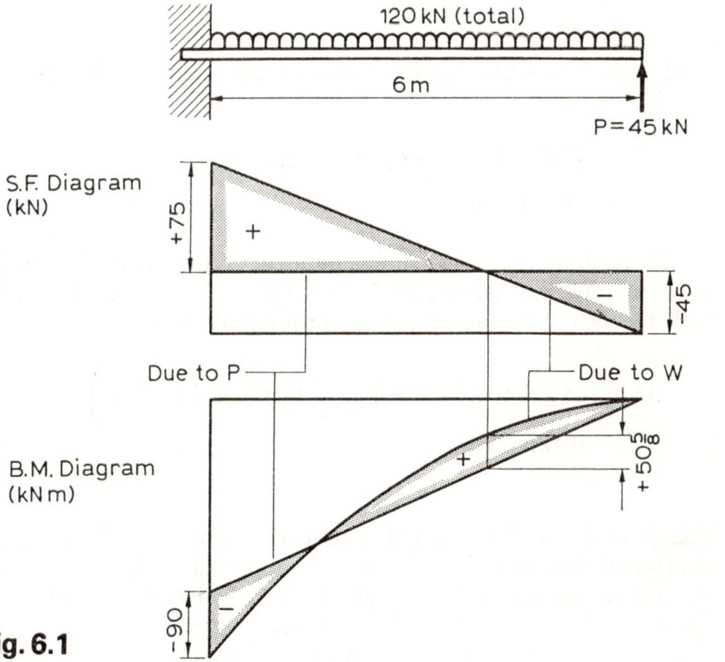

Fig. 6.1

Figure 6.1 shows the given cantilever; a prop can be regarded as a "knife edge" type of support. Its reaction on the cantilever cannot be found by taking moments since there are three unknowns involved, namely the prop reaction, the reaction at the fixed end, and the fixing moment. Only two independent moment equations can be formed, and thus one of the three unknowns must be evaluated from other considerations. (Such a problem is said to be *statically indeterminate.*)

In the present example the prop reaction P is such that the overall deflection at the prop is zero. Hence, by (i),

$$P = \tfrac{3}{8} \times 120 \text{ kN} = 45 \text{ kN} \qquad (Ans)$$

The required diagrams are obtained by the method of superposition. At the "free" end the shearing force is clearly (−) 45 kN. At the fixed end,

$$\text{Shearing force} = W - P$$
$$= 120 \text{ kN} - 45 \text{ kN}$$
$$= 75 \text{ kN}$$

The maximum hogging bending moment occurs at the fixed end and is

$$M = (45 \text{ kN} \times 6 \text{ m}) - (120 \text{ kN} \times 3 \text{ m})$$
$$= (-)\ 90 \text{ kN-m}$$

The maximum sagging bending moment occurs where the shearing force is zero, i.e. (45 kN/120 kN) × 6 m = $2\tfrac{1}{4}$ m from the "free" end. Hence the maximum sagging bending moment is

$$M = (45 \text{ kN} \times 2\tfrac{1}{4} \text{ m}) - \left(\frac{2\tfrac{1}{4} \text{ m}}{6 \text{ m}} \times 120 \text{ kN} \times 1\tfrac{1}{8} \text{ m}\right)$$
$$= 50\tfrac{5}{8} \text{ kN-m} \qquad (Ans)$$

6.2. A uniform girder, length 8 m, carries a total load of 400 kN, uniformly distributed. It is simply supported at its ends and is propped at the centre to the same level. Calculate the prop reaction and draw, to scale, the bending moment diagram.

If, however, the prop sinks 30 mm relative to the ends (due to the load) what are the prop and end reactions? The product EI = 32 MN-m².

Solution. The deflection at mid-span due to a uniformly distributed load W alone is

$$\delta_1 = 5WL^3/384EI \text{ (downwards)}$$

Similarly the deflection due to P, the prop reaction, is

$$\delta_2 = PL^3/48EI \text{ (upwards)}$$

If the three supports are at the same level, then $\delta_2 = \delta_1$, or

$$PL^3/48EI = 5WL^3/384EI$$

and

$$P = \tfrac{5}{8}W = \tfrac{5}{8} \times 400 \text{ kN} = 250 \text{ kN}$$

By symmetry, the end reactions are each

$$\tfrac{1}{2}(400 - 250) \text{ kN} = 75 \text{ kN}$$

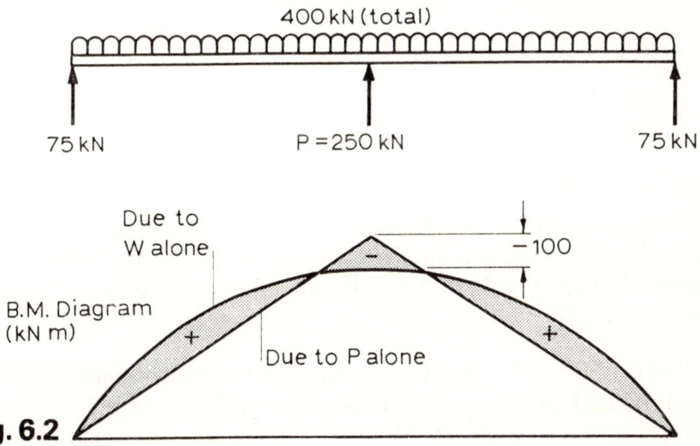

Fig. 6.2

Figure 6.2 shows the bending moment diagram which is obtained by the method of superposition.

Due to W alone (i.e. with the prop removed) the bending moment at mid-span is

$$M_1 = \tfrac{1}{8}WL = \tfrac{1}{8}(400 \text{ kN} \times 8 \text{ m}) = 400 \text{ kN-m}$$

Due to P alone the bending moment at the centre is

$$M_2 = (-)\ \tfrac{1}{4}PL = (-)\ \tfrac{1}{4}(250 \text{ kN} \times 8 \text{ m}) = (-)\ 500 \text{ kN-m}$$

The resultant bending moment at the centre is

$$M_1 + M_2 = 400 - 500 = (-)\ 100 \text{ kN-m}$$

If the prop sinks 30 mm the prop reaction P is reduced. Its new value corresponds to a deflection which is the difference of the deflection due to W and 30 mm. Therefore

$$\delta_1 - \delta_2 = 30 \times 10^{-3}\ \text{m}$$

$$\frac{5WL^3}{384EI} - \frac{PL^3}{48EI} = 30 \times 10^{-3}$$

$$\frac{5W}{384} - \frac{P}{48} = \frac{EI \times (30 \times 10^{-3})}{L^3} = \frac{(32 \times 10^6) \times (30 \times 10^{-3})}{8^3}$$

$$P = \frac{5W}{8} - \frac{48 \times 32 \times 30 \times 10^3}{512}$$

$$= \frac{5 \times 400}{8} - 90$$

$$= 160\ \text{kN} \qquad (Ans)$$

By symmetry each end reaction is

$$\tfrac{1}{2}(400 - 160) = 120\ \text{kN} \qquad (Ans)$$

6.3. A cantilever of effective length L with a concentrated load W at the free end is propped at a distance a from the fixed end to the same level as the fixed end. Find the load on the prop. Show that there is always a real point of inflexion and find its distance from the fixed end. (*U.L.*)

Solution. Figure 6.3 shows the bending moment diagrams for the load W and prop reaction P separately. Since the deflection at the prop is zero then, by Mohr's theorem, the moment about this point of the total bending moment diagram between the fixed end and the prop is zero.

Hence

$$[(\text{area pqrs}) \times \tfrac{1}{2}a] + [(\text{area srt}) \times \tfrac{2}{3}a] - [(\text{area lmn}) \times \tfrac{2}{3}a] = 0$$

$$[a \times W(L - a) \times \tfrac{1}{2}a] + [\tfrac{1}{2}a \times Wa \times \tfrac{2}{3}a] - [\tfrac{1}{2}a \times Pa \times \tfrac{2}{3}a] = 0$$

$$\tfrac{1}{2}Wa^2(L - a) + \tfrac{1}{3}Wa^3 - \tfrac{1}{3}Pa^3 = 0$$

from which

$$P = \tfrac{1}{2}W\left(\frac{3L}{a} - 1\right)$$

If there is a point of inflexion it will occur between the prop and the fixed end. The resultant bending moment will change sign between these points if Pa is numerically greater than WL (see diagram). Now

$$Pa = \tfrac{1}{2}Wa\left(\frac{3L}{a} - 1\right) = \tfrac{1}{2}W(3L - a)$$

which is clearly greater than WL for all values of a between 0 and L. Hence there is always a real point of inflexion.

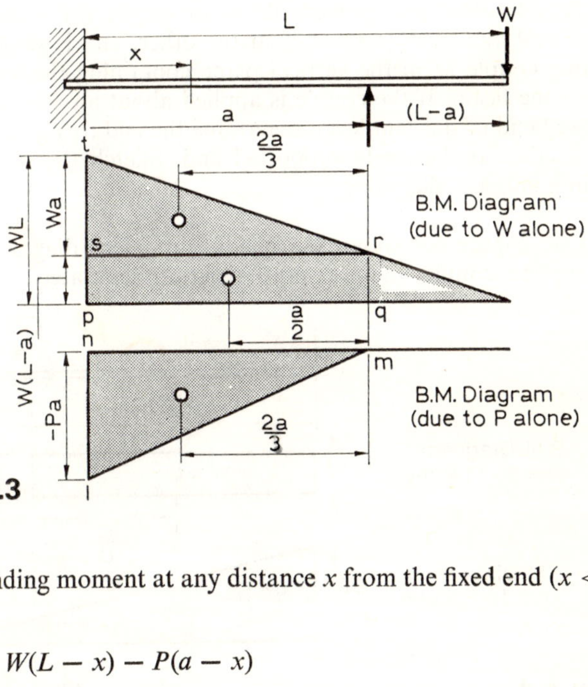

Fig. 6.3

The bending moment at any distance x from the fixed end ($x < a$) is

$$M = W(L - x) - P(a - x)$$

This is zero when

$$W(L - x) = P(a - x)$$

$$= \tfrac{1}{2}W\left(\frac{3L}{a} - 1\right)(a - x)$$

$$L - x = \tfrac{3}{2}L - \frac{3Lx}{2a} - \tfrac{1}{2}a + \tfrac{1}{2}x$$

from which

$$\frac{3x}{2}\left(\frac{L}{a} - 1\right) = \tfrac{1}{2}(L - a)$$

$$x = \frac{(L - a)}{3\left(\frac{L}{a} - 1\right)} = \frac{a}{3}$$

That is, the point of inflexion is $a/3$ from the fixed end.

6.4. A horizontal propped cantilever of length L is securely fixed at one end and freely supported at the other, and is subjected to a bending couple M in the vertical plane containing the longitudinal axis of the beam. If the couple is applied about an axis $0{\cdot}75L$ from the fixed end of the cantilever, determine the end fixing moment and the reaction at the freely supported end. Sketch the shape of the bending moment diagram. *(U.L.)*

Solution. Figure 6.4 shows the bending moment diagrams due to the applied couple and the prop reaction separately. If there is

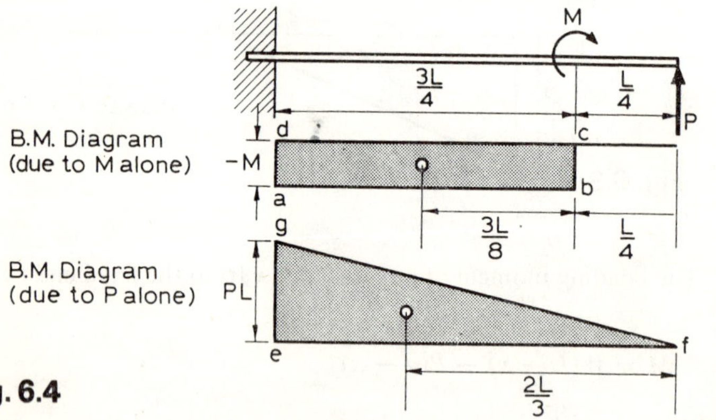

Fig. 6.4

no deflection at the propped end then the moment about this end of the total bending moment diagrams for the cantilever is zero. Hence

$$-[(\text{area abcd}) \times (\tfrac{1}{4}L + \tfrac{3}{8}L)] + [(\text{area efg}) \times \tfrac{2}{3}L] = 0$$

or

$$-(M \times \tfrac{3}{4}L \times \tfrac{5}{8}L) + (\tfrac{1}{2} \times L \times PL \times \tfrac{2}{3}L) = 0$$

from which the prop reaction is

$$P = 45M/32L$$

The fixing moment (which equals the bending moment at the fixed end) is

$$PL - M = \left(\frac{45M}{32L} \times L\right) - M = \frac{13M}{32}$$

The bending moment varies linearly from zero at the propped end to $PL/4 = 45M/128$ at the point where the couple is applied. At this point the bending moment changes to

$$\frac{45M}{128} - M = -\frac{83M}{128}$$

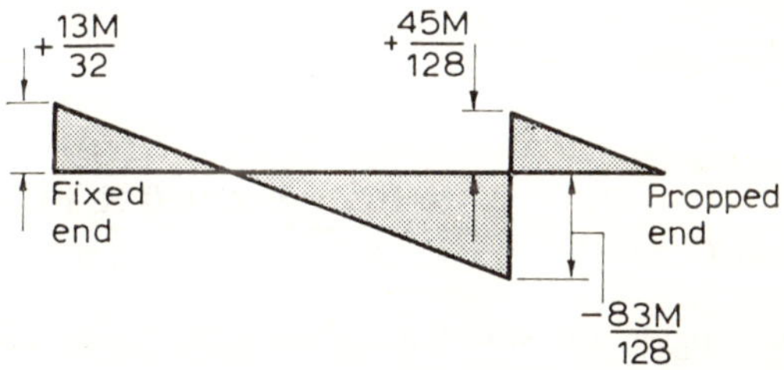

Fig. 6.5

due to the applied couple. It then varies linearly to (+) $13M/32$ at the fixed end. Figure 6.5 shows the resulting diagram for the cantilever.

6.5. A horizontal beam, $I = 8\,600\ \text{cm}^4$, carries a uniformly distributed load of 50 kN over its length of 3 m. The beam is supported by three vertical steel tie-rods, each 2 m long, one at each end and one in the middle, the end rods having a diameter of 25 mm and the centre rod a diameter of 32 mm. Calculate the deflection of the centre of the beam below the end points and the stress in each of the rods. (E for steel $= 206 \times 10^9\ \text{N/m}^2$.) (*U.L.*)

Solution. Working in newtons and metre units throughout, let P be the tensile load in the middle tie-rod.

By symmetry, the load in each end tie-rod is $\frac{1}{2}(50 \times 10^3 - P)$.

Deflection at mid-span due to the distributed load W is

$$\delta_1 = 5WL^3/384EI \text{ (downwards)}$$

Deflection at mid-span due to the load in the middle tie-rod is

$$\delta_2 = PL^3/48EI \text{ (upwards)}$$

Extension of middle tie-rod is

$$x_1 = \frac{\text{length} \times \text{stress}}{E} = \frac{2 \times P}{\frac{1}{4}\pi \times (32 \times 10^{-3})^2 \times E}$$

Extension of each end tie-rod is

$$x_2 = \frac{\text{length} \times \text{stress}}{E} = \frac{2 \times \frac{1}{2}(50 \times 10^3 - P)}{\frac{1}{4}\pi \times (25 \times 10^{-3})^2 \times E}$$

Also the resultant deflection at mid-span equals the difference in the extensions of the middle and end tie-rods.
Hence,

$$\delta_1 - \delta_2 = x_1 - x_2$$

or

$$\frac{5WL^3}{384EI} - \frac{PL^3}{48EI} = \frac{2P \times 10^6}{\frac{1}{4}\pi(32)^2E} - \frac{2 \times \frac{1}{2}(50 \times 10^3 - P) \times 10^6}{\frac{1}{4}\pi(25)^2E}$$

Cancelling E, and substituting the given values of W, L and I, we obtain

$$(2044 \times 10^3) - 65{\cdot}4P = 24{\cdot}87P - 20{\cdot}37\,(50 \times 10^3 - P)$$

from which,

$$P = 27{\cdot}7 \text{ kN}$$

The deflection of the centre of the beam below the end points is

$$\begin{aligned}\delta_1 - \delta_2 &= \frac{5WL^3}{384EI} - \frac{PL^3}{48EI} \\ &= \frac{L^3}{48EI}\left(\frac{5W}{8} - P\right) \\ &= \frac{3^3}{48 \times (206 \times 10^9) \times (8600 \times 10^{-8})} \times \left(\frac{5 \times 50 \times 10^3}{8} - 27{\cdot}7 \times 10^3\right) \\ &= 113 \times 10^{-6} \text{ m} = 0{\cdot}113 \text{ mm} \qquad (Ans)\end{aligned}$$

Stress in the middle tie-rod is

$$P/[\tfrac{1}{4}\pi \times (32 \times 10^{-3})^2]$$

$$= \frac{(27{\cdot}7 \times 10^3)}{\tfrac{1}{4}\pi \times 32^2 \times 10^{-6}} = 34{\cdot}5 \text{ MN/m}^2 \qquad (Ans)$$

Stress in the end tie-rods is

$$\frac{\tfrac{1}{2}(50 \times 10^3 - P)}{\tfrac{1}{4}\pi \times (25 \times 10^{-3})^2}$$

$$= \frac{2(50 - 27{\cdot}7) \times 10^3}{\pi \times 625 \times 10^{-6}} = 22{\cdot}7 \text{ MN/m}^2 \qquad (Ans)$$

6.6. What is an "encastré" beam? Derive an expression for the maximum deflection of such a beam when it carries a central concentrated load. A uniform beam, simply supported at its ends, carries a concentrated load of 60 kN at mid-span. Calculate the load which the same beam may carry at mid-span when it is built-in at its ends if

(*a*) the maximum deflection is unchanged,
(*b*) the maximum bending stress is unchanged.

Solution. An encastré (or built-in) beam is one which is rigidly fixed at each end in the manner of the fixed end of a cantilever. It is assumed, however, that the beam is not constrained longitudinally and a possible arrangement is shown in Fig. 6.6 where the left-hand

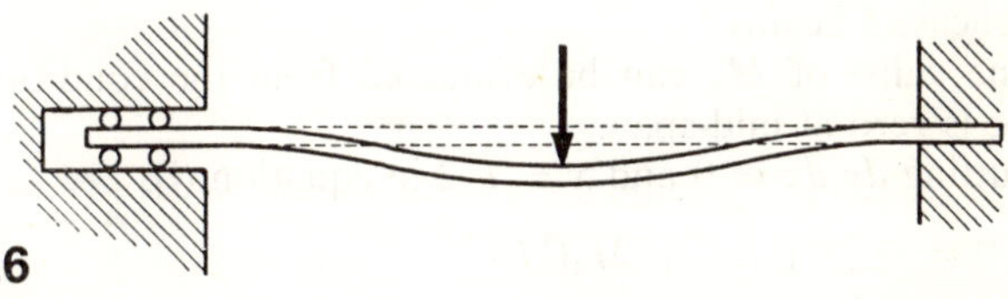

Fig. 6.6

end is held between rollers which prevent angular and vertical movement but allow each part of the beam to move longitudinally.

In addition to vertical reactions, the supports exert fixing couples on the beam. For downward (vertical) loads these couples are counter-clockwise and clockwise at the left- and right-hand ends respectively. These fixing couples must be included in all appropriate moment equations and (except for symmetrical cases) the reactions for an encastré beam are different to those of a simply supported beam with the same loading.

Figure 6.7 shows an encastré beam AB carrying a central point load W. Let M_A and M_B be the fixing couples (in magnitude only). By symmetry, each reaction is $W/2$ and the bending moment at any section XX, distance x from A, is

$$M = \tfrac{1}{2}Wx - M_A \qquad \text{(if } x \text{ is less than } \tfrac{1}{2}L\text{).}$$

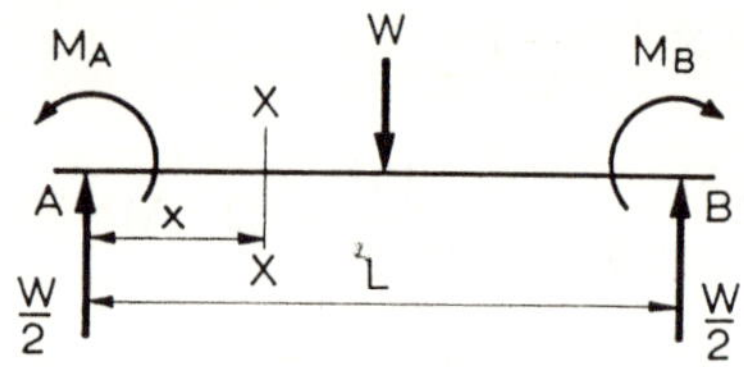

Fig. 6.7

Substituting in the differential equation of flexure,

$$EI\frac{d^2y}{dx^2} = -M = -\tfrac{1}{2}Wx + M_A$$

Integrating once,

$$EI\frac{dy}{dx} = -\tfrac{1}{4}Wx^2 + M_Ax \qquad \text{(i)}$$

(The constant of integration is zero since $dy/dx = 0$ when $x = 0$.) From the second integration,

$$EIy = -\tfrac{1}{12}Wx^3 + \tfrac{1}{2}M_Ax^2 \qquad \text{(ii)}$$

(The constant of integration is again zero, as are all such constants for encastré beams.)

The value of M_A can be evaluated from the condition that the slope is zero at mid-span.

Putting $dy/dx = 0$ and $x = L/2$ in equation (i),

$$0 = -\tfrac{1}{4}W(\tfrac{1}{2}L)^2 + M_A(\tfrac{1}{2}L)$$
$$M_A = \tfrac{1}{8}WL$$

Substituting this value in equation (ii) and putting $x = L/2$, the maximum deflection (which occurs at mid-span) is

$$y_{max} = \frac{1}{EI}[-\tfrac{1}{12}W(\tfrac{1}{2}L)^3 + \tfrac{1}{2}(\tfrac{1}{8}WL)(\tfrac{1}{2}L)^2]$$
$$= WL^3/192EI$$

(The reader should now examine the shearing force and bending moment diagrams for this case which are given in Table 3.)

In case (*a*), if W is the required central load then

$$\frac{WL^3}{192EI} = \frac{60\,(\text{kN}) \times L^3}{48EI}$$

$$W = 4 \times 60 \text{ kN} = 240 \text{ kN} \qquad (Ans)$$

In case (*b*), since the maximum bending stress is unchanged, then the maximum bending moment is unchanged. It is $WL/8$ for the built-in case (see Table 3) and hence

$$\tfrac{1}{8}WL = \frac{60 \text{ kN} \times L}{4}$$

$$W = 2 \times 60 \text{ kN} = 120 \text{ kN} \qquad (Ans)$$

6.7. Derive an expression for the maximum deflection of an encastré beam carrying a uniformly distributed load w/unit length. A uniform beam is simply supported at its ends and carries a uniformly distributed load which produces a maximum deflection of 25 mm and a maximum bending stress of 30 MN/m².

An equal beam, built-in at its ends, carries a uniformly distributed load of different magnitude.

For the latter beam calculate

(*a*) the maximum bending stress if the maximum deflections of the two beams are equal, and
(*b*) the maximum deflection if the maximum bending stresses of the two beams are equal.

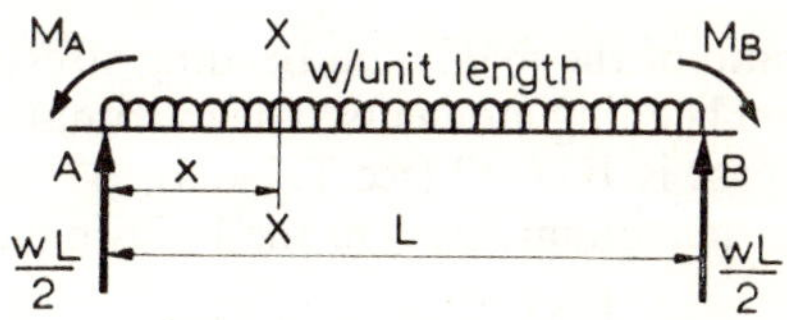

Fig. 6.8

Solution. Figure 6.8 shows the beam AB carrying the given load. By symmetry, the reactions are each $wL/2$ and the fixing moments are (numerically) equal. At any section XX, distance x from the end A, the bending moment is

$$M = \tfrac{1}{2}wLx - \tfrac{1}{2}wx^2 - M_A$$

Hence the differential equation of flexure becomes

$$EI\frac{d^2y}{dx^2} = -M = -\tfrac{1}{2}wLx + \tfrac{1}{2}wx^2 + M_A$$

Integrating twice, the deflection is given by

$$EIy = -\frac{wLx^3}{12} + \frac{wx^4}{24} + \frac{M_Ax^2}{2} \qquad \text{(i)}$$

(the constants of integration being zero).

The value of M_A can be found, from (i), using the condition that $y = 0$ when $x = L$.

Hence

$$0 = -\frac{wL(L)^3}{12} + \frac{wL^4}{24} + \frac{M_AL^2}{2}$$

$$M_A = WL^2/12$$

Substituting this value in equation (i) and putting $x = L/2$, the maximum deflection (which occurs at mid-span) is

$$y_{max} = \frac{1}{EI}\left[-\tfrac{1}{12}wL(\tfrac{1}{2}L)^3 + \tfrac{1}{24}w(\tfrac{1}{2}L)^4 + \tfrac{1}{2}(\tfrac{1}{12}wL^2)(\tfrac{1}{2}L)^2\right]$$

$$= \frac{wL^4}{384EI}$$

(The shearing force and bending moment diagrams for this case are given in Table 3.)

Let W_1 be the total load in the simply supported case and W_2 be the total load in the built-in case.

(*a*) If the deflections are equal then

$$\frac{5W_1L^3}{384EI} = \frac{W_2L^3}{384EI} \quad \text{and} \quad W_2 = 5W_1$$

The ratio of the maximum bending stresses equals the ratio of the maximum bending moments. The maximum bending moment for the built-in case is $W_2L/12$ (see Table 3).

Hence, maximum stress in the built-in case is

$$\frac{\text{max. B.M. in built-in case}}{\text{max. B.M. in simply supported case}} \times 30 \text{ MN/m}^2$$

$$= \frac{\tfrac{1}{12}W_2L}{\tfrac{1}{8}W_1L} \times 30 \text{ MN/m}^2$$

$$= \frac{W_2}{W_1} \times \frac{8}{12} \times 30 \text{ MN/m}^2$$

$$= 100 \text{ MN/m}^2 \qquad (Ans)$$

(*b*) If the maximum stresses are equal, then the maximum bending moments are equal and

$$W_1L/8 = W_2L/12 \quad \text{or} \quad W_2 = 3W_1/2$$

Therefore, the deflection in the built-in case is

$$\frac{(W_2L^3/384EI)}{(5W_1L^3/384EI)} \times 25 \text{ mm} = \frac{1}{5} \times \frac{W_2}{W_1} \times 25 \text{ mm}$$
$$= 7{\cdot}5 \text{ mm} \qquad (Ans)$$

6.8. A built-in beam AB has a clear span L and carries a concentrated load W at distances a and b from the ends A and B respectively, a being greater than b.

Establish formulae for the fixing moments and reactions. Find also the magnitude of the maximum deflection and show that for any value of a its position is within $L/6$ of the centre.

Solution. Let M_A, M_B be the fixing moments, R_A, R_B the reactions as shown in Fig. 6.9.

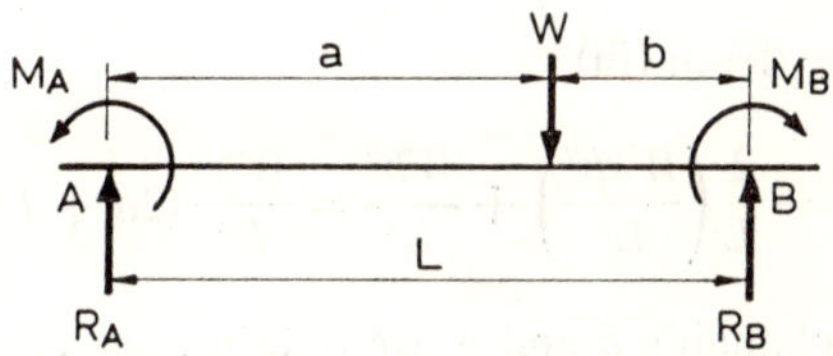

Fig. 6.9

At any distance x from A the bending moment is

$$M = R_Ax - M_A - W[x - a]$$

the term in square brackets being omitted when $x < a$.

Substituting in the differential equation of flexure,

$$EI\frac{d^2y}{dx^2} = -M = -R_Ax + M_A + W[x - a]$$

Integrating, using the Macaulay method, the slope is given by

$$EI\frac{dy}{dx} = -\tfrac{1}{2}R_Ax^2 + M_Ax + \tfrac{1}{2}W[x - a]^2 \qquad \text{(i)}$$

Integrating again to find the deflection,

$$EIy = -\tfrac{1}{6}R_Ax^3 + \tfrac{1}{2}M_Ax^2 + \tfrac{1}{6}W[x - a]^3 \qquad \text{(ii)}$$

At the right-hand end of the beam ($x = L$), the slope and deflection are both zero.

Hence, from (i),

$$0 = -\tfrac{1}{2}R_A L^2 + M_A L + \tfrac{1}{2}W[L-a]^2$$

$$R_A = \frac{2M_A}{L} + \frac{Wb^2}{L^2} \qquad \text{(iii)}$$

(since $L - a = b$).

Similarly, from (ii),

$$0 = -\tfrac{1}{6}R_A L^3 + \tfrac{1}{2}M_A L^2 + \tfrac{1}{6}W[L-a]^3$$

$$R_A = \frac{3M_A}{L} + \frac{Wb^3}{L^3} \qquad \text{(iv)}$$

Eliminating R_A from (iii) and (iv), we have

$$\frac{3M_A}{L} + \frac{Wb^3}{L^3} = \frac{2M_A}{L} + \frac{Wb^2}{L^2}$$

$$M_A = \frac{Wb^2}{L} - \frac{Wb^3}{L^2} = \frac{Wb^2}{L^2}(L-b) = \frac{Wab^2}{L^2}$$

Hence, from (iii),

$$R_A = \frac{2}{L}\left(\frac{Wab^2}{L^2}\right) + \frac{Wb^2}{L^2} = \frac{Wb^2}{L^3}(2a+L)$$

Interchanging a and b, we obtain

$$M_B = \frac{Wa^2b}{L^2} \qquad \text{and} \qquad R_B = \frac{Wa^2}{L^3}(2b+L)$$

(The formulae for M_A and M_B are useful in numerical examples and should be committed to memory.)

The position of the maximum deflection coincides with a point of zero slope and is clearly between the end A and the load. Substituting for M_A and R_A and putting $dy/dx = 0$ we have, from (i), omitting the terms in square brackets,

$$0 = -\frac{Wb^2}{L^3}(2a+L)\tfrac{1}{2}x^2 + \frac{Wab^2}{L^2}x$$

from which

$$x = 0 \quad \text{or} \quad x = 2aL/(2a+L) \qquad \text{(v)}$$

Substituting the latter value (which is clearly the required one) in (ii) and omitting the terms in square brackets, the maximum deflection is

$$y_{max} = \frac{1}{EI}\left[-\frac{Wb^2}{6L^3}(2a+L)\left(\frac{2aL}{2a+L}\right)^3 + \frac{Wab^2}{2L^2}\left(\frac{2aL}{2a+L}\right)^2\right]$$

$$= \frac{1}{EI}\left[-\frac{4Wa^3b^2}{3(2a+L)^2} + \frac{2Wa^3b^2}{(2a+L)^2}\right]$$

$$= \frac{2Wa^3b^2}{3(2a+L)^2EI}$$

From (v), the distance of the point of maximum deflection from A is

$$x = \frac{2aL}{2a+L} = \frac{2L}{2+(L/a)}$$

The maximum value of this expression clearly corresponds to $a = L$. In this case the maximum deflection occurs at a point $\frac{2}{3}L$ from A. Hence it is always within $L/6$ of the centre.

6.9. A beam spans 6 m and is rigidly built-in at each end. It carries point loads of 60 kN and 80 kN at distances of 2 m and 5 m respectively from one end. Calculate the fixing moments and reactions, and draw, to scale, the shearing force and bending moment diagrams for the beam. How far from the ends are the points of contraflexure?

Solution. The fixing moments can be calculated by the formulae of Example 6.8, using the principle of superposition.

The left-hand end (A) is taken as shown in Fig. 6.10.

For the 60 kN load alone, a = 2 m and b = 4 m,

$$M_A = \frac{Wab^2}{L^2} = \frac{60\text{ kN} \times 2\text{ m} \times (4\text{ m})^2}{(6\text{ m})^2} = 53{\cdot}3\text{ kN-m}$$

$$M_B = \frac{Wa^2b}{L^2} = \frac{60\text{ kN} \times (2\text{ m})^2 \times 4\text{ m}}{(6\text{ m})^2} = 26{\cdot}7\text{ kN-m}$$

Similarly for the 80 kN load alone

$$M_A = \frac{80\text{ kN} \times 5\text{ m} \times (1\text{ m})^2}{(6\text{ m})^2} = 11{\cdot}1\text{ kN-m}$$

$$M_B = \frac{80\text{ kN} \times (5\text{ m})^2 \times 1\text{ m}}{(6\text{ m})^2} = 55{\cdot}5\text{ kN-m}$$

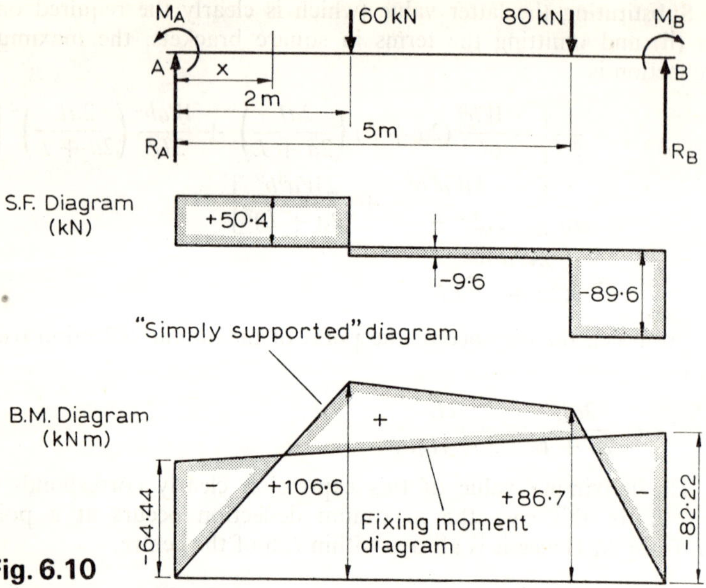

Fig. 6.10

For the two loads, therefore,

$$M_A = 53{\cdot}3 + 11{\cdot}1 = 64{\cdot}4 \text{ kN-m}$$
$$M_B = 26{\cdot}7 + 55{\cdot}5 = 82{\cdot}2 \text{ kN-m}$$

Taking moments about A to find R_B (anti-clockwise = clockwise),

$$M_A + (R_B \times 6 \text{ m}) = M_B + (60 \text{ kN} \times 2 \text{ m}) + (80 \text{ kN} \times 5 \text{ m})$$
$$(R_B \times 6 \text{ m}) = M_B - M_A + 120 \text{ kN-m} + 400 \text{ kN-m}$$

Substituting for M_A and M_B we find

$$R_B = 89{\cdot}6 \text{ kN} \qquad (Ans)$$

Similarly, taking moments about B,

$$R_A = 50{\cdot}4 \text{ kN} \qquad (Ans)$$

The shearing force diagram can now be drawn in the usual way (Fig. 6.10). The bending moment diagram is obtained by the method of superposition. The diagram due to the fixing moments alone is added to that due to the same loads acting on a similar but simply-supported beam.

In calculating values for the "simply-supported" diagram it must be remembered that the reactions obtained above apply only to the built-in case. For a similar freely supported beam, the reactions are found to be

$$R_A = 53{\cdot}3 \text{ kN} \quad \text{and} \quad R_B = 86{\cdot}7 \text{ kN}$$

With these values the "simply-supported" bending moments at the 60 kN and 80 kN loads are (+) 106·6 kN-m and (+) 86·7 kN-m respectively.

The "fixing moment" diagram is a straight line from (−) M_A at A to (−) M_B at B. (The minus signs are necessary since the bending moment due to the fixing couples is everywhere "hogging".) The diagrams are shown in Fig. 6.10.

There are clearly two points of contra-flexure (zero B.M.) Suppose the one nearer A is a distance x from this end ($x < 2$ m). For zero bending moment at this point,

$$R_A x - M_A = 0 \qquad \text{(i)}$$

(R_A being the reaction in the "fixed" case.) Therefore,

$$x = M_A/R_A = 64{\cdot}4 \text{ kN-m}/50{\cdot}4 \text{ kN} = 1{\cdot}28 \text{ m} \qquad (Ans)$$

Similarly, the distance of the other point of contra-flexure from B is

$$M_B/R_B = 82{\cdot}2 \text{ kN-m}/89{\cdot}6 \text{ kN} = 0{\cdot}92 \text{ m} \qquad (Ans)$$

(If the required point did not lie between the end and the nearest load, an extra term (or terms) would be required in equation (i).)

PROBLEMS

1. A uniform cantilever 3 m long is propped at the free end to the level of the fixed end. Calculate

(*a*) The prop reaction for a load of 160 kN uniformly distributed throughout.
(*b*) The prop reaction when the same load is concentrated at the centre.
(*c*) The position of the point of contra-flexure when a concentrated load is carried at a distance of 1 m from the fixed end.
(*d*) The maximum load which can be carried at a point 2 m from the fixed end if the prop reaction must not exceed 80 kN.

Answer. (*a*) 60 kN, (*b*) 50 kN. (*c*) 0·65 m from the fixed end. (*d*) 154·3 kN.

2. A uniform beam, 4 m long, is simply supported at its ends and is propped to the same level at mid-span. Calculate the load on the prop and the positions of the points of contra-flexure if

(*a*) The beam carries a load of 160 kN uniformly distributed throughout.
(*b*) The beam carries concentrated loads of 40 kN (each) at distances 1 m from each end.

Answer. (*a*) 100 kN; 1·5 m from each end. (*b*) 55 kN; 1·45 m from each end.

3. A uniform beam, 10 m long, is simply supported at its ends. Under a central load of 80 kN the maximum deflection is 50 mm and the maximum bending stress is 90 MN/m^2. Calculate

(*a*) The maximum deflection and bending stress for an equal beam built-in at its ends and carrying a load of 200 kN uniformly distributed throughout.
(*b*) The maximum deflection for a beam of the same flexural rigidity but 4 m long, built-in at its ends and carrying a central point load which produces a maximum bending stress of 135 MN/m^2.

Answer. (*a*) 15·63 mm; 75 MN/m^2. (*b*) 6 mm.

4. A uniform beam ($I = 8000$ cm^4) is 6 m long and carries a central point load of 50 kN. Taking $E = 200$ GN/m^2 calculate the deflection under the load if

(*a*) The beam is simply supported at its ends.
(*b*) The beam is rigidly built-in at its ends.
(*c*) The beam is built-in at one end and simply supported to the same level at the other.

Answer. (*a*) 14·06 mm. (*b*) 3·52 mm. (*c*) 6·15 mm.

5. A beam 6 m long is supported at its ends A and B on the same level and at its centre point C. When loaded with a uniformly distributed load of 64 kN/m run over the whole span, the support at C sinks. Calculate the amount of this settlement at C in order that there should be no bending stresses at this point, and also determine the three reactions. What would be the slope of the beam at C.

Sketch the bending moment diagram. $E = 200$ GN/m^2; $I = 8000$ cm^4.

(*I.Struct.E.*)

Answer. 13·5 mm; $R_A = R_B = 96$ kN, $R_C = 192$ kN; slope at C is zero; B.M. diagram is symmetrical about C, each half being a parabola. B.M. is zero at A, C, and B, and has a maximum value of 72 kN-m at 1·5 m from each end.

6. State Mohr's Theorems as applied to the flexure of beams.

A horizontal cantilever 3 m long is rigidly fixed at one end A, and supported at the same level at the other end B. There is a load of 50 kN at the centre point. Calculate

(*a*) the fixing moment at A,
(*b*) the reaction at B,
(*c*) the slope at B (in terms of *EI*).

Sketch the bending moment diagram. (*I.Struct.E.*)

Answer. (*a*) 28·1 kN-m. (*b*) 15·6 kN. (*c*) 14·06/*EI* rad. B.M. varies linearly from (−) 28·1 kN-m at A to (+) 23·4 kN-m under the load, then linearly to zero at B.

7. Figure 6.11 shows a steel strip 25 mm by 2 mm arranged as a cantilever with a load of 3 N at the free end. A roller 50 mm diameter, is placed with its axis at a distance x from the fixed end of the cantilever. Find the pressures exerted by the strip on the roller when x is 50 mm and 150 mm. Also find the deflection at the free end when $x = 100$ mm. $E = 200$ GN/m^2. (*I.Mech.E.*)

Answer. 16·5 N and 4·5 N; 0·525 mm.

8. A beam of uniform section is built-in at each end so as to have a clear span of 6 m. It carries a uniformly distributed load of 24 kN/m on the left-hand half of the beam, together with a 120 kN load at 4½ m from the left-hand end. Find the reactions and fixing moments at the ends and draw a bending moment diagram for the beam inserting the principal values. (*U.L.*)

Hint. Use the Macaulay method with an origin at the right-hand end.

Answer. At left-hand end, reaction = 77·3 kN and fixing moment = 83·3 kN-m; at right-hand end, reaction = 114·8 kN and fixing moment = 123·8 kN-m.

B.M. diagram (kN-m): Parabola from −83·3 at left-hand end to (+) 40·5 at mid-span; straight line to (+) 48·4 under the 120 kN load; straight line to −123·8 at right-hand end.

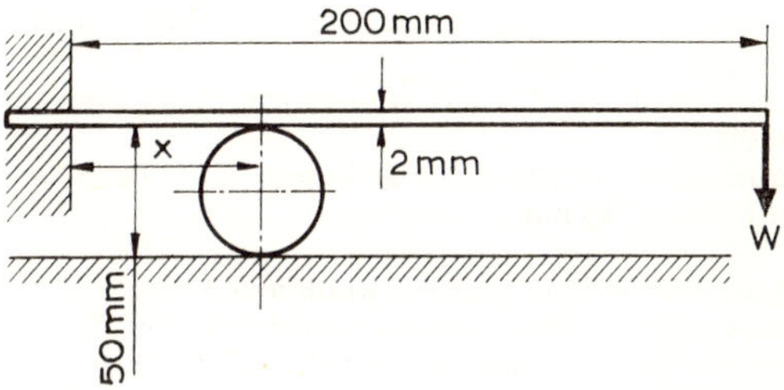

Fig. 6.11

9. A circular steel pipe 450 mm internal diameter and 6 mm thick is supported freely at each end over a span of 15 m. There is also a support at the middle of the span initially at the same level as the end supports. When the pipe is full of water the central support sinks 12 mm below the ends.

Find the pressure on each of the three supports after the settlement and draw the bending moment diagram for the pipe. Determine also the maximum bending stress in the pipe.

Specific weight of steel = 75·5 kN/m³.
Specific weight of water = 9·81 kN/m³.
E for steel = 206 GN/m². (*U.L.*)

Answer. Pressure on central support, 12·85 kN; on each end support, 10·14 kN.

B.M. diagram (kN-m): Parabola from zero at (each) end to (+) 13·9 at centre passing through a maximum of (+) 23·3 at 4·59 m from (each) end. Maximum bending stress in the steel, 24·1 MN/m².

10. A girder of 12 m span is fixed horizontally at the ends. A downward vertical load of 108 kN acts on the girder at a distance of 4 m from the left-hand end and an upward vertical force of 80 kN acts at a distance of 6 m from the right-hand end. Determine the end reactions and fixing couples and draw the bending moment and shearing force diagrams for the girder. (*U.L.*)

Answer. At left-hand end, reaction is 40 kN and fixing moment is 72 kN-m (anti-clock). At right-hand end reaction is 12 kN (downwards) and fixing moment is 24 kN-m (anti-clock). S.F. (kN): Between the left-hand end and 108 kN load, (+) 40; between 108 kN load and 80 kN force, (−) 68; between 80 kN force and right-hand end, (+) 12.

B.M. (kN-m): −72 at left-hand end, (+) 88 at 108 kN load, −48 at 80 kN force, (+) 24 at right-hand end. (Straight line in each range.)

11. A horizontal bar of 60 mm diameter is rigidly fixed at each end, the fixings being 1·2 m apart. A rigid bracket is fixed to the middle of the bar, at right angles to its axis and in the same horizontal plane. Determine the maximum radius arm of the bracket at which a vertical load of 1 kN can be suspended if the deflection of the load is not to exceed 0·5 mm.

($E = 200 \times 10^9$ N/m²; $G = 80 \times 10^9$ N/m².) (*U.L.*)

Hint. The deflections due to bending and twisting can be considered separately.
Answer. 382 mm.

12. A steel girder 248 mm deep has a span of 10 m and is rigidly built-in at both ends. The loading on the girder consists of a uniformly distributed load of 24 kN/m on the whole span and three equal point loads at the centre and the quarter-span points. Find the magnitudes of the point loads if the maximum stress due to bending is 120 MN/m².

The section of the girder is symmetrical about both the principal axes and $I_{xx} = 82000$ cm⁴.

Determine also the maximum deflection. $E = 206$ GN/m². (*U.L.*)
Answer. 63 kN; 7·59 mm.

13. A long steel strip of uniform width and 3 mm thick is laid on a level floor, but passes over a 50 mm diameter roller lying on the floor at one point. For what distance on either side of the roller will the strip be clear of the floor, and what will be the maximum stress induced, if the density of steel is 7700 kg/m³? ($E = 206 \times 10^9$ N/m²). (*U.L.*)

Hint. Treat the strip as a fixed-ended beam carrying a U.D.L. together with an upward central load. The span is the total distance for which the strip is clear of the floor and the net fixing moment at each end is zero.
Answer. 1·65 m; 68·3 MN/m².

14. Figure 6.12 shows a cantilever of uniform section and length L; it is propped at the free end and carries a load W at a distance l from the fixed end.

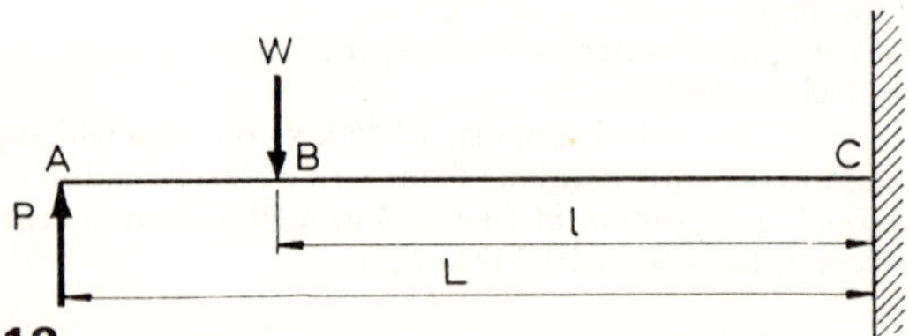

Fig. 6.12

(*a*) If the prop is removed show that the deflection at A is

$$\frac{Wl^2}{6EI}(3L - l).$$

(*b*) If the prop is in position as shown in Fig. 6.12 and the magnitude of P is such that there is no deflection at A, show that

$$P = \frac{Wl^2}{2L^3}(3L - l).$$

(*c*) In a particular case, $L = 5$ m, $l = 4$ m and $W = 75$ kN. Obtain for this case the bending moment diagram showing on it the magnitude of the maximum positive and maximum negative bending moments. (*U.L.*)

Answer. 52·8 and −36·0 kN-m; B.M. diagram is a straight line from 0 at A to +52·8 kN-m at B, then a straight line to −36·0 kN-m at C.

15. A cantilever of uniform flexural rigidity EI has a length of 3 m and is propped at a distance a from the wall. There is a uniformly distributed load of w kN per m on the overhanging length between the prop and the free end of the cantilever.

If the load on the prop is wa and the prop is level with the fixed end of the cantilever, find the value of a.

Draw, on squared paper, the bending moment diagram for the cantilever for $w = 25$ kN/m. *(U.L.)*

Answer. $a = 1{\cdot}8$ m; parabola from 0 at free end to -18 kN-m at prop, then straight line to $+9$ kN-m at fixed end.

16. A steel beam of 11 m span is built-in at both ends and carries two point loads each of 80 kN at points 3 m from the ends of the beam. The middle 3 m has a section for which the second moment of area is 24000 cm^4 and the 3 m lengths at either end have a section for which the second moment of area is 32000 cm^4. Find the fixing moments at the ends and calculate the deflection at mid-span. Take $E = 200$ GN/m^2 and neglect the weight of the beam. *(U.L.)*

Answer. 180 kN-m; 12·2 mm.

17. A beam 10 m long carrying a uniformly distributed load of 50 kN/m is lifted on three hydraulic jacks which are supplied by a common pressure line. One jack is under each end and one under the mid-point; the area of the ram of the middle jack is three times that of the outer ones.

Sketch the shearing force and bending moment diagrams for the beam showing on them all maximum values both positive and negative.

Determine also relative to the level at the ends (*a*) the level of the beam at the centre, and (*b*) the position of maximum deflection and hence the value of the maximum deflection. An approximate solution of the equation will suffice. EI for the beam = 30 $MN\text{-}m^2$. *(U.L.)*

Answer. S.F. diagram (kN): straight line from $+100$ at left-hand end to -150 at centre; straight line from $+150$ at centre to -100 at right-hand end; B.M. diagram (kN-m); parabola from 0 at left-hand end to -125 at centre, passing through a maximum of $+100$ at 2 m from the left-hand end, symmetrical about centre.

(*a*) 8·68 mm downwards; (*b*) 2·79 m from left-hand end; 11·5 mm downwards.

18. A beam 6 m long is mounted on three springs, one at each end and one at the mid-point. Initially the beam is straight and it is desired that under a uniform load of 1·2 kN per m (which includes the weight of the beam) the mid-point shall be level with the ends.

The springs used at the ends have a stiffness of 60 kN/m and the third spring a stiffness of 120 kN/m. The springs rest on a horizontal bed and their unloaded lengths are the same but the central spring is packed up to secure the above conditions.

(*a*) Find the thickness of the packing required and (*b*) sketch the shearing force and bending moment diagrams for the beam, showing on them all positive and negative maximum values.

The value of the relevant EI for the beam is 2·6 $MN\text{-}m^2$. *(U.L.)*

Answer. (*a*) 15 mm. (*b*) S.F. (kN): straight line from $+1{\cdot}35$ at left-hand end to $-2{\cdot}25$ at centre; straight line from $+2{\cdot}25$ at centre to $-1{\cdot}35$ at right-hand end. B.M. (kN-m): parabola from 0 at left-hand end to $-1{\cdot}35$ at centre passing through a maximum of $+0{\cdot}76$ at $1\frac{1}{8}$ m from left-hand end; symmetrical about centre.

19. A beam 9 m long is encastré at its end supports and carries two loads each of 40 kN symmetrically situated at $1\frac{1}{2}$ m on each side of mid-span. The left-hand end of the beam sinks below the level of the right-hand end by 2·5 mm without tilting at either end. Find the fixing moments and reactions at the ends and sketch the shearing force and bending moment diagrams for the beam. EI for the beam is 25 MN-m². (*U.L.*)
Answer. Fixing moments (kN-m) 75·4 and 84·6; reactions (kN) 39·0 and 41·0; S.F. (kN), (x in m from left-hand end): from $x = 0$ to $x = 3$, 39·0; from $x = 3$ to $x = 6$, −1·0; from $x = 6$ to $x = 9$, −41·0.

B.M. (kN-m): straight line from −75·4 at $x = 0$ to +41·5 at $x = 3$; straight line from this point to +38·5 at $x = 6$; straight line from this point to −84·6 at $x = 9$.

20. A horizontal uniform beam is firmly built-in at both ends to span a distance l and has a frictionless hinge at mid-span. The beam carries a vertical load W at a distance $l/5$ from one end. Determine the values of the fixing moments at the ends and the force acting at the hinge and sketch the bending moment and shearing force diagrams for the beam. (*U.L.*)
Answer. Fixing moments are $41Wl/250$ (nearer the load) and $9Wl/250$; force at hinge $= 9W/125$.

21. A steel beam ABCD of uniform section is firmly built-in at the ends A and D and propped at the mid-span point B so that at B the beam is level with the ends A and D. The effective span AD = 8 m; AB = 4 m, BC = 1 m and CD = 3 m. The beam carries a point load of 80 kN at C. Determine the bending moments and reactions at A, B and D and the bending moment at C. Sketch the bending moment diagram for the beam showing on it the calculated values. (*U.L.*)
Answer. $M_A = 12{\cdot}5$ kN-m, $M_B = -21{\cdot}25$ kN-m, $M_D = -26{\cdot}25$ kN-m; $R_A = -8{\cdot}4$ kN, $R_B = 67{\cdot}5$ kN, $R_D = 20{\cdot}9$ kN. $M_C = 34{\cdot}7$ kN-m.

22. A uniform section beam ABC is fixed at A and simply supported at B and C. The spans AB and BC are 20 m and 10 m long respectively. A point load of 10 kN is applied to the mid-point of the span AB and the span BC supports a uniformly distributed load of 1 kN/m over its entire length. Sketch the bending moment and shear force diagrams for the whole beam indicating the principal change points. (*U.L.*)
Answer. S.F. (kN): Between A and the 10 kN load, $+5\frac{3}{8}$; between this load and B, $-4\frac{5}{8}$; straight line from +7 at B to −3 at C.

B.M. (kN-m): Straight line from $-27\frac{1}{2}$ at A to $+26\frac{1}{4}$ at 10 kN load; straight line from $+26\frac{1}{4}$ at this load to −20 at B; parabola from −20 at B to 0 at C passing through a maximum of $+4\frac{1}{2}$ at 3 m from C.

Chapter 7

Axially-loaded Struts

In the following formulae, $L =$ length, $E =$ Young's modulus, and $I =$ second moment of area (moment of inertia) about the neutral axis.

The Euler crippling load (P_E) is:

(*a*) One end fixed, one end (position and direction) free,

$$P_E = \pi^2 EI/4L^2$$

(*b*) Both ends pin-jointed (direction-free),

$$P_E = \pi^2 EI/L^2$$

(*c*) One end pin-jointed, one end (position and direction) fixed,

$$P_E = 2\pi^2 EI/L^2$$

(*d*) Both ends (position and direction) fixed,

$$P_E = 4\pi^2 EI/L^2$$

The Rankine crippling load is

$$P_R = \frac{\sigma_c A}{1 + a(L/k)^2}$$

where $A =$ cross-sectional area (m^2),
$k =$ radius of gyration of cross-section (same units as L),
$\sigma_c =$ compressive yield stress (N/m^2),
$a =$ a numerical constant.

WORKED EXAMPLES

7.1. Define the terms "strut" and "column" and distinguish between "slender" and "stocky" struts.

Derive (approximate) formulae for the buckling load of a uniform strut assuming that it bows to a curve which is the same as that of a beam subjected to

(*a*) a central point load,
(*b*) a uniformly distributed load.

Solution. A *strut* is a bar subjected to a longitudinal compressive load (as opposed to a *tie* which is a bar in tension). The term *column* is reserved for vertical struts, as are the words *pillar* and *stanchion.*

A *slender* strut is one which is long in comparison with its cross-sectional dimensions. A *stocky* strut is one which is short in comparison with its cross-sectional dimensions.

By common experience, slender struts bend or bow under load. The bending moment at any section equals the product of the load and its eccentricity from the neutral axis of that section. Bending would not take place if the strut were perfectly straight and uniform and if the load were absolutely central. Since, however, there are always imperfections, it is necessary to investigate the stability of the strut in the bent state. The *buckling* (or crippling) load is defined as the end load which will maintain the strut in its bent form.

(*a*) Let δ be the deflection at the middle of the beam or strut,

W be the central point load for the beam,
P be the buckling load for the strut.

Then $\delta = WL^3/48EI$ and, for the internal equilibrium at mid-span, bending moment for the strut = bending moment for the beam, or

$$P\delta = WL/4$$

i.e.

$$PWL^3/48EI = WL/4$$

Hence

$$P = \frac{12EI}{L^2} \qquad \text{(Ans)}$$

This result is independent of the amount of bowing and the strut is therefore in a state of neutral equilibrium.

(*b*) In this case, $\delta = 5WL^3/384EI$, W being the total distributed load, and the maximum bending moment for the beam is $WL/8$.

Therefore,

$$P\delta = WL/8 \quad \text{or} \quad 5PWL^3/384EI = WL/8$$

$$P = \frac{9{\cdot}6EI}{L^2} \qquad \text{(Ans)}$$

7.2. Derive the Euler formula for the buckling load of a pin-ended strut, enumerating carefully the assumptions on which it is based.

Explain briefly why so little use is made of Euler's formula in practical design. (*U.L.*)

Solution. The Euler theory is based on the following assumptions:

(*a*) the strut is initially perfectly straight;
(*b*) the load is applied axially;
(*c*) the strut is very long in comparison with its cross-sectional dimensions;
(*d*) the assumptions made in the theory of bending hold good, i.e. the strut is uniform throughout and the limit of proportionality is not exceeded.

Consider a pin-ended (or pin-jointed) strut AB length L, subjected to an end compressive load P (Fig. 7.1). Suppose that, for an unspecified reason, the strut bows so that at any section XX, distance x

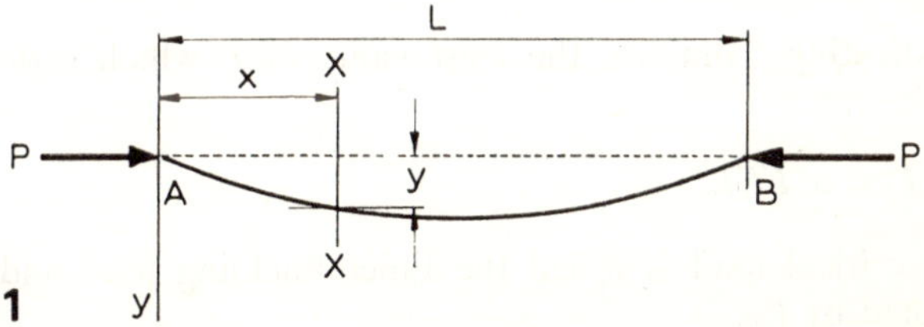

Fig. 7.1

from A, the deflection is y, as shown. The bending moment at this section is Py and substituting in the differential equation of flexure,

$$EI\frac{d^2y}{dx^2} = -M = -Py$$

$$\frac{d^2y}{dx^2} + \frac{P}{EI}y = 0$$

It is convenient to write this in the form

$$\frac{d^2y}{dx^2} + \alpha^2 y = 0 \qquad \text{(i)}$$

where

$$\alpha^2 = P/EI \qquad \text{(ii)}$$

The solution of (i) is

$$y = A\cos\alpha x + B\sin\alpha x$$

where A and B are constants.

This solution can be checked by differentiating twice to obtain d^2y/dx^2 and substituting in (i). But $y = 0$ when $x = 0$ and therefore $A = 0$. Also, $y = 0$ when $x = L$ and therefore

$$B \sin \alpha L = 0$$

This means that either $B = 0$, which is impossible since there is then no deflection at all, or

$$\sin \alpha L = 0$$

The latter alternative is the only possible one and therefore

$$\alpha L = 0, \pi, 2\pi, 3\pi, \ldots$$

The first value is rejected on the grounds that it again leads to zero deflection throughout. The *least* value which satisfies all the conditions of the problems is, therefore

$$\alpha L = \pi \qquad \text{or} \qquad \alpha^2 = \pi^2/L^2$$

Substituting from (ii), the least value of P which satisfies all the conditions is

$$P = \pi^2 EI/L^2$$

This critical load is called the Euler buckling load and is usually denoted by P_E.

The Euler formula is little used in practical design for two main reasons:

(*a*) It applies to an ideal strut only and, in practice, there is inevitably some crookedness in the strut or eccentricity of the load.

(*b*) It takes no account of the direct stress conditions. Whereas the direct stress in slender struts is small, it becomes of prime importance in stocky struts. This means that the Euler formula may give a buckling load for such struts far in excess of the load which they can withstand under direct compression. (See also Example 7.6.)

7.3. A uniform round bar, 1·8 m long, is used as a simply supported beam over a span of 1·2 m and deflects 6 mm under a central load of 6 N. Find the Euler crippling load for this bar when used as a strut with "free" ends.

Solution. For the simply-supported beam case ($L = 1{\cdot}2$ m) the maximum deflection is

$$\delta = WL^3/48EI$$

Rearranging, and substituting the values given,

$$EI = WL^3/48\delta$$
$$= \frac{(6 \text{ N}) \times (1{\cdot}2 \text{ m})^3}{48 \times (6 \times 10^{-3} \text{ m})}$$
$$= 36 \text{ N-m}^2$$

The term "free" ends means pin-jointed ends, which are "free" in direction.

The Euler crippling load (L now being $1{\cdot}8$ m) is

$$P_E = \pi^2 EI/L^2 = \frac{\pi^2 \times (36 \text{ N-m}^2)}{(1{\cdot}8 \text{ m})^2}$$
$$= 110 \text{ N} \qquad (Ans)$$

7.4. Enumerate the assumptions on which the Euler formula for a strut is based.

A straight bar of steel, 1 m long and 20 mm by 6 mm in section, is mounted in a testing-machine and loaded axially in compression till it buckles. Assuming the Euler formula for pinned ends to apply, estimate the maximum central deflection before the material attains its yield stress of 350 MN/m². Take $E = 200$ GN/m². (*U.L.*)

Solution. The assumptions are given in Example 7.2. Buckling will naturally take place about the axis having the least value of I. Hence

$$I = bd^3/12 = (20 \text{ mm}) \times (6 \text{ mm})^3/12 = 0{\cdot}36 \times 10^{-9} \text{ m}^4$$

The Euler buckling load is, therefore,

$$P_E = \pi^2 EI/L^2$$
$$= \pi^2 \times (200 \times 10^9 \text{ N/m}^2) \times (0{\cdot}36 \times 10^{-9} \text{ m}^4)/(1 \text{ m})^2$$
$$= 710 \text{ N}$$

The principle of eccentric loading can now be used. The direct stress is

$$P_E/\text{area} = 710 \text{ N}/(20 \times 10^{-3} \text{ m}) \times (6 \times 10^{-3} \text{ m})$$
$$= 5{\cdot}92 \text{ MN/m}^2 \text{ (compressive)}$$

When the yield stress is reached, the maximum bending stress is

$$350 - 5{\cdot}92 = 344{\cdot}1 \text{ MN/m}^2$$

The corresponding bending moment is

$$M = \frac{\sigma_{max}}{y_{max}} I = \frac{(344{\cdot}1 \times 10^6 \text{ N/m}^2)}{(3 \times 10^{-3} \text{ m})} \times (0{\cdot}36 \times 10^{-9} \text{ m}^4)$$
$$= 41{\cdot}3 \text{ N-m}$$

The maximum deflection equals the maximum eccentricity of the load and hence

$$\delta = M/P_E = 41{\cdot}3 \text{ N-m}/710 \text{ N} = (58{\cdot}2 \times 10^{-3}) \text{ m}$$
$$= 58{\cdot}2 \text{ mm} \qquad (Ans)$$

7.5. Explain how the Euler formula can be modified for struts having one or both ends built-in.

A 2·5 m length of tube has a crippling load of 1·2 kN when used as a strut with pin-jointed ends. Calculate the crippling load for a 3 m length of the same tube when used as a strut if

(i) both ends are "fixed",
(ii) one end is "fixed" and the other is (position and direction) "free".

Solution. In order of strength, the four cases usually considered are

(a) *One end "fixed", one end (position and direction) "free".* This is the case (Fig. 7.2a) of a strut built-in at one end but free to take up any position and direction at the other. It is

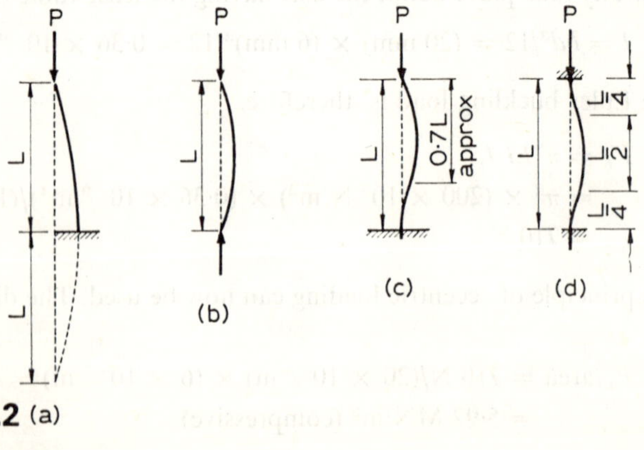

Fig. 7.2 (a)

clear from the diagram that the strut behaves in exactly the same way as a pin-jointed strut having twice the length. Replacing L by $2L$ in the Euler formula for pin-ended struts, the buckling load is

$$P = \frac{\pi^2 EI}{(2L)^2} = \frac{\pi^2 EI}{4L^2}$$

(*b*) *Both ends "free"*. This is the pin-jointed strut (Fig. 7.2b) already dealt with in Example 7.2.

(*c*) *One end "fixed", one end pin-jointed.* In this case (Fig. 7.2c) both ends are fixed in position and one is also fixed in direction. It appears that the "equivalent" pin-jointed strut is about $\frac{2}{3}$ of the length of the given strut. It can be shown that the equivalent pin-jointed length is approximately $0{\cdot}7L$, and using this value, the buckling load is

$$P = \frac{\pi^2 EI}{(0{\cdot}7L)^2} = \frac{2\pi^2 EI}{L^2} \text{ (approximately)}$$

(A formal analysis of this case is given in Example 7.8.)

(*d*) *Both ends "fixed" (in position and direction)*. In this case (Fig. 7.2d) the middle half of the strut behaves in the same way as a pin-jointed strut, the end portions each corresponding to a strut of type (*a*). The equivalent pin-jointed length is $L/2$ and the buckling load is

$$P = \frac{\pi^2 EI}{(\frac{1}{2}L)^2} = \frac{4\pi^2 EI}{L^2}$$

Since, for all cases, the crippling load is inversely proportional to the square of the length, then for a pin-jointed 3 m length of the given tube the Euler load is

$$\left(\frac{2{\cdot}5}{3}\right)^2 \times 1{\cdot}2 \text{ kN} = 833{\cdot}3 \text{ N}$$

(i) This is case (*d*) above and the crippling load is therefore four times the pin-jointed value, i.e.

$$4 \times 833{\cdot}3 \text{ N} = 3333 \text{ N} \qquad (Ans)$$

(ii) This is case (*a*) and the crippling load is one-quarter of the pin-jointed value, i.e.

$$\tfrac{1}{4} \times 833{\cdot}3 \text{ N} = 208{\cdot}3 \text{ N} \qquad (Ans)$$

7.6. Explain the term *slenderness ratio* and show that, for a given material, the Euler critical stress depends solely on this ratio.

Determine the least value of the ratio for which the Euler formula can apply in the case of a pin-jointed steel strut having a yield stress of 300 MN/m². $E = 200$ GN/m².

Solution. For a pin-jointed strut of cross-sectional area A, the stress σ_E corresponding to the Euler critical load is

$$\sigma_E = \frac{P_E}{A} = \frac{\pi^2 EI}{AL^2}$$

But $I = Ak^2$ where k is the *radius of gyration* of the cross-section (see Appendix 1). Thus the stress may be written

$$\sigma_E = \frac{\pi^2 EAk^2}{AL^2} = \frac{\pi^2 E}{(L/k)^2} \qquad \text{(i)}$$

The fraction L/k is called the *slenderness ratio* and, as the expression shows, the value of σ_E decreases as L/k increases. It should be noted that k must correspond to the appropriate plane of bending. If the strut has frictionless spherical joints at its ends then bending is possible in all planes containing the undeflected axis of the strut. In this case the least value of k for the cross-section must be used and this corresponds to the least value of I. If, however, the strut is pin-jointed then bending in only one plane is possible and this determines the value of k to be used.

For a given material, E is constant and, from (i), σ_E depends only on the value of L/k, the slenderness ratio. The variation of σ_E with L/k is illustrated by the curve of Fig. 7.3. However, the Euler analysis

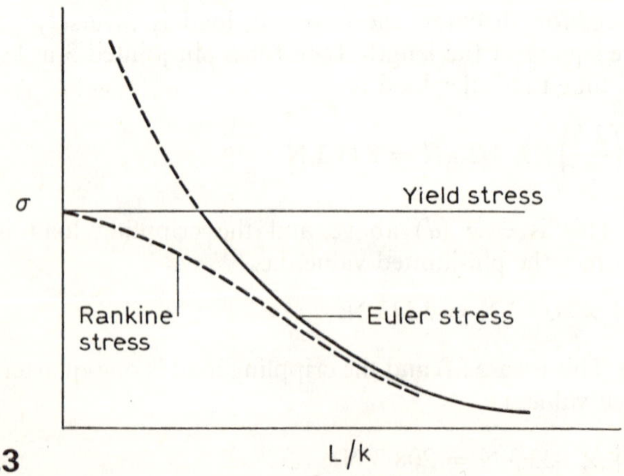

Fig. 7.3

allows for bending only and, below a certain value of L/k, it leads to values greater than the yield stress in direct compression. Clearly, the Euler formula cannot be applied below this value.

In the case of the given steel strut the value of the stress will be the given yield stress σ_Y and thus, from (i),

$$(L/k)^2 = \frac{\pi^2 E}{\sigma_Y} = \frac{\pi^2 \times 200 \times 10^9}{300 \times 10^6} = 6580$$

and therefore the least slenderness ratio for which the Euler formula can be applied is

$$L/k = \sqrt{6580} = 81{\cdot}1 \qquad (Ans)$$

This value separates the two modes of failure for the strut. Stockier struts (lower L/k) are able to resist the bending moments due to small lateral deflections or slight eccentricities of loading and ultimately fail by direct compression.

More slender struts (higher L/k), on the other hand, tend to be unstable in bending; with an end load greater than the Euler critical value they are not able to resist the bending moments due to lateral deflections or eccentricities of loading and failure is then due to excessive bending stress.

7.7. A long slender strut of length L is clamped at each end and subjected to an axial compressive force P. Derive from first principles an expression connecting the critical load and the slenderness ratio. Hence determine the slenderness ratio when the proportional limit of the steel is 210 MN/m² and $E = 200$ GN/m². (*U.L.*)

Solution. The critical load for an axially loaded strut clamped at its ends was obtained in Example 7.5 by comparison with the "ends pinned" case. Working now from first principles we must allow for fixing moments at the ends. Figure 7.4 illustrates the problem and, by symmetry,

$$M_1 = M_2 \quad \text{and} \quad R_1 = R_2 = 0$$

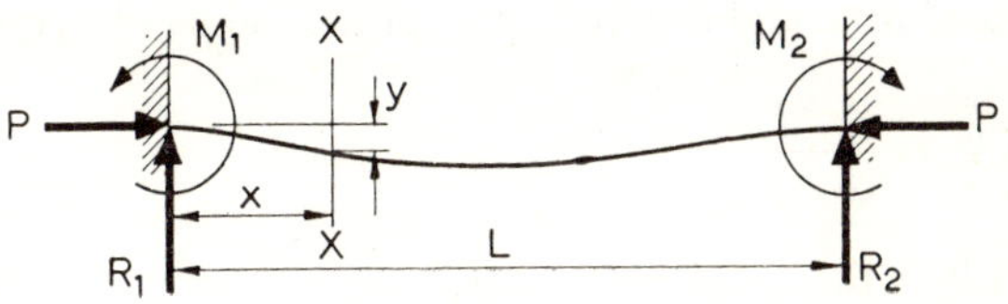

Fig. 7.4

Thus at section XX,

$$EI\frac{d^2y}{dx} = -Py + M_1$$

The solution to this differential equation is

$$y = A\cos\alpha x + B\sin\alpha x + M_1/P$$

where $\alpha^2 = P/EI$ as before. (This solution can be checked by substitution in (i)).

When $x = 0$, $y = 0$ from which $M_1/P = -A$.

Thus

$$y = A\cos\alpha x + B\sin\alpha x - A \tag{ii}$$

and, by differentiation,

$$dy/dx = -\alpha A\sin\alpha x + \alpha B\cos\alpha x$$

When $x = 0$, $dy/dx = 0$ (clamped ends) and thus $B = 0$.

Hence equation (ii) becomes

$$y = A(\cos\alpha x - 1)$$

When $x = L$, $y = 0$ and either $A = 0$ (in which case there is no deflection at all) or

$$\cos\alpha L = 1$$

$$\alpha L = 0, 2\pi, 4\pi, 6\pi \ldots$$

The first value is rejected because it again leads to zero deflection throughout. The least value satisfying all the conditions of the problem is therefore

$$\alpha L = 2\pi \qquad \alpha^2 = 4\pi^2/L^2$$

But $\alpha^2 = P/EI$ and therefore

$$P = 4\pi^2 EI/L^2$$

This is the load which will maintain the strut in its deflected form and is thus the critical load (P_{cr}). Since $I = Ak^2$ where A = cross-sectional area and k is the appropriate radius of gyration we have

$$\frac{P_{cr}}{A} = \frac{4\pi^2 E}{(L/k)^2}$$

and L/k is the *slenderness ratio.*

From the data given we can determine the value of L/k for which this result coincides with the proportionality limit. Substituting the figures gives

$$(210 \times 10^6 \text{ N/m}^2) = \frac{4\pi^2 \times (200 \times 10^9 \text{ N/m}^2)}{(L/k)^2}$$

$$(L/k)^2 = \frac{4\pi^2 \times 200 \times 10^9}{210 \times 10^6} = 3{\cdot}76 \times 10^4$$

$$L/k = 194 \qquad (Ans)$$

7.8. A straight uniform strut is both position-fixed and direction-fixed at one end and is position-fixed only at the other end where it is hinged. Derive a formula for the buckling load under ideal axial loading. You may assume the solution to the equation $\tan \theta = \theta$ is $\theta = 257\frac{1}{2}$ degrees.

Calculate the percentage error in finding the buckling load if the equivalent length for a freely hinged strut is assumed as $\frac{2}{3}L$, where L is the actual length of the strut. (*U.L.*)

Solution. Take the origin at the hinged end as shown in Fig. 7.5. There is a fixing moment at the clamped end only and, due to lack of symmetry, there are unequal reactions R_1 and R_2 at the supports.

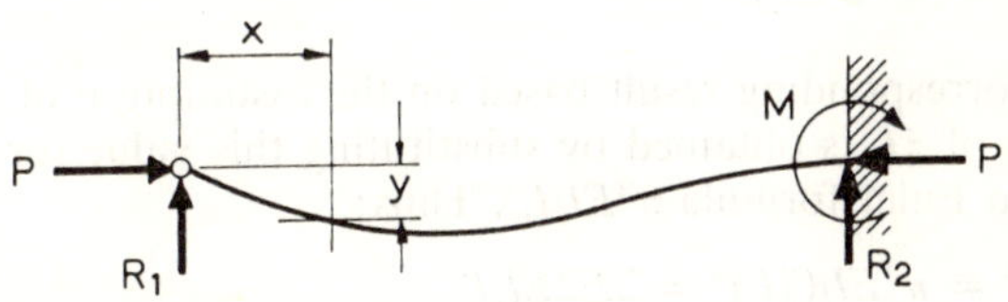

Fig. 7.5

The differential equation of flexure becomes

$$EI\frac{d^2y}{dx^2} = -M_x = -Py - R_1x$$

and the solution is

$$y = A \cos \alpha x + B \sin \alpha x - R_1x/P$$

where $\alpha^2 = P/EI$. The reader should check this solution by substitution in the differential equation.

When $x = 0$, $y = 0$ and therefore $A = 0$. Thus

$$y = B \sin \alpha x - R_1x/P \qquad \text{(i)}$$

and, by differentiation,

$$\frac{dy}{dx} = \alpha B \cos \alpha x - \frac{R_1}{P} \qquad \text{(ii)}$$

At the clamped end, $x = L$, $y = 0$ and $dy/dx = 0$. Thus from (i) and (ii),

$$B \sin \alpha L = R_1 L/P \qquad \text{(iii)}$$

and

$$\alpha B \cos \alpha L = R_1/P \qquad \text{(iv)}$$

Dividing (iii) by (iv) gives

$$\tan \alpha L = \alpha L$$

As usual, we reject the solution $\alpha L = 0$ on the grounds that it leads to no deflection whatever and the lowest value satisfying the problem is that given in the question. Thus

$$\alpha L = 257\tfrac{1}{2}° = 4{\cdot}494 \text{ rad}$$

and

$$\alpha^2 L^2 = PL^2/EI = 4{\cdot}494^2 = 20{\cdot}2$$

from which the critical load is

$$P = 20{\cdot}2\ EI/L^2 \qquad \text{(v)}$$

The corresponding result based on the assumption of an equivalent length of $\tfrac{2}{3}L$ is obtained by substituting this value for L in the pin-jointed Euler formula $\pi^2 EI/L^2$. Thus:

$$P = \pi^2 EI/(\tfrac{2}{3}L)^2 = 22{\cdot}2 EI/L^2 \qquad \text{(vi)}$$

A comparison of (v) and (vi) shows that the assumed value of the equivalent length leads to an error of

$$\frac{22{\cdot}2 - 20{\cdot}2}{20{\cdot}2} \times 100 = 9{\cdot}9 \text{ per cent} \qquad (\textit{Ans})$$

7.9. Show how a formula of the Rankine type can be developed from the Euler strut formula.

Compare the crippling loads given by the two formulae for the case of a strut 5·5 m long, the ends being "free". The minimum moment of inertia = 2400 cm^4 and the cross-sectional area = 105 cm^2. $E = 207$ GN/m^2.

In the Rankine formula take

$\sigma_Y = 310$ MN/m^2 and $a = 1/7500$

Solution. Take the case of a pin-jointed strut and let

P_E = Euler load ($\pi^2 EI/L^2$)

P_Y = the crippling load in compression for a very short strut (in which buckling is impossible). P_Y can be regarded as $\sigma_Y \times A$ where σ_Y is the yield stress (or limit of proportionality) in compression, and A the cross-sectional area.

P_R = Rankine load for a strut of any length.

It is reasonable to assume that the crippling load for a strut of any length will be given by P_R, where

$$\frac{1}{P_R} = \frac{1}{P_Y} + \frac{1}{P_E} \qquad \text{(iii)}$$

since, with this formula, P_R will always be less than either P_Y or P_E. For very short struts P_E is large and P_R will approach P_Y whereas for long struts $1/P_Y$ is small compared with $1/P_E$ and P_R will be near to the Euler value.

From (iii),

$$\frac{1}{P_R} = \frac{P_E + P_Y}{P_Y P_E} = \frac{1 + (P_Y/P_E)}{P_Y}$$

$$P_R = \frac{P_Y}{1 + (P_Y/P_E)} = \frac{\sigma_Y A}{1 + (\sigma_Y A L^2/\pi^2 EI)}$$

$$= \frac{\sigma_Y A}{1 + \dfrac{\sigma_Y}{\pi^2 E}\left(\dfrac{L}{k}\right)^2}$$

$$= \frac{\sigma_Y A}{1 + a(L/k)^2}$$

where $a = \sigma_Y/\pi^2 E$ which is a constant for a given material. The formula should be regarded as empirical and no attempt made to calculate a (using σ_Y and E). The constants σ_Y and a are found experimentally and will be given in examples where required. A similar demonstration can be given for other end conditions. For cases (*a*), (*c*), and (*d*) of Example 7.5 it will be found that the values of the constant a are respectively four times, one-half and one-quarter of its value for the pin-ended case. It is usual, however for the value given in the question to be the one appropriate to the end conditions.

The variation of the Rankine stress with L/k is indicated by the lower curve of Fig. 7.3. (Graphs of the two formulae to scale are required in Problem 5 on p. 217.)

For the strut in the question, Euler load

$$P_E = \pi^2 EI/L^2$$
$$= \pi^2 \times (207 \times 10^9\ \mathrm{N/m^2}) \times (2400 \times 10^{-8}\ \mathrm{m^4})/(5{\cdot}5\ \mathrm{m})^2$$
$$= 1621\ \mathrm{kN} \qquad (Ans)$$

$$k = \sqrt{\frac{I}{A}} = \sqrt{\frac{2400 \times 10^{-8}\ \mathrm{m^4}}{105 \times 10^{-4}\ \mathrm{m^2}}} = 47{\cdot}8\ \mathrm{mm}$$

Rankine load

$$P_R = \frac{\sigma_Y A}{1 + a(L/k)^2}$$
$$= \frac{(310 \times 10^6\ \mathrm{N/m^2}) \times (105 \times 10^{-4}\ \mathrm{m^2})}{1 + \dfrac{1}{7500}\left(\dfrac{5{\cdot}5\ \mathrm{m}}{47{\cdot}8 \times 10^{-3}\ \mathrm{m}}\right)^2}$$
$$= 1176\ \mathrm{kN} \qquad (Ans)$$

The Euler load is thus about 38 per cent greater than the Rankine.

7.10. A short length of tube 25 mm internal and 32 mm external diameter failed in compression at a load of 70 kN. When a 2·5 m length of the same tube was tested as a strut with fixed ends the failing load was 24·1 kN. Assuming that σ_Y in the Rankine formula is given by the first test, find the value of the constant *a* in the same formula.

Hence estimate the crippling load for a piece of the tube 1·5 m long when used as a strut with free ends.

Solution.

$$\text{Area of the tube} = \tfrac{1}{4}\pi(32^2 - 25^2)\ \mathrm{mm^2} = 314{\cdot}2\ \mathrm{mm^2}$$
$$I \text{ for the tube} = \tfrac{1}{64}\pi(32^4 - 25^4)\ \mathrm{mm^4} = 32290\ \mathrm{mm^4}$$
$$k = \sqrt{\frac{I}{A}} = \sqrt{\frac{32290\ \mathrm{mm^4}}{314{\cdot}2\ \mathrm{mm^2}}} = 10{\cdot}14\ \mathrm{mm}$$

If σ_Y is given by the first test then

$$\sigma_Y A = 70\ \mathrm{kN}$$

Using the Rankine formula with this value of σ_Y and the conditions of the second test,

$$P_R = \frac{\sigma_Y A}{1 + a(L/k)^2}$$
$$24{\cdot}1 \times 10^3\ \mathrm{N} = \frac{70 \times 10^3\ \mathrm{N}}{1 + a\,(2{\cdot}5\ \mathrm{m}/10{\cdot}14 \times 10^{-3}\ \mathrm{m})^2}$$

from which

$$a = 1/32000$$

(a is invariably given as a fraction).

For a pin-jointed strut the value of a is four times that of a "fixed" ended strut.

Hence, for the 1·5 m length of tube with "free" ends,

$$a = 4/32000 = 1/8000$$

and

$$P_R = \frac{\sigma_Y A}{1 + a(L/k)^2} = \frac{70 \times 10^3\ N}{1 + \dfrac{1}{8000}\left(\dfrac{1{\cdot}5\ \text{m}}{10{\cdot}14 \times 10^{-3}}\right)^2}$$

$$= 18{\cdot}74\ \text{kN} \qquad (Ans)$$

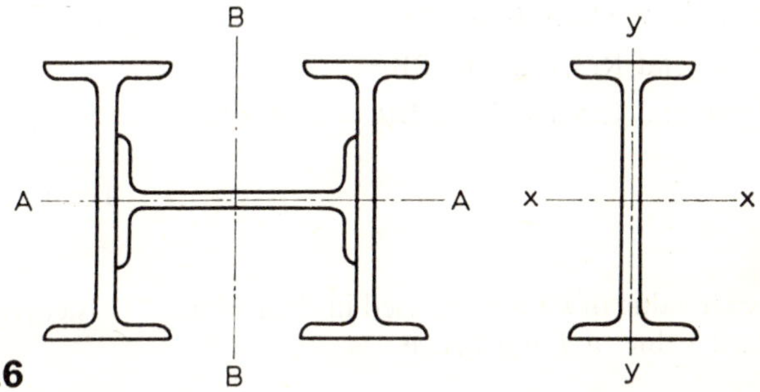

Fig. 7.6

7.11. A stanchion is built up of three 250 mm × 125 mm R.S.J.s as shown in Fig. 7.6. Calculate the second moments of area about the principal axes AA and BB. Properties of one 250 mm × 125 mm R.S.J. (see right-hand diagram of Fig. 7.6)

$I_{xx} = 6085\ \text{cm}^4$ $\qquad I_{yy} = 405\ \text{cm}^4$

Cross-sectional area = $57\ \text{cm}^2$

Web thickness = 9·15 mm

If the height of the stanchion is 6 m calculate the working load, the working stress being given by

$$\left[140 - \frac{1}{1{\cdot}3}\left(\frac{L}{k}\right)\right]\ \text{MN/m}^2$$

What factor of safety is to be used with the Rankine formula to give the same result? Take $\sigma_Y = 320$ MN/m² and $a = 1/7500$.

Solution.

I about AA:		cm⁴
I for the centre R.S.J. $= I_{yy}$	$=$	405
I for the two end R.S.J.s $= 2 \times I_{xx}$	$= 2 \times 6085$	
	$=$	12170
	Total	12575

I about BB:		cm⁴
I for the centre R.S.J. $= I_{xx}$	$=$	6085
I for one end R.S.J. about its own n.a. $= I_{yy} = 405$		
I for one end R.S.J. due to transferring to BB $= 57 \times (12{\cdot}5 + \frac{1}{2}10{\cdot}915)^2 = 9572$		
I for one end R.S.J. about BB $= 9977$		
I for two end R.S.J.s about BB $= 2 \times 9977$	$=$	19954
	Total	26039

The least value of k must be used in the formula. This corresponds to the least value of I and hence

$$k = \sqrt{\frac{I_{AA}}{\text{area}}} = \sqrt{\frac{12575 \text{ cm}^4}{3 \times 57 \text{ cm}^2}} = 7{\cdot}353 \text{ cm} = 73{\cdot}53 \text{ mm}$$

Using this value, the working stress is

$$\left[140 - \frac{1}{1{\cdot}3}\left(\frac{6\text{m}}{73{\cdot}53 \times 10^{-3}\text{m}}\right)\right] \text{MN/m}^2 = 77{\cdot}23 \text{ MN/m}^2$$

$$\begin{aligned} \text{Working load} &= \text{working stress} \times \text{area} \\ &= (77{\cdot}23 \times 10^6 \text{ N/m}^2) \times (3 \times 57 \times 10^{-4} \text{ m}^2) \\ &= 1{\cdot}32 \text{ MN} \end{aligned}$$

In practice the working load is taken as some reasonable fraction of the theoretically safe load. The ratio

safe theoretical load/working load

is called the factor of safety. It varies from 1·5–2·0 in certain "light-weight" designs up to 20–30 in structures liable to severe shock loading.

For the given stanchion, the Rankine load is

$$P_R = \frac{\sigma_Y A}{1 + a(L/k)^2} = \frac{320 \times 10^6\ \text{N/m}^2 \times (3 \times 57 \times 10^{-4}\ \text{m}^2)}{1 + \dfrac{1}{7500}\left(\dfrac{6\ \text{m}}{73{\cdot}53 \times 10^{-3}\ \text{m}}\right)^2}$$

$$= 2{\cdot}90\ \text{MN}$$

Factor of safety $= 2{\cdot}90/1{\cdot}32 = 2{\cdot}2$

PROBLEMS

1. Use the method of Example 7.1 to derive approximate formulae for the crippling load of a strut built-in at one end and free (in position and direction) at the other assuming that the deflection curve is the same as that of a cantilever subjected to

(*a*) a concentrated load at the free end,
(*b*) a uniformly distributed load.

Answer. (*a*) $3EI/L^2$; (*b*) $4EI/L^2$.

2. Calculate the Euler crippling load for a 13 mm diameter steel rod, 2 m long, if the ends are "pinned". Take Young's modulus = 200 GN/m². What would be the buckling load if one end was rigidly fixed and the other was (i) pinned, (ii) completely free, (iii) also rigidly fixed?
Answer. 692 N; (i) 1384 N; (ii) 173 N; (iii) 2768 N.

3. A round vertical bar is clamped at the lower end and is free at the other. The effective length is 1·5 m. If a horizontal force of 330 N at the top produces a (horizontal) deflection there of 13 mm what is the buckling load for the bar in the given condition?
Answer. 31·3 kN.

4. An alloy tube, 4·9 m long, extends 6 mm under a tensile load of 60 kN. Calculate the buckling load for the tube when used as a strut with pin-jointed ends. The tube diameters are 25 mm and 37 mm.
Answer. 2·51 kN.

5. Plot, on the same axes, graphs of the crippling *stress* for mild steel struts with "free" ends as given by the Euler and Rankine formulae. Take L/k as a base and use the following values of the constants

$$E = 207\ \text{GN/m}^2;\ \sigma_Y = 326\ \text{MN/m}^2;\ a = 1/7500$$

Use the following ranges of L/k: for the Rankine formula, 0 to 200; for the Euler formula, 80 to 200.

Read from your graph,

(i) the stress as given by the two formulae when $L/k = 130$;
(ii) the slenderness ratio, as given by the two formulae for a stress of 140 MN/m².

Answer. (i) By Euler, 120·9 MN/m²; by Rankine, 100 MN/m². (ii) By Euler, $L/k = 121$; by Rankine, $L/k = 100$.

6. A strut is built up of two 250 mm × 75 mm B.S.C.s placed back to back at a distance of 100 mm apart and riveted to two flange plates each 300 mm by 12 mm as shown in Fig. 7.7. Calculate the second moments of area about the principal axes AA and BB neglecting the effects of the rivets.

Properties of one B.S.C.

Cross-sectional area = 36·6 cm²

$I_{xx} = 3440$ cm⁴; $I_{yy} = 166$ cm⁴

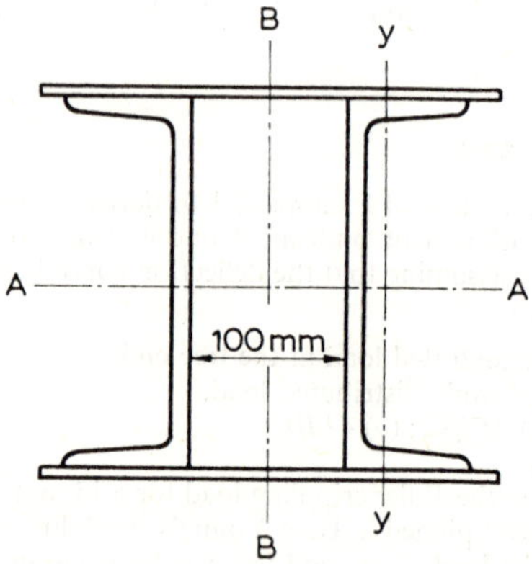

Fig. 7.7

Distance of the centroid from the back = 19 mm. (The *xx* axis for each channel coincides with AA.)

If the effective length is 4·8 m calculate the working load for the strut, using the following rules for the working stress

(i) The straight line formula,

$$\text{Working stress} = \left[90 - \frac{1}{2}\left(\frac{L}{k}\right)\right] \text{MN/m}^2$$

(ii) The parabolic formula,

$$\text{Working stress} = \left[90 - \frac{1}{120}\left(\frac{L}{k}\right)^2\right] \text{MN/m}^2$$

(iii) The Rankine formula, taking $\sigma_Y = 90$ MN/m²
$a = 1/7500$

Answer. $I_{AA} = 19244$ cm⁴. $I_{BB} = 9218$ cm⁴. (i) 870 kN; (ii) 868 kN; (iii) 881 kN.

7. The cross-member of the frame of a machine can be regarded as a strut with fixed ends. The member is 1·25 m in length and is fabricated from M.S. tube and plate with a cross-section as shown in Fig. 7.8. Taking a factor of safety of 6 and the constants "a" and "σ_Y" in the Rankine formula as 1/7500 and 320 MN/m² respectively, find the greatest thrust that can be taken by the member. Also find the thrust using the Euler formula and state why this value is not valid. (E = 207 GN/m².)
(*I.Mech.E.*)

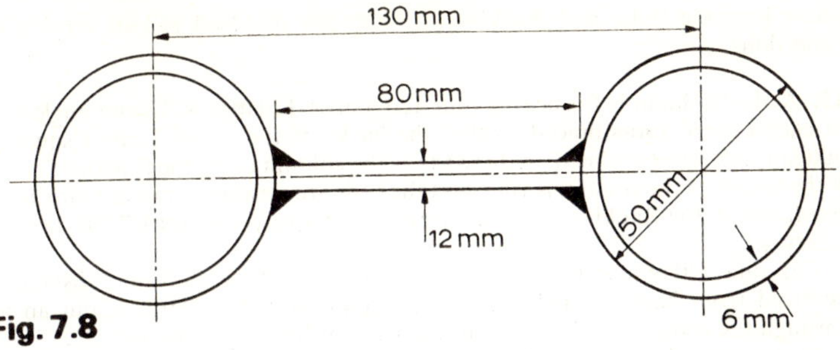

Fig. 7.8

(*Note*. The value of "a" given in the question is clearly that for the "free" ends case. As explained in Example 7.6, it is necessary to divide this by 4 to allow for the "fixing" of the ends.)
Answer. 105 kN; Euler load is 2·19 MN (without safety factor); this is invalid since it corresponds to an (average) stress of 838 MN/m² which is considerably greater than σ_Y.

8. An alloy steel tube is 80 mm external diameter and 3 mm wall thickness. A very short length of the tube was tested in compression and found to yield at a load of 392 kN. A length of 2 m when tested as a strut with hinged ends buckled at a load of 230 kN. Assuming the failing stress in the Rankine-Gordon formula to be the yield stress of the material find the value of the constant a.

A length of this tube, when tested as a beam by placing it on supports 2 m apart, had a deflection of 1·4 mm with a central load of 900 N. What is the value of the buckling load as given by the Euler formula? (*I.Mech.E.*)
Answer. $a = 1/7650$; Euler load = 264 kN.

9. A compound column is composed of two 250 mm × 200 mm *I*-section beams with their webs parallel and at 225 mm centres and a plate 450 mm × 25 mm riveted across each pair of beam flanges. Calculate the maximum axial load this section can carry on a length of 10 m assuming hinged ends, using the Rankine formula.

What would be the value of this load if one end of the column was assumed to be direction fixed? Properties of one *I*-section beam: area = 104·3 cm².

$I_{xx} = 12030$ cm⁴ $\quad I_{yy} = 2290$ cm⁴

Rankine constants: $\sigma_Y = 85$ MN/m² and $a = 1/7500$ (*I.Struct.E.*)
Answer. 2·42 MN; 3·38 MN.

10. A tube of length 1·2 m is 26 mm outer diameter and thickness 1 mm. If the material is steel ($E = 207$ GN/m²) find the load which may be applied axially so as not to exceed one-third of the Euler buckling load. Derive any formula you use. (*R.Ae.S.*)
Answer. 2·91 kN.

11. Derive the Euler formula giving the buckling load for a pin-ended slender strut carrying an axial load.

In the case of a strut of given section, let P_E be the Euler buckling load, P_c the crushing load for a short length and P the crippling load for any length of strut. Derive the Rankine formula for the crippling load assuming that

$$\frac{1}{P} = \frac{1}{P_c} + \frac{1}{P_E}$$

Show how the Euler and Rankine formulae are modified to suit varying end conditions. (*U.L.*)

12. A steel column is formed by four 120 mm × 120 mm × 12 mm angles, one at each corner, cross-braced so that the backs of the angles form a square of 500 mm side. Determine the safe road for the column using the Rankine formula if the length of the column is 10 m and the ends are position—but not direction—fixed. Use in the Rankine formula: stress 320 MN/m^2, constant 1/7 500, factor of safety 4.

For one 120 mm × 120 mm × 12 mm angle: area = 27·5 cm^2, distance of centroid from back of angle 34·0 mm, second moment of area about an axis through the centroid and parallel to the back of the angle 368 cm^4.

In the general treatment of columns comment briefly on the variations which occur in end-fixity and the method whereby allowance is made for these in strength calculations. (*U.L.*)

Hint. The column is shown in Fig. 7.9. The cross-bracing can be ignored in calculating the strength of the column.
Answer. 689 kN.

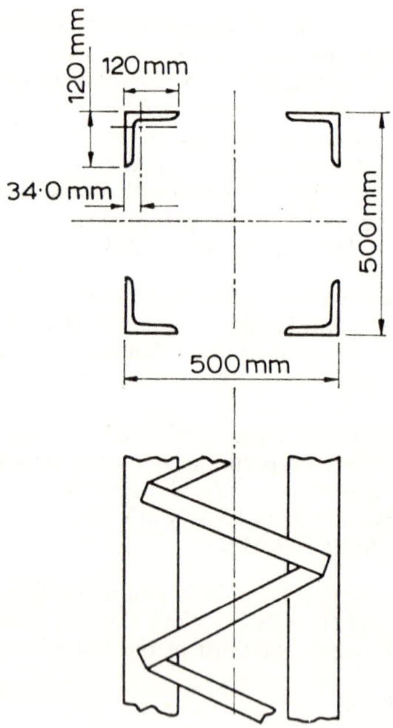

Fig. 7.9

Chapter 8

Laterally and Eccentrically Loaded Struts

The following formulae for maximum bending moment M_{max} apply to slender bars pin-jointed at their ends. In each case P is the end load, L is the length, $\alpha^2 = P/EI$ and $P_E = \pi^2 EI/L^2$, the Euler load.

(*a*) Strut, having end load P with eccentricity e.

$$M_{max} = Pe \sec \tfrac{1}{2}\alpha L = Pe \sec \tfrac{1}{2}\pi\sqrt{(P/P_E)}$$

(*b*) Strut, initially bowed to a sine wave having a maximum deflection y_0 at the centre, with end load P acting along the line joining the centres of area of the end cross-sections.

$$M_{max} = \frac{P_E}{P_E - P} P y_0$$

(*c*) Strut with an axial end load P and a central lateral point load W.

$$M_{max} = \frac{W}{2\alpha} \tan \tfrac{1}{2}\alpha L$$

(*d*) Strut with an axial end load P and a uniformly distributed lateral load w per unit length.

$$M_{max} = \frac{w}{\alpha^2} (\sec \tfrac{1}{2}\alpha L - 1)$$

In other cases the required results should be derived from first principles starting from the bending moment relationships. In the notation of Fig. 8.1 we have

$$M = Py + R_1 x - M_1 + \text{term(s) due to lateral loads} \qquad \text{(i)}$$

In the absence of lateral load terms the combination of (i) and the differential equation of flexure gives

$$EI\frac{d^2y}{dx^2} = -M = -Py - R_1x + M_1 \qquad \text{(ii)}$$

and the solution may be written

$$y = A\cos\alpha x + B\sin\alpha x - \frac{R_1x}{P} + \frac{M_1}{P} \qquad \text{(iii)}$$

where A and B are integration constants. The determination of A, B, R_1 and M_1 depends on the end conditions. At a pinned or free end

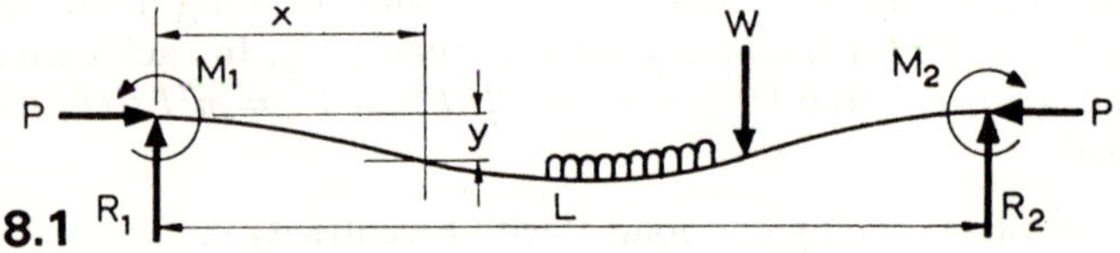

Fig. 8.1

$M = 0$. If both ends are pinned R_1 (and R_2) can be determined at the outset by taking moments. If the ends are clamped it will be necessary to derive dy/dx from (iii) and use the condition $dy/dx = 0$ at the ends.

Lateral loads give rise to an extra term or terms in (ii), normally involving powers of x. The corresponding addition to (iii) will be of the form $a + bx + cx^2$ . . ., the highest power of x being the same as that in (ii). The constants a, b, c . . . are then determined by differentiation of the result and comparison with the original differential equation.

In symmetrical cases it is often convenient to restrict the analysis to one-half of the strut and to use the conditions at mid-span for determining the integration constants.

In certain problems, where the ends of the struts are pin-jointed, the working can be simplified by differentiating equation (i) twice. This gives

$$\frac{d^2M}{dx^2} = P\frac{d^2y}{dx^2} + \text{term(s) which depend on the lateral loads}$$

and d^2y/dx^2 may be written $-M/EI$ from the differential equation of flexure. The result is an equation connecting M and x which may be solved by methods similar to those used previously for y. (See Examples 8.5 and 8.6.)

WORKED EXAMPLES

8.1. Obtain an expression for the maximum bending moment in a slender strut length L, pin-jointed at its ends and subjected to a compressive load P which acts parallel to the axis of the undeflected strut but at an eccentricity e. Show how the result leads to the formula for the Euler critical load.

A tubular strut length 1·5 m and diameter 50 mm has a wall thickness of 1 mm. An axial load of 10 kN is applied at the ends of the strut (which may be assumed as pin-jointed) and due to the load being slightly eccentric it is found that a central deflection is produced of magnitude 3 mm. Taking the eccentricity as being the same at each end, find its magnitude. $E = 70$ GN/m².

Solution. The analysis is an extension of that used in Example 7.2 of Chapter 7. In the notation of Fig. 8.2 (compare with Fig. 7.1) the bending moment at distance x from one end is

$$M = P(y + e)$$

Fig. 8.2

and the differential equation of flexure becomes

$$EI\,\frac{d^2y}{dx^2} = -M = -P(y + e)$$

Equations of this form can be solved by introducing a new variable Y such that $Y = y + e$. Since e is a constant the differential coefficients of Y and y are identical. The solution is therefore the same as that obtained on page 203 except that y is replaced by $y + e$. Thus

$$y + e = A \cos \alpha x + B \sin \alpha x$$

where $\alpha^2 = P/EI$.

When $x = 0$, $y = 0$ and therefore $A = e$. Thus

$$y + e = e \cos \alpha x + B \sin \alpha x \qquad \text{(i)}$$

When $x = L$, $y = 0$ and

$$0 + e = e \cos \alpha L + B \sin \alpha L$$

from which

$$B = \frac{e(1 - \cos \alpha L)}{\sin \alpha L} \tag{ii}$$

It is convenient to use the half-angle $\frac{1}{2}\alpha L$ and, with the relationships $\cos \alpha L = 2 \cos^2 \frac{1}{2}\alpha L - 1$ and $\sin \alpha L = 2 \sin \frac{1}{2}\alpha L \cos \frac{1}{2}\alpha L$, we obtain

$$B = e \tan \tfrac{1}{2}\alpha L$$

(This result can also be obtained from the condition $dy/dx = 0$ at mid-span.)

With this value of B the deflection at mid-span ($x = L/2$) is given by (using (ii)):

$$y_{max} + e = e \cos \tfrac{1}{2}\alpha L + e \tan \tfrac{1}{2}\alpha L \sin \tfrac{1}{2}\alpha L$$

$$= e\left(\frac{\cos^2 \frac{1}{2}\alpha L + \sin^2 \frac{1}{2}\alpha L}{\cos \frac{1}{2}\alpha L}\right)$$

$$= e \sec \tfrac{1}{2}\alpha L$$

and

$$y_{max} = e(\sec \tfrac{1}{2}\alpha L - 1) \tag{iii}$$

The maximum bending moment is therefore

$$M_{max} = P(y_{max} + e) = Pe \sec \tfrac{1}{2}\alpha L \qquad (Ans)$$

The value of $\sec \theta$ becomes infinite when $\theta = \pi/2$. Thus, if $\frac{1}{2}\alpha L = \pi/2$ the bending moment is (theoretically) infinite and the strut will collapse however small the eccentricity of the load. The corresponding value of P is given by

$$\alpha L = \pi \qquad \text{or} \qquad \alpha^2 L^2 = \pi^2$$

But $\alpha^2 = P/EI$ and hence the critical value of P is

$$P = \pi^2 EI/L^2$$

the Euler expression.

In the numerical part of the question let 50 mm be the internal diameter. Then

$$I = \frac{\pi}{64}(52^4 - 50^4) = 5{\cdot}213 \times 10^4 \text{ mm}^4$$

$$\alpha^2 = \frac{P}{EI} = \frac{(10 \times 10^3 \text{ N})}{(70 \times 10^9 \text{ N/m}^2) \times (5{\cdot}213 \times 10^4 \times 10^{-12} \text{ m}^4)}$$

$$= 2{\cdot}742 \text{ m}^{-2}$$

$$\alpha = 1{\cdot}656 \text{ m}^{-1}$$

From (ii) the eccentricity of the load is

$$e = \frac{y_{max}}{\sec \frac{1}{2}\alpha L - 1}$$

$$= \frac{3 \text{ mm}}{\sec (\frac{1}{2} \times 1{\cdot}656 \times 1{\cdot}5) - 1}$$

$$= 1{\cdot}32 \text{ mm} \qquad (Ans)$$

8.2. A hollow circular steel strut with its ends position-fixed has a length of 2·5 m, its external diameter being 100 mm and its internal

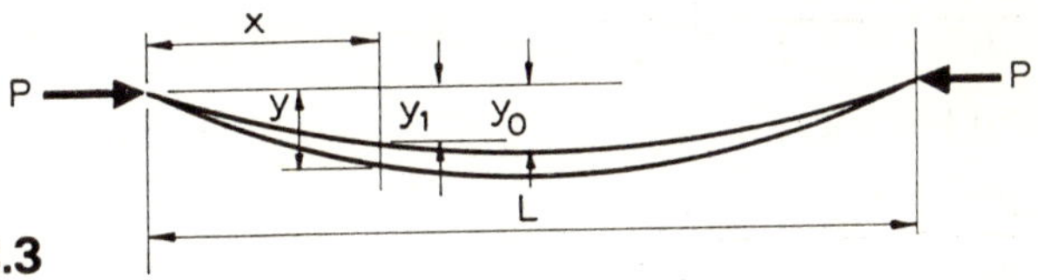

Fig. 8.3

diameter 80 mm. Before loading the strut is bent with a maximum deviation of 4 mm. Assuming that the centre line of the strut is sinusoidal determine the maximum stress due to a central compressive end load of 100 kN. ($E = 206$ GN/m².) (*U.L.*)

Solution. Figure 8.3 shows a strut, length L, subjected to an end compressive load P. Suppose that at distance x from the left-hand end, y_1 is the deflection due to the initial bending and y is the total deflection when the load is applied. Let y_0 be the maximum initial deviation. Since the initial bowing is sinusoidal we have

$$y_1 = C \sin Dx$$

where C and D are constants.

When $x = L$, $y_1 = 0$ and thus $\sin DL = 0$. The lowest value which satisfies the problem is $DL = \pi$ and hence $D = \pi/L$.

At mid-span $x = L/2$ and $y_1 = y_0$. Using the value for D found above we obtain

$$y_0 = C \sin \left(\frac{\pi}{L} \times \frac{L}{2}\right)$$

from which $C = y_0$. Thus the initial deflection is given by

$$y_1 = y_0 \sin \frac{\pi x}{L} \qquad \text{(i)}$$

At distance x from the origin the bending moment due to P is Py but the change in deflection due to the load is $y - y_1$. Hence the differential equation of flexure becomes

$$EI\frac{d^2(y - y_1)}{dx^2} = -M = -Py$$

or

$$\frac{d^2y}{dx^2} - \frac{d^2y_1}{dx^2} = -\frac{P}{EI}y = -\alpha^2 y \qquad \text{(ii)}$$

From (i),

$$\frac{d^2y_1}{dx^2} = -\frac{\pi^2 y_0}{L^2}\sin\frac{\pi x}{L}$$

and (ii) becomes

$$\frac{d^2y}{dx^2} = -\alpha^2 y - \frac{\pi^2 y_0}{L^2}\sin\frac{\pi x}{L}$$

The solution of this equation can be written

$$y = A\cos\alpha x + B\sin\alpha x + \frac{\pi^2 y_0}{\pi^2 - \alpha^2 L^2}\sin\frac{\pi x}{L} \qquad \text{(iii)}$$

and this result can be checked by successive differentiation.

When $x = 0$, $y = 0$ and therefore $A = 0$.
When $x = L$, $y = 0$ and thus $B\sin\alpha L = 0$.

For $\sin\alpha L$ to be zero, $\alpha L = \pi$ (the lowest value satisfying the problem) but this is ruled out because it would make the third term of (iii) infinite. Hence B must be zero and equation (iii) becomes

$$y = \frac{\pi^2 y_0}{\pi^2 - \alpha^2 L^2}\sin\frac{\pi x}{L}$$

The greatest deflection occurs at mid-span ($x = L/2$) and thus

$$y_{max} = \frac{\pi^2 y_0}{\pi^2 - \alpha^2 L^2} = \frac{y_0}{1 - (\alpha^2 L^2/\pi^2)}$$

But $\alpha^2 = P/EI$ and $\alpha^2 L^2/\pi^2 = P/P_E$ where P_E is the Euler load for the strut. Thus

$$y_{max} = \left(\frac{1}{1 - (P/P_E)}\right) y_0 = \left(\frac{P_E}{P_E - P}\right) y_0$$

For the given steel strut,

$$I = \frac{\pi}{64}(100^4 - 80^4)\ \text{mm}^4 = 2{\cdot}90 \times 10^6\ \text{mm}^4$$

$$P_E = \frac{\pi^2 EI}{L^2} = \frac{\pi^2 \times (206 \times 10^9\ \text{N/m}^2) \times (2{\cdot}90 \times 10^{-6}\ \text{m}^4)}{(2{\cdot}5\ \text{m})^2}$$

$$= 942\ \text{kN}$$

$$y_{max} = \left(\frac{P_E}{P_E - P}\right) y_0 = \left(\frac{942}{942 - 100}\right) \times 4\ \text{mm}$$

$$= 4{\cdot}47\ \text{mm}$$

The maximum bending moment is

$$Py_{max} = 100\ \text{kN} \times 0{\cdot}00447\ \text{m} = 447\ \text{N-m}$$

and the corresponding maximum bending stress is

$$\sigma = \frac{447\ \text{N-m} \times 0{\cdot}05\ \text{m}}{2{\cdot}90 \times 10^{-6}\ \text{m}^4} = 7{\cdot}71\ \text{MN/m}^2$$

$$\text{Direct compressive stress} = \frac{0{\cdot}1\ \text{MN}}{\frac{1}{4}\pi(100^2 - 80^2) \times 10^{-6}\ \text{m}^2}$$

$$= 35{\cdot}38\ \text{MN/m}^2$$

Maximum stress $= 7{\cdot}71 + 35{\cdot}38 = 43{\cdot}1\ \text{MN/m}^2$ (compressive)

(*Ans*)

8.3. What are the basic assumptions made in deriving the Perry–Robertson formula for struts?

For what reasons was this formula made the basis of practical design formulae?

Show that if a strut has an initial curvature in the form of a parabolic arc and is hinged at both ends (i.e. position fixed only) then the maximum compressive stress produced by a load P is

$$\sigma_{max} = \frac{P}{A}\left\{1 + \frac{es}{k^2}\frac{8P_E}{\pi^2 P}\left(\sec \tfrac{1}{2}\pi \sqrt{\left[\frac{P}{P_E}\right]} - 1\right)\right\}$$

where A is the cross-sectional area; e the initial central deflection; P_E the Eulerian crippling load; k the least radius of gyration of the section and s the distance of the extreme fibres on the compression side from the neutral axis.

Compare the maximum stress *due to bending only* derived from the given formula with that given by the Perry–Robertson assumptions when $P = \frac{1}{2}P_E$. (*U.L.*)

Solution. The Perry–Robertson formula for pin-jointed struts with end loads is based on the following assumptions:

(*a*) the strut is initially bowed to a sine wave with a maximum deviation at the centre (say e_1);
(*b*) the load is applied eccentrically (by an amount, say, e_2);
(*c*) the strut fails when the maximum compressive stress reaches the yield stress of the material in direct compression (say σ_Y).

As shown in the solution to example 8.2 assumption (*a*) leads to a maximum deflection of:

$$y_{max} = \frac{P_E}{P_E - P} e_1 \qquad \text{(i)}$$

Similarly the analysis given in Example 7.3 shows that an initial accentricity e_2 leads to a maximum deflection

$$y_{max} = e_2 (\sec \tfrac{1}{2}\alpha L - 1)$$

end the corresponding moment arm of the load P is

$$y_{max} + e_2 = e_2 \sec \tfrac{1}{2}\alpha L = e_2 \sec \tfrac{1}{2}\pi \sqrt{(P/P_E)} \qquad \text{(ii)}$$

Perry showed that the expression $\sec \frac{1}{2}\pi\sqrt{(P/P_E)}$ is very nearly equal to $1{\cdot}2P_E/(P_E - P)$ and thus (ii) becomes

$$y_{max} + e_2 = \left(\frac{P_E}{P_E - P}\right) 1{\cdot}2e_2 \qquad \text{(iii)}$$

A comparison of (i) and (iii) shows that the separate effects of e_1 and e_2 can be simultaneously allowed for by an equivalent initial bowing e given by

$$e = e_1 + 1{\cdot}2e_2$$

With the notation of the present question and assumption (*c*) above the maximum compressive stress is given by

$$\sigma_{max} = \text{direct stress} + \text{maximum compressive bending stress}$$
$$= \frac{P}{A} + \frac{Pes}{I}\left(\frac{P_E}{P_E - P}\right)$$
$$= \sigma + \frac{\sigma es}{k^2}\left(\frac{\sigma_E}{\sigma_E - \sigma}\right)$$

where $\sigma = P/A$ and σ_E = Euler stress = P_E/A

On rearrangement, the last equation becomes a quadratic

$$\sigma^2 - \sigma\left[\sigma_{max} + \sigma_E\left(1 + \frac{es}{k^2}\right)\right] + \sigma_E\sigma_{max} = 0$$

from which, taking the smaller root,

$$\sigma = \tfrac{1}{2}[\sigma_{max} + (1 + \eta)\sigma_E] - \tfrac{1}{2}\sqrt{\{[\sigma_{max} + (1 + \eta)\sigma_E]^2 - 4\sigma_E\sigma_{max}\}}$$

where $\eta = es/k^2$.

This result is known as the Perry–Robertson formula and is the basis for the calculation of safe loads for steel columns (see British Standard 449: Part 1 1970 and Part 2 1969).

This formula has been adopted for practical design because there is inevitably some initial crookedness and eccentricity of loading. The expression for σ represents closely the results obtained from experiments in which these imperfections are kept as small as possible. Robertson analysed a considerable number of results and concluded that for structural materials with a marked yield point (see Chapter 16) η could be expressed as a simple function of the slenderness ratio L/k.

Figure 8.3 illustrates an initially bowed strut, the only difference in the present case being that the maximum initial deviation is e instead of y_0. Since the initial shape is now parabolic the deflection before loading is given by

$$y_1 = a + bx + cx^2 \tag{iv}$$

where a, b and c are constants. When $x = 0$, $y_1 = 0$ and $a = 0$. At $x = \frac{1}{2}L$ the slope $dy_1/dx = 0$ and hence $b = -cL$.

In addition, when $x = \frac{1}{2}L$, $y_1 = e$. This gives $b = 4e/L$ and $c = -4e/L^2$. Thus, from (iv),

$$y_1 = 4e\left(\frac{x}{L} - \frac{x^2}{L^2}\right) \tag{v}$$

The differential equation of flexure is now

$$EI\frac{d^2(y - y_1)}{dx^2} = -M = -Py$$

$$\text{or } \frac{d^2y}{dx^2} - \frac{d^2y_1}{dx^2} = -\alpha^2 y \tag{vi}$$

From (v),

$$\frac{d^2y_1}{dx^2} = -\frac{8e}{L^2}$$

and equation (vi) becomes

$$\frac{d^2y}{dx^2} = -\alpha^2 y - \frac{8e}{L^2}$$

The solution of this equation is

$$y = A\cos\alpha x + B\sin\alpha x - \frac{8e}{\alpha^2L^2} \qquad \text{(vii)}$$

When $x = 0$, $y = 0$ and $A = 8e/\alpha^2L^2$.
When $x = L$, $y = 0$ and

$$B = \frac{8e}{\alpha^2L^2}\operatorname{cosec}\alpha L(1 - \cos\alpha L)$$

But cosec αL and cos αL can be written in terms of the half-angles $\frac{1}{2}\alpha L$ so that

$$B = \frac{8e}{\alpha^2L^2}\,\frac{1}{2\sin\frac{1}{2}\alpha L\cos\frac{1}{2}\alpha L}\,(2\sin^2\tfrac{1}{2}\alpha L)$$

$$= \frac{8e}{\alpha^2L^2}\tan\tfrac{1}{2}\alpha L$$

The maximum deflection occurs at mid-span ($x = L/2$) and substituting for A and B, equation (vii) gives

$$y_{max} = \frac{8e}{\alpha^2L^2}\cos\tfrac{1}{2}\alpha L + \frac{8e}{\alpha^2L^2}\tan\tfrac{1}{2}\alpha L\sin\tfrac{1}{2}\alpha L - \frac{8e}{\alpha^2L^2}$$

$$= \frac{8e}{\alpha^2L^2}\left(\frac{\cos^2\frac{1}{2}\alpha L + \sin^2\frac{1}{2}\alpha L}{\cos\frac{1}{2}\alpha L} - 1\right)$$

$$= \frac{8e}{\alpha^2L^2}(\sec\tfrac{1}{2}\alpha L - 1)$$

$$= \frac{8e}{\pi^2}\frac{P_E}{P}\left(\sec\tfrac{1}{2}\pi\sqrt{(P/P_E)} - 1\right)$$

since

$$\alpha^2 = \frac{P}{EI} = \frac{\pi^2}{L^2}\frac{P}{P_E}$$

The maximum bending moment is Py_{max} and the maximum compressive stress is therefore

Direct stress + maximum bending stress

$$= \frac{P}{A} + \frac{M_{max}s}{I}$$

$$= \frac{P}{A} + \frac{Ps}{I}\frac{8e}{\pi^2}\frac{P_E}{P}\left(\sec\tfrac{1}{2}\pi\sqrt{\left[\frac{P}{P_E}\right]} - 1\right)$$

$$= \frac{P}{A}\left\{1 + \frac{es}{k^2}\frac{8P_E}{\pi^2P}\left(\sec\tfrac{1}{2}\pi\sqrt{\left[\frac{P}{P_E}\right]} - 1\right)\right\}$$

since $I = Ak^2$.

If $P = \frac{1}{2}P_E$ the maximum bending moment will be

$$Py_{max} = \frac{8e}{\pi^2} \times 2P\,(\sec \tfrac{1}{2}\pi\sqrt{\tfrac{1}{2}} - 1) = 2{\cdot}03eP \qquad \text{(viii)}$$

With the basic Perry–Robertson assumption of sinusoidal bowing we have

$$M_{max} = \left(\frac{P_E}{P_E - P}\right) eP = \left(\frac{1}{1 - \frac{1}{2}}\right) eP = 2eP \qquad \text{(ix)}$$

Comparing (viii) and (ix) we see that the formula in the question gives a slightly higher result than the Perry–Robertson assumptions.

8.4. A slender strut of uniform section and of length L has pin-jointed ends and it is initially straight and vertical. It carries an axial

Fig. 8.4

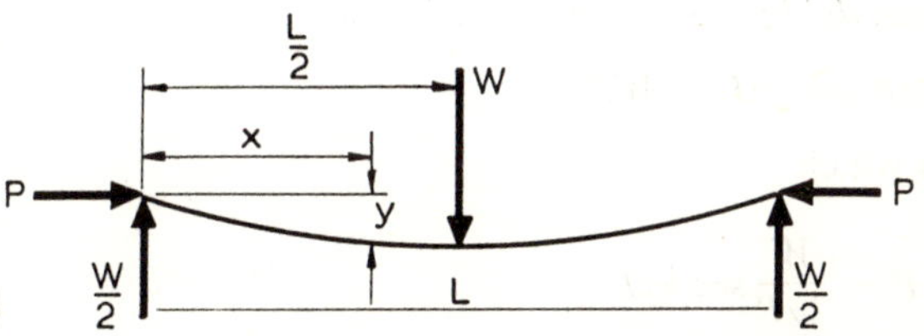

load P and also a horizontal lateral load W applied at the middle of its length and acting in the plane in which P would cause bending to occur. Show that the maximum deflection is

$$\Delta = \frac{W}{2\alpha P}\tan \tfrac{1}{2}\alpha L - \frac{WL}{4P}$$

where

$$\alpha = \sqrt{(P/EI)}$$

In the case of a given strut the magnitude of P is $P = P_E/4$ where P_E is the Euler critical load for the strut. Find the ratio of the maximum deflection produced by P and the lateral load W acting together, to that produced by W acting alone. *(U.L.)*

Solution. Figure 8.4 gives the notation, the strut being shown horizontal for ease of comparison with other examples. By symmetry the lateral reactions at the ends are each $W/2$. The differential equation of flexure is given by

$$EI\frac{d^2y}{dx^2} = -M = -Py - \tfrac{1}{2}Wx$$

for the range $0 \leqslant x \leqslant L/2$.

The solution of this equation is

$$y = A\cos\alpha x + B\sin\alpha x - Wx/2P \tag{i}$$

where $\alpha^2 = P/EI$ and A and B are constants.

When $x = 0$, $y = 0$ and thus, from (i), $A = 0$.

By differentiation the slope is

$$dy/dx = \alpha B\cos\alpha x - W/2P \tag{ii}$$

Since equations (i) and (ii) apply only to the left-hand half of the strut the constant B must be determined from the zero slope condition at mid-span.

Thus $dy/dx = 0$ when $x = L/2$ and

$$\alpha B\cos\tfrac{1}{2}\alpha L = W/2P$$

from which

$$B = \frac{W}{2\alpha P}\sec\tfrac{1}{2}\alpha L$$

Using the values obtained for A and B, the maximum deflection, which occurs at mid-span, is given by (i) with $x = L/2$. Hence

$$\Delta\ (\text{or } y_{max}) = \frac{W}{2\alpha P}\sec\tfrac{1}{2}\alpha L\sin\tfrac{1}{2}\alpha L - \frac{W}{2P}\frac{L}{2}$$

$$= \frac{W}{2\alpha P}\tan\tfrac{1}{2}\alpha L - \frac{WL}{4P}$$

as required. Although the question does not call for the maximum bending moment it occurs at mid-span and is easily found from this last result to be

$$M_{max}\frac{W}{2\alpha}\tan\tfrac{1}{2}\alpha L$$

The Euler load $P_E = \pi^2EI/L^2$ and hence $P = \pi^2EI/4L^2$ in the present example.

Also

$$\alpha^2 = P/EI = \pi^2/4L^2 \quad \text{and} \quad \alpha = \pi/2L$$

The maximum deflection with P and W acting together is therefore

$$\Delta = \frac{W}{2\alpha P}\tan \tfrac{1}{2}\alpha L - \frac{WL}{4P}$$

$$= \frac{W}{2(\pi/2L)(\pi^2 EI/4L^2)}\tan\frac{\pi}{2L}\frac{L}{2} - \frac{WL}{4(\pi^2 EI/4L^2)}$$

$$= \frac{4WL^3}{\pi^3 EI}\tan \tfrac{1}{4}\pi - \frac{WL^3}{\pi^2 EI}$$

$$= \frac{WL^3}{EI}\left(\frac{4}{\pi^3} - \frac{1}{\pi^2}\right)$$

The maximum deflection with W acting alone is

$$\Delta = WL^3/48EI$$

The required ratio is therefore $\left(\frac{4}{\pi^3} - \frac{1}{\pi^2}\right) \Big/ \frac{1}{48}$ which equals 1·33.

8.5. Show that

$$M = \frac{wEI}{P}\left(\sec \tfrac{1}{2}l\sqrt{\left[\frac{P}{EI}\right]} - 1\right)$$

represents the maximum bending moment in a horizontal strut of length l, having pin-jointed ends, if it weighs w per unit length and is subjected to an axial load P.

A horizontal pin-ended strut 5 m long is formed from a standard rolled steel Tee section (150 mm × 100 mm × 12·5 mm @ 24·13 kg/m). The axial compressive load is 100 kN. Determine the maximum stress in the steel if the XX axis is horizontal and the table of the Tee forms the compression face. The centroid of the section is 2·5 cm below the top.

(I_{XX} = 253 cm^4; cross-sectional area = 30·8 cm^2).
E= 206 GN/m^2. (*U.L.*)

Solution. The weight of the strut is equivalent to a uniformly distributed load and Fig. 8.5 shows the effective loading. The bending moment at a distance x from one end is

$$M = Py + \tfrac{1}{2}wlx - \tfrac{1}{2}wx^2 \qquad \text{(i)}$$

An expression for the deflection is not required and the following method enables us to determine M in terms of x without first finding the deflection equation.

Differentiating equation (i) twice with respect to x,

$$\frac{d^2M}{dx^2} = P\frac{d^2y}{dx^2} - w = -\frac{P}{EI}M - w$$

since $EI\,(d^2y/dx^2) = -M$.

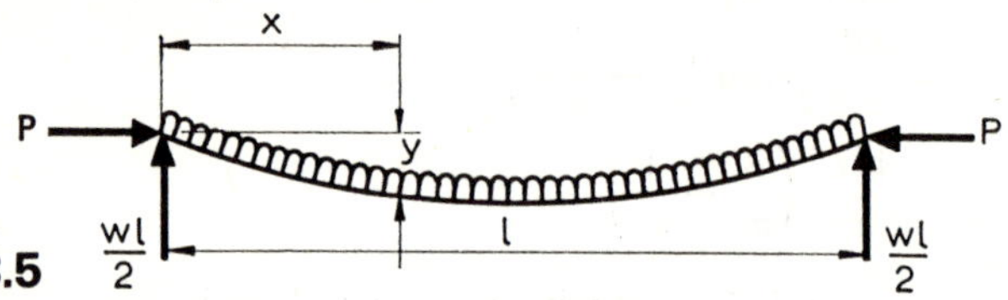

Fig. 8.5

The solution of this equation is

$$M = A\cos\alpha x + B\sin\alpha x - w/\alpha^2 \qquad \text{(ii)}$$

where $\alpha^2 = P/EI$.

The strut is pin-jointed at the ends and, when $x = 0$, $M = 0$. From (i), therefore, $A = w/\alpha^2$.

With this value and the condition $x = l$, $M = 0$, we have

$$\frac{w}{\alpha^2}\cos\alpha l + B\sin\alpha l = \frac{w}{\alpha^2}$$

from which

$$B = \frac{w}{\alpha^2}\frac{(1-\cos\alpha l)}{\sin\alpha l}$$

Using the trigonometrical identities $\cos\alpha l = 1 - 2\sin^2\frac{1}{2}\alpha l$ and $\sin\alpha l = 2\sin\frac{1}{2}\alpha l\cos\frac{1}{2}\alpha l$ this becomes

$$B = \frac{w}{\alpha^2}\tan\tfrac{1}{2}\alpha l$$

The maximum bending moment occurs at mid-span where $x = l/2$ and with the values obtained for A and B, equation (ii) gives

$$M_{max} = \frac{w}{\alpha^2}\cos\tfrac{1}{2}\alpha l + \frac{w}{\alpha^2}\tan\tfrac{1}{2}\alpha l\sin\tfrac{1}{2}\alpha l - \frac{w}{\alpha^2}$$

$$= \frac{w}{\alpha^2}\left(\frac{\cos^2\frac{1}{2}\alpha l + \sin^2\frac{1}{2}\alpha l}{\cos\frac{1}{2}\alpha l} - 1\right)$$

$$= \frac{w}{\alpha^2}\left(\sec\tfrac{1}{2}\alpha l - 1\right)$$

This is the required result since $\alpha^2 = P/EI$.

For the given Tee section,

$$\text{Direct compressive stress} = \frac{100 \times 10^3\ \text{N}}{30{\cdot}8 \times 10^{-4}\ \text{m}^2}$$

$$= 32{\cdot}5\ \text{MN/m}^2$$

$$\alpha^2 = \frac{P}{EI} = \frac{(100 \times 10^3)\ \text{N}}{(206 \times 10^9)\ \text{N/m}^2 \times (253 \times 10^{-8})\ \text{m}^4}$$

$$= 0{\cdot}1919\ \text{m}^{-2} \qquad \alpha = 0{\cdot}438\ \text{m}^{-1}$$

The reference "@ 24·15 kg/m" means that the strut weighs 24·15 kgf per metre and thus $w = 24{\cdot}15 \times 9{\cdot}81 = 237$ N/m.

Also $l = 5$ m and the maximum bending moment is

$$M_{max} = \frac{w}{\alpha^2}(\sec\tfrac{1}{2}\alpha l - 1)$$

$$= \frac{237\ \text{N/m}}{0{\cdot}1919\ \text{m}^{-2}}[\sec(\tfrac{1}{2} \times 0{\cdot}438 \times 5) - 1]$$

$$= 1461\ \text{N-m}$$

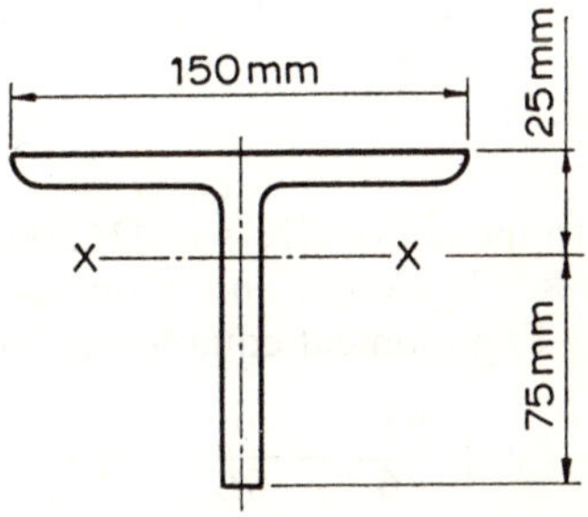

Fig. 8.6

The maximum compressive stress due to bending occurs at the top edge (see Fig. 8.6). If y_{max} is now the distance from the centroid of the section to this edge, maximum compressive bending stress is

$$\frac{My_{max}}{I} = \frac{(1{\cdot}461 \times 10^3\ \text{N-m}) \times (25 \times 10^{-3}\ \text{m})}{253 \times 10^{-8}\ \text{m}^4}$$

$$= 14{\cdot}4\ \text{MN/m}^2$$

Combining this with the direct compressive stress gives a maximum compressive stress of

$$32{\cdot}5 + 14{\cdot}4 = 46{\cdot}9\ \text{MN/m}^2$$

At the bottom edge of the section, $y_{max} = 7{\cdot}5$ cm, the maximum tensile stress is found to be 43·2 MN/m² and the resultant of this and the direct compressive stress is 10·7 MN/m² (tensile)

8.6. A straight strut of length L and of uniform section is hinged at both ends and is loaded along its axis with a thrust P. It also carries a transverse distributed load which varies uniformly in intensity from w per unit length at one end A to zero at the other end B.

Show that the distance x from the end B to the section at which the maximum bending moment occurs is given by

$$\cos \alpha x = \frac{\sin \alpha L}{\alpha L}$$

where $\alpha^2 = P/EI$.

If the thrust P is 81 per cent of the Eulerian crippling load, find the position and value of the maximum bending moment. (*U.L.*)

Solution. Take the origin at the end B. Then, with the notation of Fig. 8.7 the intensity of loading at a section XX is wx/L, the average

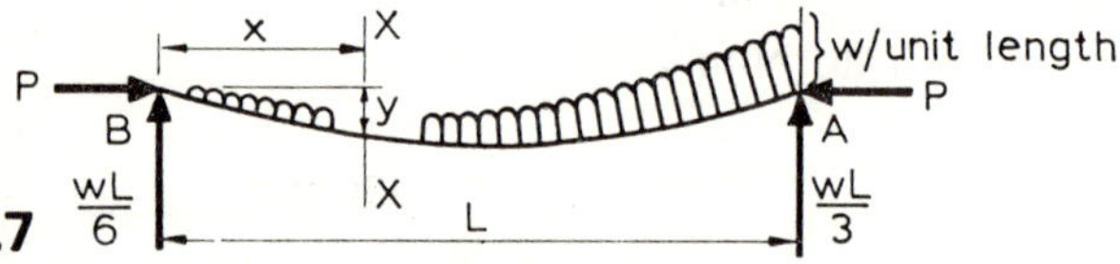

Fig. 8.7

intensity to the left of XX is $wx/2L$, the total lateral load on the portion B to XX is $wx^2/2L$ and its moment arm about XX is $x/3$.

The bending moment equation is therefore

$$M = Py + \frac{wLx}{6} - \frac{wx^3}{6L}$$

and, by differentiating twice with respect to x,

$$\frac{d^2M}{dx^2} = P\frac{d^2y}{dx^2} - \frac{wx}{L} = -\alpha^2 M - \frac{wx}{L}$$

(since $d^2y/dx^2 = -M/EI$)
The solution of this equation is

$$M = A \cos \alpha x + B \sin \alpha x - wx/\alpha^2 L \qquad \text{(i)}$$

When $x = 0$, $M = 0$ (pin-jointed) and thus $A = 0$.
Also when $x = L$, $M = 0$ and $B = (w/\alpha^2) \operatorname{cosec} \alpha L$
For maximum (or minimum) $dM/dx = 0$ and this gives

$$\alpha B \cos \alpha x - \frac{w}{\alpha^2 L} = 0$$

Using the value obtained above for B we have

$$\alpha\left(\frac{w}{\alpha^2}\operatorname{cosec}\alpha L\right)\cos\alpha x = \frac{w}{\alpha^2 L}$$

$$\cos\alpha x = \frac{\sin\alpha L}{\alpha L}$$

as required. From physical considerations this is a maximum rather than a minimum.

The Euler load is $\pi^2 EI/L^2$ and hence

$$P = 0{\cdot}81\pi^2 EI/L^2$$

Therefore

$$\alpha^2 = 0{\cdot}81\pi^2/L^2 \quad \text{and} \quad \alpha = 0{\cdot}9\pi/L$$

The position of maximum bending moment is given by

$$\cos\alpha x = (\sin\alpha L)/\alpha L = (\sin 0{\cdot}9\pi)/0{\cdot}9\pi = 0{\cdot}121$$

$$\alpha x = 83^\circ\ 3' \quad \text{or} \quad 1{\cdot}45 \text{ rad}$$

and the position of the maximum bending moment is given by

$$x = \frac{1{\cdot}45}{0{\cdot}9\pi/L} = 0{\cdot}513\,L \qquad (Ans)$$

With $\alpha = 0{\cdot}9\pi/L$ and $x = 0{\cdot}513L$ the maximum bending moment can be found from (i). It is

$$M_{max} = \frac{w}{\alpha^2}\operatorname{cosec}\alpha L\sin\alpha x - \frac{wx}{\alpha^2 L}$$

$$= \frac{wL^2}{0{\cdot}81\pi^2}\operatorname{cosec}\left(\frac{0{\cdot}9\pi}{L}L\right)\sin\left(\frac{0{\cdot}9\pi}{L}\times 0{\cdot}513L\right) - \frac{w\times 0{\cdot}513L\times L^2}{0{\cdot}81\pi^2\times L}$$

$$= \frac{wL^2}{0{\cdot}81\pi^2}(\operatorname{cosec}162^\circ\sin 82{\cdot}1^\circ - 0{\cdot}513)$$

$$= 0{\cdot}330wL^2 \qquad (Ans)$$

8.7. Obtain expressions for the bending moments at the ends and centre of a uniform strut, built in at both ends, and subjected to a uniform lateral load of intensity w. The strut length is L, the end thrust P, and the elastic properties EI. Take $\alpha^2 = P/EI$. Show, without elaborate analysis, from the expressions derived that for practical struts the end moments are greater numerically than the central moment and of opposite sign. (*U.L.*)

Solution. Figure 8.8 illustrates the notation. By symmetry the end reactions $R_1 = R_2 = wL/2$ and $M_1 = M_2$. The differential equation of flexure gives

$$EI\frac{d^2y}{dx^2} = -M = -Py - R_1x + M_1 + \tfrac{1}{2}wx^2 \qquad \text{(i)}$$

and the solution may be written

$$y = A\cos\alpha x + B\sin\alpha x + (a + bx + cx^2) \qquad \text{(ii)}$$

where the constants a, b and c can be determined by substituting (ii) in (i) and equating coefficients of powers of x.

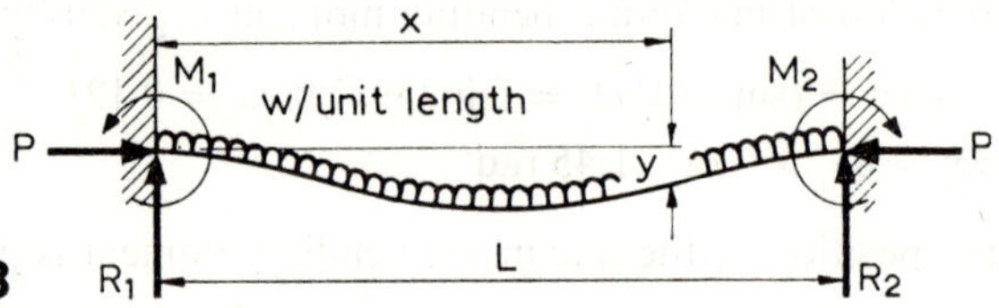

Fig. 8.8

From (ii),

$$\frac{d^2y}{dx^2} = -\alpha^2(A\cos\alpha x + B\sin\alpha x) + 2c$$

$$= -\alpha^2(y - a - bx - cx^2) + 2c$$

and

$$EI\frac{d^2y}{dx^2} = -P(y - a - bx - cx^2) + 2cEI$$

On comparing coefficients of x in this expression with those in (i) we find:

for x^2 $\quad cP = w/2$ and $c = w/2P$

for x $\quad bP = -R_1$ and $b = -R_1/P$

for number term,

$$Pa + 2cEI = M_1$$

$$a = \frac{M_1 - 2cEI}{P} = \frac{M_1}{P} - \frac{w}{\alpha^2 P}$$

(since $c = w/2P$).

Thus the deflection equation (ii) becomes

$$y = A\cos\alpha x + B\sin\alpha x + \frac{M_1}{P} - \frac{R_1 x}{P} + \frac{wx^2}{2P} - \frac{w}{\alpha^2 P} \qquad \text{(iii)}$$

When $x = 0$, $y = 0$ and from (iii)

$$A = \frac{w}{\alpha^2 P} - \frac{M_1}{P}$$

By differentiating (iii) the slope is given by

$$\frac{dy}{dx} = -\alpha A\sin\alpha x + \alpha B\cos\alpha x - \frac{R_1}{P} + \frac{wx}{P} \qquad \text{(iv)}$$

At $x = 0$, $dy/dx = 0$ and $B = R_1/\alpha P$.

To find M_1 it is necessary to use a further "end" condition. It is convenient to use $dy/dx = 0$ when $x = L/2$, (from considerations of symmetry), and, from (iv) we have, noting that $R_1 = wL/2$,

$$0 = -\alpha A\sin\tfrac{1}{2}\alpha L + \alpha B\cos\tfrac{1}{2}\alpha L - \frac{wL}{2P} + \frac{wL}{2P}$$

$$\cot\tfrac{1}{2}\alpha L = \frac{A}{B} = \frac{2\alpha}{wL}\left(\frac{w}{\alpha^2} - M_1\right)$$

The end moment at $x = 0$ is therefore

$$M_1 = \frac{w}{\alpha^2} - \frac{wL}{2\alpha}\cot\tfrac{1}{2}\alpha L$$

and with this result the constant A is

$$\frac{wL}{2\alpha P}\cos\tfrac{1}{2}\alpha L$$

Combining (i) and (iii) and substituting for A, B, M_1 and R_1 the expression for bending moment becomes

$$\begin{aligned} M &= Py + R_1 x - M_1 - \tfrac{1}{2}wx^2 \\ &= P\left(A\cos\alpha x + B\sin\alpha x - \frac{w}{\alpha^2 P}\right) \\ &= \frac{wL}{2\alpha}\left(\cot\tfrac{1}{2}\alpha L\cos\alpha x + \sin\alpha x\right) - \frac{w}{\alpha^2} \end{aligned}$$

The end moment M_0 is obtained by putting $x = 0$ and thus

$$M_0 = \frac{wL}{2\alpha}\cot\tfrac{1}{2}\alpha L - \frac{w}{\alpha^2} = \frac{wL}{2\alpha}\left(\cot\tfrac{1}{2}\alpha L - \frac{2}{\alpha L}\right)$$

Similarly the bending moment M_c at mid-span ($x = L/2$) is

$$M_c = \frac{wL}{2\alpha}\left(\frac{\cos \frac{1}{2}\alpha L}{\sin \frac{1}{2}\alpha L}\cos \tfrac{1}{2}\alpha L + \sin \tfrac{1}{2}\alpha L\right) - \frac{w}{\alpha^2}$$

$$= \frac{wL}{2\alpha}\operatorname{cosec} \tfrac{1}{2}\alpha L - \frac{w}{\alpha^2} = \frac{wL}{2\alpha}\left(\operatorname{cosec} \tfrac{1}{2}\alpha L - \frac{2}{\alpha L}\right)$$

For practical struts $\frac{1}{2}\alpha L$ must be less than $\frac{1}{2}\pi$. Denoting $\frac{1}{2}\alpha L$ by θ the two expressions become

$$M_0 = \frac{wL}{2\alpha}\left(\frac{1}{\tan\theta} - \frac{1}{\theta}\right)$$

and

$$M_c = \frac{wL}{2\alpha}\left(\frac{1}{\sin\theta} - \frac{1}{\theta}\right)$$

For practical values of θ, $\sin\theta < \theta < \tan\theta$ so that M_0 is negative and M_c positive. In addition, a consideration of corresponding values of θ, $\sin\theta$ and $\tan\theta$ shows that M_0 is numerically greater than M_c.

PROBLEMS

1. Show that the maximum bending moment M_{max} for a strut, pin-jointed at its ends, with an end load which is parallel to the undeflected line of the strut but eccentric to it is given by

$$\frac{M_{max}}{M_0} = \sec \tfrac{1}{2}\pi\sqrt{(P/P_E)}$$

where M_0 is the bending moment in the strut if its deflection is ignored and P_E = Euler load.

Plot M_{max}/M_0 against P/P_E for values of P/P_E from 0 to 0·81. (It is convenient to take intermediate values of 0·09, 0·16, 0·25, 0·36, 0·49 and 0·64.)
Answer. For the suggested values of P/P_E the values of M_{max}/M_0 are: 1·00, 1·12, 1·24, 1·41, 1·70, 2·20, 3·24, 6·39.

2. Show, with the notation of the previous example, that if the end load is axial but the strut has an initial sinusoidal bowing the maximum bending moment is given by

$$\frac{M_{max}}{M_0} = \frac{1}{1 - P/P_E}$$

where M_0 is now the maximum bending moment due to the initial bowing (ignoring the further deflection due to bending).

Plot M_{max}/M_0 against P/P_E for the same range of values as before.
Answer. 1·00, 1·10, 1·19, 1·33, 1·56, 1·96, 2·78, 5·26.

3. If, in Problem 1, the strut is straight and the end load is axial but there is a central lateral point load W, show that

$$M_{max}/M_0 = \tan\theta/\theta$$

where M_0 is now the maximum bending moment when $P = 0$ (i.e. $M_0 = WL/4$) and $\theta = \frac{1}{2}\pi\sqrt{(P/P_E)}$.

Plot M_{max}/M_0 against P/P_E for the same range of values as before.

If, however, the lateral load is uniformly distributed over the length of the strut, show that

$$\frac{M_{max}}{M_0} = \frac{2}{\theta^2}(\sec\theta - 1)$$

Plot M_{max}/M_0 against P/P_E on the same axes as for the central point load case. (*Note*: $M_0 = WL/8$, and, for small angles $\sec\theta \simeq 1 + \frac{1}{2}\theta^2$.)

Answer.

	P/P_E	0	0·09	0·16	0·25	0·36	0·49	0·64	0·81
$\frac{M_{max}}{M_0}$	Point load	1·00,	1·08,	1·16,	1·27,	1·46,	1·79,	2·45,	4·47
	UDL	1·00,	1·10,	1·20,	1·34,	1·58,	1·99,	2·83,	5·39

4. A pin-jointed strut length L carries an axial end load P. In addition there are couples (each M_0) applied at the ends, clockwise at one end and counterclockwise at the other. If $\alpha^2 = P/EI$ where EI is the flexural rigidity, show that the maximum bending moment in the strut is $M_0 \sec \alpha L/2$.

Obtain expressions for the maximum deflection (in terms of M_0, L, E and I) (*a*) when P is one-half of the Euler critical load and (*b*) when P is zero.

Answer. (*a*) $0{\cdot}254\ M_0L^2/EI$; (*b*) $M_0L^2/8EI$ (compare with Example 5.1 of Chapter 5).

5. A tubular strut is 6 cm external and 5 cm internal diameter. It is 2 m long and has hinged ends. The load is parallel to the axis but eccentric to it. Find the maximum allowable eccentricity for a crippling load 0·75 × Euler load, the yield stress being 300 MN/m². Work from first principles. $E = 206$ GN/m². (*U.L.*)

Answer. 2·83 mm.

6. A long slender strut, originally straight and securely fixed at one end and (completely) free at the other end, is loaded at the free-end with an eccentric load whose line of action is parallel to the original axis of the strut. Deduce an expression for the deviation of the free end from its original position.

Determine this deviation and the greatest compressive stress for a steel strut complying with the above conditions; length 3 m; circular cross-section 5 cm external diameter and 2·5 cm internal diameter; load 3·6 kN and original eccentricity 7·5 cm. ($E = 206$ GN/m².) (*U.L.*)

Answer. $e(\sec \alpha L - 1)$ where $\alpha^2 = P/EI$; 26·4 mm; 34·1 MN/m².

7. Derive a formula for the maximum compressive stress induced in an initially straight, slender, uniform strut when loaded along an axis having an eccentricity e at both ends, which are pin-jointed.

A straight steel pin-jointed strut is 5 cm diameter and 1·25 m long. Calculate:

(*a*) the (Euler) crippling load when loaded along the central axis;
(*b*) the eccentricity which will cause failure at 75 per cent of this load if the yield-point stress of the material is 270 MN/m². $E = 206$ GN/m². (*U.L.*)

Answer. (*a*) 399 kN, (*b*) 1·01 mm.

8. A slender strut of uniform section and length L has ends which are fixed in position but free in direction. A force P acts at each end of the strut in a direction parallel with the axis of the strut and at a distance e from the axis but the two forces act on opposite sides of the neutral layer, i.e. their distance apart is $2e$. Show that

(*a*) the condition for maximum bending moment is given by
$$\tan mx = -\cot \tfrac{1}{2}mL$$
where $m = \sqrt{(P/EI)}$ and x is the distance from an end to the section where maximum bending moment occurs, and

(*b*) maximum bending moment occurs at the ends if
$$P < \pi^2 EI/L^2$$
(*U.L.*)

Hint. There are equal and opposite lateral reactions at the supports of $2Pe/L$. Since deflections are not required it is convenient to differentiate the bending moment equation twice and obtain d^2M/dx^2 as in Examples 8.7 and 8.8.

9. A long strut of constant section is initially straight. A thrust is applied eccentrically at both ends and on the same side of the centre-line with the eccentricity at one end twice that at the other. The eccentricities are small compared with the length of the strut.

If the length of the strut is L and the thrust P, show that the maximum bending-moment occurs at a distance x from the end with the smaller eccentricity where
$$\tan mx = \frac{2 - \cos mL}{\sin mL} \quad \text{and} \quad m = \sqrt{(P/EI)}$$

If in the above problem $L = 0{\cdot}75$ m and the strut is circular and 2·5 cm diam., calculate the value of the eccentricities which will produce a maximum stress of 300 MN/m² with a thrust P of 35 kN. Take $E = 200$ GN/m².

Hint. Differentiate bending moment equation twice as in Examples 8.7 and 8.8.
Answer. 2·8 mm and 5·6 mm.

10. (*a*) An initially straight, slender strut of uniform flexural rigidity EI and length l is built-in at both ends and loaded with an axial compressive load P. One end of the strut, and the force P acting at that end, are displaced laterally an amount e, so that they remain parallel to their original direction. If the end fixing moments, both of the same sense, are each denoted M, prove that
$$M = \frac{Pe(1 - \cos ml)}{2 - 2\cos ml - ml \sin ml}$$
where $m = \sqrt{(P/EI)}$.

(*b*) A steel tube 2·5 m long, having outside and inside diameters of 5 cm and $3\frac{1}{2}$ cm respectively, is used as a strut in the manner described in (*a*). If $P = 7000$ N and $e = 16$ mm find the value of M and also the maximum normal stress in the material. $E = 200$ GN/m². (*U.L.*)
Answer. 702 N-m; 78·5 MN/m².

11. A hollow circular steel strut with its ends position-fixed has a length of 2·5 m, its external diameter being 10 cm and its internal diameter 8 cm. Before loading the strut is bent with a maximum deviation of 5 mm. Assuming that the centre line of the strut is sinusoidal, determine the maximum stress due to a central compressive end load of 100 kN. $E = 200$ GN/m². (*U.L.*)
Answer. 45·1 MN/m².

12. A slender strut 1·5 m long and of rectangular section 2·5 cm × 1 cm transmits a longitudinal thrust P acting at the centre at each end. The strut was slightly bent about its minor principal axis before loading.

If P is increased from 200 N to 600 N the deflection at the middle of the length increases by 1·4 mm; determine the amount of deflection before loading.

Find also the total deflection and the maximum stress when P is 1200 N. $E = 205$ GN/m². (*U.L.*)

Answer. 4·0 mm; 11·1 mm and 36·8 MN/m² (compressive).

13. A straight circular section strut of length L has an applied axial compressive load P. It is loaded at the centre with a load W acting at right angles to its axis.

Prove that the maximum bending moment is

$$-\frac{W}{2m}\tan\tfrac{1}{2}mL \quad \text{where } m = \sqrt{(P/EI)}$$

and derive a formula for the central deflection.

If the strut is of steel 2·5 cm dia. and 1·25 m long with an axial load of 16 kN calculate the value of W which will cause collapse if the yield point stress is 280 MN/m² and $E = 206$ GN/m². (*U.L.*)

Answer. $\dfrac{W}{2mP}\tan\tfrac{1}{2}mL - \dfrac{WL}{4P}$; 496 N

14. A straight horizontal rod of steel 3 m long and 3 cm in diameter is freely supported at its ends. An axial thrust of P is applied to each end. Find the greatest value if the maximum stress is not to exceed 36 MN/m². Density of steel 7·83 g/cm³; $E = 200$ GN/m². (*U.L.*)

Hint. A "trial-and-error" solution is required; calculate P_E and take values of P between 0 and P_E.

Answer. $P = 2\,450$ N.

15. A slender strut of uniform section and length L and with pin-jointed ends, is subjected to an axial thrust P and to a distributed lateral load of intensity increasing uniformly from zero at each end to a maximum of w per unit length at the mid-section of the strut. Show that the maximum bending moment is

$$M_{max} = \frac{w}{m^2}\left(\frac{2}{mL}\tan\tfrac{1}{2}mL - 1\right)$$

where $m^2 = P/EI$.

Further, show by using the approximation

$$\tan\theta = \theta + \theta^3/3 + 2\theta^5/15$$

that the value of M_{max} may be obtained by using the approximate formula

$$M_{max} = \mu + P\Delta$$

where μ is the maximum bending moment and Δ the maximum deflection due to the lateral load only. ($\Delta = wL^4/120\,EI$). (*U.L.*)

16. In the notation of the previous example show that the deflection at mid-span is

$$\frac{w}{P}\left(\frac{2}{m^3L}\tan\tfrac{1}{2}mL - \frac{L^2}{12} - \frac{1}{m^2}\right)$$

17. A pinned strut AB of constant section carries an axial load P and terminal couples M_A and M_B both deflecting the strut in the same direction. Obtain a formula for the maximum stress in the strut. (*U.L.*)
Answer.

$$\frac{P}{A} + \frac{\sqrt{[M_A^2 + M_B^2 - 2M_AM_B \cos \alpha L]}}{Z \sin \alpha L}$$

where A and Z are the section area and modulus respectively and $\alpha^2 = P/EI$.

18. An initially straight slender strut of uniform section and length l has hinged ends through which it is axially loaded by a force P. The magnitude of P is between $\pi^2 EI/l^2$ and $4\pi^2 EI/l^2$. Equal opposite couples M_0 are applied to the ends to maintain equilibrium, the slope at each end then being ϕ radians. Show that

(*a*) $M_0 = \dfrac{P\phi}{m}(-\cot \tfrac{1}{2}ml)$

(*b*) the maximum deflection $\Delta = \dfrac{\phi}{m} \tan \tfrac{1}{4}ml$

where $m = \sqrt{(P/EI)}$.

If the strut is a steel bar 1 m long and of rectangular section 6 cm × 2 cm; $P = 2\pi^2 EI/l^2$ and $\phi = 0{\cdot}01$ radian, find the values of M_0 and Δ. $E = 206$ GN/m². (*U.L.*)

Answer. (*a*) 280 N-m; (*b*) 4·54 mm.

19. A strut of length $2a$ has each end fixed in an elastic material which exerts a restraining moment μ per unit angular displacement (one radian). Prove that the critical load P is given by the equation

$$\mu n \tan na + P = 0$$

where $n^2 = P/EI$.

Such a strut, 2·5 m in length, has a theoretical critical load of 15 500 N on the assumption that the ends are pinned, fixed in position but not in direction. Determine the percentage increase in the critical load if the constraint offered at the ends is 170 N-m per degree of rotation. (*U.L.*)

Hint. A "trial-and-error" solution is required; take values of P between 1 and 4 times the critical load for pinned ends.
Answer. 79·1 per cent.

20. An initially straight strut, L m long, has lateral loading w per metre and a longitudinal load P applied with eccentricity e m at both ends. If the strut has area A, relevant second moment of area I and section modulus Z, and the end moments and lateral loading have opposing effects, find an expression for the central bending moment, and show that the maximum stress at the centre will be equal to

$$\frac{P}{A} + \frac{1}{Z}\left[\left(Pe - \frac{wEI}{P}\right) \sec \tfrac{1}{2}L\sqrt{\frac{P}{EI}} + \frac{wEI}{P}\right] \qquad (U.L.)$$

Answer. Maximum bending moment is the numerator of the second term in the given expression.

21. A horizontal bar of uniform section and of length L, is simply supported at its ends. In addition to the uniform load w per unit length due to its own weight, the bar is subjected to longitudinal thrusts F acting at points on the vertical centre lines of the end sections at a distance e below the centres. Show that the resultant maximum bending moment in the beam will have its least possible value if

$$e = \frac{w(\sec \frac{1}{2}mL - 1)}{Fm^2 (\sec \frac{1}{2}mL + 1)}$$

where $m^2 = F/EI$.

If the bar is of steel 2·5 m long of rectangular cross-section 8 cm wide and 2·5 cm deep and weighs 15·7 kgf/m and if the end thrust is 13 kN find the eccentricity e as already defined and also the corresponding maximum deflection. E = 206 GN/m². (*U.L.*)

Hint. The required condition is obtained when the moment at the end is equal and opposite to that at mid-span. Equate $-Pe$ to the maximum bending moment as obtained in the previous example.

Answer. 0·00547 m or 5·47 mm; 0·59 mm (upwards).

22. A slender strut of length L is encastré at one end and pin-jointed at the other. It carries an axial load P and a couple M at the pinned end. If its flexural rigidity is EI and $P/EI = \mu^2$ show that the magnitude of the couple at the fixed end is

$$M \frac{\mu L - \sin \mu L}{\mu L \cos \mu L - \sin \mu L}$$

What is the value of this couple when P is one quarter of the Euler critical load and also when P is zero? (*U.L.*)

Hint. The limiting value of $(\theta - \sin \theta)/(\theta \cos \theta - \sin \theta)$ as $\theta \to 0$ is found by expanding the trigonometrical functions as infinite series and considering the first few terms.

Answer. $0{\cdot}571M$; $\frac{1}{2}M$.

Chapter 9

Complex Stresses

In a two-dimensional stress system having perpendicular normal stresses σ_x and σ_y, together with complementary shear stresses τ_{xy} acting on the same planes, the normal stress σ_n in a direction making an angle θ with that of σ_x is given by

$$\sigma_n = \tfrac{1}{2}(\sigma_x + \sigma_y) + \tfrac{1}{2}(\sigma_x - \sigma_y)\cos 2\theta + \tau_{xy}\sin 2\theta$$

(the sign convention being that of Figs. 9.2 and 9.3).

The corresponding shear stress τ is given by

$$\tau = \tfrac{1}{2}(\sigma_x - \sigma_y)\sin 2\theta + \tau_{xy}\cos 2\theta$$

The principal stresses σ_1 and σ_2 are the roots of the equation

$$(\sigma - \sigma_x)(\sigma - \sigma_y) = {\tau_{xy}}^2$$

from which

$$\sigma_1 = \tfrac{1}{2}(\sigma_x + \sigma_y) + \tfrac{1}{2}\sqrt{[(\sigma_x - \sigma_y)^2 + 4{\tau_{xy}}^2]}$$

and

$$\sigma_2 = \tfrac{1}{2}(\sigma_x + \sigma_y) - \tfrac{1}{2}\sqrt{[(\sigma_x - \sigma_y)^2 + 4{\tau_{xy}}^2]}$$

The directions of the principal stresses relative to that of σ_x are given by

$$\tan 2\theta = \frac{2\tau_{xy}}{\sigma_x - \sigma_y}$$

The maximum shear stress is

$$\tau_{max} = \tfrac{1}{2}(\sigma_1 - \sigma_2) = \tfrac{1}{2}\sqrt{[(\sigma_x - \sigma_y)^2 + 4{\tau_{xy}}^2]}$$

and it acts on planes which are inclined at 45° to the principal planes.

The normal (direct) stress σ_n and tangential (shear) stress τ on a plane making an angle θ with the principal plane on which σ_1 acts are

$$\sigma_n = \tfrac{1}{2}(\sigma_1 + \sigma_2) + \tfrac{1}{2}(\sigma_1 - \sigma_2) \cos 2\theta$$
$$\tau = (-)\,\tfrac{1}{2}(\sigma_1 - \sigma_2) \sin 2\theta$$

the minus sign in front of the expression for τ being necessary for sign convention consistency (see Example 9.3).

For a circular shaft subjected to bending and torsion simultaneously (but without axial load) the equivalent bending moment M_E which, acting alone, would produce the same maximum normal stress is

$$M_E = \tfrac{1}{2}[M + \sqrt{(M^2 + T^2)}]$$

and the equivalent twisting moment T_E which, acting alone, would produce the same maximum shear stress is

$$T_E = \sqrt{(M^2 + T^2)}$$

where M and T are the applied bending and twisting moments respectively. (See also Chapter 16.)

In a three-dimensional stress system there are three mutually perpendicular stresses σ_1, σ_2 and σ_3 and, if $\sigma_1 > \sigma_2 > \sigma_3$ the true maximum shear stress is $\tfrac{1}{2}(\sigma_1 - \sigma_3)$.

WORKED EXAMPLES

9.1. Deduce expressions for the normal and shear stresses on an oblique plane within a bar subjected to simple tension. Calculate the maximum shear stress on any oblique plane of a 50 mm diameter bar subjected to a pull of 100 kN. What planes have a shear stress of 16 MN/m^2? What are the normal stresses on these planes?

Solution. The equilibrium of a bar in tension was first mentioned in Chapter 1 (Example 1.1). There, however, the stress on a cross-section perpendicular to the axis was considered. In order to establish the conditions on an oblique plane it is necessary to consider the equilibrium of a small wedge within the bar, such as ABC (Fig. 9.1a). AB represents a plane perpendicular to the axis of the bar and AC is an oblique plane (or "interface") making an angle θ with AB.

Let the direct tensile stress acting on AB (Fig. 9.1b) be σ and suppose the triangular block ABC is kept in equilibrium by stresses σ_n and τ which are respectively normal and tangential to the plane AC.

If, for convenience, we assume the block to have unit depth

perpendicular to the plane of the paper, then the *forces* acting on it are

$\sigma \times$ AB horizontally to the right,

$\sigma_n \times$ AC perpendicular to the plane AC,

$\tau \times$ AC parallel to (i.e. along) the plane AC.

These forces and the angles between them are shown in Fig. 9.1c.

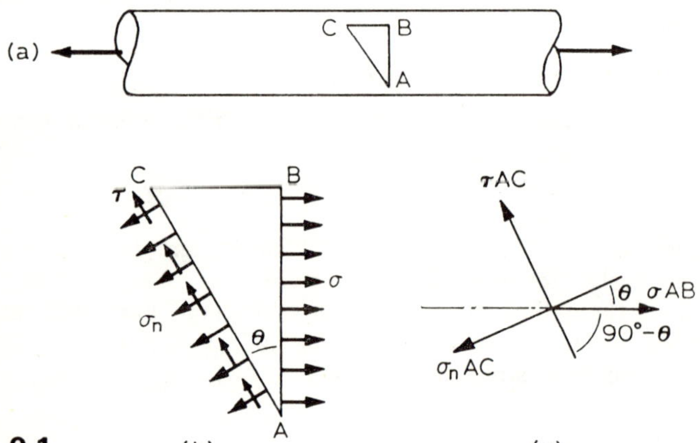

Fig. 9.1 (b) (c)

For equilibrium, the sum of the resolved parts of all the forces in any direction is zero. In all examples it is best to resolve in directions perpendicular and parallel to the inclined plane.

Hence, resolving perpendicular to AC,

$$\sigma_n \times \mathrm{AC} = \sigma \times \mathrm{AB} \cos \theta$$

$$\sigma_n = \sigma \frac{\mathrm{AB}}{\mathrm{AC}} \cos \theta = \sigma \cos^2 \theta \qquad \text{(i)}$$

Resolving in the direction AC,

$$\tau \times \mathrm{AC} = \sigma \times \mathrm{AB} \cos (90^\circ - \theta) = \sigma \times \mathrm{AB} \sin \theta$$

$$\tau = \sigma \frac{\mathrm{AB}}{\mathrm{AC}} \sin \theta = \sigma \cos \theta \sin \theta = \tfrac{1}{2}\sigma \sin 2\theta \qquad \text{(ii)}$$

The expressions (i) and (ii) are independent of the size of the block ABC and, in cases where the stress σ is not uniform, can be considered as applying at a point.

The maximum value of σ_n is clearly σ itself and occurs when

$\theta = 0$ (or 180°), i.e. when AC coincides with AB. σ_n is zero when $\theta = 90°$.

The maximum value of τ is (from (ii)) $\frac{1}{2}\sigma$ and this occurs when $\sin 2\theta = 1$, i.e. when $2\theta = 90°$ and $\theta = 45°$.

τ is zero when $\theta = 0°$ or $90°$.

Similar results are obtained for a bar in compression.

For the bar in the question the direct stress is

$$\sigma = \frac{\text{load}}{\text{area}} = \frac{100\ \text{kN}}{\frac{1}{4}\pi \times (50\ \text{mm})^2} = 50{\cdot}9\ \text{MN/m}^2\ \text{(tensile)}$$

Thus, the maximum shear stress is

$$\tau = \tfrac{1}{2}\sigma = \tfrac{1}{2} \times 50{\cdot}9\ \text{MN/m}^2 = 25{\cdot}45\ \text{MN/m}^2 \qquad (Ans)$$

For a shear stress of 16 MN/m² we have, from (ii),

$$\sin 2\theta = \frac{2\tau}{\sigma} = \frac{2 \times 16\ \text{MN/m}^2}{50{\cdot}9\ \text{MN/m}^2} = 0{\cdot}628$$

$$2\theta = 38°\ 54' \text{ or } (180° - 38°\ 54') = 141°\ 6'$$

$$\theta = 19°\ 27' \text{ or } 70°\ 33'$$

When $\theta = 19°\ 27'$,

$$\begin{aligned}\text{Normal stress } \sigma_n &= \sigma \cos^2 \theta \\ &= 50{\cdot}9\ \text{MN/m}^2 \times (\cos 19°\ 27')^2 \\ &= 45{\cdot}2\ \text{MN/m}^2\ \text{(tensile)} \qquad (Ans)\end{aligned}$$

When $\theta = 70°\ 33'$

$$\begin{aligned}\text{Normal stress } \sigma_n &= 50{\cdot}9\ \text{MN/m}^2 \times (\cos 70°\ 33')^2 \\ &= 5{\cdot}64\ \text{MN/m}^2\ \text{(tensile)} \qquad (Ans)\end{aligned}$$

9.2. Derive formulae for the normal and tangential stresses on an oblique plane within a material subjected to two perpendicular direct stresses.

A piece of steel plate is subjected to perpendicular stresses of 80 and 50 MN/m², both tensile. Calculate the normal and tangential stresses and the magnitude and direction of the resultant stress on the interface whose normal makes an angle of 30° with the axis of the second stress.

Solution. The method is similar to that of Example 9.1. Let P_x and P_y be the forces (Fig. 9.2a) and σ_x and σ_y the corresponding stresses. These are shown as tensile (Fig. 9.2b) and if the sign convention of

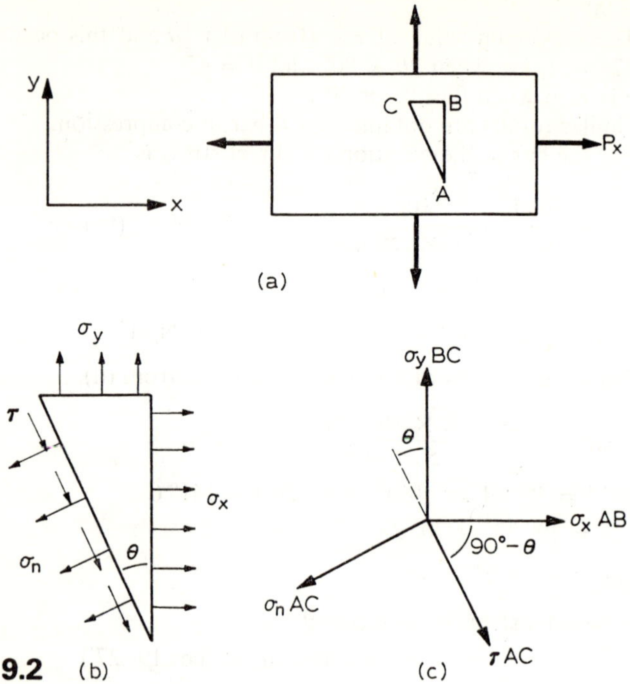

Fig. 9.2

"tensile stresses are positive, compressive are negative" is adopted then the results are applicable to all cases.

The forces acting on the wedge ABC are shown in Fig. 9·2c (again assuming unit depth).

Resolving perpendicular to AC,

$$\sigma_n \times \text{AC} = \sigma_x \times \text{AB} \cos\theta + \sigma_y \times \text{BC} \cos(90° - \theta)$$

or

$$\begin{aligned}\sigma_n &= \sigma_x \frac{\text{AB}}{\text{AC}} \cos\theta + \sigma_y \frac{\text{BC}}{\text{AC}} \sin\theta \\ &= \sigma_x \cos^2\theta + \sigma_y \sin^2\theta \\ &= \sigma_x \left(\frac{1 + \cos 2\theta}{2}\right) + \sigma_y \left(\frac{1 - \cos 2\theta}{2}\right) \\ &= \tfrac{1}{2}(\sigma_x + \sigma_y) + \tfrac{1}{2}(\sigma_x - \sigma_y)\cos 2\theta \qquad \text{(i)}\end{aligned}$$

(since $\cos 2\theta = 2\cos^2\theta - 1 = 1 - 2\sin^2\theta$).

Resolving in the direction AC,

$$\tau \times AC + \sigma_x \times AB \cos(90° - \theta) = \sigma_y \times BC \cos\theta$$

or

$$\tau = -\sigma_x \frac{AB}{AC} \sin\theta + \sigma_y \frac{BC}{AC} \cos\theta$$
$$= -\sigma_x \cos\theta \sin\theta + \sigma_y \sin\theta \cos\theta$$
$$= -(\sigma_x - \sigma_y) \sin\theta \cos\theta$$
$$= -\tfrac{1}{2}(\sigma_x - \sigma_y) \sin 2\theta \qquad \text{(ii)}$$

Figure 9.3 shows the relationship between the forces on the block ABC for the interface given in the problem.

The angle between the normal of the interface and the axis of σ_y is 30°.

Hence

$$\theta = 90° - 30° = 60°$$
$$\sigma_x = 80 \text{ MN/m}^2 \quad \text{and} \quad \sigma_y = 50 \text{ MN/m}^2$$
$$\tfrac{1}{2}(\sigma_x + \sigma_y) = 65 \text{ MN/m}^2 \quad \text{and} \quad \tfrac{1}{2}(\sigma_x - \sigma_y) = 15 \text{ MN/m}^2$$

Using formulae (i) and (ii), the normal and tangential stresses on the interface are

$$\sigma_n = \tfrac{1}{2}(\sigma_x + \sigma_y) + \tfrac{1}{2}(\sigma_x - \sigma_y)\cos 2\theta$$
$$= 65 + (15 \times \cos 120°) = 57{\cdot}5 \text{ MN/m}^2 \qquad (\textit{Ans})$$

and

$$\tau = -\tfrac{1}{2}(\sigma_x - \sigma_y)\sin 2\theta$$
$$= -15 \times \sin 120° = (-)\ 13 \text{ MN/m}^2 \qquad (\textit{Ans})$$

Since the stresses σ_n and τ apply to the same area (AC) then their resultant σ_r (also based on AC) is, from Fig. 9.3,

$$\sigma_r = \sqrt{[\sigma_n^2 + \tau^2]}$$
$$= \sqrt{[57{\cdot}5^2 + (-13)^2]}$$
$$= 58{\cdot}95 \text{ MN/m}^2 \qquad (\textit{Ans})$$

The angle made by this resultant with the normal is given by

$$\tan\phi = \tau/\sigma_n = -13/57{\cdot}5 = -0{\cdot}226$$
$$\phi = -12°\,44'$$

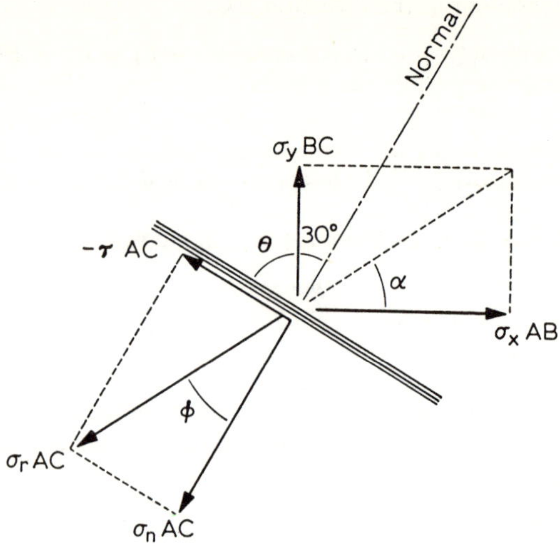

Fig. 9.3

Also, since the three *forces* $\sigma_r \times$ AC, $\sigma_x \times$ AB and $\sigma_y \times$ BC must together be in equilibrium, then (again using Fig. 9.3) the angle made by the resultant with the axis of σ_x is α, where

$$\tan \alpha = \frac{\sigma_y \times \text{BC}}{\sigma_x \times \text{AB}} = \frac{\sigma_y}{\sigma_x} \tan \theta \tag{iii}$$

With the values given,

$$\tan \alpha = \frac{50}{80} \times \tan 60° = 1{\cdot}08$$

$$\alpha = 47° \, 16' \qquad (Ans)$$

As a check, $\alpha - \phi = \theta$ (from the geometry of Fig. 9.3) and, using the results obtained,

$$47° \, 16' - (-12° \, 44') = 60°$$

9.3. What are "complementary shear stresses"? Derive expressions for the normal and tangential stresses on an interface of a piece of material in pure shear.

What are the numerical values of the stresses on planes making angles of 15° and 65° with the directions of the applied stresses, which are 46 MN/m^2?

Solution. In this and later examples it is convenient to adopt the following sign conventions. Perpendicular axes x and y are taken with the positive directions being those of Fig. 9.4a. Normal stresses are positive when tensile, i.e. when they are directed outwards from the planes on which they act. A shear stress is positive if it and the outward normal to the plane on which it acts have the same signs in respect to the positive directions of the x and y co-ordinate axes. These definitions are used in the following analysis.

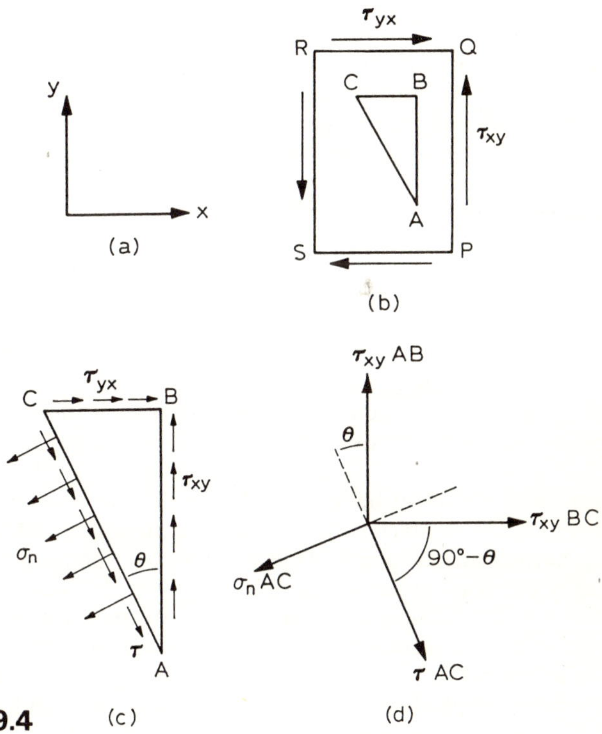

Fig. 9.4

Consider a rectangular block PQRS (Fig. 9.4b) so small that the stresses on its faces can be considered uniform. A shear stress is applied to the faces PQ and RS. The outward normal to the face PQ is in the positive x-direction. The shear stress on PQ is therefore positive when acting in the positive direction of y, i.e. vertically upwards. It is denoted by τ_{xy}. (Notice that the absence of a normal stress on PQ does not affect this definition.)

The shear stresses on PQ and RS produce an anti-clockwise couple on the block. For equilibrium, therefore, there must be an equal and opposite couple. This can only be provided by shear stresses along the faces PS and RQ as shown, since uniform normal stresses cannot produce a couple. The stress which is "called out" on the faces PS and RQ is said to be complementary to the applied stress. This is denoted by τ_{yx} and the reader should check that the directions of the stresses shown on the faces of PQRS are all positive in accordance with the sign convention now adopted.

With unit depth of material the forces along the four faces are

$$\tau_{xy} \times \mathrm{PQ} \qquad \tau_{yx} \times \mathrm{RQ} \qquad \tau_{xy} \times \mathrm{SR} \qquad \tau_{yx} \times \mathrm{PS}$$

The couple due to the stresses τ_{xy} is $(\tau_{xy} \times \mathrm{PQ}) \times \mathrm{PS}$ (anti-clockwise) and that due to the stresses τ_{yx} is $(\tau_{yx} \times \mathrm{PS}) \times \mathrm{PQ}$ (clockwise) since PQ = SR and PS = RQ.

Equating these two couples for equilibrium

$$\tau_{xy} \times \mathrm{PQ} \times \mathrm{PS} = \tau_{yx} \times \mathrm{PS} \times \mathrm{PQ}$$

Hence

$$\tau_{xy} = \tau_{yx}$$

The stresses τ_{xy} and τ_{yx} are called complementary shear stresses. In all cases, whether there are normal stresses acting or not, the intensities of shear stresses across two planes at right angles are equal.

A material is said to be in pure (or simple) shear when it is subjected to complementary shear stresses only as in Fig. 9.4a.

The normal and tangential (shear) stresses on oblique planes are found by considering the wedge ABC, Fig. 9.4c. In accordance with the sign convention now adopted the shear stress τ on the plane AC is positive in the direction C to A. The forces acting on the wedge are shown in Fig. 9.4d.

Resolving perpendicular to AC in the usual way and replacing τ_{yx} by τ_{xy}, to which it is equal,

$$\sigma_n \mathrm{AC} = \tau_{xy} \times \mathrm{BC} \cos\theta + \tau_{xy} \times \mathrm{AB} \cos(90° - \theta)$$

$$\sigma_n = \tau_{xy} \frac{\mathrm{BC}}{\mathrm{AC}} \cos\theta + \tau_{xy} \frac{\mathrm{AB}}{\mathrm{AC}} \sin\theta$$

$$= \tau_{xy} \sin\theta \cos\theta + \tau_{xy} \cos\theta \sin\theta$$

$$= 2\tau_{xy} \sin\theta \cos\theta$$

$$= \tau_{xy} \sin 2\theta \qquad \text{(i)}$$

Resolving in the direction AC

$$\tau \times AC = \tau_{xy}\, AB \cos\theta - \tau_{xy}\, BC \cos(90° - \theta)$$

$$\tau = \tau_{xy}\frac{AB}{AC}\cos\theta - \tau_{xy}\frac{BC}{AC}\sin\theta$$

$$= \tau_{xy}\cos^2\theta - \tau_{xy}\sin^2\theta$$

$$= \tau_{xy}\cos 2\theta \qquad \text{(ii)}$$

(since $\cos 2\theta = \cos^2\theta - \sin^2\theta$).

If $\tau = 46$ MN/m² and $\theta = 15°$ then, using (i) and (ii),

$$\sigma_n = 46 \sin 30° = 23 \text{ MN/m}^2 \qquad (Ans)$$

$$\tau = 46 \cos 30° = 39{\cdot}8 \text{ MN/m}^2 \qquad (Ans)$$

When $\theta = 65°$

$$\sigma_n = 46 \sin 130° = 35{\cdot}3 \text{ MN/m}^2 \qquad (Ans)$$

$$\tau = 46 \cos 130° = -29{\cdot}6 \text{ MN/m}^2 \qquad (Ans)$$

9.4. Derive expressions for the normal and tangential stresses on an interface for the case of two perpendicular direct stresses together with complementary shear stresses on the same planes.

When a certain thin-walled tube is subjected to internal pressure and torque the stresses in the tube wall are

(*a*) 60 MN/m² tensile,
(*b*) 30 MN/m² tensile in a direction at right angles to (*a*),
(*c*) complementary shear stresses of 45 MN/m² in the directions (*a*) and (*b*).

Calculate the normal and tangential stresses on the two planes which are equally inclined to (*a*) and (*b*).

What are the results if, due to an end thrust, (*b*) is compressive, (*a*) and (*c*) being unchanged?

Solution. The required formulae can be obtained from the results established in the previous examples.

From equation (i) of Example 9.2, and equation (i) of Example 9.3, the normal stress on any interface is

$$\sigma_n = \tfrac{1}{2}(\sigma_x + \sigma_y) + \tfrac{1}{2}(\sigma_x - \sigma_y)\cos 2\theta + \tau_{xy}\sin 2\theta \qquad \text{(i)}$$

Similarly the tangential stress is (using equation (ii) of Example 9.2, and equation (ii) of Example 9.3).

$$\tau = -\tfrac{1}{2}(\sigma_x - \sigma_y)\sin 2\theta + \tau_{xy}\cos 2\theta \qquad \text{(ii)}$$

(The reader is advised to obtain these expressions for himself from first principles, using an extension of the method shown in Examples 9.1, 2 and 3.)

The interface is equally inclined to the directions (*a*) and (*b*) when $\theta = 45°$, 135°, 225° and 315°. It is clear, however, that the last two values give the same planes as the first two. For this reason, only 45° and 135° are considered.

For all angles, $\sigma_x = 60$ MN/m²; $\sigma_y = 30$ MN/m²

$$\tau_{xy} = 45 \text{ MN/m}^2$$
$$\tfrac{1}{2}(\sigma_x + \sigma_y) = 45 \text{ MN/m}^2; \ \tfrac{1}{2}(\sigma_x - \sigma_y) = 15 \text{ MN/m}^2$$

When $\theta = 45°$,

$$\begin{aligned}\sigma_n &= \tfrac{1}{2}(\sigma_x + \sigma_y) + \tfrac{1}{2}(\sigma_x - \sigma_y)\cos 2\theta + \tau_{xy}\sin 2\theta \\ &= 45 + (15 \times \cos 90°) + (45 \times \sin 90°) \\ &= 90 \text{ MN/m}^2 \qquad (Ans)\end{aligned}$$

and

$$\begin{aligned}\tau &= -\tfrac{1}{2}(\sigma_x - \sigma_y)\sin 2\theta + \tau_{xy}\cos 2\theta \\ &= -(15 \times \sin 90°) + (45 \times \cos 90°) \\ &= -15 \text{ MN/m}^2 \qquad (Ans)\end{aligned}$$

Similarly, when $\theta = 135°$,

$$\begin{aligned}\sigma_n &= 45 + (15 \times \cos 270°) + (45 \times \sin 270°) \\ &= 0 \qquad (Ans)\end{aligned}$$

and

$$\begin{aligned}\tau &= -(15 \times \sin 270°) + (45 \times \cos 270°) \\ &= 15 \text{ MN/m}^2 \qquad (Ans)\end{aligned}$$

If σ_x and σ_y are taken to be positive when tensile and negative when compressive, then the same convention will apply to the results for the normal stress (σ_n). In the case with end thrust, therefore,

$$\sigma_x = 60 \text{ MN/m}^2; \ \sigma_y = -30 \text{ MN/m}^2$$
$$\tfrac{1}{2}(\sigma_x + \sigma_y) = +15 \text{ MN/m}^2 \text{ and } \tfrac{1}{2}(\sigma_x - \sigma_y) = +45 \text{ MN/m}^2$$

When $\theta = 45°$,

$$\sigma_n = 15 + (45 \times \cos 90°) + (45 \times \sin 90°)$$
$$= 60 \text{ MN/m}^2 \text{ (tensile)} \qquad (Ans)$$

and

$$\tau = -(45 \times \sin 90°) + (45 \times \cos 90°)$$
$$= -45 \text{ MN/m}^2 \qquad (Ans)$$

Similarly, when $\theta = 135°$,

$$\sigma_n = 15 + (45 \times \cos 270°) + (45 \times \sin 270°)$$
$$= -30 \text{ MN/m}^2\text{, i.e. } 30 \text{ MN/m}^2 \text{ (compressive)} \qquad (Ans)$$

and

$$\tau = -(45 \times \sin 270°) + (45 \times \cos 270°)$$
$$= 45 \text{ MN/m}^2 \qquad (Ans)$$

9.5. What are "principal stresses"? Obtain expressions for the principal stresses in the case of two perpendicular direct stresses together with complementary shear stresses on the same planes. Calculate the principal stresses for the thin-walled tube of Example 9.4.

Solution. In any two-dimensional stress system there are two planes (which will be shown to be mutually perpendicular) on which the shear (or tangential) stress τ is zero. These planes are called *principal planes* and the corresponding values of σ_n are called the *principal stresses*. It is clear that the principal stresses are the resultant stresses for the principal planes.

Suppose that AC in Fig. 9.5 is a principal plane and σ is the corresponding principal stress. Let θ be the angle made by the principal

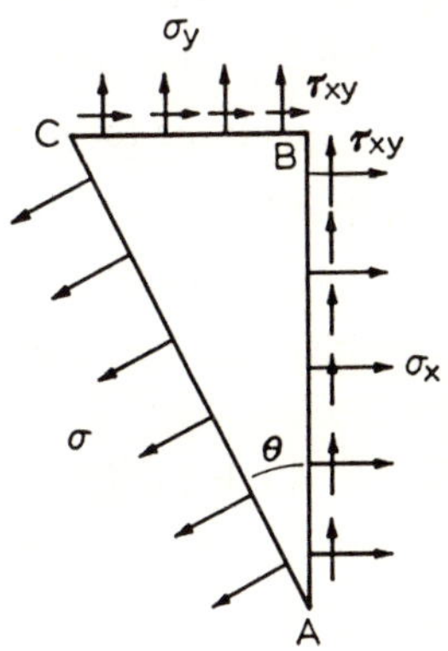

Fig. 9.5

plane with the plane on which σ_x acts. The *forces* acting on ABC are (considering unit depth of material):

σ_xAB + τ_{xy}BC horizontally to the right
σ_yBC + τ_{xy}AB vertically upwards
σAC perpendicular to AC (i.e. at an angle θ to the horizontal)

Resolving horizontally

$$\sigma \mathrm{AC} \cos\theta = \sigma_x \mathrm{AB} + \tau_{xy}\,\mathrm{BC}$$

$$\sigma\cos\theta = \sigma_x\frac{\mathrm{AB}}{\mathrm{AC}} + \tau_{xy}\frac{\mathrm{BC}}{\mathrm{AC}}$$

$$= \sigma_x\cos\theta + \tau_{xy}\sin\theta$$

Hence, dividing by cos θ and re-arranging,

$$\sigma - \sigma_x = \tau_{xy}\tan\theta \qquad \text{(i)}$$

Similarly, resolving vertically,

$$\sigma \mathrm{AC}\cos(90^\circ - \theta) = \sigma_y\,\mathrm{BC} + \tau_{xy}\,\mathrm{AB}$$

$$\sigma\sin\theta = \sigma_y\frac{\mathrm{BC}}{\mathrm{AC}} + \tau_{xy}\frac{\mathrm{AB}}{\mathrm{AC}}$$

$$= \sigma_y\sin\theta + \tau_{xy}\cos\theta$$

from which

$$\sigma - \sigma_y = \tau_{xy}\cot\theta \qquad \text{(ii)}$$

Multiplying (i) by (ii)

$$(\sigma - \sigma_x)(\sigma - \sigma_y) = (\tau_{xy}\tan\theta \times \tau_{xy}\cot\theta) = \tau_{xy}^2$$

Removing brackets and rearranging,

$$\sigma^2 - (\sigma_x + \sigma_y)\sigma + (\sigma_x\sigma_y - \tau_{xy}^2) = 0$$

and, solving this quadratic for σ, the principal stresses are

$$\sigma = \tfrac{1}{2}\{(\sigma_x + \sigma_y) \pm \sqrt{[(\sigma_x + \sigma_y)^2 - 4(\sigma_x\sigma_y - \tau_{xy}^2)]}\}$$

$$= \tfrac{1}{2}(\sigma_x + \sigma_y) \pm \tfrac{1}{2}\sqrt{[(\sigma_x - \sigma_y)^2 + 4\tau_{xy}^2]}$$

Taking the positive and negative square roots separately, the two principal stresses are

$$\sigma_1 = \tfrac{1}{2}(\sigma_x + \sigma_y) + \tfrac{1}{2}\sqrt{[(\sigma_x - \sigma_y)^2 + 4\tau_{xy}^2]}$$

and

$$\sigma_2 = \tfrac{1}{2}(\sigma_x + \sigma_y) - \tfrac{1}{2}\sqrt{[(\sigma_x - \sigma_y)^2 + 4\tau_{xy}^2]}$$

With the original values of Example 9.4,

$$\sigma_x + \sigma_y = +90 \qquad \sigma_x - \sigma_y = +30 \qquad \tau_{xy} = 45$$

$$\sigma_1 = \tfrac{90}{2} + \tfrac{1}{2}\sqrt{[30^2 + (4 \times 45^2)]} = 45 + 47{\cdot}44$$

$$= (+)\ 92{\cdot}44\ \text{MN/m}^2\ \text{(tensile)} \qquad (\textit{Ans})$$

$$\sigma_2 = 45 - 47{\cdot}44 = -2{\cdot}44\ \text{MN/m}^2,\ \text{i.e. } 2{\cdot}44\ \text{MN/m}^2\ \text{(compressive)} \qquad (\textit{Ans})$$

After the end thrust is applied,

$$\sigma_x + \sigma_y = +30 \qquad \sigma_x - \sigma_y = +90 \qquad \tau_{xy} = 45$$

$$\sigma_1 = \tfrac{30}{2} + \tfrac{1}{2}\sqrt{[90^2 + (4 \times 45^2)]} = 15 + 63{\cdot}65$$

$$= (+)\ 78{\cdot}65\ \text{MN/m}^2\ \text{(tensile)} \qquad (\textit{Ans})$$

$$\sigma_2 = 15 - 63{\cdot}65 = -48{\cdot}65\ \text{MN/m}^2,\ \text{i.e. } 48{\cdot}65\ \text{MN/m}^2\ \text{(compressive)} \qquad (\textit{Ans})$$

9.6. Show that the principal stresses are the extreme values of the normal stress for any interface under conditions of complex stress. A 50 mm diameter bar is subjected to a pull of 70 kN and a torque of 1·25 kN-m. Calculate the principal stresses for a point on the surface of the bar and show by a diagram the relation between the principal planes and the axis of the bar.

Solution. From equations (i) and (ii) of Example 9.4, the normal and tangential stresses on any interface are

$$\sigma_n = \tfrac{1}{2}(\sigma_x + \sigma_y) + \tfrac{1}{2}(\sigma_x - \sigma_y)\cos 2\theta + \tau_{xy}\sin 2\theta \qquad \text{(i)}$$

$$\tau = -\tfrac{1}{2}(\sigma_x - \sigma_y)\sin 2\theta + \tau_{xy}\cos 2\theta \qquad \text{(ii)}$$

The principal planes correspond to $\tau = 0$, and hence they make angles with the plane on which σ_x acts, given by

$$0 = -\tfrac{1}{2}(\sigma_x - \sigma_y)\sin 2\theta + \tau_{xy}\cos 2\theta$$

$$\tan 2\theta = 2\tau_{xy}/(\sigma_x - \sigma_y) \qquad \text{(iii)}$$

Differentiating the expression for σ_n to find its maximum and minimum values,

$$\frac{d\sigma_n}{d\theta} = \tfrac{1}{2}(\sigma_x - \sigma_y) \times (-2\sin 2\theta) + 2\tau_{xy}\cos 2\theta$$

For a maximum or minimum $d\sigma_n/d\theta$ is zero, and θ is given by

$$0 = -(\sigma_x - \sigma_y)\sin 2\theta + 2\tau_{xy}\cos 2\theta$$

$$\tan 2\theta = 2\tau_{xy}/(\sigma_x - \sigma_y) \qquad \text{(iv)}$$

Since the values of θ found from (iii) and (iv) are identical then the principal planes correspond to the planes of maximum and minimum normal stress. The words "maximum" and "minimum" are here used in the mathematical sense. The principal stresses may be both tensile, both compressive or one tensile and one compressive according to the nature and magnitude of the applied stresses.

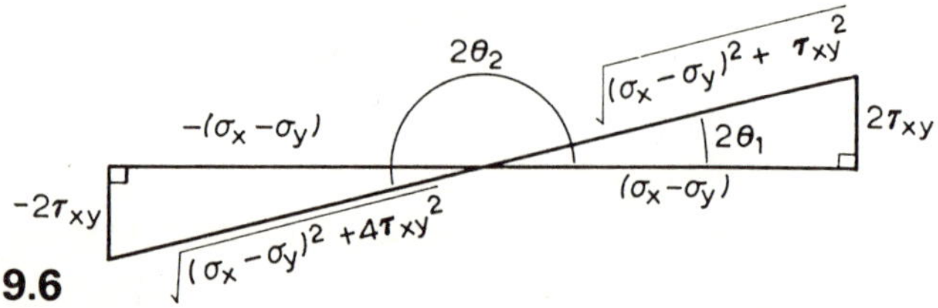

Fig. 9.6

As shown in Fig. 9.6, there are two values of 2θ which satisfy (iii) and (iv). Denoting these values by $2\theta_1$ and $2\theta_2$ we have

$$\sin 2\theta_1 = 2\tau_{xy}/\sqrt{[(\sigma_x - \sigma_y)^2 + 4\tau_{xy}^2]}$$
$$\cos 2\theta_1 = (\sigma_x - \sigma_y)/\sqrt{[(\sigma_x - \sigma_y)^2 + 4\tau_{xy}^2]}$$
$$\sin 2\theta_2 = -2\tau_{xy}/\sqrt{[(\sigma_x - \sigma_y)^2 + 4\tau_{xy}^2]}$$
$$\cos 2\theta_2 = -(\sigma_x - \sigma_y)/\sqrt{[(\sigma_x - \sigma_y)^2 + 4\tau_{xy}^2]}$$

(assuming the positive square root in each case).

It is clear that the angles $2\theta_1$ and $2\theta_2$ differ by 180°. Hence, θ_1 and θ_2 differ by 90° so that the two principal planes are mutually perpendicular.

Substituting the values for $\sin 2\theta_1$ and $\cos 2\theta_1$ found above in equation (i), the corresponding principal stress is

$$\sigma_1 = \tfrac{1}{2}(\sigma_x + \sigma_y) + \tfrac{1}{2}(\sigma_x - \sigma_y)\left\{\frac{(\sigma_x - \sigma_y)}{\sqrt{[(\sigma_x - \sigma_y)^2 + 4\tau_{xy}^2]}}\right\}$$
$$+ \tau_{xy}\left\{\frac{2\tau_{xy}}{\sqrt{[(\sigma_x - \sigma_y)^2 + 4\tau_{xy}^2]}}\right\}$$
$$= \tfrac{1}{2}(\sigma_x + \sigma_y) + \tfrac{1}{2}\left[\frac{(\sigma_x - \sigma_y)^2 + 4\tau_{xy}^2}{\sqrt{[(\sigma_x - \sigma_y)^2 + 4\tau_{xy}^2]}}\right]$$
$$= \tfrac{1}{2}(\sigma_x + \sigma_y) + \tfrac{1}{2}\sqrt{[(\sigma_x - \sigma_y)^2 + 4\tau_{xy}^2]}$$

This result has been obtained by a different method in Example 9.5. The reader should verify that the ratios of $2\theta_2$ lead to the appropriate result for σ_2.

For the bar in the question, suppose σ_x is the direct tensile stress.

$$\sigma_x = \text{load/area} = 70\ \text{kN}/\tfrac{1}{4}\pi \times (50\ \text{mm})^2 = 35{\cdot}7\ \text{MN/m}^2$$

$\sigma_y = 0$ (since there is no other applied direct stress) and, from the torsion equation,

$$\tau = \frac{T}{J}r = \frac{T}{\frac{1}{16}\pi d^3} = \frac{1{\cdot}25\ \text{kN-m}}{\frac{1}{16}\pi \times (50\ \text{mm})^3} = 50{\cdot}9\ \text{MN/m}^2$$

Using these values, the principal stresses are

$$\sigma_1 = \tfrac{1}{2}(35{\cdot}7 + 0) + \tfrac{1}{2}\sqrt{[(35{\cdot}7 - 0)^2 + (4 \times 50{\cdot}9^2)]}$$

$$= 17{\cdot}85 + 53{\cdot}95 = 71{\cdot}8\ \text{MN/m}^2\ \text{(tensile)} \qquad (Ans)$$

$$\sigma_2 = 17{\cdot}85 - 53{\cdot}95 = 36{\cdot}1\ \text{MN/m}^2\ \text{(compressive)} \qquad (Ans)$$

The angles made by the principal planes with the plane on which σ_x acts are given by

$$\tan 2\theta = 2\tau_{xy}/(\sigma_x - \sigma_y) = 2 \times 50{\cdot}9/(35{\cdot}7 - 0) = 2{\cdot}8515$$

$$\theta_1 = 35^\circ\ 20' \qquad \theta_2 = \theta_1 + 90^\circ = 125^\circ\ 20'$$

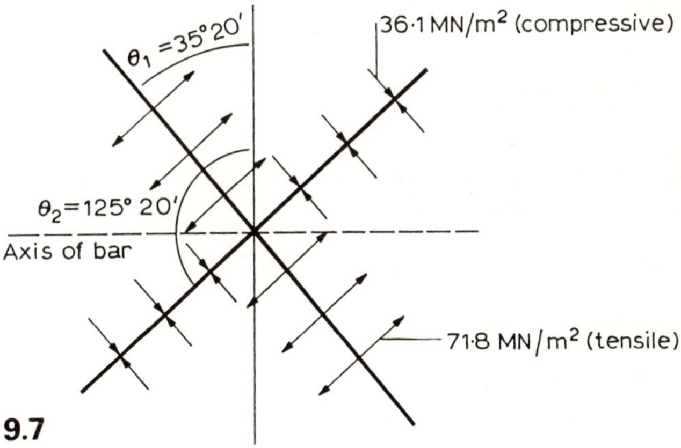

Fig. 9.7

θ_1 and θ_2 are the angles made by the principal planes with a cross-section of the bar which is perpendicular to the axis. They are thus the angles made by the normals of the principal planes with the axis of the bar (see Fig. 9.7).

9.7. Derive a formula for the maximum shear stress on any interface under conditions of complex stress. Discuss the cases of complex stress when one (or both) of the direct applied stresses is zero and when there is no applied shear stress.

At a point in the web of a girder the bending stress is 80 MN/m² (tensile) and the shearing stress at the same point is 30 MN/m². Calculate:

(*a*) the principal stresses,
(*b*) the maximum shear stress,
(*c*) the tensile stress which, when acting alone, would produce the same maximum shear stress.
(*d*) the shearing stress which, when acting alone, would produce the same maximum principal stress.

Draw a diagram showing the relationship between the principal planes and the planes of maximum shear stress.

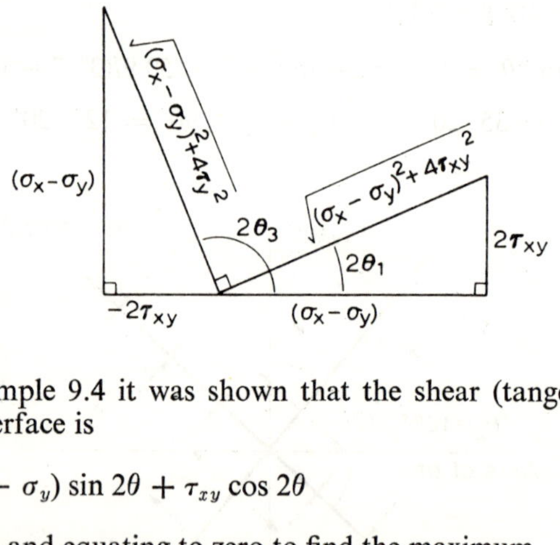

Fig. 9.8

Solution. In Example 9.4 it was shown that the shear (tangential) stress on any interface is

$$\tau = -\tfrac{1}{2}(\sigma_x - \sigma_y)\sin 2\theta + \tau_{xy}\cos 2\theta \qquad \text{(i)}$$

Differentiating (i) and equating to zero to find the maximum,

$$d\tau/d\theta = -\tfrac{1}{2}(\sigma_x - \sigma_y)(2\cos 2\theta) + \tau_{xy}(-2\sin 2\theta)$$

$$0 = -(\sigma_x - \sigma_y)\cos 2\theta - 2\tau_{xy}\sin 2\theta$$

$$\tan 2\theta = (\sigma_x - \sigma_y)/-2\tau_{xy} \qquad \text{(ii)}$$

From Fig. 9.8 the trigonometrical ratios for $2\theta_3$ are

$$\sin 2\theta_3 = (\sigma_x - \sigma_y)/\sqrt{[(\sigma_x - \sigma_y)^2 + 4\tau_{xy}{}^2]}$$

$$\cos 2\theta_3 = -2\tau_{xy}/\sqrt{[(\sigma_x - \sigma_y)^2 + 4\tau_{xy}{}^2]}$$

Substituting these values in (i), the maximum shear stress is

$$\tau_{max} = \tfrac{1}{2}(\sigma_x - \sigma_y)\left[\frac{(\sigma_x - \sigma_y)}{\sqrt{[(\sigma_x - \sigma_y)^2 + 4\tau_{xy}^2]}}\right] - \tau_{xy}\left[\frac{-2\tau_{xy}}{\sqrt{[(\sigma_x - \sigma_y)^2 + 4\tau_{xy}^2]}}\right]$$

$$= \tfrac{1}{2}\left[\frac{(\sigma_x - \sigma_y)^2 + 4\tau_{xy}^2}{\sqrt{[(\sigma_x - \sigma_y)^2 + 4\tau_{xy}^2]}}\right]$$

$$= \tfrac{1}{2}\sqrt{[(\sigma_x - \sigma_y)^2 + 4\tau_{xy}^2]} \qquad \text{(iii)}$$

The other possible value of θ which satisfies (ii) is given by

$$\sin 2\theta_3 = -(\sigma_x - \sigma_y)/\sqrt{[(\sigma_x - \sigma_y)^2 + 4\tau_{xy}^2]}$$
$$\cos 2\theta_3 = 2\tau_{xy}/\sqrt{[(\sigma_x - \sigma_y)^2 + 4\tau_{xy}^2]}$$

The reader can easily verify that the value of τ in this case is the same as (iii), except for a change in sign.

The strength of a material in shear is unaffected by the sign of the shearing stress acting on it—in contrast to normal stresses where the strength in compression can be very different to that in tension. It is important, nevertheless, to apply a sign convention consistently in the working of examples to ensure the correct numerical values at the end.

It can be seen that the expression (iii) is the final term in the formulae for the principal stresses given by equations (iii) and (iv) in Example 9.5. Taking the difference between these two expressions gives

$$\sigma_1 - \sigma_2 = \sqrt{[(\sigma_x - \sigma_y)^2 + 4\tau_{xy}^2]}$$

so that, by comparison with (iii) above,

$$\tau_{max} = \tfrac{1}{2}(\sigma_1 - \sigma_2)$$

i.e. half the difference between the principal stresses.

N.B. It must be emphasized that τ_{max} in this case is the maximum shear stress in the *xy* plane. It is possible to obtain greater values in three-dimensional cases as indicated in Example 9.13.

It should be noted that in general, the normal stress is not zero on the planes of maximum shear stress.

The expressions which have been obtained for the maximum shear stress at (iii) and for the principal stresses (see Example 9.5) can be applied to all examples. They are simplified when one (or two) of the three stresses σ_x, σ_y and τ_{xy} is zero.

If σ_y is zero, the principal stresses are

$$\left.\begin{aligned}\sigma_1 &= \tfrac{1}{2}\sigma_x + \tfrac{1}{2}\sqrt{[\sigma_x^2 + 4\tau_{xy}^2]}\\ \sigma_2 &= \tfrac{1}{2}\sigma_x - \tfrac{1}{2}\sqrt{[\sigma_x^2 + 4\tau_{xy}^2]}\end{aligned}\right\} \qquad \text{(iv)}$$

and the maximum shear stress is

$$\tau_{max} = \tfrac{1}{2}\sqrt{[\sigma_x^2 + 4\tau_{xy}^2]} \qquad \text{(v)}$$

The principal planes are given by

$$\tan 2\theta = 2\tau_{xy}/\sigma_x \qquad \text{(vi)}$$

If σ_x and σ_y are both zero, the principal stresses are

$$\left.\begin{aligned}\sigma_1 &= \tfrac{1}{2}\sqrt{(4\tau_{xy}^2)} = \tau_{xy}\\ \sigma_2 &= -\tau_{xy}\end{aligned}\right\} \qquad \text{(vii)}$$

and the maximum shear stress is

$$\tau_{max} = \tfrac{1}{2}\sqrt{(4\tau_{xy}^2)} = \tau_{xy} \qquad \text{(viii)}$$

The principal planes are given by

$$\tan 2\theta = 2\tau_{xy}/0 \rightarrow \infty$$

or

$$2\theta_1 = 90° \quad \text{and} \quad \theta_1 = 45°$$

i.e. the principal stresses are both equal to τ_{xy}, but one is tensile and the other is compressive; the principal planes bisect the planes on which the applied shear stress acts and this stress is itself the maximum shear stress.

If τ_{xy} (alone) is zero, the principal stresses are

$$\left.\begin{aligned}\sigma_1 &= \tfrac{1}{2}(\sigma_x + \sigma_y) + \tfrac{1}{2}(\sigma_x - \sigma_y) = \sigma_x\\ \sigma_2 &= \tfrac{1}{2}(\sigma_x + \sigma_y) - \tfrac{1}{2}(\sigma_x - \sigma_y) = \sigma_y\end{aligned}\right\} \qquad \text{(ix)}$$

and the maximum shear stress is

$$\tau_{max} = \tfrac{1}{2}(\sigma_x - \sigma_y) \qquad \text{(x)}$$

The principal planes are given by

$$\tan 2\theta = 0/(\sigma_x - \sigma_y) = 0$$

or

$$2\theta_1 = 0 \quad \text{and} \quad \theta_1 = 0$$

i.e. the applied stresses are themselves the principal stresses.

If τ_{xy} and σ_y are both zero, the principal stresses are

$$\sigma_1 = \sigma_x \quad \text{and} \quad \sigma_2 = 0 \qquad \text{(xi)}$$

and the maximum shear stress is

$$\tau_{max} = \tfrac{1}{2}\sigma_x \qquad \text{(xii)}$$

Using the values for the girder, $\sigma_x = 80$ MN/m², $\sigma_y = 0$ and $\tau_{xy} =$ 30 MN/m².

(*a*) By (iv), the principal stresses are

$$\sigma_1 = (\tfrac{1}{2} \times 80) + \tfrac{1}{2}\sqrt{[80^2 + (4 \times 30^2)]} = 90 \text{ MN/m}^2 \text{ (tensile)} \qquad (Ans)$$

$$\sigma_2 = (\tfrac{1}{2} \times 80) - \tfrac{1}{2}\sqrt{[80^2 + (4 \times 30^2)]} = 10 \text{ MN/m}^2 \text{ (compressive)} \qquad (Ans)$$

(*b*) By (v), the maximum shear stress is

$$\tau_{max} = \tfrac{1}{2}\sqrt{[80^2 + (4 \times 30^2)]} = 50 \text{ MN/m}^2 \qquad (Ans)$$

(*c*) By (xii), the required tensile stress is

$$\sigma_x = 2\tau_{max} = 2 \times 50 \text{ MN/m}^2 = 100 \text{ MN/m}^2 \qquad (Ans)$$

(*d*) By (vii), the required shear stress is

$$\tau = \sigma_1 = 90 \text{ MN/m}^2 \qquad (Ans$$

By (vi), the principal planes are given by

$$\tan 2\theta_1 = 2 \times 30/80 = 0{\cdot}75$$

Hence

$$2\theta_1 = 36°\ 52' \quad \text{and} \quad \theta_1 = 18°\ 26'$$
$$\theta_2 = \theta_1 + 90° = 108°\ 26'$$

Figure 9.9 shows the relationship between the principal planes, the planes of maximum shear stress and the plane on which σ_x acts.

9.8. Explain the "Mohr stress circle" method for obtaining the stresses on any interface.

Draw the diagram for principal stresses of 120 MN/m² (tensile) and 80 MN/m² (compressive) and estimate the magnitude and direction of the resultant stresses on planes making angles of 20° and 65° with the plane of the first principal stress. Find also the normal and tangential stresses on these planes.

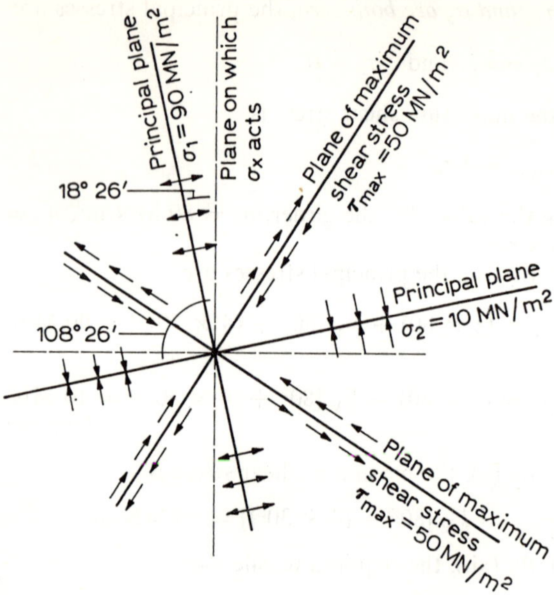

Fig. 9.9

Solution. Since principal planes are planes of zero shear stress then the results obtained in Example 9.2 for two perpendicular direct stresses are immediately applicable to principal stresses. Using equations (i) and (ii) of that question (putting $\sigma_x = \sigma_1$ and $\sigma_y = \sigma_2$) the normal and tangential stresses on an interface, which makes an angle θ with the "σ_1" principal plane, are

$$\sigma_n = \tfrac{1}{2}(\sigma_1 + \sigma_2) + \tfrac{1}{2}(\sigma_1 - \sigma_2)\cos 2\theta \qquad \text{(i)}$$

and

$$\tau = -\tfrac{1}{2}(\sigma_1 - \sigma_2)\sin 2\theta \qquad \text{(ii)}$$

The Mohr circle is a geometrical construction by which these results are represented graphically. The method of finding σ_n and τ is as follows.

From a point O mark off along a horizontal axis OA and OB proportional to σ_2 and σ_1 respectively. Draw a circle on AB as diameter (Fig. 9.10). For the interface required, draw a line AP making an angle θ with the horizontal. Then the horizontal and vertical co-ordinates of P represent the numerical values of the normal and tangential stresses on the interface.

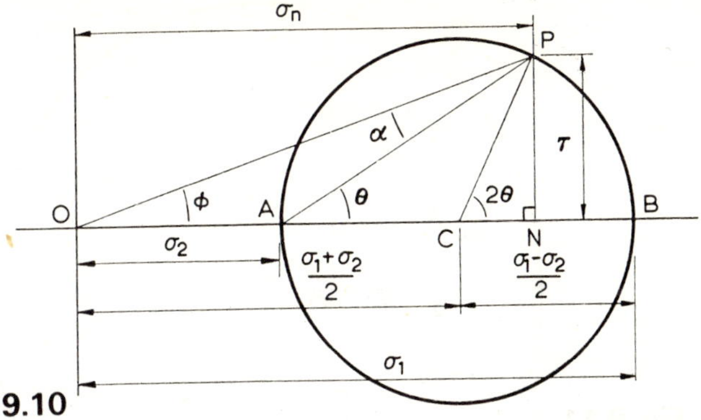

Fig. 9.10

This is proved by noting that

$OC = \frac{1}{2}(OB + OA)$ and thus represents $\frac{1}{2}(\sigma_1 + \sigma_2)$,
radius $CB = \frac{1}{2}(OB - OA)$ and thus represents $\frac{1}{2}(\sigma_1 - \sigma_2)$,
and $\angle PCB = 2 \times \angle PAB = 2\theta$

Hence, if PN is the perpendicular from P on to the horizontal axis

$$ON = OC + CN = OC + (\text{radius } CP \times \cos PCN)$$
$$= \tfrac{1}{2}(\sigma_1 + \sigma_2) + \tfrac{1}{2}(\sigma_1 - \sigma_2)\cos 2\theta$$

By comparison with (i), ON represents σ_n. Also

$$NP = \text{radius } CP \times \sin PCN$$
$$= \tfrac{1}{2}(\sigma_1 - \sigma_2)\sin 2\theta$$

By comparison with (ii), NP represents τ, except for a change in sign.

Furthermore, OP, which is the vector sum of ON and NP, represents the magnitude of the resultant stress, and its direction relative to the normal is given by ϕ (the angle POA). From the geometry of Fig. 9.3 the angle made by this resultant with the axis of σ_1 is

$$\alpha = \theta - \phi$$

In Fig. 9.10, angle OPA clearly equals α.

Figure 9.10 illustrates the case when σ_1 and σ_2 are both positive. If σ_2 is zero, A coincides with O. If σ_2 is negative then A is to the left of O, and O lies inside the circle. In this last case, normal (and resultant) stresses will be negative for points on the circle to the left of the vertical axis through O.

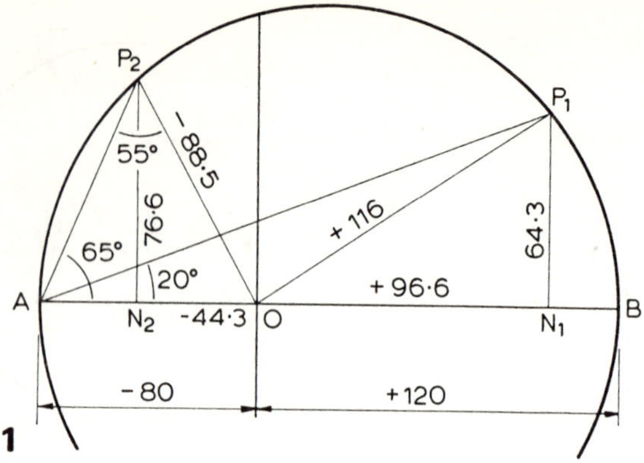

Fig. 9.11

Figure 9.11 shows the circle for the principal stresses given in the question. P_1 and P_2 correspond to the 20° and 65° planes respectively. The results are given on the lines which represent them.

9.9. At a point in a bracket the stresses on two mutually perpendicular planes are 60 MN/m² tensile and 30 MN/m² tensile. The shear stress across these planes is 15 MN/m². Find, using the Mohr stress circle, the principal stresses and maximum shear stress at the point.

Solution. The diagram is shown in Fig. 9.12. The points F and E are marked off on the horizontal axis to represent the given direct stresses (+60 and +30 MN/m²). The shear stress is 15 MN/m² on both the planes on which these stresses act. Hence the centre of the circle C is the mid-point of EF. The circle must pass through

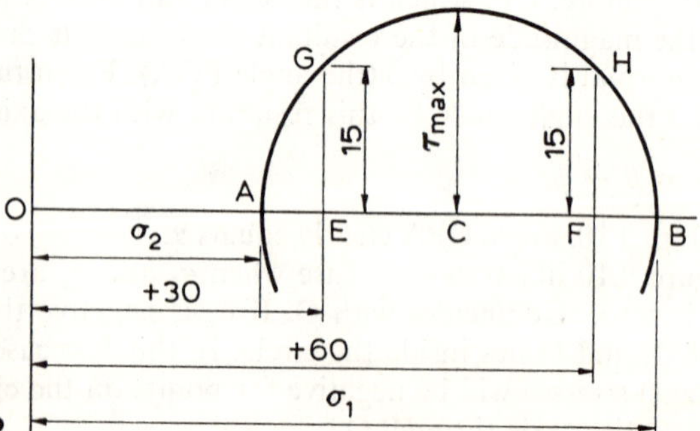

Fig. 9.12

G and H where EG and FH are verticals representing the given shear stress of 15 MN/m².

By measurement, from the diagram, the principal stresses are

$$\sigma_1 \ (= \text{OB}) = 66{\cdot}2 \text{ MN/m}^2 \qquad (Ans)$$

$$\sigma_2 \ (= \text{OA}) = 23{\cdot}8 \text{ MN/m}^2 \qquad (Ans)$$

The maximum shear stress, which is obviously represented by the radius, is

$$\tau_{max} = 21{\cdot}2 \text{ MN/m} \qquad (Ans)$$

N.B. The usual sign convention must be observed in the location of points E and F.

9.10. A solid shaft 125 mm diameter transmits 600 kW at 300 rev/min. It is also subjected to a bending moment of 9 kN-m and to an end thrust. If the maximum principal stress is limited to 80 MN/m², determine the permissible end thrust.

Determine the position of the plane in which the principal stress acts and draw a diagram showing the position of the plane relative to the torque and the plane of the bending moment applied to the shaft. (*U.L.*)

Solution. The torque transmitted by the shaft is

$$T = \frac{\text{power}}{\text{angular speed}} = \frac{600 \times 10^3 \text{ W}}{300 \times 2\pi/60 \text{ rad/s}} = 19{\cdot}1 \text{ kN-m}$$

The maximum shear stress in the shaft is

$$\tau_{xy} = \frac{Tr}{J} = \frac{T}{\frac{1}{16}\pi d^3} = \frac{16 \times 19{\cdot}1 \text{ kN-m}}{\pi \times (0{\cdot}125 \text{ m})^3}$$

$$= 49{\cdot}8 \text{ MN/m}^2$$

The direct stresses due to the bending moment and the end thrust act in the same direction. At the point where the bending stress is a maximum we have, since there is no direct stress at right angles to the bending stress,

$$\sigma_x = \text{max. bending stress} + \text{stress due to end thrust}$$

$$\sigma_y = 0 \quad \text{and} \quad \tau_{xy} = 49{\cdot}8 \text{ MN/m}^2$$

In examples where the principal stress is given, the relationship

$$(\sigma - \sigma_x)(\sigma - \sigma_y) = \tau_{xy}^2$$

is more convenient than the formulae for principal stresses given by equations (iii) and (iv) in Example 9.5. In the present case put $\sigma = -80$ (the negative sign signifying compressive) and thus

$$(-80 - \sigma_x)(-80 - 0) = 49{\cdot}8^2$$

$$6400 + 80\sigma_x = 2480$$

from which

$$\sigma_x = (2480 - 6400)/80 = (-)\ 49\ \text{MN/m}^2$$

the minus sign again signifying compression.

From thc bending equation, the maximum bending stress is

$$\frac{My_{max}}{I} = \frac{M}{\frac{1}{32}\pi d^3} = \frac{32 \times 9\ \text{kN-m}}{\pi \times (0{\cdot}125\ \text{m})^3} = 46{\cdot}95\ \text{MN/m}^2$$

Hence the maximum permissible stress due to the end thrust is

$$(\sigma_x - \text{max. bending stress}) = (49 - 46{\cdot}95)\ \text{MN/m}^2$$
$$= 2{\cdot}05\ \text{MN/m}^2$$

and

$$\text{Permissible end thrust} = \text{stress} \times \text{area}$$
$$= 2{\cdot}05\ \text{MN/m}^2 \times [\tfrac{1}{4}\pi \times (0{\cdot}125\ \text{m})^2]$$
$$= 25{\cdot}2\ \text{kN} \qquad (Ans)$$

The principal plane is most easily found by using equation (i) of Example 9.5. This gives

$$\tan\theta = \frac{\sigma - \sigma_x}{\tau_{xy}} = \frac{-80 - (-49)}{49{\cdot}8} = -0{\cdot}6226$$

and

$$\theta = -31°\ 54' \quad \text{or} \quad (180° - 31°\ 54')$$

Figure 9.13 shows the position of the principal plane relative to the axis of the bar.

The value of θ can also be found from the relationship

$$\tan 2\theta = 2\tau_{xy}/(\sigma_x - \sigma_y)$$

but this gives two results corresponding to the two principal planes.

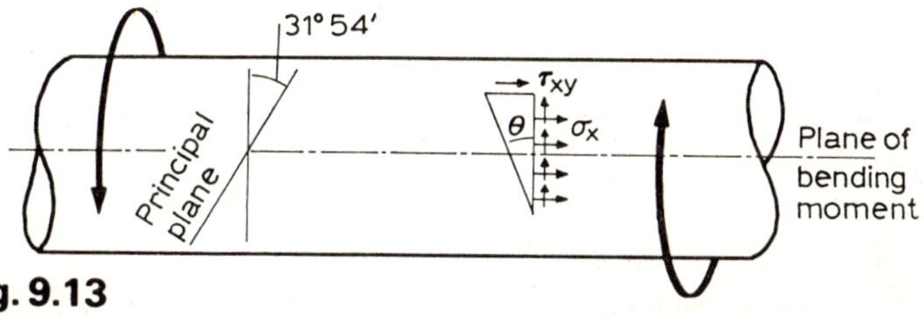

Fig. 9.13

9.11. Show that when a bending moment M and a twisting moment T act simultaneously on a circular shaft the maximum principal stress is equal to the maximum bending stress due to a simple bending moment M_E, where

$$M_E = \tfrac{1}{2}[M + \sqrt{(M^2 + T^2)}]$$

A hollow shaft is subjected to a bending moment of 2·5 kN-m and a twisting moment of 3 kN-m. Calculate the maximum principal stress in the shaft, the diameters being 100 mm external and 75 mm internal.

Solution. If D is the outside diameter of the shaft (solid or hollow) and I and J have their usual meanings, then at a point on the surface where the bending stress has its greatest value,

$$\text{Bending stress } \sigma_x = \frac{M}{I} y_{max} = \frac{MD}{2I}$$

$$\text{Shear stress } \tau_{xy} = \frac{Tr}{J} = \frac{TD}{2J} = \frac{TD}{4I}$$

(since $J = 2I$ for a circular shaft).

As in the previous example, $\sigma_y = 0$ and the maximum principal stress is (substituting for σ_x and τ_{xy})

$$\sigma_1 = \tfrac{1}{2}\sigma_x + \tfrac{1}{2}\sqrt{[\sigma_x^2 + 4\tau_{xy}^2]}$$

$$= \frac{1}{2}\left(\frac{MD}{2I}\right) + \frac{1}{2}\sqrt{\left[\left(\frac{MD}{2I}\right)^2 + 4\left(\frac{TD}{4I}\right)^2\right]}$$

$$= \frac{1}{2}\left(\frac{D}{2I}\right)(M + \sqrt{[M^2 + T^2]}) \qquad \text{(i)}$$

If M_E is the bending moment which, when acting alone, would cause the same maximum direct stress then

$$\sigma_1 = \frac{M_E\, y_{max}}{I} = \frac{M_E D}{2I} \qquad \text{(ii)}$$

From (i) and (ii),

$$\frac{M_E D}{2I} = \frac{1}{2}\left(\frac{D}{2I}\right)(M + \sqrt{[M^2 + T^2]})$$

$$M_E = \tfrac{1}{2}\sqrt{(M^2 + T^2)}$$

M_E is called the *equivalent bending moment.*

With the values given,

$$M_E = \tfrac{1}{2}(2{\cdot}5 + \sqrt{[2{\cdot}5^2 + 3^2]}) = 3{\cdot}2 \text{ kN-m}$$

$$I = \tfrac{1}{64}\pi\,(100^4 - 75^4) \text{ mm}^4 = 3{\cdot}355 \times 10^6 \text{ mm}^4$$

$$= 3{\cdot}355 \times 10^{-6} \text{ m}^4$$

Hence, using (ii), the maximum principal stress is

$$\sigma_1 = \frac{(3{\cdot}2 \times 10^3 \text{ N-m}) \times (50 \times 10^{-3} \text{ m})}{3{\cdot}355 \times 10^{-6} \text{ m}^4} = 47{\cdot}7 \text{ MN/m}^2 \quad (Ans)$$

9.12. Assuming a formula for the maximum shear stress under conditions of complex stress in two dimensions, obtain an expression for an "equivalent twisting moment" which, if acting alone, would produce the same maximum shear stress in a circular shaft as a bending moment M and twisting moment T acting together.

A solid circular shaft is to transmit 900 kW at 500 rev/min. It is supported in bearings 1·8 m apart and carries a flywheel weighing 20 kN midway between them. Find the minimum diameter of the shaft if the maximum permissible shear stress is 75 MN/m².

Solution. With the notation of Example 9.11, the maximum bending stress

$$\sigma_x = \frac{M}{I}\, y_{max} = \frac{MD}{2I} = \frac{MD}{J}$$

and the applied shear stress

$$\tau_{xy} = \frac{T}{J}\, r = \frac{TD}{2J}$$

Substituting in the expression for the maximum shear stress on any interface,

$$\tau_{max} = \tfrac{1}{2}\sqrt{[\sigma_x^2 + 4\tau_{xy}^2]}$$

$$= \frac{1}{2}\sqrt{\left[\left(\frac{MD}{J}\right)^2 + 4\left(\frac{TD}{2J}\right)^2\right]}$$

$$= \left(\frac{D}{2J}\right)\sqrt{[M^2 + T^2]} \qquad \text{(i)}$$

Let T_E be the "equivalent twisting moment" which, if acting alone, would produce a maximum shear stress equal to that due to M and T combined. Then, from the torsion equation,

$$\tau_{max} = \frac{T_E r}{J} = \frac{T_E D}{2J} \qquad \text{(ii)}$$

From (i) and (ii),

$$\frac{T_E D}{2J} = \frac{D}{2J}\sqrt{[M^2 + T^2]} \quad \text{or} \quad T_E = \sqrt{[M^2 + T^2]}$$

For the shaft in the question,

$$T = \frac{\text{power}}{\text{angular speed}}$$

$$= \frac{900 \times 10^3 \text{ W}}{(2\pi/60) \times 500 \text{ rad/s}} = 17{\cdot}18 \text{ kN-m}$$

Assuming the shaft to be simply-supported, the maximum bending moment is

$$M = \tfrac{1}{4}WL = \tfrac{1}{4} \times 20 \text{ kN} \times 1{\cdot}8 \text{ m} = 9 \text{ kN-m}$$

Using these values, the equivalent twisting moment is

$$T_E = \sqrt{[9^2 + 17{\cdot}18^2]} = 19{\cdot}4 \text{ kN-m}$$

From the torsion equation

$$T_E/J = \tau_{max}/\tfrac{1}{2}d \qquad 2J/d = T_E/\tau_{max}$$

or

$$\pi d^3/16 = T_E/\tau_{max}$$

$$d^3 = (16 \times 19{\cdot}4 \text{ kN-m})/(\pi \times 75 \text{ MN/m}^2) = 1{\cdot}318 \times 10^{-3} \text{ m}^3$$

Hence

Minimum diameter $d = \sqrt[3]{(1{\cdot}318 \times 10^{-3})}$ m $= 109{\cdot}6$ mm

(*Ans*)

9.13. At a point in the wall of a circular tube subjected to internal pressure and torque the stresses on three mutually perpendicular planes are as follows:

(*a*) A tensile stress of 120 MN/m^2 acting in a tangential direction.
(*b*) A tensile stress of 60 MN/m^2 acting parallel to the axis of the tube.
(*c*) A compressive stress of 20 MN/m^2 acting radially.
(*d*) Complementary shear stresses of +40 MN/m^2 acting on the same planes as (*a*) and (*b*).

There is no shear stress on the plane on which stress (*c*) acts. Determine the principal stresses, the three relative maximum shear stresses and the true maximum shear stress.

Solution. Consider a small element of the wall of the tube in the form of a cube as shown in Fig. 9.14. Let σ_x, σ_y and σ_z be the normal

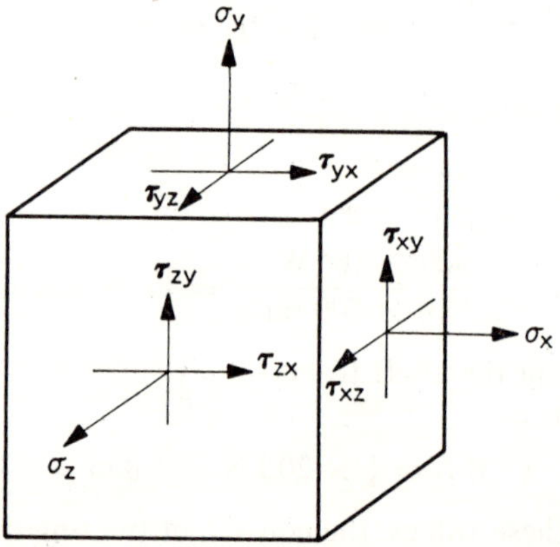

Fig. 9.14

stresses on such an element the stresses being positive (tensile) in the directions shown. With the sign convention adopted earlier the shear stresses acting on the faces are positive in the directions shown and, since complementary shear stresses are equal,

$$\tau_{xy} = \tau_{yx} \qquad \tau_{yz} = \tau_{zy} \qquad \tau_{zx} = \tau_{xz}$$

Using the numerical values in the question and working in MN/m^2 throughout, we can put

$$\sigma_x = 120 \qquad \sigma_y = 60 \qquad \sigma_z = -20$$

$$\tau_{xy} = 40 \qquad \tau_{yz} = \tau_{zx} = 0$$

Since $\tau_{yz} = 0$, σ_z is a principal stress. The other principal stresses must lie within a plane at right angles to it, i.e. the plane xy. Their values are:

$$\sigma_1 = \tfrac{1}{2}(\sigma_x + \sigma_y) + \tfrac{1}{2}\sqrt{[(\sigma_x - \sigma_y)^2 + 4\tau_{xy}{}^2]}$$
$$= \tfrac{1}{2}(120 + 60) + \tfrac{1}{2}\sqrt{[(120 - 60)^2 + 4 \times 40^2]}$$
$$= 90 + 50 = 140$$

and

$$\sigma_2 = 90 - 50 = 40$$

Putting $\sigma_3 = \sigma_z = -20$ the three principal stresses are 140 MN/m^2 (tensile), 40 MN/m^2 (tensile) and 20 MN/m^2 (compressive). (*Ans*)

There are three two-dimensional stress systems in all and they are completely defined by the three principal stresses taken in pairs.

For the xy plane

$$\tau_{max} = \tfrac{1}{2}(\sigma_1 - \sigma_2) = \tfrac{1}{2}(140 - 40) = 50$$

For the yz plane

$$\tau_{max} = \tfrac{1}{2}(\sigma_2 - \sigma_3) = \tfrac{1}{2}(40 + 20) = 30$$

For the xz plane

$$\tau_{max} = \tfrac{1}{2}(\sigma_1 - \sigma_3) = \tfrac{1}{2}(140 + 20) = 80$$

The true maximum shear stress is therefore 80 MN/m^2 (*Ans*)

PROBLEMS

1. A 32 mm diameter bar is subjected to a pull of 54 kN. Calculate the normal and tangential stresses on planes making angles of 5°, 37° and 79° with the axis of the bar. What is the maximum shear stress on any interface and what is the normal stress on this plane?

Answer. (All stresses in MN/m^2.)

For 5° plane, $\theta = 85°$; $\sigma_n = 0{\cdot}51$; $\tau = -5{\cdot}82$.
For 37° plane, $\theta = 53°$; $\sigma_n = 24{\cdot}3$; $\tau = -32{\cdot}25$.
For 79° plane, $\theta = 11°$; $\sigma_n = 64{\cdot}66$; $\tau = -12{\cdot}57$.
Max. shear stress ($\theta = 45°$) is $-33{\cdot}55$.
Normal stress on same plane = 33·55.

2. Calculate the minimum diameter of a bar to carry a tensile load of 80 kN if

(i) the tensile stress is limited to 70 MN/m^2;
(ii) the shear stress is limited to 70 MN/m^2.

Answer. (i) 38·2 mm, (ii) 27·0 mm.

3. In each of the following cases σ_x and σ_y are perpendicular direct stresses (in MN/m^2). Calculate the normal, tangential and resultant stresses on planes making angles of 19°, 42° and 79° with the axis of σ_x and also for the plane having the maximum shear stress. Find also the obliquity of the resultant stress (i.e. its inclination to the normal).

(*a*) $\sigma_x = +60, \sigma_y = +15.$
(*b*) $\sigma_x = +60, \sigma_y = -15.$
(*c*) $\sigma_x = +30, \sigma_y = -75.$
(*d*) $\sigma_x = -45, \sigma_y = -90.$

Answer. (All stresses in MN/m^2; ϕ is the angle of obliquity.)

(*a*) For 19°, $\theta = 71°, \sigma_n = +19{\cdot}77; \tau = -13{\cdot}86$
$\sigma_r = 24{\cdot}09; \phi = -35°$

For 42°, $\theta = 48°, \sigma_n = +35{\cdot}14; \tau = -22{\cdot}38$
$\sigma_r = 41{\cdot}4; \phi = -32{\cdot}5°$

For 79°, $\theta = 11°, \sigma_n = +58{\cdot}36; \tau = -8{\cdot}43$
$\sigma_r = 58{\cdot}98; \phi = -8{\cdot}2°$

For τ_{max}, $\theta = 45°, \sigma_n = +37{\cdot}5; \tau = -22{\cdot}5$
$\sigma_r = 43{\cdot}72; \phi = -31°$

(*b*) For 19°, $\theta = 71°, \sigma_n = -7{\cdot}05; \tau = -23{\cdot}08$
$\sigma_r = (-)\ 24{\cdot}09; \phi = 73°$

For 42°, $\theta = 48°, \sigma_n = +18{\cdot}59; \tau = -37{\cdot}29;$
$\sigma_r = 41{\cdot}67; \phi = -63{\cdot}5°$

For 79°, $\theta = 11°, \sigma_n = +57{\cdot}27; \tau = -14{\cdot}04$
$\sigma_r = 58{\cdot}98; \phi = -13{\cdot}8°$

For τ_{max}, $\theta = 45°, \sigma_n = +22{\cdot}5; \tau = -37{\cdot}5$
$\sigma_r = 43{\cdot}72, \phi = -59°$

(*c*) For 19°, $\theta = 71°, \sigma_n = -63{\cdot}87; \tau = -32{\cdot}32$
$\sigma_r = (-)\ 71{\cdot}56; \phi = 26{\cdot}8°$

For 42°, $\theta = 48°, \sigma_n = -27{\cdot}99; \tau = -52{\cdot}22$
$\sigma_r = (-)\ 59{\cdot}2; \phi = 61{\cdot}8°$

For 79°, $\theta = 11°, \sigma_n = +26{\cdot}77; \tau = -19{\cdot}66$
$\sigma_r = 33{\cdot}21; \phi = -36{\cdot}3°$

For τ_{max}, $\theta = 45°, \sigma_n = -22{\cdot}5; \tau = -52{\cdot}5$
$\sigma_r = (-)\ 57{\cdot}12; \phi = 66{\cdot}8°$

(*d*) For 19°, $\theta = 71°, \sigma_n = -85{\cdot}23; \tau = -13{\cdot}86;$
$\sigma_r = (-)\ 86{\cdot}35; \phi = 9{\cdot}2°$

For 42°, $\theta = 48°, \sigma_n = -69{\cdot}85; \tau = -22{\cdot}38;$
$\sigma_r = (-)\ 73{\cdot}36; \phi = 17{\cdot}8°$

For 79°, $\theta = 11°, \sigma_n = -46{\cdot}65; \tau = -8{\cdot}43$
$\sigma_r = (-)\ 47{\cdot}4; \phi = 10{\cdot}2°$

For τ_{max}, $\theta = 45°, \sigma_n = -67{\cdot}5; \tau = -22{\cdot}5;$
$\sigma_r = (-)\ 71{\cdot}14; \phi = 18{\cdot}3°.$

4. Calculate the principal stresses and the maximum shear stress for the following cases of two perpendicular direct stresses together with complementary shear stresses in the same directions. Draw diagrams showing the positions of the principal planes and planes of maximum shear stress relative to the planes of the applied stresses.

(*a*) $\sigma_x = 120$ MN/m² (tensile); $\sigma_y = 45$ MN/m² (tensile) $\tau_{xy} = +45$ MN/m²
(*b*) $\sigma_x = 30$ MN/m² (tensile); $\sigma_y = 75$ MN/m² (compressive); $\tau_{xy} = +15$ MN/m²
(*c*) $\sigma_x = 0$; $\sigma_y = 45$ MN/m² (compressive); $\tau_{xy} = -75$ MN/m²

Answer. (All stresses are in MN/m², positive values being tensile. θ_1 is the angle between the planes on which σ_1 and σ_x act. The angle of the "σ_2" principal plane is $\theta_2 = \theta_1 + 90°$. Planes of maximum shear stress are 45° from the principal planes.)

(*a*) $\sigma_1 = +141{\cdot}1$; $\sigma_2 = +23{\cdot}9$; $\tau_{max} = 58{\cdot}6$; $\theta_1 = 25{\cdot}1°$
(*b*) $\sigma_1 = +32{\cdot}1$; $\sigma_2 = -77{\cdot}1$; $\tau_{max} = 54{\cdot}6$; $\theta_1 = 7{\cdot}9°$
(*c*) $\sigma_1 = +55{\cdot}8$; $\sigma_2 = -100{\cdot}8$; $\tau_{max} = 78{\cdot}3$; $\theta_1 = -36{\cdot}7°$

5. Solve Problem 3 (*a*)–(*d*) by the Mohr stress circle construction.

6. Draw the Mohr stress circle for the three cases given in Problem 4 and find, by measurement, the principal and maximum shearing stresses.

7. Verify the following statements from the geometry of the Mohr stress circle (Fig. 9.10) σ_1 and σ_2 being the principal stresses.

(i) The maximum shear stress is $\tau_{max} = \frac{1}{2}(\sigma_1 - \sigma_2)$
(ii) The normal and resultant stresses on the plane of maximum shear stress are

$$\sigma_n = \tfrac{1}{2}(\sigma_1 + \sigma_2) \quad \text{and} \quad \sigma_r = \sqrt{[\tfrac{1}{2}(\sigma_1{}^2 + \sigma_2{}^2)]}$$

(iii) The extreme values of the resultant stress are σ_1 and σ_2
(iv) The maximum obliquity of the resultant stress is given by
$\sin\phi = (\sigma_1 - \sigma_2)/(\sigma_1 + \sigma_2)$
(*Hint.* ϕ is a maximum when OP is a tangent.)
(v) The maximum obliquity occurs when $\theta = \phi + 45°$.

8. A solid circular shaft transmits 2240 kW at 400 rev/min and is also subjected at one section to a bending moment of 30 kN-m. Find the equivalent bending moment and equivalent twisting moment for this section, and hence calculate the minimum diameter of the shaft if the maximum shear stress is to be 60 MN/m².
Answer. $M_E = 45{\cdot}67$ kN-m; $T_E = 61{\cdot}33$ kN-m; minimum $d = 173$ mm.

9. A hollow propeller shaft, having 250 mm and 150 mm external and internal diameters respectively, transmits 1200 kW with a thrust of 400 kN. Find the speed of the shaft if the maximum principal stress is not to exceed 60 MN/m². What is the value of the maximum shear stress at this speed? (*I.Mech.E.*)
Answer. 80·5 rev/min; 53·63 MN/m².

10. Explain the meaning of the terms "Equivalent Bending Moment" and "Equivalent Twisting Moment" in a shaft subject to both bending and torsion.

A solid shaft 76 mm diameter and spanning 3 m between bearings, has a pulley weighing 225 N mounted at the centre of the span. The maximum speed is to be 200 rev/min when transmitting 30 kW. Find the value of the maximum principal stress in the shaft. (Neglect weight of shaft.) (*I.Struct.E.*)
Answer. 4·445 MN/m².

11. Explain the meaning of Principal Stress. At a point in a beam there are tensile stresses of 75 and 45 MN/m² respectively at right angles to one another together with a shearing stress of 37·5 MN/m².

Calculate the maximum and minimum principal stresses in the material and the direction of the planes on which these stresses occur. (*I.Struct.E.*)
Answer. 100·4 and 19·6 MN/m² (tensile) on planes making angles of 34·1° and 124·1° respectively with the plane on which the 75 MN/m² stress acts.

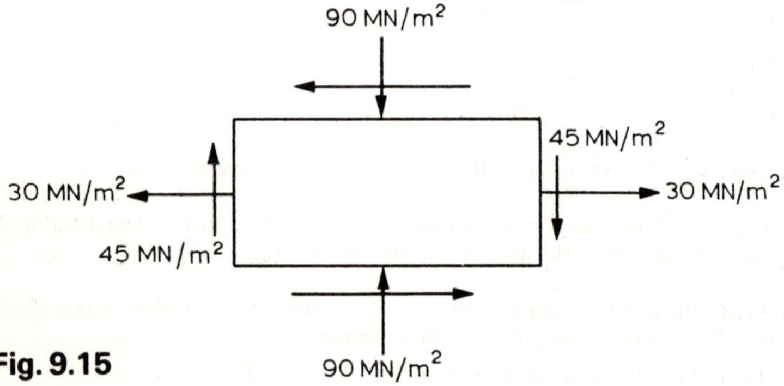

Fig. 9.15

12. Prove that in any material in a state of shear strain a shear stress on any plane must be accompanied by a shear stress of equal intensity on any plane at right angles thereto.

At a point in the web of a beam there is a shear stress of 37·5 MN/m² and a tensile stress of 90 MN/m².

Find the normal and tangential components of stress on a plane making 30° with a plane at right angles to the direction of the tensile stress. (*I.Struct.E.*)
Answer. σ_n = 100·0 MN/m² (tensile); τ = 20·2 MN/m².

13. The following data apply to an electric motor driving a line shaft: output power = 7·5 W; speed = 950 rev/min; effective diameter of motor pulley = 120 mm; diameter of motor shaft = 38 mm; distance from line of action of belt pull on driving pulley to centre of bearing = 125 mm, and ratio of belt tensions 2·5. For a point on the surface of the shaft at the centre of the bearing find the principal stresses and the maximum shear stress. (*I.Mech.E.*)
Answer. 68·6 and (−) 0·72 MN/m²; 34·66 MN/m².

14. It is calculated that the stresses on a small rectangular element of a control lever are as indicated in the diagram (Fig. 9.15). Determine the principal stresses and the maximum shear stress at the element, explaining carefully your method of estimation. (*R.Ae.S.*)
Answer. 45 MN/m² (tensile) and 105 MN/m² (compressive); 75 MN/m².

15. A solid circular shaft is subjected to an axial torque T and to a bending moment M. If $M = kT$ determine in terms of k, the ratio of the maximum principal stress to the maximum shear stress. Find the power transmitted by a 50 mm diameter shaft, at a speed of 300 rev/min, when $k = 0{\cdot}4$ and the maximum shear stress is 75 MN/m². (*I.Mech.E.*)

Answer. Ratio $= \dfrac{k + \sqrt{(1 + k^2)}}{\sqrt{(1 + k^2)}}$; 53·7 kW.

16. In a two-dimensional stress system the normal stresses at a point are σ_x and σ_y and the shear stress is τ_{xy}. Show by means of the Mohr circle diagram how the principal stresses at the point may be determined when (*a*) σ_x and σ_y are tensile stresses and (*b*) σ_x and σ_y are of opposite sign. When σ_x and σ_y are of the same sign one of the principal stresses may be zero. Illustrate this by a diagram and show how τ_{xy} is then related to σ_x and σ_y. (*I.Mech.E.*)
Answer. One principal stress is zero if ${\tau_{xy}}^2 = \sigma_x\sigma_y$.

17. A horizontal shaft 76 mm in diameter and 1·2 m long is held rigidly at the ends. A torque of 4·5 kN-m is applied to the shaft, in a plane perpendicular to the axis of the shaft, at a distance of 0·9 m from one end. Find the magnitude of the fixing torques and the maximum shear stress in the shaft. If a concentrated load of 9 kN is now hung from the centre of the shaft find the maximum principal stress at a point on the top of the shaft, 0·15 m from the end where the lesser fixing couple occurs. (*I.Mech.E.*)
Answer. Fixing torques are 3·375 kN-m (at end nearest applied torque) and 1·125 kN-m; maximum shear stress = 39·2 MN/m²; after load is applied required principal stress = 23·05 MN/m² (tensile).

18. In a circular shaft subjected to an axial twisting moment T and a bending moment M show that when $M = 1{\cdot}2T$, the ratio of the maximum shearing stress to the greater principal stress is approximately 0·566. (*U.L.*)

19. The principal stresses at a point in a material are 45 MN/m² tension and 75 MN/m² tension. Working from first principles, determine, for a plane inclined at 40° to the plane on which the latter stress acts:

(*a*) the magnitude and angle of obliquity of the resultant stress;
(*b*) the normal and tangential component stresses. (*U.L.*)

Answer. (*a*) $\sigma_r = 64{\cdot}35$ MN/m²; $\phi = 13{\cdot}3°$. (*b*) $\sigma_n = 62{\cdot}61$ MN/m² (tensile); $\tau = 14{\cdot}78$ MN/m².

20. Draw and describe Mohr's Circle of Stress and prove that it may be used to represent the state of stress at a point within a stressed material. Illustrate your answer by sketches. If, at a point within a material, the minimum and maximum principal stresses are 30 MN/m² and 90 MN/m² respectively (both tension), determine the shearing stress and normal stress on a plane passing through the point and making an angle of $\tan^{-1} 0{\cdot}25$ with the plane on which the maximum principal stress acts. (*U.L.*)
Answer. $\sigma_n = 86{\cdot}48$ MN/m² (tensile); $\tau = 14{\cdot}12$ MN/m².

21. At a certain point in a piece of elastic material there are normal stresses of 45 MN/m² tension and 30 MN/m² compression on two planes at right angles to one another together with shearing stresses of 22·5 MN/m² on the same planes. If the loading on the material is increased so that the stresses reach values of k times those given, find the maximum value of k if the maximum direct stress in the material is not to exceed 120 MN/m^2 and the maximum shearing stress is not to exceed 75 MN/m^2. (*U.L.*)

Answer. $k = 2{\cdot}342$ for direct stress condition and 1·715 for the shearing stress condition.

22. At a certain point in a piece of material there are two planes at right angles to one another on which there are shearing stresses of 36 MN/m^2 together with normal stresses of 120 MN/m^2 tension on one plane and 56 MN/m^2 tension on the other plane. Determine for the given point:

(*a*) the magnitudes of the principal stresses,
(*b*) the inclinations of the principal planes,
(*c*) the maximum shearing stresses and the inclinations of the planes on which they act,
(*d*) the maximum strain if $E = 204$ GN/m^2 and Poisson's ratio $= 0{\cdot}29$.

Make a diagram to show clearly the quantities and directions found in relationship to the given planes and stresses. (*U.L.*)

Answer. (*a*) 136·2 and 39·8 MN/m^2 tension;
(*b*) 24·2° and 114·2° to plane on which 120 MN/m^2 acts;
(*c*) ±48·2 MN/m^2 on planes at 45° to principal planes;
(*d*) 0·000611 (this result depends on the theory given in Chapter 10).

23. A 250 mm diameter solid shaft drives a screw propeller with an output of 6 MW. When the forward speed of the vessel is 20 knots the speed of revolution of the propeller is 240 rev/min. Find the maximum shearing stress due to the torque and the axial compressive stress due to the thrust in the shaft; hence find for a point on the surface of the shaft (*a*) the principal stresses and (*b*) the directions of the principal planes relative to the shaft axis. Make a diagram to show clearly the direction of the principal planes and stresses relative to the shaft axis.

If a graphical solution is used a clear explanation must be given of the method by which answers to both (*a*) and (*b*) were obtained. 1 knot = 0·5148 m/s. (*U.L.*)

Answer. 77·8 MN/m^2; 11·9 MN/m^2; (*a*) 83·9 MN/m^2 (compressive) and 72 MN/m^2 (tensile); (*b*) 42°50′, and 132°50′ to the plane on which the axial compressive stress acts.

24. A 76 mm diameter steel shaft transmits a torque of 4 kN-m and at the same time is subjected to an axial thrust of 44 kN. If the greatest allowable compressive stress in the shaft is 100 MN/m^2 find

(*a*) the greatest bending moment to which the shaft may be subjected in addition to the torque and thrust,
(*b*) the smaller principal stress and the maximum shearing stress at the point where the 100 MN/m^2 stress occurs. (*U.L.*)

Answer. (*a*) 2·96 kN-m. (*b*) 21·5 MN/m^2 (tensile) and 60·8 MN/m^2.

25. In a two-dimensional stress system (Fig. 9.16) there are two planes AB and BC at right angles on which there are normal tensile stresses of 50 MN/m^2 and 60 MN/m^2 respectively together with shearing stresses of 30 MN/m^2. AC is any plane inclined at θ to the plane BC.

(*a*) Determine the normal and tangential components of the stress on AC for $\theta = 30$ degrees.

(*b*) Find the magnitude of θ which will make AC the principal plane on which the larger principal stress acts and determine the magnitude of this stress. Analytical or graphical methods may be used. (*U.L.*)

Answer. (*a*) 83·5 MN/m² (tensile); 10·67 MN/m²; (*b*) 40°16′; and 85·4 MN/m² (tensile).

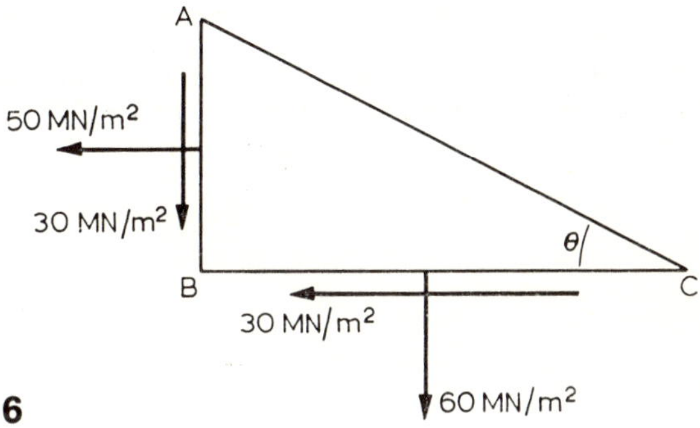

Fig. 9.16

26. The stress conditions at a certain point are represented in Fig. 9.17. AB and BC are two planes at right angles on which there are shearing stresses of 35 MN/m² acting towards B; there are also normal stresses of σ_x on AB and σ_y on BC. The resultant stress on AC is 75 MN/m² tension and it has an angle of obliquity of 18° as shown.

(*a*) Determine the magnitude of the normal stresses on AB and BC and state whether they are tension or compression.
(*b*) Determine the principal stresses at the point and inclinations relative to BC, of the planes on which they act.

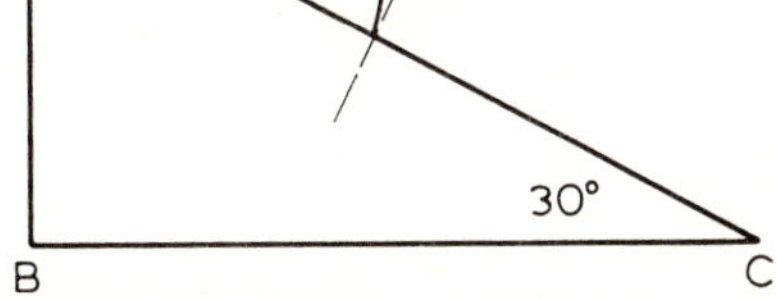

Fig. 9.17

Make a diagram to show clearly the stresses and planes found. (*U.L.*)
Answer. (*a*) −29·5, +64·5 MN/m²; (*b*) +64·5, −29·5 MN/m², 18°21′, 108°21′.

27. At a section of a rotating shaft there is a bending moment which produces a maximum direct stress of ±75 MN/m² and a torque which produces a maximum shearing stress of 45 MN/m².

Consider a certain point on the surface of the shaft where the bending stress is initially 75 MN/m² tension and find the principal stresses at the point in magnitude and direction:

(a) when the point is at the initial position;
(b) when the shaft has turned through 45°;
(c) when the shaft has turned through 90°, i.e. the point is at the neutral axis.

Make sketches to show the changes in the principal planes and stresses.

(U.L.)

Answer. (a) +96·2, −21·2 MN/m², 25° 6′ and 115° 6′ (to axis of shaft). (b) +78·8, −25·6 MN/m², 29° 46′ and 119° 46′. (c) +45, −45 MN/m², 45° and 135°.

Chapter 10

Analysis of Strain

The following definitions are additional to those given at the head of Chapter 1.

$$\text{Poisson's ratio } \nu = \frac{\text{Lateral strain}}{\text{Longitudinal strain}}$$

$$\text{Volumetric strain} = \frac{\text{Change in volume}}{\text{Original volume}}$$

$$\text{Bulk modulus } K = \frac{\text{Volumetric stress}}{\text{Volumetric strain}}$$

If E = Young's modulus, G = modulus of rigidity

$$E = 2G(1 + \nu) = 3K(1 - 2\nu)$$

In a two-dimensional stress system the principal stresses (σ_1 and σ_2) and principal strains (ϵ_1 and ϵ_2) are related by the equations:

$$\epsilon_1 = \frac{1}{E}(\sigma_1 - \nu\sigma_2) \quad \text{and} \quad \epsilon_2 = \frac{1}{E}(\sigma_2 - \nu\sigma_1)$$

$$\sigma_1 = \frac{E}{1 - \nu^2}(\epsilon_1 + \nu\epsilon_2) \quad \text{and} \quad \sigma_2 = \frac{E}{1 - \nu^2}(\epsilon_2 + \nu\epsilon_1)$$

WORKED EXAMPLES

10.1. Define Poisson's ratio. A rectangular bar, 75 mm wide and 50 mm thick, extends 1·90 mm in a length of 1·5 m under an axial tensile load of 1 MN. If the corresponding decrease in width is 27·5 μm calculate Young's modulus and Poisson's ratio for the material of the bar.

What would be the change in *thickness* of the bar due to a pull of 800 kN?

Solution. When a bar is loaded in tension it increases in length but decreases in all cross-sectional dimensions (Fig. 10.1a). Similarly, a

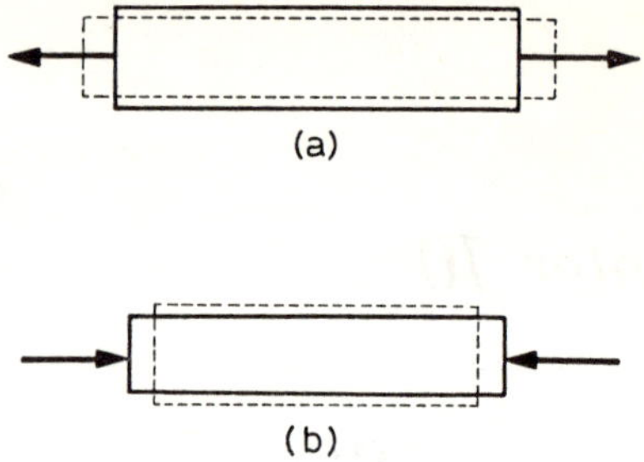

Fig. 10.1

compressive load produces a decrease in length but an increase in the cross-sectional dimensions (Fig. 10.1b).

Lateral strain is defined as

$$\frac{\text{change in width (thickness or diameter)}}{\text{original width (thickness or diameter)}}$$

It is found, by experiment, that lateral strain is proportional to longitudinal strain within the limit of proportionality.

Poisson's ratio is defined as

$$\frac{\text{lateral strain}}{\text{longitudinal strain}}$$

and is denoted by ν (or $1/m$). For most metals its value lies between 0·25 and 0·35. Although Poisson's ratio is always given as a positive number it is important to note that tensile longitudinal strains are accompanied by compressive lateral strains and vice versa.

For the bar in the question,

$$\text{Longitudinal stress} = \frac{(1 \times 10^6\ \text{N})}{(75 \times 50 \times 10^{-6}\ \text{m}^2)} = 266{\cdot}7\ \text{MN/m}^2$$

$$\text{Longitudinal strain} = \frac{\text{extension}}{\text{original length}} = \frac{1{\cdot}9 \times 10^{-3}\ \text{m}}{1{\cdot}5\ \text{m}}$$
$$= 0{\cdot}001\,267$$

$$\text{Lateral strain} = \frac{\text{change in width}}{\text{original width}} = \frac{(27{\cdot}5 \times 10^{-6}\ \text{m})}{(75 \times 10^{-3}\ \text{m})}$$
$$= 0{\cdot}000\,3667$$

$$\text{Young's modulus } E = \frac{\text{longitudinal stress}}{\text{longitudinal strain}}$$
$$= (266{\cdot}7 \times 10^6\ \text{N/m}^2)/0{\cdot}001\,267$$
$$= 210\ \text{GN/m}^2 \qquad (Ans)$$

$$\text{Poisson's ratio } \nu = \frac{\text{lateral strain}}{\text{longitudinal strain}} = \frac{0{\cdot}000\,3667}{0{\cdot}001\,267}$$
$$= 0{\cdot}2895$$

For the 800 kN load,

$$\text{Longitudinal stress} = \frac{800 \times 10^3 \text{ N}}{75 \times 50 \times 10^{-6} \text{ m}^2} = 213 \text{ MN/m}^2$$

$$\text{Longitudinal strain} = \frac{\text{stress}}{E} = \frac{213 \times 10^6 \text{ N/m}^2}{210 \times 10^9 \text{ N/m}^2}$$

$$= 0{\cdot}001\,014$$

$$\text{Lateral strain} = \nu \times \text{longitudinal strain}$$

$$= 0{\cdot}2895 \times 0{\cdot}001\,014 = 0{\cdot}000\,293$$

Thus,

$$\text{Change in thickness} = \text{thickness} \times \text{lateral strain}$$

$$= 50 \text{ mm} \times 0{\cdot}000\,293$$

$$= 14{\cdot}6\ \mu\text{m (decrease)}$$

10.2. Derive expressions for the principal strains under conditions of complex stress in two dimensions, the principal stresses being σ_1 and σ_2.

Calculate the change in the area of a rectangular steel plate, 300 mm by 200 mm when it is subjected to tensile (direct) stresses of 120 MN/m² and 80 MN/m² on its longer and shorter edges respectively.

$E = 200 \text{ GN/m}^2 \qquad \nu = 0{\cdot}3$

Solution. The principal strains are the strains in the directions of the principal stresses. Due to the Poisson's ratio effect each principal stress produces a strain in the direction of the other.

Regarding tensile strains as positive, the stress σ_1 produces a strain of σ_1/E in its own direction and a (lateral) strain of $-\nu\sigma_1/E$ in the direction of the axis of σ_2.

Similarly, σ_2 produces a strain of σ_2/E in its own direction and $-\nu\sigma_2/E$ along the axis of σ_1.

Thus the total strain along the axis of σ_1 is

$$\frac{\sigma_1}{E} - \frac{\nu\sigma_2}{E}$$

and the total strain along the axis of σ_2 is

$$\frac{\sigma_2}{E} - \frac{\nu\sigma_1}{E}$$

(It is clear from these expressions that a strain can exist along an axis of zero stress or that, if $\sigma_1 = \nu\sigma_2$ or $\sigma_2 = \nu\sigma_1$ there is a stress along an axis of zero strain.)

If a rectangle, length l and breadth b, has these dimensions increased to $(l + \delta l)$ and $(b + \delta b)$ respectively, then the increase in its area is

$$\begin{aligned}\delta A &= \text{final area} - \text{original area}\\ &= (l + \delta l)(b + \delta b) - lb\\ &= b\delta l + l\delta b + \delta l\delta b\end{aligned}$$

If δl and δb are small quantities then the product $\delta l\delta b$ will be negligible and, if A is the original area,

$$\begin{aligned}\delta A &= b\delta l + l\delta b\\ &= lb\left(\frac{\delta l}{l} + \frac{\delta b}{b}\right)\\ &= A \times [\text{strain in the direction of the length}\\ &\qquad + \text{strain in the direction of the breadth}]\end{aligned}$$

When the directions of the principal strains coincide with the axes of the rectangle,

$$\delta A/A = \text{sum of the two principal strains}$$

For the given steel plate, putting $\sigma_1 = 120$ MN/m² and $\sigma_2 = 80$ MN/m², the principal strains are (in terms of E)

$$\frac{\sigma_1}{E} - \frac{\nu\sigma_2}{E} = \frac{120}{E} - \frac{0{\cdot}3 \times 80}{E} = \frac{96}{E}$$

(in the direction of the length) and

$$\frac{\sigma_2}{E} - \frac{\nu\sigma_1}{E} = \frac{80}{E} - \frac{0{\cdot}3 \times 120}{E} = \frac{44}{E}$$

(in a direction perpendicular to the length).

Therefore, the change in area is

$$\begin{aligned}\delta A &= A \times (\text{sum of the two principal strains})\\ &= (300\text{ mm} \times 200\text{ mm}) \times \left(\frac{96}{E} + \frac{44}{E}\right)\\ &= \frac{(6 \times 10^{-2}\text{ m}^2) \times (140 \times 10^6\text{ N/m}^2)}{(200 \times 10^9\text{ N/m}^2)} = 42\text{ mm}^2 \qquad (\textit{Ans}\end{aligned}$$

(The reader should note that the same result would be obtained if the directions of the two given stresses were interchanged.)

10.3. A rectangular plate of steel is stressed as shown in Fig. 10.2 and the strains in the *X* and *Y* directions are $11{\cdot}85 \times 10^{-5}$ and $9{\cdot}47 \times 10^{-5}$ respectively. Find the stresses σ_x and σ_y and hence determine the normal and shear stresses on the interface AB. ($E = 207$ GN/m²; $1/m$ or $\nu = 0{\cdot}28$). *(I.Mech.E.)*

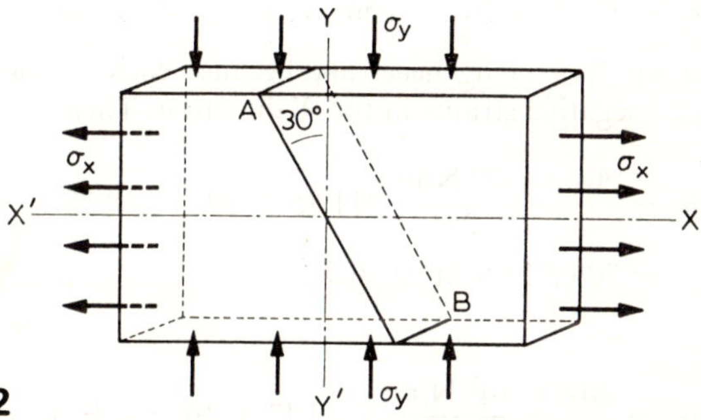

Fig. 10.2

Solution. The strains in the directions of the principal stresses are called the *principal strains.* It is convenient to have formulae for the principal stresses in terms of these strains and, in the two-dimensional case, these are derived as follows.

Let ϵ_x and ϵ_y be the strains in the X and Y directions respectively. Then, if tensile stresses and strains are taken as positive,

$$\epsilon_x = \frac{\sigma_x}{E} - \frac{\nu\sigma_y}{E} \tag{i}$$

$$\epsilon_y = \frac{\sigma_y}{E} - \frac{\nu\sigma_x}{E} \tag{ii}$$

Multiplying (ii) by ν we obtain

$$\nu\epsilon_y = \frac{\nu\sigma_y}{E} - \frac{\nu^2\sigma_x}{E}$$

and adding to (i),

$$\epsilon_x + \nu\epsilon_y = \frac{\sigma_x}{E}(1 - \nu^2)$$

from which

$$\sigma_x = \frac{E}{1 - \nu^2}(\epsilon_x + \nu\epsilon_y)$$

Similarly

$$\sigma_y = \frac{E}{1 - \nu^2}(\epsilon_y + \nu\epsilon_x)$$

With the numerical data given in the question

$$\epsilon_x = 11{\cdot}85 \times 10^{-5} \quad \text{and} \quad \epsilon_y = -9{\cdot}45 \times 10^{-5},$$

the minus sign being necessary because each of the given stresses leads to negative strains in the Y direction. Then

$$\sigma_x = \frac{207 \times 10^9 \text{ N/m}^2}{1 - 0{\cdot}28^2}(11{\cdot}85 \times 10^{-5} - 0{\cdot}28 \times 9{\cdot}47 \times 10^{-5})$$
$$= 20{\cdot}65 \text{ MN/m}^2 \text{ (tensile)} \qquad (Ans)$$

and

$$\sigma_y = \frac{207 \times 10^9 \text{ N/m}^2}{1 - 0{\cdot}28^2}(-9{\cdot}47 \times 10^{-5} + 0{\cdot}28 \times 11{\cdot}85 \times 10^{-5})$$
$$= -13{\cdot}81 \text{ MN/m}^2 \text{ (i.e. compressive)} \qquad (Ans)$$

With these results the normal and shear stresses on the given interface are

$$\sigma_\theta = \tfrac{1}{2}(\sigma_x + \sigma_y) + \tfrac{1}{2}(\sigma_x - \sigma_y)\cos 2\theta$$
$$= \tfrac{1}{2}(20{\cdot}65 - 13{\cdot}81) + \tfrac{1}{2}(20{\cdot}65 + 13{\cdot}81)\cos 60°$$
$$= 12{\cdot}04 \text{ MN/m}^2 \text{ (tensile)} \qquad (Ans)$$

and

$$\tau = \tfrac{1}{2}(\sigma_x - \sigma_y)\sin 2\theta$$
$$= \tfrac{1}{2}(20{\cdot}65 + 13{\cdot}81)\sin 60°$$
$$= 14{\cdot}92 \text{ MN/m}^2 \qquad (Ans)$$

10.4. Derive expressions for the principal strains under conditions of three-dimensional complex stress in terms of the principal stresses.

A bar of metal 100 × 50 mm in cross-section is 250 mm long. It carries a tensile load of 400 kN in the direction of its length, a compressive load of 4000 kN on its 100 mm × 250 mm faces and a tensile load of 2000 kN on its 50 mm × 250 mm faces. If $E = 200$ GN/m^2 and Poisson's ratio is 0·25 determine the change in volume of the bar.

What change must be made in the 4000 kN load in order that there shall be no change in the volume of the bar?

Solution. Under conditions of complex stress in three dimensions there are always three mutually perpendicular planes on which the shear stress is zero. These planes are the principal planes and the normal stresses on them are the principal stresses (σ_1, σ_2 and σ_3). The strains in the directions of the principal stresses are the principal strains. Since each principal stress causes a lateral strain in the directions of *both* of the others the principal strains are, by an extension of the method of Example 10.2,

$$\left(\frac{\sigma_1}{E} - \frac{\nu\sigma_2}{E} - \frac{\nu\sigma_3}{E}\right) \text{ in the direction in which } \sigma_1 \text{ acts,}$$

$$\left(\frac{\sigma_2}{E} - \frac{\nu\sigma_3}{E} - \frac{\nu\sigma_1}{E}\right) \text{ in the direction in which } \sigma_2 \text{ acts,}$$

$$\left(\frac{\sigma_3}{E} - \frac{\nu\sigma_1}{E} - \frac{\nu\sigma_2}{E}\right) \text{ in the direction in which } \sigma_3 \text{ acts.}$$

If a rectangular block, length l and cross-section $b \times d$, has these dimensions increased by small amounts δl, δb, and δd respectively the increase in volume is

$$\begin{aligned}\delta V &= (\text{final volume}) - (\text{original volume})\\ &= (l + \delta l)(b + \delta b)(d + \delta d) - lbd\\ &= (lbd + lb\delta d + ld\delta b + bd\delta l + l\delta b\delta d\\ &\quad + b\delta l\delta d + d\delta l\delta b + \delta l\delta b\delta d) - lbd\end{aligned}$$

Neglecting $\delta l\delta b\delta d$ and the products $\delta b\delta d$, $\delta l\delta d$ and $\delta l\delta b$ we have, if V is the original volume,

$$\begin{aligned}\delta V &= lb\delta d + ld\delta b + bd\delta l\\ &= lbd\left(\frac{\delta d}{d} + \frac{\delta b}{b} + \frac{\delta l}{l}\right)\\ &= V \times \text{(sum of the three strains in the directions of the length, width and breadth)}\end{aligned}$$

If the principal stresses act in the directions of the length, width, and breadth, then

$$\delta V/V = \text{sum of the three principal strains}$$

($\delta V/V$ is called the *volumetric strain*.)

Using the values given in the question,

$$\sigma_1 = (400 \times 10^3 \text{ N})/(100 \times 50 \times 10^{-6} \text{ m}^2) = 80 \text{ MN/m}^2$$
$$\sigma_2 = -(4000 \times 10^3 \text{ N})/(100 \times 250 \times 10^{-6} \text{ m}^2) = -160 \text{ MN/m}^2$$
$$\sigma_3 = (2000 \times 10^3 \text{ N})/(50 \times 250 \times 10^{-6} \text{ m}^2) = 160 \text{ MN/m}^2$$

The volumetric strain is

$$\frac{\delta V}{V} = \left(\frac{\sigma_1}{E} - \frac{\nu\sigma_2}{E} - \frac{\nu\sigma_3}{E}\right) + \left(\frac{\sigma_2}{E} - \frac{\nu\sigma_3}{E} - \frac{\nu\sigma_1}{E}\right) + \left(\frac{\sigma_3}{E} - \frac{\nu\sigma_1}{E} - \frac{\nu\sigma_2}{E}\right)$$
$$= \frac{1}{E}[\sigma_1 + \sigma_2 + \sigma_3 - 2\nu(\sigma_1 + \sigma_2 + \sigma_3)]$$
$$= \frac{1}{E}(\sigma_1 + \sigma_2 + \sigma_3)(1 - 2\nu) \qquad \text{(i)}$$
$$= 1/(200 \times 10^9 \text{ N/m}^2)[(80 - 160 + 160) \times 10^6 \text{ N/m}^2] \times (1 - 2 \times 0{\cdot}2)5$$
$$= 1/5000$$

Hence the change in volume is

$$\delta V = V/5000 = (250 \text{ mm} \times 100 \text{ mm} \times 50 \text{ mm})/5000 = 250 \text{ mm}^3 \qquad (Ans)$$

If there is no change in the volume of the bar, then from (i)

$$\sigma_1 + \sigma_2 + \sigma_3 = 0$$
$$\sigma_2 = -(\sigma_1 + \sigma_3) = -(80 + 160) = -240 \text{ MN/m}^2$$

The corresponding load (on the 100 mm × 250 mm face) is

240 MN/m² × 100 mm × 250 mm = 6000 kN (compression) *(Ans)*

Thus the 4000 kN load must be increased by 2000 kN (compression) *(Ans)*

10.5. A cylindrical bar 150 mm long by 75 mm diameter is subjected to an axial compressive load of 450 kN. Taking $E = 200$ GN/m² and $\nu = 0{\cdot}3$ calculate the change in volume of the bar. If all lateral swelling of the bar is prevented what is the pressure on the side of the bar and the change in volume?

Solution. If a cylindrical bar, length l and diameter d, has these dimensions increased to $(l + \delta l)$ and $(d + \delta d)$ respectively, the change in volume is

$$\begin{aligned}\delta V &= \tfrac{1}{4}\pi(d + \delta d)^2(l + \delta l) - \tfrac{1}{4}\pi d^2 l \\ &= \tfrac{1}{4}\pi(d^2 l + 2dl\delta d + l\delta d^2 + d^2\delta l + 2d\delta d\delta l + \delta l\delta d^2 - d^2 l) \\ &= \tfrac{1}{4}\pi(2dl\delta d + d^2\delta l)\end{aligned}$$

(neglecting, as usual, the products of small quantities).

Hence

$$\delta V = \tfrac{1}{4}\pi d^2 l\left(2\,\frac{\delta d}{d} + \frac{\delta l}{l}\right)$$

$$\frac{\delta V}{V} = \text{(twice the diametral strain + the longitudinal strain)}$$

$$\begin{aligned}\text{Longitudinal stress} &= \frac{450 \times 10^3\ \text{N}}{\tfrac{1}{4}\pi \times (75 \times 10^{-3}\ \text{m})^2} \\ &= 102\ \text{MN/m}^2\ \text{(compressive)}\end{aligned}$$

$$\begin{aligned}\text{Longitudinal strain} = \text{stress}/E &= -\frac{102 \times 10^6\ \text{N/m}^2}{200 \times 10^9\ \text{N/m}^2} \\ &= -0{\cdot}000\,510\end{aligned}$$

$$\begin{aligned}\text{The diametral (lateral) strain} &= -\nu \times \text{longitudinal strain} \\ &= -0{\cdot}3\,(-0{\cdot}000\,510) \\ &= 0{\cdot}000\,153\end{aligned}$$

Thus, the change in volume is

$$\begin{aligned}\delta V &= V\,\text{(twice the diametral strain + the longitudinal strain)} \\ &= \left[\frac{\pi}{4}(75\ \text{mm})^2(150\ \text{mm})\right][2(0{\cdot}000\,153) + (-\,0{\cdot}000\,510)] \\ &= 135\ \text{mm}^3\end{aligned}$$

Let p MN/m² be the pressure on the surface of the bar necessary to prevent all lateral swelling. By symmetry, it is the same at all points on the surface. Hence, both lateral principal stresses are equal to $-p$, and if σ_1 is the longitudinal stress then

$$\sigma_1 = -102 \qquad \sigma_2 = \sigma_3 = -p$$

Since the strain is zero in the direction of any diameter,

$$\frac{\sigma_2}{E} - \frac{\nu\sigma_3}{E} - \frac{\nu\sigma_1}{E} = 0$$

or

$$\frac{(-p)}{E} - \frac{\nu(-p)}{E} = \frac{\nu\sigma_1}{E}$$

$$p(\nu - 1) = \nu\sigma_1 \quad \text{or} \quad p = \nu\sigma_1/(\nu - 1)$$

The required pressure is therefore

$$\nu\sigma_1 \times \left(\frac{1}{\nu - 1}\right) = \frac{0{\cdot}3 \times (-102)}{(0{\cdot}3 - 1)} = 43{\cdot}7 \text{ MN/m}^2 \qquad (Ans)$$

since the diametral strain is zero, the change in volume is

$$\begin{aligned} \delta V &= V \times \text{longitudinal strain} \\ &= V \times \left(\frac{\sigma_1}{E} - \frac{\nu\sigma_2}{E} - \frac{\nu\sigma_3}{E}\right) \\ &= \frac{V}{E} \times (-102 + 2\nu p) \\ &= \left[\frac{\frac{1}{4}\pi \times (75 \text{ mm})^2 \times 150 \text{ mm}}{200 \text{ GN/m}^2}\right] \times \\ &\qquad [-102 + (2 \times 0{\cdot}3 \times 43{\cdot}7)] \text{ MN/m}^2 \\ &= -251 \text{ mm}^3 \qquad (Ans) \end{aligned}$$

i.e. a decrease of 251 mm³.

10.6. Define bulk modulus and derive an expression relating it to Young's modulus.

Calculate the change in volume of a 15 cm cube of steel when it is immersed to a depth of 300 m in sea-water which weighs 10 kN/m³

$$E = 200 \text{ GN/m}^2 \qquad \nu = 0{\cdot}29$$

Solution. When a body is subjected to the same pressure at all points on its surface the volumetric strain is proportional to this stress (or pressure). The *bulk modulus* is defined as

$$\frac{\text{stress}}{\text{volumetric strain}}$$

and is usually denoted by K.

Since the volumetric strain can also be obtained as the sum of the three principal strains (which involve E and ν) then a relationship between K, E and ν can be deduced.

Suppose a rectangular block is subjected to a pressure p on all faces. The principal stresses are

$$\sigma_1 = \sigma_2 = \sigma_3 = -p$$

and each principal strain is

$$\frac{\sigma_1}{E} - \frac{\nu\sigma_2}{E} - \frac{\nu\sigma_3}{E} = -\left(\frac{p}{E} - \frac{2\nu p}{E}\right)$$

The volumetric strain is thus

$$-3 \times \left(\frac{p}{E} - \frac{2\nu p}{E}\right)$$

$$= -\frac{3p}{E}(1 - 2\nu) \qquad \text{(i)}$$

By definition, however,

$$\text{volumetric strain} = \frac{\text{stress}}{K} = \frac{-p}{K} \qquad \text{(ii)}$$

Equating (i) and (ii),

$$-\frac{p}{K} = -\frac{3p}{E}(1 - 2\nu)$$

$$E = 3K(1 - 2\nu) \qquad (\textit{Ans})$$

With the values given in the question,

$$K = \frac{E}{3(1 - 2\nu)} = \frac{200 \times 10^9 \text{ N/m}^2}{3[1 - (2 \times 0{\cdot}29)]}$$

$$= 159 \text{ GN/m}^2$$

The pressure at the given depth is

$$p = 300 \text{ m} \times 10 \text{ kN/m}^3 = 3 \text{ MN/m}^2$$

The volumetric strain $\delta V/V = \text{stress}/K$

$$= \frac{-3 \times 10^6 \text{ N/m}^2}{159 \times 10^9 \text{ N/m}^2}$$

$$= -1{\cdot}89 \times 10^{-5}$$

The change in volume $\delta V = V \times 1{\cdot}89 \times 10^{-5}$

$$= (15 \times 10^{-2} \text{ m})^3 \times (-1{\cdot}89 \times 10^{-5})$$

$$= -63{\cdot}8 \text{ mm}^3 \qquad (\textit{Ans})$$

10.7. Prove the relationship between Young's modulus, the modulus of rigidity and Poisson's ratio.

When a bar of certain material, 50 mm diameter, is subjected to an axial pull of 250 kN the extension on a gauge length of 200 mm is 338 μm and the decrease in diameter is 26 μm. Calculate Young's modulus, Poisson's ratio, the shear modulus and bulk modulus for this material.

Solution. Consider a cube ABCD which is distorted by (pure) shear stress τ to a position AB′C′D as shown in Fig. 10.3. For pure shear the principal stresses are $\sigma_1 = +\tau$ and $\sigma_2 = -\tau$ acting along the diagonals AC and BD respectively.

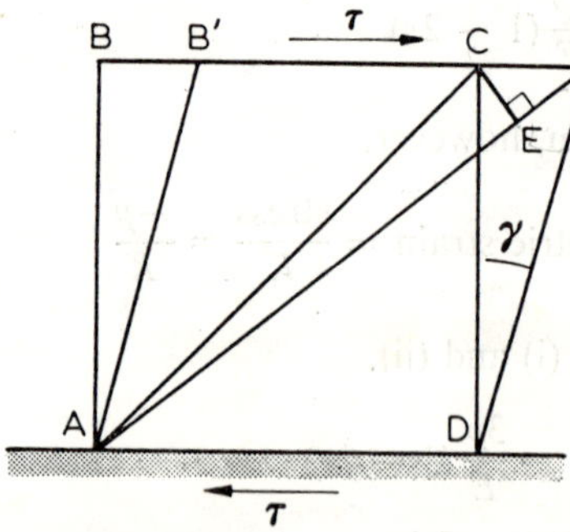

Fig. 10.3

Since the strain along the diagonal AC can be expressed in terms of the shear modulus or, alternatively, in terms of Young's modulus and Poisson's ratio, then a relationship between these three quantities can be deduced.

If CE is a perpendicular from C on to the new position of the diagonal AC′ then, since the distortion is small, the strain along AC′ is

$$\frac{EC'}{AC} = \frac{CC' \cos EC'C}{AC} = \frac{CC' \cos 45°}{AC} \text{ (approximately)}$$

$$= \frac{CC' \cos 45°}{(DC/\cos 45°)} \qquad \text{(since } DC/AC = \cos 45°\text{)}$$

$$= \frac{CC'}{DC} \cos^2 45°$$

$$= \frac{1}{2}\left(\frac{CC'}{DC}\right)$$

However CC′/DC is the shear strain γ where

$$\gamma = \text{shear stress}/G = \tau/G$$

Hence, strain along AC′ is

$$\tau/2G \qquad \text{(i)}$$

In terms of the principal stresses $\sigma_1(=\tau)$ and $\sigma_2(= -\tau)$,

$$\text{Diagonal strain} = \frac{\sigma_1}{E} - \frac{\nu\sigma_2}{E} = \frac{\tau}{E} - \frac{\nu(-\tau)}{E}$$

$$= \frac{\tau}{E}(1 + \nu) \qquad \text{(ii)}$$

Since (i) and (ii) represent the same strain,

$$\frac{\tau}{2G} = \frac{\tau}{E}(1 + \nu)$$

$$E = 2G(1 + \nu)$$

Combining this with the result obtained in Example 10.6 (p. 293),

$$E = 2G(1 + \nu) = 3K(1 - 2\nu)$$

and, if any two of the four elastic constants E, G, K, ν are given, then the others can be calculated.

From the data given in the question,

$$E = \frac{\text{stress}}{\text{strain}} = \frac{250 \times 10^3\ \text{N}}{\frac{1}{4}\pi \times (50 \times 10^{-3}\ \text{m})^2} \times \frac{200 \times 10^{-3}\ \text{m}}{338 \times 10^{-6}\ \text{m}}$$

$$= 75{\cdot}4\ \text{GN/m}^2 \qquad (Ans)$$

$$\text{Poisson's ratio } \nu = \text{lateral strain/longitudinal strain}$$

$$= \frac{26 \times 10^{-6}\ \text{m}}{50 \times 10^{-3}\ \text{m}} \times \frac{200 \times 10^{-3}\ \text{m}}{338 \times 10^{-6}\ \text{m}}$$

$$= 0{\cdot}3078 \qquad (Ans)$$

$$\text{Shear modulus } G = E/2(1 + \nu)$$

$$= \frac{75{\cdot}4 \times 10^9\ \text{N/m}^2}{2(1 + 0{\cdot}3078)} = 28{\cdot}8\ \text{GN/m}^2 \qquad (Ans)$$

$$\text{Bulk modulus } K = E/3(1 - 2\nu)$$

$$= \frac{75{\cdot}4 \times 10^9\ \text{N/m}^2}{3[1 - (2 \times 0{\cdot}3078)]} = 65{\cdot}4\ \text{GN/m}^2 \qquad (Ans)$$

10.8. A hollow round bar deflects 6 mm under a central load of 10 kN when it is used as a simply-supported beam over a span of 3·6 m. Calculate the angle of twist in this length when the bar is subjected to a torque of 5 kN-m. Poisson's ratio = $\frac{3}{11}$.

Solution. Since $E = 2G(1 + \nu)$

$$G = \frac{E}{2(1+\nu)} = \frac{E}{2(1+\frac{3}{11})} = \frac{11E}{28}$$

Also, for all circular sections, $J = 2I$.

Using the "beam" data, since the central deflection Δ is $WL^3/48EI$ then

$$EI = \frac{WL^3}{48\Delta} = \frac{(10 \times 10^3\ \text{N}) \times (3{\cdot}6\ \text{m})^3}{48 \times (6 \times 10^{-3}\ \text{m})} = 1{\cdot}62\ \text{MN-m}^2$$

Hence

$$GJ = \frac{11E}{28} \times 2I = \frac{11}{14}EI = \frac{11 \times 1{\cdot}62}{14} = 1{\cdot}27\ \text{MN-m}^2$$

From the torsion equation,

$$\theta = \frac{TL}{GJ} = \frac{(5 \times 10^3\ \text{N-m}) \times (3{\cdot}6\ \text{m})}{1{\cdot}27\ \text{MN-m}^2}$$

$$= 0{\cdot}01414\ \text{rad or } 0{\cdot}81^\circ \qquad (Ans)$$

10.9. Prove that the principal stresses σ_1 and σ_2 at a point in a material can be derived from the strains ϵ_1 and ϵ_2 in the direction of the principal stresses by the equations

$$\sigma_1 = \frac{E}{1-\nu^2}(\epsilon_1 + \nu\epsilon_2)$$

$$\sigma_2 = \frac{E}{1-\nu^2}(\epsilon_2 + \nu\epsilon_1)$$

where ν = Poisson's ratio.

In order to determine the principal stresses at a point in a structural member two strain gauges are fixed, their directions being at 30° to the known directions of the principal stresses, The measured strains in these two directions are $+455 \times 10^{-6}$ and -32×10^{-6} respectively. If $E = 200\ \text{GN/m}^2$ and $\nu = 0{\cdot}3$ find the magnitudes of the principal stresses. *(U.L.)*

Solution. The derivation of the equations in the question is given in Example 10.3 (x and y being replaced by 1 and 2). A note on strain gauges is given in the solution to Example 10.10.

For the second part of the question suppose (Fig. 10.4) that σ_θ and $\sigma_{(\theta+90°)}$ are the normal stresses in directions each making an angle θ with the directions of the principal stresses σ_1 and σ_2. Then from the analysis given in Chapter 9,

$$\sigma_\theta = \tfrac{1}{2}(\sigma_1 + \sigma_2) + \tfrac{1}{2}(\sigma_1 - \sigma_2)\cos 2\theta$$

and

$$\begin{aligned}\sigma_{(\theta+90°)} &= \tfrac{1}{2}(\sigma_1 + \sigma_2) + \tfrac{1}{2}(\sigma_1 - \sigma_2)\cos(2\theta + 180°)\\ &= \tfrac{1}{2}(\sigma_1 + \sigma_2) - \tfrac{1}{2}(\sigma_1 - \sigma_2)\cos 2\theta\end{aligned}$$

The strain in the direction of σ_θ is (allowing for the Poisson's ratio effect)

$$\begin{aligned}\epsilon_\theta &= \frac{\sigma_\theta}{E} - \frac{\nu\sigma_{(\theta+90°)}}{E}\\ &= \frac{1}{2E}[(\sigma_1 + \sigma_2) + (\sigma_1 - \sigma_2)\cos 2\theta - \nu(\sigma_1 + \sigma_2)\\ &\qquad + \nu(\sigma_1 - \sigma_2)\cos 2\theta]\\ &= \frac{1}{2}\left(\frac{\sigma_1}{E} - \frac{\nu\sigma_2}{E}\right) + \frac{1}{2}\left(\frac{\sigma_2}{E} - \frac{\nu\sigma_1}{E}\right)\\ &\qquad + \frac{1}{2}\left[\left(\frac{\sigma_1}{E} - \frac{\nu\sigma_2}{E}\right) - \left(\frac{\sigma_2}{E} - \frac{\nu\sigma_1}{E}\right)\right]\cos 2\theta\end{aligned}$$

But from the expressions derived in Example 10.2 the principal strains are

$$\epsilon_1 = \frac{\sigma_1}{E} - \frac{\nu\sigma_2}{E} \quad \text{and} \quad \epsilon_2 = \frac{\sigma_2}{E} - \frac{\nu\sigma_1}{E}$$

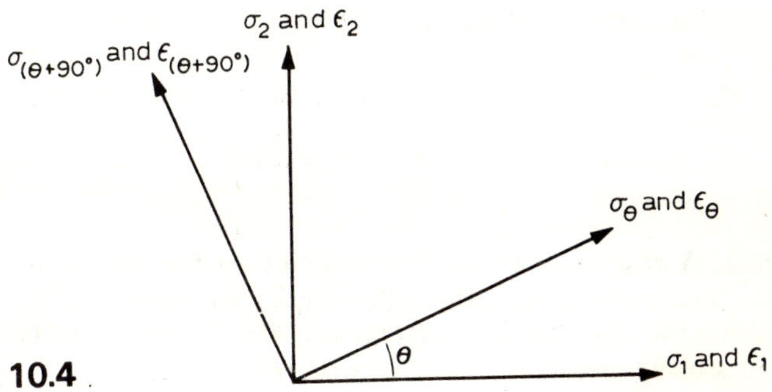

Fig. 10.4

The strain ϵ_θ can therefore be expressed in terms of the principal strains as

$$\epsilon_\theta = \tfrac{1}{2}(\epsilon_1 + \epsilon_2) + \tfrac{1}{2}(\epsilon_1 - \epsilon_2) \cos 2\theta$$

A comparison with the expression for σ_θ obtained in Chapter 9 shows that ϵ_θ is related to the principal strains ϵ_1 and ϵ_2 by an expression which has the same form as that relating σ_θ to the principal stresses (σ_1 and σ_2).

With the numerical data of the question $\theta = 30°$ and

$$\tfrac{1}{2}(\epsilon_1 + \epsilon_2) + \tfrac{1}{2}(\epsilon_1 - \epsilon_2) \cos 60° = 455 \times 10^{-6} \qquad \text{(i)}$$

For the perpendicular direction ($\theta = 120°$)

$$\tfrac{1}{2}(\epsilon_1 + \epsilon_2) - \tfrac{1}{2}(\epsilon_1 - \epsilon_2) \cos 60° = -32 \times 10^{-6} \qquad \text{(ii)}$$

Adding (i) and (ii) gives

$$\epsilon_1 + \epsilon_2 = 422 \times 10^{-6} \qquad \text{(iii)}$$

Subtracting (ii) from (i)

$$\epsilon_1 - \epsilon_2 = 487 \times 10^{-6} \times 2 = 974 \times 10^{-6} \qquad \text{(iv)}$$

since $\cos 60° = \tfrac{1}{2}$.
Adding (iii) and (iv)

$$\epsilon_1 = \tfrac{1}{2}(422 + 974) \times 10^{-6} = 698 \times 10^{-6}$$

Subtracting (iv) from (iii)

$$\epsilon_2 = \tfrac{1}{2}(422 - 974) \times 10^{-6} = -276 \times 10^{-6}$$

Using the equations given in the question

$$\sigma_1 = \frac{200 \times 10^9 \text{ N/m}^2}{1 - 0{\cdot}3^2} (698 \times 10^{-6} - 0{\cdot}3 \times 276 \times 10^{-6})$$

$$= 135 \text{ MN/m}^2 \text{ (tensile)} \qquad (Ans)$$

$$\sigma_2 = \frac{200 \times 10^9 \text{ N/m}^2}{1 - 0{\cdot}3^2} (-276 \times 10^{-6} + 0{\cdot}3 \times 698 \times 10^{-6})$$

$$= -14{\cdot}7 \text{ MN/m}^2 \text{ (i.e. compressive)} \qquad (Ans)$$

10.10. A rosette of three strain gauges on the surface of a metal plate under stress gave the following strain readings: No. 1 at 0°, +0·000592; No. 2 at 45° +0·000308; No. 3 at 90,° −0·000432, the angles being measured anti-clockwise from gauge No. 1.

Determine the magnitude of the principal strains and their directions relative to the axis of gauge No. 1. If $E = 203$ GN/m^2 and Poisson's ratio is 1/3 find the principal stresses.

Prove any formula used. (*U.L.*)

Solution. There are a number of instruments for measuring the small changes in length which occur when a metal bar or plate is stressed. Most of them depend on systems of mechanical or optical levers and they are known as *extensometers*. Their main use is in the determination of the mechanical properties of materials and typical results are given in Chapter 16.

Nowadays the *electrical resistance strain gauge* is widely used to measure the strains in machine and structural elements, particularly under conditions of complex stress. The principle of this device is that the electrical resistance of a wire changes when the wire is strained. The basic gauge (Fig. 10.5a) consists of a grid of wire or foil

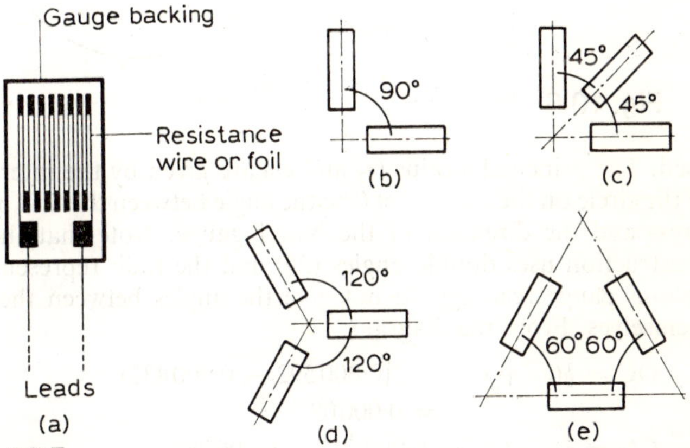

Fig. 10.5

mounted on a special thin backing. The gauge is usually less than 2 cm long. It is stuck to the surface of the metal at the point and in the direction for which the strain is required.

In order to measure strains under conditions of complex stress two or more basic gauges are used simultaneously and they are manufactured in various configurations. Fig. 10.5b shows two gauges mounted at right angles and this arrangement is suitable when the directions of the principal stresses are known (as in Example 10.9). Otherwise it is necessary to measure strains in at least three directions and this is done by a *rosette* of gauges mounted at convenient angles to one another. Figure 10.5c illustrates the configuration described in the present example, which is known as a rectangular rosette.

The mathematical relationships between the strains can be derived from the Mohr circle and Fig. 10.6 gives the notation to be

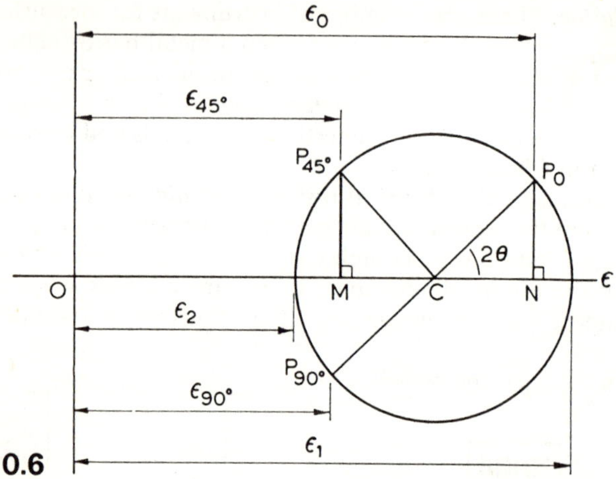

Fig. 10.6

used. The principal strains (ϵ_1 and ϵ_2) are given by the intersections of the circle on the ϵ axis. Let θ be the angle between the first principal stress and the direction of the No. 1 gauge. Note that the Mohr construction uses double angles (2θ) and the radii representing the various gauge readings do not give the angles between the gauges themselves. From the diagram

$$OC = \tfrac{1}{2}(\epsilon_0 + \epsilon_{90}) = \tfrac{1}{2}(0{\cdot}000\,592 - 0{\cdot}000\,432)$$
$$= 0{\cdot}000\,08$$

$$CN = \epsilon_0 - OC = 0{\cdot}000\,592 - 0{\cdot}000\,08$$
$$= 0{\cdot}000\,512$$

$$P_0N = MC = OC = \varepsilon_{45} = 0{\cdot}000\,08 - 0{\cdot}000\,308$$
$$= -0{\cdot}000\,228$$

$$\text{Radius } P_0C = \sqrt{[CN^2 + P_0N^2]} = \sqrt{[0{\cdot}000\,512^2 + (-0{\cdot}000\,228)^2]}$$
$$= 0{\cdot}000\,575$$

The principal strains are, therefore,

$$\epsilon_1 = OC + \text{radius} = 0{\cdot}000\,08 + 0{\cdot}000\,575$$
$$= 0{\cdot}000\,655$$

$$\epsilon_2 = OC - \text{radius} = 0{\cdot}000\,08 - 0{\cdot}000\,575$$
$$= -0{\cdot}000\,495$$

Also from the geometry of the Mohr circle the angle between the axis of gauge No. 1 and the greater principal strain is given by

$$\tan 2\theta = \frac{P_0N}{CN} = \frac{-0{\cdot}000228}{0{\cdot}000512} = -0{\cdot}445$$

$$2\theta = -24^\circ \quad (\text{or } 156^\circ)$$

Since the No. 1 gauge strain is closer to ϵ_1 than ϵ_2 it is the (numerically) smaller angle which is required and thus

$$\theta = -12^\circ$$

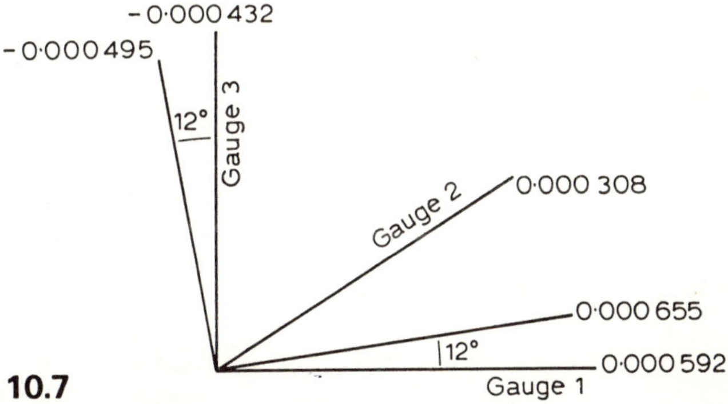

Fig. 10.7

The directions of the strains are shown in Fig. 10.7. The principal stresses are found as in the previous example.

$$\sigma_1 = \frac{E}{1-\nu^2}(\epsilon_1 + \nu\epsilon_2)$$

$$= \frac{203 \times 10^9}{1-(\frac{1}{3})^2}(0{\cdot}000655 - \tfrac{1}{3} \times 0{\cdot}000495)$$

$$= 112 \text{ MN/m}^2 \text{ (tensile)} \qquad (\textit{Ans})$$

$$\sigma_2 = \frac{203 \times 10^9}{1-(\frac{1}{3})^2}(-0{\cdot}000495 + \tfrac{1}{3} \times 0{\cdot}000655)$$

$$= -63{\cdot}3 \text{ MN/m}^2 \text{ (i.e. compressive)} \qquad (\textit{Ans})$$

10.11. A strain gauge rosette is used for the determination of the stress condition at a point on the surface of a steel plate subjected to plane stress. The axes of the three gauges are denoted OA, OB and OC and these are 120° from one to the next. The observed strains are $\epsilon_1 = +0{\cdot}000554$ along OA, $\epsilon_2 = -0{\cdot}000456$ along OB, $\epsilon_3 = +0{\cdot}000064$ along OC.

Determine the inclinations of the principal planes at O relative to OA and the magnitudes of the principal stresses. Determine also the strain in the direction at right angles to OA. Either graphical or analytical methods may be used. Take $E = 200$ GN/m² and Poisson's ratio = 0·3.

Make a diagram showing the directions OA, OB and OC and the principal strains and stresses in their correct relative positions.

(*U.L.*)

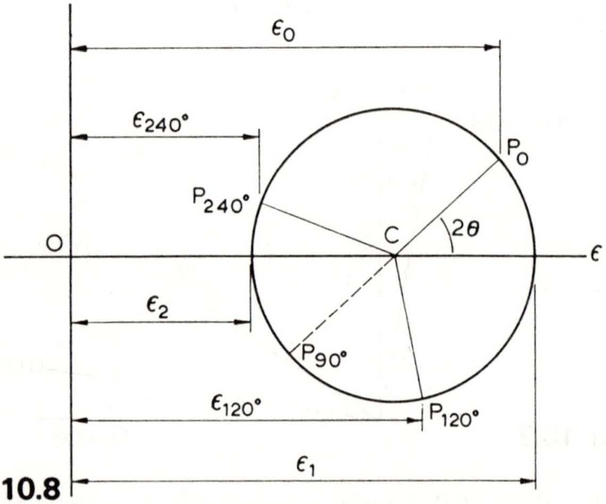

Fig. 10.8

Solution. An alternative to the rectangular rosette is one in which the three gauges are mounted symmetrically. This can be achieved by a "star" configuration (Fig. 10.5d) or by making the gauges the sides of an equilateral triangle, Fig. 10.5e, in which case it is known as a *delta rosette.*

Again the analysis can be made by reference to the Mohr circle Fig. 10.8, and for consistency with earlier examples, let ϵ_1 and ϵ_2 be the principal strains, ϵ_0, ϵ_{120} and ϵ_{240} the given strains.

Then from the geometry of Fig. 10.8, if r = radius,

$$\epsilon_0 = OC + r\cos 2\theta \quad \text{(i)}$$

$$\begin{aligned}\epsilon_{120} &= OC + r\cos(2\theta + 240^\circ)\\ &= OC + r(\cos 2\theta\cos 240^\circ - \sin 2\theta \sin 240^\circ)\\ &= OC - r\cos 2\theta\cos 60^\circ + r\sin 2\theta\sin 60^\circ \quad \text{(ii)}\end{aligned}$$

$$\begin{aligned}\epsilon_{240} &= OC + r\cos(2\theta + 480^\circ)\\ &= OC + r(\cos 2\theta\cos 120^\circ - \sin 2\theta\sin 120^\circ)\\ &= OC - r\cos 2\theta\cos 60^\circ - r\sin 2\theta\sin 60^\circ \quad \text{(iii)}\end{aligned}$$

Adding (i), (ii) and (iii)

$$\epsilon_0 + \epsilon_{120} + \epsilon_{240} = 3 \times OC + r\cos 2\theta - 2r\cos 2\theta\cos 60^\circ$$
$$= 3 \times OC$$

(since $\cos 60^\circ = \frac{1}{2}$).
With the values given in the question

$$OC = \tfrac{1}{3}(\epsilon_0 + \epsilon_{120} + \epsilon_{240})$$
$$= \tfrac{1}{3}(0{\cdot}000554 - 0{\cdot}000456 + 0{\cdot}000064)$$
$$= 0{\cdot}000054$$

Also, subtracting (iii) from (ii)

$$\epsilon_{120} - \epsilon_{240} = 2r\sin 2\theta\sin 60^\circ = \sqrt{3}r\sin 2\theta$$

(since $\sin 60^\circ = \sqrt{3}/2$).
Substituting the numerical values,

$$r\sin 2\theta = (\epsilon_{120} - \epsilon_{240})/\sqrt{3} = (-0{\cdot}000456 - 0{\cdot}000064)/\sqrt{3}$$
$$= -0{\cdot}0003 \qquad \text{(iv)}$$

From (i)

$$r\cos 2\theta = \epsilon_0 - OC$$
$$= 0{\cdot}000554 - 0{\cdot}000054$$
$$= 0{\cdot}0005 \qquad \text{(v)}$$

Dividing (iv) by (v) we have

$$\tan 2\theta = -0{\cdot}0003/0{\cdot}0005 = -0{\cdot}6$$
$$2\theta = -31^\circ \text{ (or } 149^\circ)$$
$$\theta = -15\tfrac{1}{2}^\circ \qquad (\textit{Ans})$$

The principal planes are therefore $15\frac{1}{2}^\circ$ and $105\frac{1}{2}^\circ$ anti-clockwise from OA. (See Fig. 10.9).
From (i) the radius of the Mohr circle is given by

$$r = \frac{\epsilon_0 - OC}{\cos 2\theta} = \frac{0{\cdot}000554 - 0{\cdot}000054}{\cos(-31^\circ)} = 0{\cdot}000583$$

The principal strains are therefore

$$\epsilon_1 = OC + r = 0{\cdot}000054 + 0{\cdot}000583$$
$$= 0{\cdot}000637$$
$$\epsilon_2 = OC - r = 0{\cdot}000054 - 0{\cdot}000583$$
$$= -0{\cdot}000529$$

Therefore

$$\sigma_1 = \frac{E}{1-\nu^2}(\epsilon_1 + \nu\epsilon_2)$$

$$= \frac{200}{1-0{\cdot}3^2}(0{\cdot}000\,637 - 0{\cdot}3 \times 0{\cdot}000\,529)$$

$$= 105 \text{ MN/m}^2 \text{ (tensile)} \qquad (Ans)$$

$$\sigma_2 = \frac{E}{1-\nu^2}(\epsilon_2 + \nu\epsilon_1)$$

$$= \frac{200}{1-0{\cdot}3^2}(-0{\cdot}000\,529 + 0{\cdot}3 \times 0{\cdot}000\,637)$$

$$= -74{\cdot}3 \text{ MN/m}^2 \text{ (i.e. compressive)} \qquad (Ans)$$

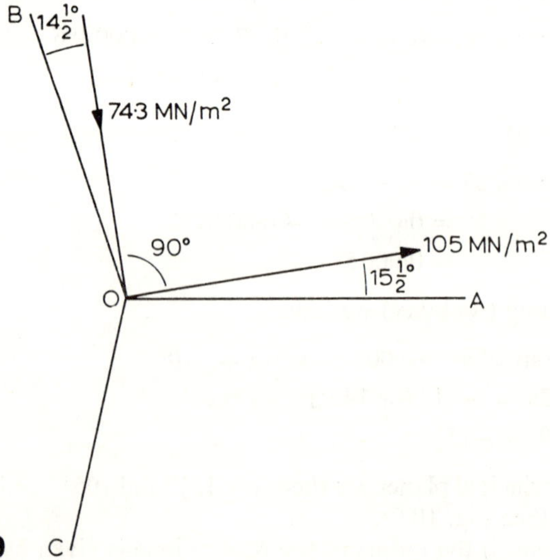

Fig. 10.9

If θ is now measured from the direction of ϵ_1 its value for the direction at right angles to OA is $(90 - 15\frac{1}{2})° = 74\frac{1}{2}°$. The strain in this direction is

$$\epsilon = \tfrac{1}{2}(\epsilon_1 + \epsilon_2) + \tfrac{1}{2}(\epsilon_1 - \epsilon_2)\cos 2\theta$$
$$= \tfrac{1}{2}(0{\cdot}000\,637 - 0{\cdot}000\,529)$$
$$+ \tfrac{1}{2}(0{\cdot}000\,637 + 0{\cdot}000\,529)\cos 149°$$
$$= -0{\cdot}000\,446 \qquad (Ans)$$

Alternatively, from Fig. 10·8, by symmetry,

$$OC - \epsilon_{90} = \epsilon_0 - OC = 0{\cdot}0005 \quad \text{(from (iv))}$$

Hence

$$\epsilon_{90} = OC - 0{\cdot}0005 = 0{\cdot}000054 - 0{\cdot}0005$$
$$= -0{\cdot}000446 \qquad \textit{(Ans)}$$

PROBLEMS

1. Calculate the change in the diameter of a short cast-iron column when it carries a compressive load of 650 kN. The original diameter is 76 mm. $E = 100$ GN/m^2 and $\nu = \frac{1}{4}$.
Answer. Increase of 0·0272 mm.

2. Under a certain axial tensile load the diameter of a steel bar is reduced from 75 mm to 74·975 mm. Calculate the change in the cross-sectional area due to this load and the value of the tensile load.

$E = 206$ GN/m^2 and $\nu = 0{\cdot}28$.
Answer. 2·944 mm^2; 1·086 MN.

3. A 150 mm × 100 mm × 75 mm block carries (axial) loads as follows:

(*a*) 720 kN (tension) on the 100 mm × 75 mm faces
(*b*) 1200 kN (compression) on the 150 mm × 100 mm faces
(*c*) 1080 kN (tension) on the 150 mm × 75 mm faces

If Poisson's ratio is 0·3, calculate, in terms of E, the strain in the direction in which each load acts.

Taking $E = 180$ GN/m^2, find the change in the volume of the block.
Answer. (*a*) $+0{\cdot}0912/E$; (*b*) $-0{\cdot}1376/E$; (*c*) $+0{\cdot}0912/E$; 280 mm^3 increase.

4. What is meant by the term "principal stress"?

A tube, 20 mm internal diameter and wall thickness 5 mm, is subjected to an axial tensile load of 30 kN and a twisting moment of 750 N-m. Calculate the maximum principal stress in the tube. What stress acting alone would produce a strain equal to the maximum principal strain? Take Poisson's ratio $\nu = 0{\cdot}3$.
Answer. Maximum principal stress = 218 MN/m^2; the maximum principal strain is equal to that caused by a simple tension of 257 MN/m^2.

5. A rectangular block 250 mm × 100 mm × 75 mm is subjected to axial loads as follows:

(*a*) 480 kN tensile in the direction of its length
(*b*) 1000 kN compressive on the 250 mm × 100 mm faces
(*c*) 900 kN tensile on the 250 mm × 75 mm faces

Taking Poisson's ratio ν as $\frac{1}{4}$, find in terms of E the strain in the direction in which each stress acts. If $E = 206$ GN/m^2, calculate the change in the volume of the block due to the above loading, the shear modulus, and bulk modulus for the material of the block.

Answer. (*a*) $+62/E$; (*b*) $-68/E$; (*c*) $+42/E$. Increase of 330 mm³; $G =$ 82 GN/ m²; $K = 136{\cdot}7$ GN/m².

6. Find the values of ν if (*a*) $E = 2\frac{1}{2}G$; (*b*) $E = K$; (*c*) $E = 1\frac{1}{2}K$; (*d*) $K = 2G$.
Answer. (*a*) 0·25; (*b*) 0·333; (*c*) 0·25; (*d*) 0·286.

7. Establish the following results, assuming the relationships proved in Examples 10.6 and 10.7.

(*a*) $G = \dfrac{3K}{2}\left(\dfrac{1-2\nu}{1+\nu}\right)$

(*b*) $\nu = \dfrac{E-2G}{2G} = \dfrac{3K-E}{6K} = \dfrac{3K-2G}{6K+2G}$

(*c*) $E = 9GK/(3K+G)$
(*d*) $G = 3KE/(9K-E)$
(*e*) $K = GE/(9G-3E)$

8. A bar of magnesium alloy, 32 mm diameter, is tested on a gauge length of 200 mm in tension and also in torsion. A tensile load of 50 kN produces an extension of 0·27 mm and a torque of 200 N-m produces a twist of 1·22°. Calculate Young's modulus, the shear modulus, the bulk modulus and Poisson's ratio for this material.
Answer. $E = 46{\cdot}0$ GN/m²; $G = 18{\cdot}3$ GN/m²; $K = 32{\cdot}0$ GN/m²; $\nu = 0{\cdot}26$.

9. Explain what is meant by the terms Young's modulus, Shear modulus, Poisson's ratio.

Derive a relationship between these quantities, and give their respective dimensions. (*R.Ae.S.*)
Answer. Poisson's ratio is a pure number: the two moduli have the dimensions of stress, i.e. force/area = $ML^{-1}T^{-2}$.

10. Define Poisson's Ratio. A steel bar 4 m long and 50 mm square in section is elongated by an axial tensile load of 100 kN. If $E = 200$ GN/m² and Poisson's ratio is 0·25, find the increase in volume of the bar. (*I.Struct.E.*)
Answer. 1000 mm³.

11. Working from first principles find
(*a*) the normal stress,
(*b*) the tangential stress,

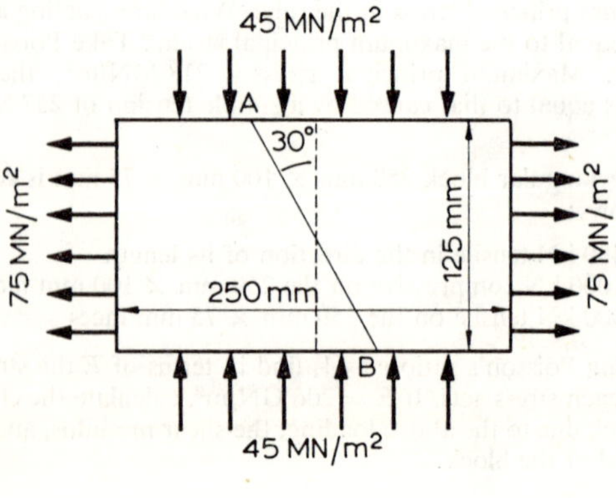

Fig. 10.10

(*c*) the resultant stress, on the interface AB of the mild steel plate stressed as shown in Fig. 10.10.

If the dimensions of the plate were 250 mm by 125 mm before straining, estimate the change in area of the plate.

$E = 206\ \mathrm{GN/m^2}$; $\nu = 0{\cdot}28$ (*I.Mech.E.*)

Answer. (*a*) 45 $\mathrm{MN/m^2}$ (tensile); (*b*) 52·0 $\mathrm{MN/m^2}$; (*c*) 68·7 $\mathrm{MN/m^2}$, change in area, 3·28 $\mathrm{mm^2}$ (increase).

12. A circle 30 cm in diameter is scribed on a mild steel plate before it is stressed as shown in Fig. 10.11. After stressing, the circle deforms to an ellipse. Calculate the lengths of the major and minor axes of the ellipse and also find their directions.

$E = 206\ \mathrm{GN/m^2}$; $\nu = 0{\cdot}28$ (*I.Mech.E.*)

Answer. Major axis, 30·0109 cm; minor axis, 29·9979 cm; these axes are 18° 26′ and 108° 26′ from the vertical centre-line (measured in a clockwise sense).

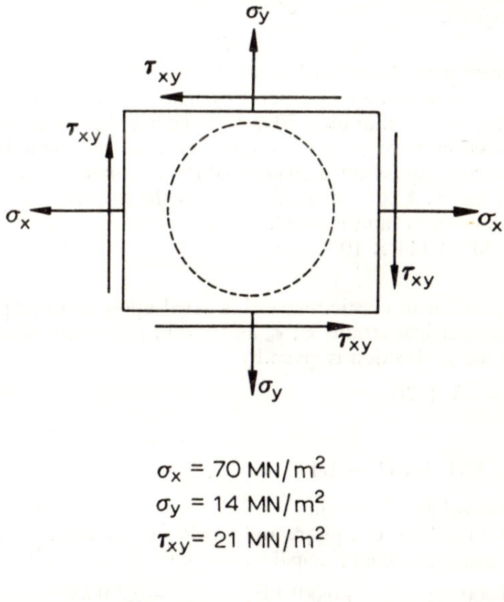

$\sigma_x = 70\ \mathrm{MN/m^2}$
$\sigma_y = 14\ \mathrm{MN/m^2}$
$\tau_{xy} = 21\ \mathrm{MN/m^2}$

Fig. 10.11

13. Derive an expression showing the relationship between the following elastic constants: Young's modulus, modulus of rigidity and Poisson's ratio.

A vertical rod of length L and diameter d is fixed at its upper end and carries a load W at the lower end. The extension of the rod due to W is x. If a torque T, applied in a horizontal plane at the lower end of the rod, gives an angle of twist θ radians show that Poisson's ratio may be obtained from the formula

$$\nu \text{ or } \frac{1}{m} = \frac{Wd^2\theta}{16Tx} - 1$$ (*I.Mech.E.*)

14. Deduce the relation between the modulus of elasticity, the modulus of rigidity and Poisson's ratio.

A hollow shaft of 75 mm external diameter and 50 mm internal diameter twists through an angle of 0·6° in a length of 1·2 m when subjected to an axial twisting moment of 1 kN-m. Estimate the deflection at the centre of the shaft due to its own weight when placed in a horizontal position on supports 1·2 m apart.

Weight of shaft 190 N per metre length. Poissons ratio 0·3. (*U.L.*)

Answer. 0·0344 mm.

15. For an elastic material, express Poisson's ratio in terms of the bulk modulus and the modulus of rigidity, and prove the derivation of the expression.

Determine the percentage change in volume of a steel bar 8 cm square in section and 1 m long when subjected to an axial compressive load of 20 kN. What change in volume would a 10 cm cube of steel suffer at a depth of 5 km in sea-water?

For steel, modulus of elasticity is 200 GN/m² and modulus of rigidity is 80 GN/m². (*U.L.*)

Answer. Poisson's ratio $\nu = (3K - 2G)/(6K + 2G)$; volume of bar decreases by 0·000781 per cent; cube decreases by 377 mm³ (assuming sea-water weighs 10·06 kN/m³).

16. A brass plate 3 mm thick is stretched in the plane of the plate in two directions at right angles. An extensometer arranged in the x-direction gave an extension of 36×10^{-3} mm on a 50 mm gauge length, and another extensometer, in the y-direction, gave an extension of 17×10^{-3} mm on a 100 mm gauge length. If the normal stress on a plane making an angle θ with the y-axis is 59·33 MN/m², find θ. Also find the decrease in the thickness of the plate. Take E for brass as 80 GN/m² and $\nu = 0{\cdot}3$. (*I.Mech.E.*)

Answer. 30°; $1{\cdot}144 \times 10^{-3}$ mm.

17. An element of elastic material is acted upon by three principal stresses and the three principal strains ϵ_x, ϵ_y and ϵ_z are measured. Show that the principal stress in the x-direction is given by

$$\sigma_x = \alpha\Delta + 2G\epsilon_x$$

where

$$\alpha = E/(1 + \nu)(1 - 2\nu)$$

Δ is the volumetric strain;
G is the modulus of rigidity and ν is Poisson's ratio.
In a certain test the principal strains were found to be

$$+0{\cdot}00071; \qquad +0{\cdot}00140; \qquad -0{\cdot}00185$$

Determine the three principal stresses. Take $E = 200$ GN/m² and Poisson's ratio 0·3. (*U.L.*)

Answer. 209 MN/m² (tensile); 315 MN/m² (tensile) and 184 MN/m² (compressive).

18. If, under two-dimensional conditions of stress, a mean stress σ_m and shear stress σ_s are defined as

$$\sigma_m = \tfrac{1}{2}(\sigma_1 + \sigma_2) \quad \text{and} \quad \sigma_s = \tfrac{1}{2}(\sigma_1 - \sigma_2)$$

show that σ_s and σ_m are respectively the radius of the Mohr circle of stress and the distance of its centre from the origin.

Show also that the principal stresses are given by $\sigma_m \pm \sigma_s$ If ϵ_m and ϵ_s are similarly defined in terms of the principal strains ϵ_1 and ϵ_2 prove that

$$\sigma_m = \left(\frac{E}{1-\nu}\right)\epsilon_m \quad \text{and} \quad \sigma_s = \left(\frac{E}{1+\nu}\right)\epsilon_s$$

The principal strains at a point on the surface of an aircraft structural component are measured as 0·001 15 and −0·000 65. Determine ϵ_m and ϵ_s and hence calculate the principal stresses. $E = 91$ GN/m^2 and Poisson's ratio is 0·3.
Answer. 0·000 25 and 0·000 90; 95·5 MN/m^2 (tensile) and 30·5 MN/m^2 (compressive).

19. In a rectangular rosette of three strain gauges the second and third gauges make angles of 45° and 90° respectively with the first. If the strains in these directions are denoted by ϵ_0, ϵ_{45} and ϵ_{90} show that the principal strains are given by

$$\tfrac{1}{2}(\epsilon_0 + \epsilon_{90}) \pm \tfrac{1}{2}\sqrt{[(\epsilon_0 - \epsilon_{90})^2 + (\epsilon_0 + \epsilon_{90} - 2\epsilon_{45})^2]}$$

and their directions relative to that of the first gauge are given by

$$\tan 2\theta = (\epsilon_0 + \epsilon_{90} - 2\epsilon_{45})/(\epsilon_0 - \epsilon_{90})$$

In a particular case $\epsilon_0 = 0{\cdot}001\,00$, $\epsilon_{45} = 0{\cdot}000\,15$ and $\epsilon_{90} = -0{\cdot}000\,20$. Determine the principal strains and principal stresses taking $E = 200$ GN/m^2 and Poisson's ratio $= \frac{1}{3}$.
Make a diagram showing the directions of the principal stresses relative to the axes of the three gauges.
Answer. +0·001 05 and −0·000 25; 217·6 MN/m^2 (tensile) and 22·5 MN/m^2 (tensile).

20. Show, with the notation of the previous question, that the principal stresses can be determined from the readings of a rectangular strain gauge rosette by the formula

$$\frac{E}{2(1-\nu)}(\epsilon_0 + \epsilon_{90}) \pm \frac{E}{2(1+\nu)}\sqrt{[(\epsilon_0 - \epsilon_{90})^2 + (\epsilon_0 + \epsilon_{90} - 2\epsilon_{45})^2]}$$

Obtain the corresponding expression for a delta rosette.
Answer.

$$\frac{E}{3(1-\nu)}(\epsilon_0 + \epsilon_{60} + \epsilon_{120}) \pm \frac{E}{3(1+\nu)}\sqrt{[(2\epsilon_0 - \epsilon_{60} - \epsilon_{120})^2 + 3(\epsilon_{120} - \epsilon_{60})^2]}$$

Chapter 11

Thin Shells

The hoop (circumferential) and longitudinal stresses in a thin, cylindrical shell subjected to an internal pressure P are respectively

$$\sigma_H = \frac{Pd}{2t} \quad \text{and} \quad \sigma_L = \frac{Pd}{4t}$$

where

$d =$ cylinder diameter, $t =$ wall thickness.

The corresponding proportional increase in the capacity of the cylindrical shell is

$$\frac{Pd}{4tE}(5 - 4\nu)$$

where $\nu =$ Poisson's ratio.

The circumferential stress in a thin spherical shell is (with the same notation)

$$\sigma_c = \frac{Pd}{4t}$$

The corresponding proportional increase in the capacity of the spherical shell is

$$\frac{3Pd}{4tE}(1 - \nu)$$

The circumferential stress σ (N/m^2) in a thin rotating circular rim is given by

$$\sigma = \rho v^2$$

where $\rho =$ density of material in the rim (kg/m^3)

$v =$ circumferential velocity (m/s).

WORKED EXAMPLES

11.1. Derive formulae for the longitudinal and circumferential stresses in a thin, cylindrical shell subjected to an internal fluid pressure.

Calculate the minimum wall thickness of a thin cylinder, 1·2 m diameter, if it is to withstand an internal pressure of 1·75 MN/m^2 and

(*a*) the longitudinal stress must not exceed 28 MN/m^2,
(*b*) the circumferential stress must not exceed 42 MN/m^2.

Solution. In the following theory it is assumed that the radial stresses in the cylinder wall are small compared to the circumferential and longitudinal-stresses and that there are no longitudinal stays in the cylinder. In addition, the stresses are assumed to be uniformly distributed through the wall (Figs. 11.1 and 11.2).

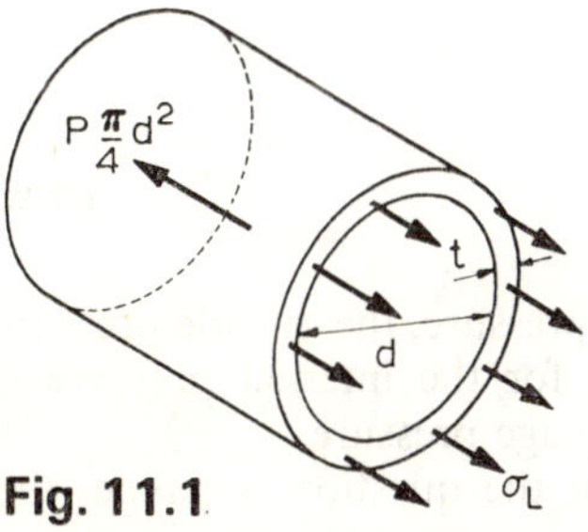

Fig. 11.1

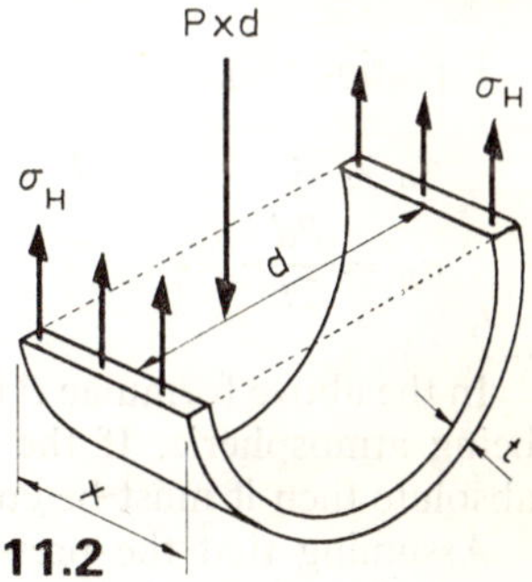

Fig. 11.2

Due to the pressure on the ends of the cylinder there is a longitudinal tensile stress in the cylinder wall. If P is the fluid pressure, the total force on the end of the cylinder is

$$\text{Pressure} \times \text{area} = P \times \tfrac{1}{4}\pi d^2 \qquad \text{(i)}$$

where d is the internal diameter.

The force corresponding to the longitudinal stress σ_L is (Fig. 11.1),

$$\text{Stress} \times \text{area} = \text{stress} \times \text{mean circumference} \times \text{thickness}$$
$$= \sigma_L \times \pi d \times t \text{ (approx.)} \qquad \text{(ii)}$$

(*N.B.* If d is taken to be the *mean* diameter then the area of the wall of the tube is πdt exactly. In this case, however, expression (i) is an approximation.)

For equilibrium, the forces given by (i) and (ii) must be equal. Thus,

$$\sigma_L \times \pi d \times t = P \times \tfrac{1}{4}\pi d^2$$

$$\sigma_L = \frac{Pd}{4t} \qquad (Ans)$$

Consider the equilibrium of the wall and contents of a length x of half the cylinder (Fig. 11.2). Due to the fluid pressure there is a (downward) force of

$$\text{Pressure} \times \text{area} = P \times x \times d \qquad \text{(iii)}$$

This force must be balanced by the (upward) force due to the circumferential (or hoop) stress σ_H, which is

$$\text{Stress} \times \text{area} = \sigma_H \times 2(x \times t) \qquad \text{(iv)}$$

Equating (iii) and (iv),

$$\sigma_H \times 2(x \times t) = P \times x \times d$$

$$\sigma_H = \frac{Pd}{2t} \qquad (Ans)$$

In the above formulae P is the gauge pressure, the outside pressure being atmospheric. If the value given for the internal pressure is absolute then it must be corrected to gauge pressure.

Assuming that the pressure given in the question is gauge, the required thickness is

$$(a)\ t = \frac{Pd}{4\sigma_L} = \frac{1{\cdot}75\ \text{MN/m}^2 \times 1{\cdot}2\ \text{m}}{4 \times 28\ \text{MN/m}^2} = 18{\cdot}8\ \text{mm} \qquad (Ans)$$

$$(b)\ t = \frac{Pd}{2\sigma_H} = \frac{1{\cdot}75\ \text{MN/m}^2 \times 1{\cdot}2\ \text{m}}{2 \times 42\ \text{MN/m}^2} = 25\ \text{mm} \qquad (Ans)$$

11.2. A boiler shell is to be made of plates 12 mm thick and is to withstand a steam pressure of 1·3 MN/m^2 abs. The efficiencies of the longitudinal and circumferential joints are respectively 75 and 35 per cent. If the tensile stress in the plating must not exceed 105 MN/m^2 find the maximum permissible diameter.

Solution. The efficiency of the joint is the ratio of the stress it can withstand to that of the plating on either side of it. Also, the longitudinal joints correspond to the circumferential stress and vice versa. Hence

$$\sigma_L = 105\ \text{MN/m}^2 \times 0{\cdot}35 = 36{\cdot}75\ \text{MN/m}^2$$
$$\sigma_H = 105\ \text{MN/m}^2 \times 0{\cdot}75 = 78{\cdot}75\ \text{MN/m}^2$$

Taking atmospheric pressure as $102\ \text{kN/m}^2$ abs., the internal pressure is

$$P = 1300 - 102 = 1198\ \text{kN/m}^2\ \text{(gauge)}$$

From longitudinal stress considerations, the maximum diameter is

$$d = \frac{4t\sigma_L}{P} = \frac{4 \times (12 \times 10^{-3}\ \text{m}) \times (36{\cdot}75 \times 10^6\ \text{N/m}^2)}{1{\cdot}198 \times 10^6\ \text{N/m}^2}$$
$$= 1{\cdot}47\ \text{m}$$

From considerations of circumferential stress, the maximum diameter is

$$d = \frac{2t\sigma_H}{P} = \frac{2 \times (12 \times 10^{-3}\ \text{m}) \times (78{\cdot}75 \times 10^6\ \text{N/m}^2)}{1{\cdot}198 \times 10^6\ \text{N/m}^2}$$
$$= 1{\cdot}58\ \text{m}$$

To satisfy both conditions, the diameter must not exceed 1·47 m (*Ans*)

11.3. In a certain experiment on combined stresses, a mild steel tube, 25 mm internal diameter and 1·5 mm wall thickness, was closed at the ends and subjected to an internal fluid pressure of $840\ \text{kN/m}^2$. At the same time the tube was subjected to an axial pull of 886 N and to pure torsion by means of a couple, the axis of which coincided with the axis of the tube. If for the purposes of the experiment a maximum principal stress of $36\ \text{MN/m}^2$ were required in the material at the outer surface of the tube find the applied torque in N-m. (*I.Mech.E.*)

Solution. Longitudinal stress due to the axial pull is

$$\frac{\text{load}}{\text{area}} = \frac{886\ \text{N}}{\frac{1}{4}\pi[(28\ \text{mm})^2 - (25\ \text{mm})^2]} = 7{\cdot}0\ \text{MN/m}^2\ \text{(tensile)}$$

(The hoop stress varies slightly across the wall of the cylinder. The formula for σ_H derived in Example 11.1 under-estimates the stress at the inside edge and over-estimates that at the outside edge. For values of t/d less than 1/20, however, the maximum deviation from that given by the formula for σ_H is less than 5 per cent if d is the internal diameter.)

Treating the cylinder as thin,

$$\sigma_H = \frac{Pd}{2t} = \frac{(840 \times 10^3\ \text{N/m}^2) \times 25\ \text{mm}}{2 \times 1{\cdot}5\ \text{mm}} = 7\ \text{MN/m}^2\ \text{(tensile)}$$

$$\sigma_L = \frac{Pd}{4t} = \frac{(840 \times 10^3\ \text{N/m}^2) \times 25\ \text{mm}}{4 \times 1{\cdot}5\ \text{mm}} = 3{\cdot}5\ \text{MN/m}^2\ \text{(tensile)}$$

Since there is no pressure on the outside edge of the tube the radial stress there is zero. The stresses at a point on the outer surface are

$$\sigma_x = \sigma_L + \text{(longitudinal stress due to the axial pull)}$$
$$= 3{\cdot}5\ \text{MN/m}^2 + 7\ \text{MN/m}^2 = (+)\ 10{\cdot}5\ \text{MN/m}^2$$
$$\sigma_y = \sigma_H = (+)\ 7\ \text{MN/m}^2$$

and τ_{xy} = shear stress due to the torque.

Substituting these values in the principal stress relationship

$$(\sigma - \sigma_x)(\sigma - \sigma_y) = \tau_{xy}{}^2$$

we have

$$\tau_{xy}{}^2 = (36 - 10{\cdot}5)(36 - 7) = 739{\cdot}5$$
$$\tau_{xy} = 27{\cdot}2\ \text{MN/m}^2$$
$$J \text{ for the section} = \tfrac{1}{32}\pi[(28\ \text{mm})^4 - (25\ \text{mm})^4]$$
$$= 2{\cdot}23 \times 10^4\ \text{mm}^4$$

Thus, the applied torque is

$$T = \frac{\tau_{xy}}{r} \times J = \frac{(27{\cdot}2 \times 10^6\ \text{N/m}^2) \times (2{\cdot}23 \times 10^{-8}\ \text{m}^4)}{14 \times 10^{-3}\ \text{m}}$$
$$= 43{\cdot}3\ \text{N-m} \qquad \textit{(Ans)}$$

11.4. Derive an expression for the proportional increase of capacity of a thin, cylindrical shell subjected to an internal fluid pressure.

Calculate the increase in the volume enclosed by a boiler shell, 2·4 m long and 0·9 m in diameter, when it is subjected to an internal pressure of 1·8 MN/m². The wall thickness is such that the maximum tensile stress in the shell is 21 MN/m² under this pressure. Take $E = 200\ \text{GN/m}^2$ and $\nu = 0{\cdot}28$.

Solution. At the inner edge of the cylinder wall the three principal stresses are σ_L, σ_H, and the radial compressive stress which is (numerically) equal to the fluid pressure. The last of these is generally negligible in comparison with σ_L and σ_H (for *thin* shells). Using the method of Chapter 10, Example 10.2,

$$\text{Longitudinal strain} = \frac{\sigma_L}{E} - \frac{\nu\sigma_H}{E}$$

$$= \frac{Pd}{4tE} - \frac{\nu Pd}{2tE} = \frac{Pd}{4tE}(1 - 2\nu)$$

(where E is Young's modulus and ν is Poisson's ratio).

Similarly, circumferential strain

$$= \frac{\sigma_H}{E} - \frac{\nu\sigma_L}{E}$$

$$= \frac{Pd}{2tE} - \frac{\nu Pd}{4tE}$$

$$= \frac{Pd}{4tE}(2 - \nu)$$

The diameter increases in the same proportion as the circumference (since circumference $= \pi d$) and, in addition, the expression obtained in Chapter 10, Example 10.5, for the volumetric strain of a cylindrical bar is applicable without alteration to the capacity of the shell.

Thus the proportional increase of the volume enclosed by the shell is

$$\frac{\delta V}{V} = (\text{twice the circumferential strain} + \text{the longitudinal strain})$$

$$= \frac{Pd}{2tE}(2 - \nu) + \frac{Pd}{4tE}(1 - 2\nu)$$

$$= \frac{Pd}{4tE}(5 - 4\nu)$$

(The reader can verify that if the radial stress is considered the above expression is increased by $3\nu P/E$. This term causes an increase of less than 4 per cent when $t/d = 1/20$ and $\nu = 0{\cdot}25$. In either case, no allowance has been made for the constraint or distortion of the ends of the cylinder.)

Since the hoop stress is greater than the lon gitudinal stress then, for the cylinder given,

$$t = \frac{Pd}{2\sigma_H} = \frac{(1{\cdot}8 \times 10^6\ \text{N/m}^2) \times 0{\cdot}9\ \text{m}}{2 \times (21 \times 10^6\ \text{N/m}^2)}$$

$$= 38{\cdot}6\ \text{mm}$$

The increase in volume is

$$\delta V = V \times \frac{Pd}{4tE}(5 - 4\nu)$$

$$= [\tfrac{1}{4}\pi \times (0{\cdot}9\ \text{m})^2 \times (2{\cdot}4\ \text{m})]$$

$$\times \left[\frac{(1{\cdot}8 \times 10^6\ \text{N/m}^2) \times (0{\cdot}9\ \text{m})}{4 \times (38{\cdot}6 \times 10^{-3}\ \text{m}) \times (200 \times 10^9\ \text{N/m}^2)}(5 - 1{\cdot}12)\right]$$

$$= 311\ \text{cm}^3$$

11.5. Derive a formula for the hoop (or tangential) stress in a thin spherical shell under internal pressure.

Calculate the safe working pressure for a spherical vessel, 1·2 m diameter and 12 mm wall thickness, if the tensile stress is limited to 42 MN/m².

Solution. Consider the equilibrium of a hemisphere and its contents has sown in Fig. 11.3. The fluid pressure causes an upward force of (pressure × area) = $P\frac{1}{4}\pi d^2$, in the usual notation. The total downward force due to the hoop (or tangential) stress σ_c in the wall is

$$\sigma_c \times \text{area} = \sigma_c \times \text{circumference} \times \text{thickness}$$

$$= \sigma_c \pi d t$$

where t is the wall thickness.

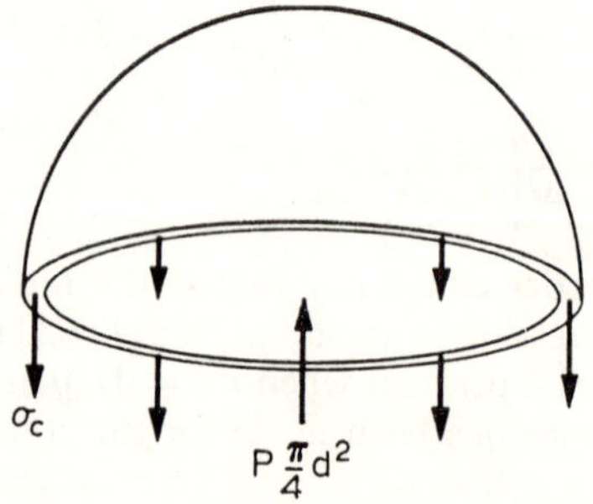

Fig. 11.3

For equilibrium, these two forces are equal and thus,

$$\sigma_c \pi d t = P \tfrac{1}{4}\pi d^2$$

$$\sigma_c = \frac{Pd}{4t} \qquad (Ans)$$

(This expression is identical with that obtained in Example 10.1 for the longitudinal stress in a thin cylinder.)

The stress σ_c is the same in every direction tangential to the spherical shell.

With the values given,

$$P = \frac{4t\sigma_c}{d} = \frac{4 \times (12 \times 10^{-3}\ \text{m}) \times (42 \times 10^6\ \text{N/m}^2)}{1{\cdot}2\ \text{m}}$$

$$= 1{\cdot}68\ \text{MN/m}^2 \qquad (Ans)$$

11.6. Derive a formula for the proportional increase of capacity of a thin spherical shell due to an internal pressure.

Calculate the increase in volume of a spherical shell, 1 m diameter and 10 mm thick, when it is subjected to an internal pressure of 1·4 MN/m².

$$E = 200\ \text{GN/m}^2 \quad \text{and} \quad \nu = 0{\cdot}3.$$

Solution. At any point in the shell the tangential stress in every direction is σ_c, and thus the circumferential strain in any direction is

$$\frac{\sigma_c}{E} - \frac{\nu\sigma_c}{E} = \frac{\sigma_c}{E}(1 - \nu) = \frac{Pd}{4tE}(1 - \nu)$$

using the result and notation of Example 10.5.

If δd is the increase in the diameter, the proportional increase in volume is

$$\frac{\delta V}{V} = \frac{\text{final volume} - \text{original volume}}{\text{original volume}}$$

$$= \frac{\frac{1}{6}\pi(d + \delta d)^3 - \frac{1}{6}\pi d^3}{\frac{1}{6}\pi d^3}$$

$$= \frac{d^3 + 3d^2\delta d + 3d\delta d^2 + \delta d^3 - d^3}{d^3}$$

$$= 3\frac{\delta d}{d} \quad \text{(neglecting products of small quantities)}$$

$$= 3 \times \text{diametral (or circumferential) strain}$$

$$= \frac{3Pd}{4tE}(1 - \nu) \qquad (Ans)$$

(If the radial stress is considered the term $3\nu P/E$ is added to this expression. The difference amounts to nearly 7 per cent when $t/d = 1/20$ and $\nu = 0{\cdot}25$.)

For the values given in the question, the increase in volume is

$$\delta V = \tfrac{1}{6}\pi d^3 \times \frac{3Pd}{4tE}(1 - \nu)$$

$$= \frac{\pi P d^4}{8tE}(1 - \nu)$$

$$= \frac{\pi \times (1{\cdot}4 \times 10^6 \text{ N/m}^2) \times (1 \text{ m})^4}{8 \times (10 \times 10^{-3} \text{ m}) \times (200 \times 10^9 \text{ N/m}^2)} \times (1 - 0{\cdot}3)$$

$$= 193 \text{ cm}^3 \qquad (Ans)$$

11.7. A copper tube 50 mm internal diameter, 1 m long and 1 mm thick, has closed ends and is filled with water under pressure. Neglecting any distortion of the end plates, determine the alteration in pressure when an additional 3 cm³ of water is pumped into the tube.

Modulus of elasticity for copper = 100 GN/m²
Poisson's ratio = 0·3
Bulk modulus for water = 2 GN/m² (*U.L.*)

Solution. Let P (N/m²) be the increase in pressure. Then the proportional increase in the capacity of the cylinder is

$$\frac{Pd}{4tE}(5 - 4\nu) = \frac{P \text{ N/m}^2 \times 0{\cdot}05 \text{ m}}{4 \times 0{\cdot}001 \text{ m} \times (100 \times 10^9 \text{ N/m}^2)}(5 - 1{\cdot}2)$$

$$= 4{\cdot}75 \times 10^{-10} \times P \qquad \text{(i)}$$

The proportional decrease in the volume of water originally in the cylinder is

$$\text{Volumetric strain} = \frac{\text{pressure}}{\text{bulk modulus}} = \frac{P \text{ N/m}^2}{2 \times 10^9 \text{ N/m}^2}$$

$$= 5 \times 10^{-10} \times P \qquad \text{(ii)}$$

The sum of expressions (i) and (ii) equals the ratio of the additional volume of water to the original capacity of the cylinder. Hence

$$(4{\cdot}75 \times 10^{-10} \times P) + (5 \times 10^{-10} \times P)$$

$$= \frac{3 \times 10^{-6} \text{ m}^3}{\tfrac{1}{4}\pi \times (0{\cdot}05 \text{ m})^2 \times (1 \text{ m})}$$

$$= 1{\cdot}53 \times 10^{-3}$$

and, multiplying through by 10^{10},

$$9{\cdot}75\,P = 1{\cdot}53 \times 10^7 \qquad P = 1{\cdot}57 \times 10^6 \text{ N/m}^2$$

The alteration in pressure is 1·57 MN/m² (or 15·7 bar) (*Ans*)

11.8. A bronze sleeve of 200 mm internal diameter and 6 mm thick is pressed over a steel liner of 200 mm external diameter and 95 mm thick with a force fit allowance of 0·075 mm on the common diameter. Treating both as thin cylinders, find:

(*a*) the radial pressure at the common surface;
(*b*) the hoop stresses;
(*c*) the respective percentages of the fit allowance met by the expansion of the sleeve and by the compression of the liner.

For bronze, modulus of elasticity = 110 GN/m²
Poisson's ratio = 0·33
For steel, modulus of elasticity = 200 GN/m²
Poisson's ratio = 0·304 (*U.L.*)

Solution. Let P (N/m²) be the radial pressure at the common surface. Using thin cylinder formulae, the hoop stress in the bronze sleeve, σ_B, is

$$\sigma_B = \frac{Pd}{2t} = \frac{P\text{ N/m}^2 \times 0{\cdot}2\text{ m}}{2 \times 0{\cdot}006\text{ m}}$$

$$= 16{\cdot}67P\text{ N/m}^2\text{ (tensile)} \qquad \text{(i)}$$

Similarly the hoop stress in the steel liner, σ_s, is

$$\sigma_s = \frac{P\text{ N/m}^2 \times 0{\cdot}2\text{ m}}{2 \times 0{\cdot}015} = 6{\cdot}67P\text{ N/m}^2\text{ (compressive)} \qquad \text{(ii)}$$

(Note that thin cylinder theory may be inadequate for the design of tubes subjected to external pressure since they may collapse with longitudinal corrugations, i.e. by elastic instability as in the case of slender struts.)

The diameter/thickness ratio of the liner is less than 14 and the radial stresses should be allowed for in calculating the hoop strains.

At the common surface the radial stresses in the sleeve and liner equal P N/m², the common pressure between the two.

$$\text{Tensile hoop strain in sleeve} = \frac{\sigma_B}{E_B} + \frac{\nu_B P}{E_B}$$

$$= \frac{16{\cdot}67P + 0{\cdot}33P}{110 \times 10^9} = 1{\cdot}545 \times 10^{-10}P$$

$$\text{Compressive hoop strain in liner} = \frac{\sigma_s}{E_s} - \frac{\nu_s P}{E_s}$$

$$= \frac{6{\cdot}67P - 0{\cdot}304P}{200 \times 10^9} = 3{\cdot}18 \times 10^{-11}P$$

The sum of these strains must equal the ratio of the force fit allowance to the common diameter. Hence

$$1{\cdot}545 \times 10^{-10}P + 3{\cdot}18 \times 10^{-11}P = \frac{0{\cdot}075}{200}$$

from which

$$P = 2{\cdot}01 \times 10^6 \text{ N/m}^2$$

The radial pressure at the common surface is therefore 2·01 MN/m² (or 20·1 bar). (If the radial stresses are ignored the result is 2·03 MN/m².) (*Ans*)

(*b*) From (i) the hoop stress in the sleeve is

$$\sigma_B = 16{\cdot}67 \times 2{\cdot}01 = 33{\cdot}5 \text{ MN/m}^2 \text{ (tensile)} \qquad (Ans)$$

Similarly, from (ii), the hoop stress in the liner is

$$\sigma_s = 6{\cdot}67 \times 2{\cdot}01 = 13{\cdot}4 \text{ MN/m}^2 \text{ (compressive)} \qquad (Ans)$$

(*c*) The percentage of the total fit allowance met by the expansion of the sleeve is given by

$$\frac{\text{Sleeve strain}}{\text{Total strain}} \times 100 = \frac{1{\cdot}545 \times 10^{-10} \times P}{(1{\cdot}545 \times 10^{-10} + 3{\cdot}18 \times 10^{-11})P} \times 100$$

$$= 82{\cdot}9 \text{ per cent} \qquad (Ans)$$

Similarly the percentage of the fit allowance met by the compression of the liner is 17·1 per cent. (*Ans*)

11.9. A cast-iron pipe having an internal diameter of 300 mm has walls 6 mm thick and is closely wound with a single layer of steel wire 3 mm diameter under a stress of 8 MN/m^2.

Calculate the stresses in the pipe and the wire when the internal pressure in the pipe is 10 bar.

E for steel = 206 GN/m^2.

E for cast iron = 103 GN/m^2.

Poisson's ratio for cast iron = 0·3. (*U.L.*)

Solution. Before the internal pressure is applied there is a compressive hoop stress in the tube due to the tension of the wire. Consider the equilibrium of a length x metres of half the tube as shown in Fig. 11.4. Since the wire is closely wound there are $x/0{\cdot}003$ coils in this

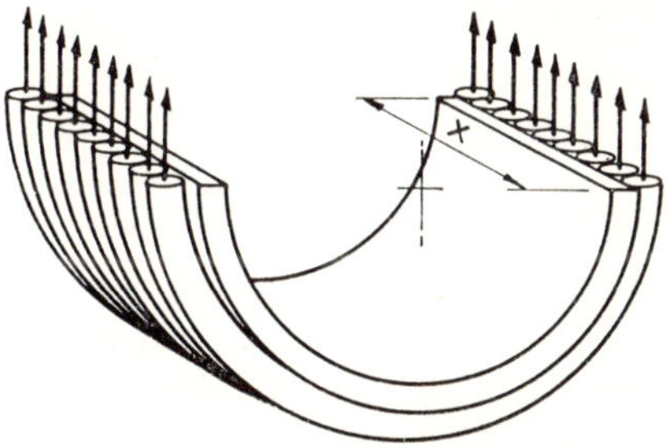

Fig. 11.4

length and the total upward force is

(Area of wire) × (stress in wire) × (twice the number of coils)

$$= \tfrac{1}{4}\pi \times (0{\cdot}003\ \text{m})^2 \times (8 \times 10^6\ \text{N/m}^2) \times \frac{2x}{0{\cdot}003}$$

$$= 3{\cdot}77 \times 10^4 x\ \text{N}$$

If σ N/m^2 is the compressive hoop stress in the tube, then for the same length x metres, there is a downward force of

$$(\sigma\ \text{N/m}^2) \times 2x\ \text{m} \times 0{\cdot}006\ \text{m} = 0{\cdot}012x\sigma\ \text{N}$$

Equating these two forces, we have

$$0{\cdot}012x\sigma = 3{\cdot}77 \times 10^4 x \qquad \sigma = 3{\cdot}14 \times 10^6$$

The hoop stress in the pipe due to the initial wire tension is therefore 3·14 MN/m^2 (compressive).

Let σ_c and σ_s be the (hoop) stresses (in N/m^2) in the pipe and wire due to the internal pressure, tension being positive. If we assume that the pipe bears the longitudinal force due to the pressure then there is a longitudinal stress σ_L given by

$$\sigma_L = \frac{Pd}{4t} = \frac{10 \times 10^5 \times 300}{4 \times 6} = 12{\cdot}5 \times 10^6 \text{ N/m}^2 \text{ (tensile)}$$

Allowing for this longitudinal stress the hoop strain in the pipe due to the internal pressure is

$$\frac{\sigma_c}{E_c} - \frac{\nu\sigma_L}{E_c} = \frac{\sigma_c - 0{\cdot}3 \times 12{\cdot}5 \times 10^6}{103 \times 10^9}$$

The strain in the wire is

$$\frac{\sigma_s}{E_s} = \frac{\sigma_s}{206 \times 10^9}$$

For practical purposes these two strains are equal and hence

$$\frac{\sigma_s}{206 \times 10^9} = \frac{\sigma_c - 0{\cdot}3 \times 12{\cdot}5 \times 10^6}{103 \times 10^9}$$

$$\sigma_s = 2\sigma_c - 7{\cdot}5 \times 10^6 \qquad \text{(i)}$$

Also for the length x metres the force due to the fluid pressure on a diametral plane must be balanced by the forces in the pipe wall and wire. Hence

$$(10 \times 10^5 \text{ N/m}^2) \times x \text{ m} \times 0{\cdot}3 \text{ m}$$

$$= \sigma_s \text{ N/m}^2 \times \tfrac{1}{4}\pi \times (0{\cdot}003 \text{ m})^2 \times \frac{2x}{0{\cdot}003} + \sigma_c \times 2x \text{ m} \times 0{\cdot}006 \text{ m}$$

$$3 \times 10^5 = \tfrac{1}{2} \times 0{\cdot}003\, \pi\sigma_s + 0{\cdot}012\sigma_c$$

and on rearranging,

$$\sigma_c = 2{\cdot}5 \times 10^7 - 0{\cdot}125\pi\sigma_s \qquad \text{(ii)}$$

Solving the simultaneous equations (i) and (ii) we obtain

$$\sigma_s = 23{\cdot}8 \times 10^6 \quad \text{and} \quad \sigma_c = 15{\cdot}6 \times 10^6$$

Thus, after the internal pressure is applied the resultant hoop stress in the pipe is

$$\sigma_c - 3{\cdot}14 \times 10^6 = 15{\cdot}6 \times 10^6 - 3{\cdot}14 \times 10^6$$

$$= 12{\cdot}5 \text{ MN/m}^2 \text{ (tensile)}$$

$$\text{Resultant stress in wire} = \sigma_s + 8 \times 10^6$$

$$= 23{\cdot}8 \times 10^6 + 8 \times 10^6$$

$$= 31{\cdot}8 \text{ MN/m}^2 \text{ (tensile)} \qquad (Ans)$$

(It is sometimes assumed that the longitudinal force due to the internal pressure is not borne by the pipe at all. In this case, equation (i) becomes $\sigma_s = 2\sigma_c$ and the solution of the simultaneous equations is

$$\sigma_s = 28{\cdot}0 \times 10^6 \quad \text{and} \quad \sigma_c = 14{\cdot}0 \times 10^6)$$

11.10. A thin circular hoop, made of material whose density is ρ kg/m³ is revolving about its axis with a circumferential (tangential) velocity v m/s. Find an expression for the circumferential stress in the rim.

Calculate this stress in the case of a thin steel rim, 1·5 m diameter, which is revolving about its axis at 800 rev/min. Density of steel, 7·83 Mg/m³.

Solution. Consider a small part (AB) of the hoop which subtends an angle θ (radians) at the centre as shown in Fig. 11.5. The length of this

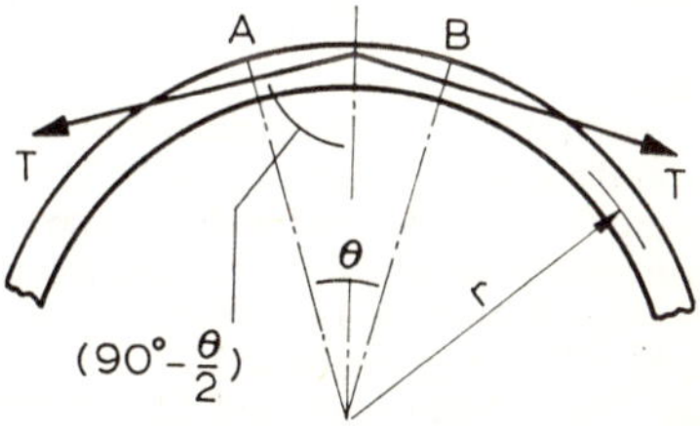

Fig. 11.5

part (measured round the circumference) is θr (where r is the radius) and, if a is its cross-sectional area, its mass is $\rho a\theta r$. Due to centripetal action this mass has an acceleration towards the centre of v^2/r. The force producing this acceleration is the sum of the components of the tensions T at the ends of AB towards the centre.

Using the relationship force = mass × acceleration, we have

$$2T\cos(90° - \tfrac{1}{2}\theta) = \rho a\theta r \times \frac{v^2}{r}$$

$$2T\sin\tfrac{1}{2}\theta = \rho a\theta v^2$$

For small angles $\sin\frac{1}{2}\theta = \frac{1}{2}\theta$ (approximately) and

$$T\theta = \rho a\theta v^2$$

Thus, the hoop stress $T/a = \rho v^2$. (This analysis assumes that the rim is free to expand and is unrestrained by spokes.)

With the values given in the question,

$$\begin{aligned} v &= \text{radius} \times \text{angular velocity} \\ &= (0{\cdot}75 \text{ m}) \times (2\pi \times 800/60 \text{ rad/s}) \\ &= 62{\cdot}8 \text{ m/s} \end{aligned}$$

Hence

$$\begin{aligned} \text{Hoop stress} &= \rho v^2 \\ &= (7{\cdot}83 \times 10^3 \text{ kg/m}^3) \times (62{\cdot}8 \text{ m/s})^2 \\ &= 30{\cdot}9 \text{ MN/m}^2 \end{aligned} \qquad (Ans)$$

(since 1 kg-m/s² is 1 N).

PROBLEMS

1. Solve the following problems relating to thin cylinders.

(*a*) Find the safe working pressure if the tensile stress (longitudinal or circumferential) must not exceed 46 MN/m², the diameter being 40 × wall thickness.

(*b*) Solve part (*a*) if the efficiencies of the longitudinal and circumferential joints in the cylinder are respectively 72 and 37 per cent.

(*c*) Calculate the diameter and wall thickness of a flat-ended compressed air drum to give a volume of 14 m³ and a maximum stress of 90 MN/m² for an absolute pressure of 1·8 MN/m².
The length/diameter ratio is 3.

(*d*) A steel plate, 1·8 m square and 12 mm thick, is bent to form the curved surface of a drum with one longitudinal joint whose efficiency can be taken as 55 per cent. If the tensile stress in the plate is to be limited to 100 MN/m², find the maximum safe internal pressure.

(*e*) A thin cylinder closed at its ends has an internal cross-sectional area of 25 cm². If the internal pressure is 2 MN/m², what axial pull must be applied to the ends of the cylinder so that the longitudinal and circumferential stresses in the cylinder wall are eqnal.

Answer. (*a*) 2·30 MN/m²; (*b*) 1·66 MN/m²; (*c*) $d = 1{\cdot}81$ m, $t = 18{\cdot}1$ mm; (*d*) 2·3 MN/m²; (*e*) 5 kN.

2. What pressure will increase the capacity of a mild steel cylinder having a diameter/thickness ratio of 30 by 1/2000 of its original value? $E = 200$ GN/m² and Poisson's ratio = 0·3.
Answer. 3·51 MN/m².

3. A thin cylinder, 50 mm internal diameter and 1 mm wall thickness, is closed at its ends and subjected to an internal pressure of 560 kN/m². If the cylinder is now subjected to a torque of 45 N-m, the axis of which coincides with the axis of the cylinder, calculate the principal and maximum shear stress for a point on the outside surface of the cylinder.
Answer. Principal stresses are 21·9 MN/m² (tensile) and 0·885 MN/m² (compressive); maximum shear stress, 11·39 MN/m².

4. The following problems relate to thin spherical shells.

(*a*) Calculate the hoop stress when the diameter is 1·2 m, the wall thickness is 25 mm and the internal pressure is 2 MN/m^2.
(*b*) Calculate the safe working pressure if the volume is to be 28 m^3, the wall thickness is 75 mm and the maximum permissible tensile stress is 45 MN/m^2.
(*c*) Find the maximum permissible volume if the wall thickness is 12 mm, the maximum tensile stress is to be 80 MN/m^2, and the internal pressure is 1·2 MN/m^2.

Answer. (*a*) 24 MN/m^2; (*b*) 3·58 MN/m^2; (*c*) 17·15 m^3.

5. Compare (*a*) the maximum tensile stresses, and (*b*) the proportional increases in volume of a thin cylinder and a thin spherical shell having the same internal pressure and same diameter/wall thickness ratio ($\nu = 0{\cdot}25$).
Answer. (*a*) Stress in cylinder = 2 × stress in sphere. (*b*) Proportional volume change for cylinder is 16/9 times that for the sphere.

6. In each of the following rotating rim problems, the material is steel of density 7·8 Mg/m^3.

(*a*) Calculate the stress for a radius of 0·9 m and a speed of 500 rev/min.
(*b*) What is the maximum speed if the radius is 1·8 m and the tensile stress must not exceed 30 MN/m^2?
(*c*) If the tensile stress is limited to 45 MN/m^2, find the diameter of the largest rim which may revolve at 450 rev/min.

Answer. (*a*) 17·3 MN/m^2; (*b*) 329 rev/min; (*c*) 3·22 m.

7. Show that when a thin-walled spherical vessel of diameter d and thickness t is subjected to an internal fluid pressure p the increase in volume is equal to

$$\frac{\pi p d^4}{8tE}(1 - \nu)$$

To what depth would a copper float, 250 mm diameter and 3 mm thick, have to be sunk in sea-water in order that its diameter is decreased by 0·03 m?

E for copper = 100 GN/m^2
ν for copper = 0·27.

Weight of sea-water = 10·06 kN/m^3. (*I.Mech.E.*)
Answer. 78·5 m.

8. A long cylindrical welded pressure vessel with plane ends is to be 1·2 m internal diameter and is to hold gas at a pressure of 2 MN/m^2. If the maximum allowable tensile stress is 140 MN/m^2, find the thickness of plate required for the shell. Give the thickness to the nearest mm and prove any formula used. Neglect end effects.

What modifications would require to be made to the calculations if the circumferential and longitudinal joints were riveted instead of welded? (*U.L.*)
Answer. 9 mm; if the joints were riveted allowance would have to be made for their efficiency being less than 100 per cent.

9. A gunmetal tube, 25 mm internal diameter and 1·5 mm wall thickness, is closed at its ends by plugs which are 0·4 m apart. The tube is subjected to a

constant internal fluid pressure of 2 MN/m^2, and at the same time pulled in an axial direction with a force of 2 kN. Find

(*a*) the change in length between the plugs and
(*b*) the change in the external diameter of the tube.

For gunmetal take $E = 100$ GN/m^2
Poisson's ratio $= 0{\cdot}35$. (*I.Mech.E.*)
Answer. (*a*) 0·0753 mm; (*b*) 0·00274 mm.

10. Derive formulae to give the longitudinal and circumferential tensions in a thin boiler shell, stating the assumptions made in your argument.

A cylindrical compressed air drum is 1·7 m in diameter with plates 12 mm thick. The efficiencies of the longitudinal and circumferential joints are respectively 85 and 45 per cent. If the tensile stress in the plating is to be limited to 100 MN/m^2, find the maximum safe air pressure. (*U.L.*)
Answer. 1·2 MN/m^2 (hoop stress is limiting condition).

11. A thin cylindrical tube, 76 mm internal diameter and wall thickness 5 mm, is closed at the ends and subjected to an internal pressure of 5·5 MN/m^2. A torque of 500π N-m is also applied to the tube.

Determine the maximum and minimum principal stresses and also the maximum shearing stress in the wall of the tube. (*U.L.*)
Answer. Principal stresses, 65·2 MN/m^2 (tensile) and 2·54 MN/m^2 (compressive); maximum shearing stress, 33·9 MN/m^2.

12. A thin-walled cylinder of length l and internal diameter d is made of plates of thickness t and subjected to an internal pressure of intensity p. Show that, if the effect of the ends is neglected, the increase in volume due to the pressure is

$$\pi p d^3 l(5 - 4\nu)/16tE$$

where E is the modulus of elasticity and ν is Poisson's ratio.

A steel cylinder 0·9 m long, 150 mm internal diameter, plate thickness 5 mm is subjected to an internal pressure of 7 MN/m^2; the increase in volume due to the pressure is 15·45 cm^3. Find the values of Poisson's ratio and the modulus of rigidity. Assume $E = 206$ GN/m^2. (*U.L.*)
Answer. $\nu = 0{\cdot}292$; $G = 80$ GN/m^2.

13. A thin-walled steel cylinder with flat ends is 0·9 m diameter and 2·5 m long. It is filled with water and the pressure is increased to 3·5 MN/m^2 (gauge). Find:

(*a*) the necessary thickness of the shell if the working tensile stress is 105 MN/m^2,
(*b*) the extra volume of water which has to be pumped into the cylinder to produce the given pressure. Allow for the elasticity of the cylinder and the compressibility of the water but neglect the distortion of the ends. Bulk modulus for water 2 GN/m^2.

Modulus of elasticity for steel = 200 GN/m^2. Poisson's ratio = 0·3.
Answer. (*a*) 15 mm; (*b*) 4370 cm^3.

14. Derive a formula for the hoop stress in a thin cylinder having a mean radius R and made of material of density ρ when rotating at ω radians per second about its axis. What is the most important assumption you make?

Apply this theory to find the maximum allowable speed in rev/min for a fly-wheel 1·5 m external diameter and 50 mm thick. The material has a density of 7·3 g/cm³ and the hoop stress is limited to 15 MN/m². (*U.L.*)
Answer. $\rho\omega^2R^2$; 597 rev/min.

15. A copper tube 40 mm external diameter and 37 mm internal diameter is closely wound with steel wire 1 mm diameter. Stating clearly the assumptions made, estimate the tension at which the wire must have been wound if an internal gauge pressure of 20 bar produces a tensile circumferential stress of 7 MN/m² in the copper tube.

E for steel = 1·6 × E for copper. (*U.L.*)

Hint. Since Poisson's ratio is not given assume that there is no longitudinal stress in the tube.
Answer. 9·63 N.

16. A cylindrical vessel, 400 mm long and 100 mm diameter internally, is made of brass 3 mm thick. The ends are flat and connected by a steel bolt 12 mm diam., which passes centrally through the ends for which it serves as a stay, nuts being screwed on the outside at each end. The vessel is filled with water after which more water is pumped in to raise the pressure in the vessel. Determine the quantity of water added to produce a rise in pressure of 6 bar. Assume that the ends remain flat.

For steel: $E = 206$ GN/m².
For brass: $E = 96$ GN/m²; $\nu = 0{\cdot}33$.
For water: $K = 2$ GN/m².

Answer. 1·54 cm³.

17. A brass cylinder with closed ends has an inside diameter of 50 mm and a wall thickness of 3 mm. The cylinder is wound with a single layer of steel wire 1 mm diameter wound under a pull of 16 N and with the coils just touching each other. If the cylinder is subsequently subjected to an internal pressure of 100 bar determine the final hoop stress in the cylinder and the tensile stress in the wire,

(*a*) if the temperature remains constant;
(*b*) if the temperature is raised 100°C.

Young's moduli: steel, 204 GN/m².
brass, 96 GN/m².
Poisson's ratio: brass, 0·333.
Coefficients of expansion: steel, 12×10^{-6} per °C,
brass, 19×10^{-6} per °C. (*U.L.*)

Answer. All tensile (*a*) 53·2 MN/m² (brass); 115·4 MN/m² (steel). (*b*) 29·2 MN/m² (brass); 207 MN/m² (steel).

Chapter 12

Thick Cylinders and Rotating Discs

The stress formulae used in this Chapter are based on the convention "*tensile is positive, compressive is negative*". These formulae involve two constants A and B whose values, in a given example, depend on the conditions at the inner and outer surfaces.

At radius r in the wall of a cylinder subject to internal or external pressure the hoop (or circumferential) stress σ_H and radial stress σ_R are given by

$$\sigma_H = A + \frac{B}{r^2} \quad \text{and} \quad \sigma_R = A - \frac{B}{r^2}$$

The + and − signs in these formulae may be interchanged without affecting the final results; only the sign of B will be altered. The radial stress is always compressive and hence σ_R is negative.

If the external pressure is zero and the internal pressure is P, the maximum hoop stress is

$$\sigma_{Hmax} = P\left(\frac{R_2^{\,2} + R_1^{\,2}}{R_2^{\,2} - R_1^{\,2}}\right) = P\left(\frac{k^2 + 1}{k^2 - 1}\right)$$

where R_1 and R_2 are the internal and external radii respectively and $k = R_2/R_1$.

At all points in a solid shaft subjected to an external pressure P,

$$\sigma_H = \sigma_R = -P$$

At radius r in a thin circular disc of uniform density ρ rotating at angular velocity ω about its axis, the hoop and radial stresses are

$$\sigma_H = A + \frac{B}{r^2} - \tfrac{1}{8}\rho r^2\omega^2(1 + 3\nu)$$

$$\sigma_R = A - \frac{B}{r^2} - \tfrac{1}{8}\rho r^2\omega^2(3 + \nu)$$

where ν = Poisson's ratio.

It will be seen that when $\omega = 0$ these formulae reduce to those for a thick cylinder subjected to pressure.

WORKED EXAMPLES

12.1. Discuss the three principal stresses at any point in a cylindrical shell subjected to internal (or external) pressure. State and comment on the assumption usually made in thick cylinder theory regarding the radial stress σ_R and the circumferential stress σ_H at any point in the cylinder wall.

A thick-walled cylinder, 150 mm internal and 200 mm external diameter, is subjected to an internal pressure of 280 bar. Calculate the longitudinal stress assuming that the cylinder bears all the longitudinal tension. If the hoop stress at the inside surface is 100 MN/m^2 (tensile) calculate its value at the outside surface.

Solution. The three principal stresses are:

(*a*) A longitudinal stress σ_L (provided that the cylinder bears the longitudinal tension due to the internal pressure);
(*b*) A hoop or circumferential stress σ_H;
(*c*) A radial stress σ_R.

All stresses will be regarded as positive when tensile. The radial stress (*c*) is invariably compressive and therefore negative.

In thin cylinder theory (*c*) is neglected and (*a*) and (*b*) are assumed to be constant through the cylinder wall. In thick cylinders, the variation of the hoop stress σ_H is considerable and the radial stress σ_R is no longer negligible.

It is assumed that plane sections of the cylinder perpendicular to the longitudinal axis remain plane under the pressure. As a consequence, the longitudinal strain is the same at all points in the cylinder wall, i.e. it is independent of the radius. If no longitudinal tension is borne by the cylinder, the (tensile) longitudinal strain at any radius is

$$-\frac{\nu\sigma_R}{E} - \frac{\nu\sigma_H}{E}$$

Hence, to satisfy the requirements of uniform longitudinal strain,

$$\sigma_R + \sigma_H = \text{constant} \qquad (1)$$

If the longitudinal tension is borne by the cylinder then the longitudinal strain at any point is

$$\frac{\sigma_L}{E} - \frac{\nu\sigma_R}{E} - \frac{\nu\sigma_H}{E} = \frac{\sigma_L}{E} - \nu\left(\frac{\sigma_R + \sigma_H}{E}\right)$$

To satisfy the requirements of uniform longitudinal strain, and assuming that (1) still holds, we have

$$\sigma_L = \text{constant} \tag{2}$$

since ν and E are constants.

The relationship (1) is the assumption on which thick cylinder theory is based.

For the values given in the question, using the assumption (2), the longitudinal stress is

$$\sigma_L = \frac{\text{longitudinal force due to the pressure}}{\text{cross-sectional area of the cylinder wall}}$$

$$= \frac{(280 \times 10^5\ \text{N/m}^2) \times \frac{1}{4}\pi \times (0{\cdot}15\ \text{m})^2}{\frac{1}{4}\pi[(0{\cdot}2\ \text{m})^2 - (0{\cdot}15\ \text{m})^2]} = 36\ \text{MN/m}^2\ \text{(tensile)}$$

(*Ans*)

At the inside surface, the radial stress in the wall is numerically equal to the internal pressure (i.e. 28 MN/m²) but opposite in sign. Thus, using (1) the value of the constant is

$$\sigma_R + \sigma_H = -28 \times 10^6 + 100 \times 10^6 = 72\ \text{MN/m}^2$$

At the outside surface the radial stress σ_R is zero (atmospheric pressure) and hence

$$\sigma_R + \sigma_H = 0 + \sigma_H = 72\ \text{MN/m}^2$$

The hoop or circumferential stress at the outside edge is therefore

$$\sigma_H = 72\ \text{MN/m}^2\ \text{(tensile)}$$ (*Ans*)

12.2. Develop formulae for the radial and circumferential stresses in a thick-walled cylinder subjected to internal and external pressure. Use the formulae to check the hoop stress given in the question of Example 12.1.

Solution. Let σ_R be the radial stress and σ_H the circumferential stress at radius r within the cylinder wall. The following theory (Lamé's) leads to generalized equations relating σ_R, σ_H and r which involve numerical constants whose values, in particular examples, are determined from boundary conditions such as the pressures on the inner and outer surfaces. The radial stress is always compressive as shown in Fig. 12.1 and the equations can be deduced from the equilibrium of a small element ABCD shown in that diagram. The work is simplified, however, by taking an annular ring of the cylinder having internal radius r and thickness δr as shown in Fig. 12.2. In this diagram all stresses are taken as positive if tensile and the radial stress σ_R can be regarded as a negative internal pressure on the ring. Similarly there is a negative external pressure $\sigma_R + \delta\sigma_R$ at radius $r + \delta r$.

The conditions for the equilibrium of one-half of this ring are similar to those for a thin cylinder (Chapter 11, Example 11.1).

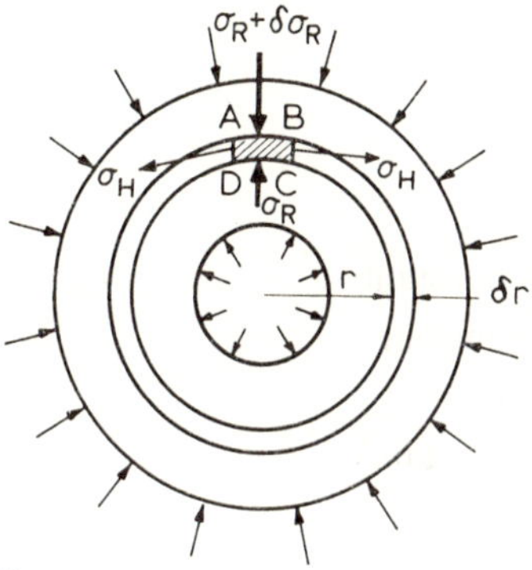

Fig. 12.1

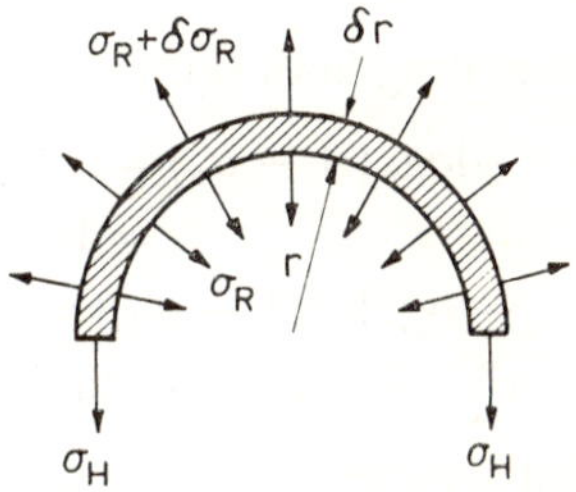

Fig. 12.2

For unit axial length, the total downward force (Fig. 12.2) is

$$2r\sigma_R + 2\sigma_H\delta r$$

and the total upward force is

$$2(r + \delta r)(\sigma_R + \delta\sigma_R)$$

Equating these two quantities

$$2(r + \delta r)(\sigma_R + \delta\sigma_R) = 2r\sigma_R + 2\sigma_H\delta r$$

$$r\sigma_R + \sigma_R\delta r + r\delta\sigma_R + \delta r\delta\sigma_R = r\sigma_R + \sigma_H\delta r$$

$$r\delta\sigma_R + \sigma_R\delta r - \sigma_H\delta r = 0$$

neglecting products of small quantities. Dividing through by δr and proceeding to the limit

$$r\frac{d\sigma_R}{dr} + \sigma_R - \sigma_H = 0 \qquad \text{(i)}$$

A second relationship between σ_R and σ_H was discussed in Example 12.1. As stated there, $\sigma_R + \sigma_H$ is assumed to be a constant and it is convenient in the working which follows to denote this constant by $2A$.

Thus

$$\sigma_R + \sigma_H = 2A \qquad \text{(ii)}$$

Substituting the value of σ_H from (ii) in (i) we obtain

$$r\frac{d\sigma_R}{dr} + \sigma_R - (2A - \sigma_R) = 0$$

$$r\frac{d\sigma_R}{dr} = 2(A - \sigma_R)$$

Separating differentials

$$\frac{d\sigma_R}{A - \sigma_R} = 2\frac{dr}{r}$$

Integrating both sides

$$-\log_e (A - \sigma_R) = 2 \log_e r + \text{constant}$$
$$= \log_e r^2 + \text{constant}$$
$$-\log_e [r^2(A - \sigma_R)] = \text{constant}$$
$$r^2(A - \sigma_R) = B \quad \text{(a different constant)}$$

Rearranging,

$$\sigma_R = A - \frac{B}{r^2} \tag{iii}$$

and, using (ii),

$$\sigma_H = 2A - \sigma_R = 2A - \left(A - \frac{B}{r^2}\right) = A + \frac{B}{r^2} \tag{iv}$$

The constants A and B are found, in a given example, by using known boundary conditions, e.g. the radial stress is taken as zero at a surface exposed to atmospheric conditions.

For the cylinder of Example 12.1 the radial stress at the inner surface is numerically equal to the internal pressure (but is negative).

Hence, when $r = 0{\cdot}075$ m, $\sigma_R = -280 \times 10^5$ N/m^2 and, from (iii),

$$-280 \times 10^5 = A - \frac{B}{0{\cdot}075^2} \tag{v}$$

Also, when $r = 0{\cdot}1$ m, $\sigma_R = 0$ and

$$0 = A - \frac{B}{0{\cdot}1^2} \tag{vi}$$

Solving the simultaneous equations (v) and (vi),

$$A = 36 \times 10^6 \quad \text{and} \quad B = 36 \times 10^4$$

in newton and metre units.

Using these values of the constants the hoop stress at the inner surface is, putting $r = 0{\cdot}075$,

$$\sigma_H = A + \frac{B}{r^2} = 36 \times 10^6 + \frac{36 \times 10^4}{0{\cdot}075^2}$$

$$= 100 \text{ MN/m}^2 \qquad (Ans)$$

(Putting $r = 0{\cdot}1$ m the stress at the outside surface can be evaluated.)

12.3. Obtain an expression for the maximum hoop stress for a cylinder subjected to an internal pressure P in terms of k the ratio of the external to the internal diameter.

Find the value of k if the maximum hoop stress is to be

(*a*) 3 × internal pressure;
(*b*) 20 × internal pressure.
(*c*) Solve case (*b*) using thin cylinder theory.

Solution. Let R_1 and R_2 be the internal and external radii respectively. The radial and hoop stresses at radius r are, as shown in Example 12.2,

$$\sigma_R = A - \frac{B}{r^2} \quad \text{and} \quad \sigma_H = A + \frac{B}{r^2}$$

At the inside surface $\sigma_R = -P$ (the internal pressure). Thus, putting $r = R_1$,

$$-P = A - \frac{B}{R_1^2} \qquad \text{(i)}$$

Also, at the outside surface the radial stress is zero (atmospheric pressure) and, putting $r = R_2$,

$$0 = A - \frac{B}{R_2^2} \qquad \text{(ii)}$$

Subtracting (i) from (ii),

$$P = \frac{B}{R_1^2} - \frac{B}{R_2^2} = B\left(\frac{R_2^2 - R_1^2}{R_1^2 R_2^2}\right)$$

$$B = P\left(\frac{R_1^2 R_2^2}{R_2^2 - R_1^2}\right) \qquad \text{(iii)}$$

Substituting this value of B in (ii), we obtain

$$A = \frac{B}{R_2^2} = P\left(\frac{R_1^2}{R_2^2 - R_1^2}\right) \quad \text{(iv)}$$

Using (iii) and (iv) the hoop stress at any radius r is

$$\sigma_H = A + \frac{B}{r^2} = P\left(\frac{R_1^2}{R_2^2 - R_1^2}\right) + \frac{P}{r^2}\left(\frac{R_1^2 R_2^2}{R_2^2 - R_1^2}\right)$$

$$= \frac{PR_1^2}{R_2^2 - R_1^2}\left(\frac{R_2^2}{r^2} + 1\right) \quad \text{(v)}$$

The maximum value of σ_H clearly corresponds to the smallest value of r, i.e. R_1 (at the inside surface). Hence, the maximum hoop stress is

$$\sigma_{H(max)} = \frac{PR_1^2}{R_2^2 - R_1^2}\left(\frac{R_2^2}{R_1^2} + 1\right)$$

$$= \frac{PR_1^2}{R_2^2 - R_1^2}\left(\frac{R_2^2 + R_1^2}{R_1^2}\right)$$

$$= P\left(\frac{R_2^2 + R_1^2}{R_2^2 - R_1^2}\right)$$

$$= P\left(\frac{k^2 + 1}{k^2 - 1}\right) \quad \text{(vi)}$$

since the ratio of the radii equals the ratio of the diameters.

(*a*) If $\sigma_{H(max)} = 3P$, then

$$3 = \left(\frac{k^2 + 1}{k^2 - 1}\right)$$

$$3(k^2 - 1) = k^2 + 1 \qquad 2k^2 = 4$$

$$k = \sqrt{2} = 1{\cdot}414 \quad (Ans)$$

(*b*) If $\sigma_{H(max)} = 20P$,

$$20 = \left(\frac{k^2 + 1}{k^2 - 1}\right)$$

$$19k^2 = 21$$

$$k = \sqrt{(21/19)} = 1{\cdot}051 \quad (Ans)$$

(*c*) Using thin cylinder theory to solve (*b*), the hoop stress is $\sigma_H = Pd/2t$ and, taking d as the internal diameter, $\sigma_H = 20P$. Thus

$$20 = d/2t \qquad d/t = 40$$

$$k = \frac{\text{external diameter}}{\text{internal diameter}} = \frac{d + 2t}{d}$$

$$= 1 + \frac{2t}{d} = 1 + \frac{2}{40}$$

$$= 1{\cdot}05 \qquad \textit{(Ans)}$$

12.4. A steel sleeve of length 12 cm and 12 cm external diameter is shrunk on to a steel shaft of diameter 8 cm such that the maximum tensile hoop stress induced in the sleeve is 130 MN/m². Determine the resulting normal pressure, in MN/m², at the mating surface. Plot graphs showing the distributions of the radial and hoop stresses throughout the thickness of the sleeve. (*I.Mech.E.*)

Solution. In the notation of Example 12.3, $k = 3/2$. Hence using equation (vi) of that example

$$130 \text{ MN/m}^2 = P\left[\frac{(3/2)^2 + 1}{(3/2)^2 - 1}\right]$$

and the normal pressure at the mating surface is

$$P = 130 \text{ MN/m}^2 \times \left[\frac{(3/2)^2 - 1}{(3/2)^2 + 1}\right]$$

$$= 50 \text{ MN/m}^2 \qquad \textit{(Ans)}$$

Using the Lamé formulae

$$\sigma_R = A - \frac{B}{r^2} \quad \text{and} \quad \sigma_H = A + \frac{B}{r^2}$$

we have, working in MN/m² for stress and cm for radius,

$$\sigma_H = 130 \text{ when } r = 4 \quad \text{or} \quad 130 = A + \frac{B}{4^2} \qquad \text{(i)}$$

Also

$$\sigma_R = 0 \text{ when } r = 6 \quad \text{or} \quad 0 = A - \frac{B}{6^2} \qquad \text{(ii)}$$

Solving the simultaneous equations (i) and (ii),

$$A = 40 \quad \text{and} \quad B = 1440$$

Using these values the radial and hoop stresses are given by

$$\sigma_R = 40 - \frac{1440}{r^2} \quad \text{and} \quad \sigma_H = 40 + \frac{1440}{r^2}$$

The results for σ_R are negative, indicating compressive stress and, tabulating values,

r cm	σ_R MN/m² (compressive)	σ_H MN/m² (tensile)
4	50	130
$4\frac{1}{2}$	31	111
5	17·6	97·6
$5\frac{1}{2}$	7·5	87·5
6	0	80

Figure 12.3 shows the distribution of the radial and hoop stresses throughout the thickness of the sleeve.

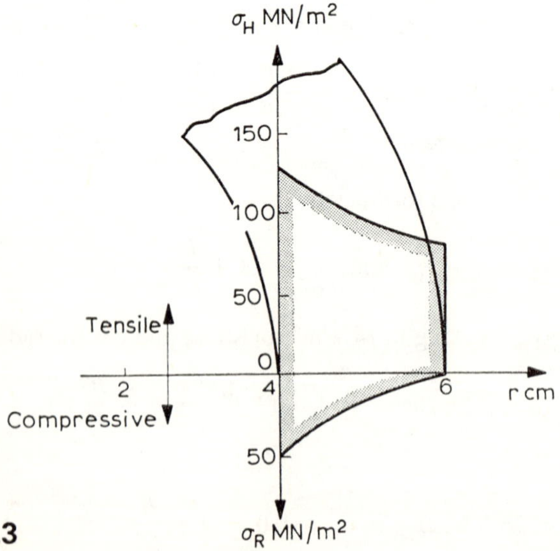

Fig. 12.3

12.5. A compound tube is made by shrinking one tube on another. The common diameter is 6 cm, the shrinkage allowance (based on diameter) is 0·01 cm and each tube is 1 cm thick. If both tubes are of steel ($E = 200$ GN/m^2) calculate the radial pressure between the tubes.

Draw a diagram to show the distribution of the hoop stress throughout the wall:

(*a*) due to the shrinkage, and
(*b*) due to an internal pressure of 60 MN/m^2

Show also the resultant hoop stress distribution.

Solution. Working in MN/m^2 for stress and cm for radius throughout, let P be the radial pressure between the tubes due to shrinkage.

For the outer tube, the diameter ratio $k = \frac{4}{3}$ and the hoop stress at the common radius is

$$P\left(\frac{k^2+1}{k^2-1}\right) = \frac{25P}{7} \qquad \text{(i)}$$

For the inner tube, take

$$\sigma_R = A - \frac{B}{r^2}$$

$$\sigma_H = A + \frac{B}{r^2}$$

When $r = 3$, $\sigma_R = -P$, and

$$-P = A - \frac{B}{3^2} \qquad \text{(ii)}$$

Also, when $r = 2$, $\sigma_R = 0$, and

$$0 = A - \frac{B}{2^2} \qquad \text{(iii)}$$

Subtracting (ii) from (iii),

$$P = \frac{B}{9} - \frac{B}{4} = -\frac{5B}{36}$$

Hence

$$B = -36P/5$$

and, substituting in (iii),

$$A = -9P/5$$

Using these values of A and B, the hoop stress in the inner tube at the common radius ($r = 3$) is

$$-\frac{9P}{5} - \frac{36P}{5 \times 3^2} = -\frac{13P}{5} \qquad \text{(iv)}$$

Thus, at the common radius, tensile hoop strain in the outer tube is

$$\frac{\sigma_H}{E} + \frac{\nu P}{E} = \frac{25P}{7E} + \frac{\nu P}{E}$$

Tensile hoop strain in inner tube is

$$-\frac{13P}{5E} + \frac{\nu P}{E}$$

Total strain difference

$$\begin{aligned} &= \text{(tensile hoop strain in outer tube)} \\ &\quad - \text{(tensile hoop strain in inner tube)} \\ &= \left(\frac{25P}{7E} + \frac{\nu P}{E}\right) - \left(-\frac{13P}{5E} + \frac{\nu P}{E}\right) \\ &= \frac{P}{E}\left(\frac{25}{7} + \frac{13}{5}\right) = \frac{216P}{35E} \end{aligned}$$

This strain difference equals the ratio of the shrinkage allowance to the common diameter, since the proportional change in diameter equals the proportional change in circumference. Thus

$$\frac{216P}{35E} = \frac{0{\cdot}01 \times 10^{-1}}{6}$$

$$P = \frac{0{\cdot}001}{6} \times \frac{35 \times 200 \times 10^9}{216} = 5{\cdot}4 \times 10^6 \text{ N/m}^2$$

The radial pressure between the tubes is 5·4 MN/m^2. (*Ans*)
Using this value of P the constants for the inner tube are

$$A = -9P/5 = -9 \times 5{\cdot}4/5 = -9{\cdot}72$$
$$B = -36P/5 = -36 \times 5{\cdot}4/5 = -38{\cdot}88$$

With these values, the hoop stress at any radius r in the inner tube is

$$\sigma_H = A + \frac{B}{r^2} = -9{\cdot}72 - \frac{38{\cdot}88}{r^2} \qquad \text{(v)}$$

Since these values of the constants A and B apply to the inner tube only, we must for the outer tube put

$$\sigma_R = A' - \frac{B'}{r^2} \qquad \sigma_H = A' + \frac{B'}{r^2}$$

When $r = 3$, $\sigma_R = -P = -5{\cdot}4$, and thus

$$-5{\cdot}4 = A' - \frac{B'}{3^2} \tag{vi}$$

When $r = 4$, $\sigma_R = 0$, and thus

$$0 = A' - \frac{B'}{4^2} \tag{vii}$$

Solving the simultaneous equations (vi) and (vii)

$$A' = 6{\cdot}944 \quad \text{and} \quad B' = 111{\cdot}1$$

With these values the hoop stress at any radius in the outer tube is

$$\sigma_H = 6{\cdot}944 + \frac{111{\cdot}1}{r^2} \tag{viii}$$

The stresses due to the internal pressure can be calculated by regarding the two tubes as a single cylinder (provided they are of the same material) with internal and external diameters 4 cm and 8 cm respectively. Hence, due to the internal pressure

$$\sigma_R = A'' - \frac{B''}{r^2} \quad \text{and} \quad \sigma_H = A'' + \frac{B''}{r^2}$$

throughout both tubes.

When $r = 2$, $\sigma_R = -60$. Thus

$$-60 = A'' - \frac{B''}{2^2} \tag{ix}$$

Also, when $r = 4$, $\sigma_R = 0$. Thus

$$0 = A'' - \frac{B''}{4^2} \tag{x}$$

Solving the simultaneous equations (ix) and (x)

$$A'' = 20 \quad \text{and} \quad B'' = 320$$

and the hoop stress at any radius r is

$$\sigma_H = 20 + \frac{320}{r^2} \tag{xi}$$

Using equations (v), (viii) and (xi) the hoop stresses in MN/m² are:

r cm	Inner tube 2	$2\frac{1}{2}$	3	Outer tube 3	$3\frac{1}{2}$	4
Due to shrinkage						
Equation (v)	−19·44	−15·94	−14·04			
Equation (viii)				19·29	16·01	13·89
Due to pressure Equation (xi)	100	71·2	55·56	55·56	46·12	40
Resultant stress	80·56	55·26	41·52	74·85	62·13	53·89

Figure 12.4 shows the distribution of the hoop stress due to the shrinkage and the internal pressure as well as the resultant stress.

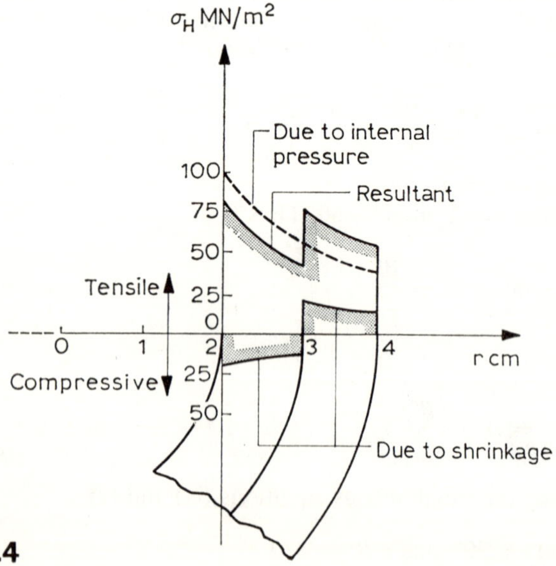

Fig. 12.4

(The reader should note that the maximum resultant stress in the compound tube is 80·56 MN/m² compared with 100 MN/m² for a single tube of the same overall dimensions and internal pressure.)

12.6. How can the thick cylinder formulae be applied to a solid shaft subjected to an external pressure?

Calculate the change in diameter of a 100 mm solid shaft subected to an external pressure of 80 MN/m².

$E = 200$ GN/m² and $\nu = 0{\cdot}28$.

Solution. The formulae

$$\sigma_R = A - \frac{B}{r^2} \quad \text{and} \quad \sigma_H = A + \frac{B}{r^2}$$

are valid for a shaft subjected to external pressure since the shaft can be regarded as a cylinder of zero internal radius. Since the formulae must apply when $r = 0$ and the stress is not infinite at the centre, then $B = 0$.

At all points in the shaft, therefore $\sigma_R = \sigma_H = A$. At the outside surface $\sigma_R = -P$ where P is the external pressure and hence, for all radii,

$$\sigma_R = \sigma_H = -P \qquad \text{(i)}$$

For the shaft given in the question, the circumferential strain at the outside surface is

$$\left(\frac{\sigma_H}{E} - \frac{\nu\sigma_R}{E}\right) = \left(\frac{-P}{E} + \frac{\nu P}{E}\right)$$
$$= -\frac{P}{E}(1-\nu)$$
$$= -\left(\frac{80 \times 10^6 \text{ N/m}^2}{200 \times 10^9 \text{ N/m}^2}\right) \times (1 - 0{\cdot}28)$$
$$= -2{\cdot}88 \times 10^{-4}$$

Hence the change in diameter is

$$100 \text{ mm} \times 2{\cdot}88 \times 10^{-4} = 0{\cdot}0288 \text{ mm (decrease)} \qquad (Ans)$$

12.7. Determine the driving allowance for a cast-iron hub, 6 cm external diameter, on a steel shaft 4 cm diameter, if the maximum circumferential stress in the hub is to be 60 MN/m² and draw a diagram showing the variation of circumferential and radial stresses across the section of the hub when it is driven on the shaft.

E for steel = 200 GN/m². E for cast-iron = 100 GN/m². Poisson's ratio = 0·3 for both.

Solution. The diameter ratio for the hub is $k = 6/4 = 3/2$. If $\sigma_{H(max)}$ is the maximum circumferential stress in the hub and P is the radial pressure at the common surface, then

$$\sigma_{H(max)} = P\left(\frac{k^2+1}{k^2-1}\right)$$

$$P = \sigma_{H(max)}\left(\frac{k^2-1}{k^2+1}\right) = 60 \text{ MN/m}^2\left[\frac{(3/2)^2-1}{(3/2)^2+1}\right]$$

$$= 300/13 \text{ MN/m}^2$$

Hence, working in MN/m² units, at all points in the shaft

$$\sigma_R = \sigma_H = -300/13$$

At the common diameter, if suffixes s and c refer to steel and cast iron respectively, we have

Tensile hoop strain in hub

$$\frac{(\sigma_H)_c}{E_c} - \frac{\nu(\sigma_R)_c}{E_c}$$

$$= \frac{60}{100\times10^3} - \frac{0{\cdot}3\times(-300/13)}{100\times10^3} = \frac{66\frac{12}{13}}{10^5}$$

Tensile hoop strain in shaft:

$$\frac{(\sigma_H)_s}{E_s} - \frac{\nu(\sigma_R)_s}{E_s}$$

$$= \frac{-300/13}{200\times10^3} - \frac{0{\cdot}3\times(-300/13)}{200\times10^3} = -\frac{8\frac{1}{13}}{10^5}$$

Total strain difference

$$= (\text{hoop strain in hub}) - (\text{hoop strain in shaft})$$

$$= \frac{66\frac{12}{13}}{10^5} - \frac{-8\frac{1}{13}}{10^5} = 75\times10^{-5}$$

Required driving allowance

= total strain difference × diameter

= $(75 \times 10^{-5}) \times 4$ cm

= 0·003 cm or 0·03 mm (*Ans*)

At any radius r in the hub, the radial and circumferential stresses are

$$\sigma_R = A - \frac{B}{r^2} \quad \text{and} \quad \sigma_H = A + \frac{B}{r^2}$$

When $r = 2$, $\sigma_H = 60$, and thus $60 = A + \dfrac{B}{2^2}$ (i)

Also, when $r = 3$, $\sigma_R = 0$, and thus $0 = A - \dfrac{B}{3^2}$ (ii)

Solving the simultaneous equations (i) and (ii)

$A = 18{\cdot}46$ and $B = 166{\cdot}1$

Using these values,

$$\sigma_R = 18{\cdot}46 - \frac{166{\cdot}1}{r^2} \quad \text{and} \quad \sigma_H = 18{\cdot}46 + \frac{166{\cdot}1}{r^2}$$

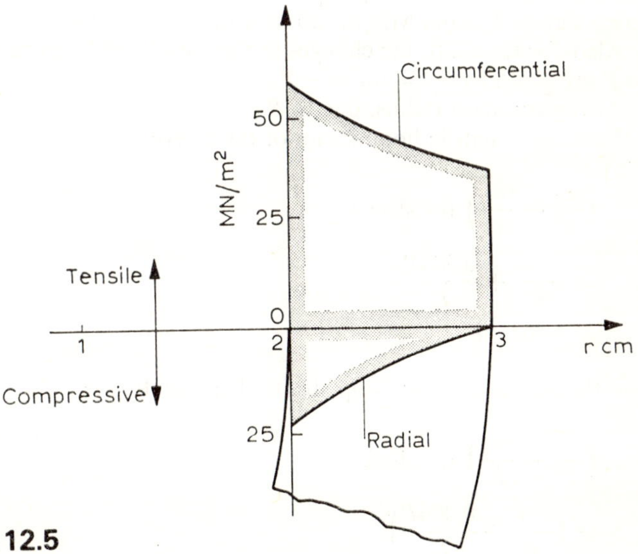

Fig. 12.5

The stress distributions drawn from these equations are shown in Fig. 12.5, the principal values being:

r cm	2	$2\frac{1}{2}$	3
Radial compressive stress, MN/m^2	23·1	8·1	0
Circumferential tensile stress, MN/m^2	60	45	36·9

12.8. A steel sleeve is pressed on to a solid steel shaft of 5 cm diameter. The radial pressure between the shaft and the sleeve is 20 MN/m^2 and the hoop stress at the inner surface of the sleeve is 48 MN/m^2. If an axial compressive load of 60 kN is now applied to the shaft determine the alteration of radial pressure.

Poisson's ratio = 0·304. *(U.L.)*

Solution. Let P be the increase in radial pressure due to the compressive load.

Since the hoop stress at the inner surface of the sleeve bears a constant ratio to the radial pressure P, the increase in the hoop stress due to P is $(48\ MN/m^2/20\ MN/m^2) \times P = 2{\cdot}4P$.

Also, for the shaft, the changes in the radial and hoop stresses due to P are given by $\sigma_R = \sigma_H = -P$.

At the common radius, due to P,

Increase in tensile hoop strain of the sleeve is

$$\left(\frac{\sigma_H}{E} - \frac{\nu\sigma_R}{E}\right) \text{ for sleeve}$$

$$= \frac{2{\cdot}4P}{E} + \frac{0{\cdot}304P}{E}$$

$$= 2{\cdot}704P/E \tag{i}$$

and the increase in the hoop strain of the shaft is

$$\left(\frac{\sigma_H}{E} - \frac{\nu\sigma_R}{E}\right) \text{ for shaft}$$

$$= \left(\frac{-P}{E} + \frac{0{\cdot}304P}{E}\right)$$

$$= \frac{-0{\cdot}696P}{E} \tag{ii}$$

Also, the proportional increase in the diameter of the shaft due to the compressive load is

$$\nu \times \text{longitudinal strain} = 0{\cdot}304 \times \text{stress}/E$$

$$= \frac{0{\cdot}304 \times (60 \times 10^3\ \text{N})}{\frac{1}{4}\pi \times (5 \times 10^{-2}\ \text{m})^2 \times E} \qquad \text{(iii)}$$

Since the change in (internal) diameter of the sleeve must equal the change in diameter of the shaft, then
Tensile hoop strain of sleeve due to P

= Proportional increase in shaft diameter due to load
+ Hoop strain in the shaft

or, from (i), (ii) and (iii),

$$\frac{2{\cdot}704P}{E} = \frac{0{\cdot}304 \times (60 \times 10^3\ \text{N})}{\frac{1}{4}\pi\,(5 \times 10^{-2}\ \text{m})^2 \times E} - \frac{0{\cdot}696P}{E}$$

$$2{\cdot}704P = 9{\cdot}292 \times 10^6 - 0{\cdot}696P$$

$$P = 2{\cdot}73 \times 10^6\ \text{N/m}^2$$

The alteration in radial pressure is $2{\cdot}73$ MN/m² (increase) (*Ans*)

12.9. A thin rotating disc of uniform density ρ rotates at angular velocity ω. Show that the radial stress at any radius r is given by

$$\sigma_R = A - \frac{B}{r^2} - \tfrac{1}{8}(3 + \nu)\rho\omega^2 r^2$$

where ν is Poisson's ratio and A and B are constants.

If the disc has an inner radius of $7\frac{1}{2}$ cm and an outer radius of 30 cm find the maximum radial stress at a speed of 3600 rev/min. Take $E = 200$ GN/m², $\nu = 0{\cdot}33$ and $\rho = 7{\cdot}5$ Mg/m³. (*U.L.*)

Solution. Suppose (Fig. 12.6) ABCD is a small element of a rotating disc at radius r, having radial width δr, and subtending a small angle θ at the centre 0. Let σ_R and $(\sigma_R + \delta\sigma_R)$ be the radial stresses at radii r and $(r + \delta r)$, and σ_H the hoop or circumferential stress at radius r, tensile being positive in all cases. For unit thickness of material the forces on this element are as shown in the diagram. These forces are not in equilibrium and resolving towards O,

$$\text{Resultant force} = \sigma_R r\theta - (\sigma_R + \delta\sigma_R)(r + \delta r)\theta + 2\sigma_H \delta r \cos(90^\circ - \tfrac{1}{2}\theta)$$

$$= -\sigma_R \delta r\theta - \delta\sigma_R r\theta + \sigma_H \delta r\theta$$

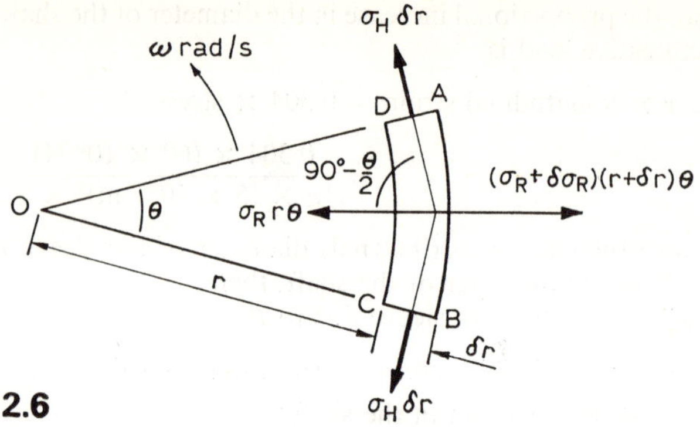

Fig. 12.6

ignoring products of small quantities and noting that $\sin \frac{1}{2}\theta = \frac{1}{2}\theta$ for small angles.

The mass of the element is $\rho \times$ volume $= \rho r\theta\delta r$ and due to the rotational speed ω it has an acceleration $\omega^2 r$ towards O. Using the relationship, force = mass × acceleration,

$$-\sigma_R\delta r\theta - \delta\sigma_R r\theta + \sigma_H\delta r\theta = (\rho r\theta\delta r) \times \omega^2 r$$

$$-\sigma_R - \frac{\delta\sigma_R}{\delta r} r + \sigma_H = \rho\omega^2 r^2$$

and, in the limit,

$$\sigma_H - \sigma_R = \rho\omega^2 r^2 + r\frac{d\sigma_R}{dr} \qquad \text{(i)}$$

A second relationship between σ_H and σ_R is obtained from considerations of strain. If a point at radius r in the unstressed disc moves radially outwards by an amount u the circumferential strain is

$$\frac{\text{Increase in circumference}}{\text{Original circumference}} = \frac{2\pi(r + u) - 2\pi r}{2\pi r} = \frac{u}{r}$$

and relating this to the stresses we have

$$\frac{u}{r} = \frac{\sigma_H}{E} - \frac{\nu\sigma_R}{E} \qquad \text{(ii)}$$

Similarly a point at radius $(r + \delta r)$ moves radially outwards by an amount $(u + \delta u)$ and the radial strain of an element such as ABCD in Fig. 12.6 is

$$\frac{\text{Increase in radial width}}{\text{Original radial width}} = \frac{[(r + \delta r + u + \delta u) - (r + u)] - \delta r}{\delta r}$$

$$= \frac{\delta u}{\delta r}$$

and, in terms of the corresponding stresses,

$$\frac{du}{dr} = \frac{\sigma_R}{E} - \frac{\nu\sigma_H}{E} \tag{iii}$$

Obtaining du/dr from (ii) and substituting in (iii) we obtain:

$$\left(\frac{\sigma_H}{E} - \frac{\nu\sigma_R}{E}\right) + \frac{r}{E}\left(\frac{d\sigma_H}{dr} - \frac{\nu d\sigma_R}{dr}\right) = \frac{\sigma_R}{E} - \frac{\nu\sigma_H}{E}$$

$$(\sigma_H - \sigma_R)(1 + \nu) = r\left(\nu\,\frac{d\sigma_R}{dr} - \frac{d\sigma_H}{dr}\right)$$

Substituting for $\sigma_H - \sigma_R$ from (i),

$$\left(\rho\omega^2 r^2 + r\,\frac{d\sigma_R}{dr}\right)(1 + \nu) = r\left(\nu\,\frac{d\sigma_R}{dr} - \frac{d\sigma_H}{dr}\right)$$

from which

$$\frac{d\sigma_H}{dr} + \frac{d\sigma_R}{dr} = -\rho\omega^2 r(1 + \nu)$$

Integrating with respect to r,

$$\sigma_H + \sigma_R = -\tfrac{1}{2}\rho\omega^2 r^2\,(1 + \nu) + 2A \tag{iv}$$

where the constant of integration is denoted by $2A$ for convenience later. Subtracting (i) from (iv) to eliminate σ_H,

$$2\sigma_R = -\tfrac{1}{2}\rho\omega^2 r^2\,(1 + \nu) + 2A - \rho\omega^2 r^2 - r\,\frac{d\sigma_R}{dr}$$

$$2\sigma_R + r\,\frac{d\sigma_R}{dr} = -\tfrac{1}{2}\rho\omega^2 r^2\,(3 + \nu) + 2A$$

The left-hand side can be written as $\dfrac{1}{r}\dfrac{d}{dr}(r^2\sigma_R)$ and hence

$$\frac{d}{dr}(r^2\sigma_R) = -\tfrac{1}{2}\rho\omega^2 r^3(3 + \nu) + 2Ar$$

Integrating with respect to r

$$r^2\sigma_R = -\tfrac{1}{8}\rho\omega^2 r^4(3 + \nu) + Ar^2 - B \tag{v}$$

where the second integration constant is denoted by $-B$.

On rearranging,

$$\sigma_R = A - \frac{B}{r^2} - \tfrac{1}{8}(3 + \nu)\rho\omega^2 r^2$$

which is the required result. If it is substituted in (iv) the circumferential or hoop stress is found to be

$$\sigma_H = A + \frac{B}{r^2} - \tfrac{1}{8}(1 + 3\nu)\rho\omega^2 r^2$$

The radial stress σ_R is zero at a free surface and hence $\sigma_R = 0$ when $r = R_1$ and also when $r = R_2$. Thus

$$0 = A - \frac{B}{R_1^2} - \tfrac{1}{8}(3 + \nu)\rho\omega^2 R_1^2$$

and

$$0 = A - \frac{B}{R_2^2} - \tfrac{1}{8}(3 + \nu)\rho\omega^2 R_2^2$$

Solving these simultaneous equations for A and B we obtain

$$A = \tfrac{1}{8}\rho\omega^2(3 + \nu)(R_1^2 + R_2^2)$$
$$B = \tfrac{1}{8}\rho\omega^2(3 + \nu)(R_1^2 R_2^2)$$

Substituting in the expression for σ_R the radial stress is given by

$$\sigma_R = \tfrac{1}{8}\rho\omega^2(3 + \nu)\left(R_1^2 + R_2^2 - \frac{R_1^2 R_2^2}{r^2} - r^2\right)$$

Differentiating the expression in the last bracket with respect to r, and equating to zero to find the maximum,

$$2R_1^2 R_2^2/r^3 = 2r \quad \text{or} \quad r^2 = R_1 R_2$$

and the corresponding value of the expression in the bracket is

$$R_1^2 + R_2^2 - \frac{R_1^2 R_2^2}{R_1 R_2} - R_1 R_2 = (R_2^2 - 2R_1 R_2 + R_1^2)$$
$$= (R_2 - R_1)^2$$

Thus

$$\sigma_{R(max)} = \tfrac{1}{8}\rho\omega^2(3 + \nu)(R_2 - R_1)^2$$

With the numerical data given in the question,

$$\omega = 2\pi N/60 = 2\pi \times 3600/60 = 120\pi \text{ rad/s}$$

and

$$\sigma_{R(max)} = \tfrac{1}{8} \times (7{\cdot}5 \times 10^3 \text{ kg/m}^3) \times (120\pi \text{ rad/s})^2 \times (3 + 0{\cdot}33) \times [(30 - 7{\cdot}5) \times 10^{-2} \text{ m}]^2$$
$$= 2{\cdot}25 \times 10^7 \text{ N/m}^2 \quad \text{or} \quad 22{\cdot}5 \text{ MN/m}^2 \qquad (Ans)$$

12.10. The general expressions for radial and hoop stresses at radius r in a thin circular rotating disc of uniform thickness are respectively

$$A + \frac{B}{r^2} - \tfrac{1}{8}(3 + \nu)\rho\omega^2 r^2 \quad \text{and} \quad A - \frac{B}{r^2} - \tfrac{1}{8}(1 + 3\nu)\rho\omega^2 r^2$$

where A and B are constants, ν = Poisson's ratio, ρ = density of the material and ω the angular velocity.

A thin disc of uniform thickness is 1 m external diameter and has a very small central hole. It rotates at 3000 rev/min.

Calculate the maximum hoop stress deriving any formula used from those given above.

Take $\rho = 7{\cdot}83$ Mg/m³ and $\nu = \frac{1}{3}$. *(U.L.)*

(Note: The signs of B in the equations in the question are opposite to those of the previous problem. As explained at the beginning of the chapter only the intermediate working is affected; the final results are unaltered.)

Solution. As in the previous solution, $\sigma_R = 0$ when $r = R_1$ and also when $r = R_2$. From these boundary conditions,

$$A = \tfrac{1}{8}\rho\omega^2(3 + \nu)(R_1^2 + R_2^2)$$

and

$$B = -\tfrac{1}{8}\rho\omega^2(3 + \nu)(R_1^2 R_2^2)$$

With these values of A and B the hoop stress is

$$\sigma_H = \tfrac{1}{8}\rho\omega^2\left[(3 + \nu)\left(R_1^2 + R_2^2 + \frac{R_1^2 R_2^2}{r^2}\right) - (1 + 3\nu)r^2\right]$$

From the form of this expression σ_H is greatest when r has its least value, i.e. R_1. Thus

$$\sigma_{H(max)} = \tfrac{1}{4}\rho\omega^2[(1 - \nu)R_1^2 + (3 + \nu)R_2^2]$$

If the central hole is very small ($R_1 \simeq 0$) the maximum stress tends to the value

$$\sigma_{H(max)} = \tfrac{1}{4}\rho\omega^2 R_2^2(3 + \nu)$$

With the numerical data given in the question,

$$\omega = 2\pi\,\text{N}/60 = 2\pi \times 3000/60 = 100\pi \text{ rad/s}$$

and

$$\sigma_{H(max)} = \tfrac{1}{4} \times (7{\cdot}83 \times 10^3 \text{ kg/m}^3) \times (100\pi \text{ rad/s})^2 \times (1 \text{ m})^2 \times (3 + \tfrac{1}{3})$$

$$= 6{\cdot}44 \times 10^8 \text{ N/m}^2 = 644 \text{ MN/m}^2 \qquad (Ans)$$

PROBLEMS

1. Solve the following thick cylinder problems using the assumptions of Example 12.1.

(*a*) If there is an internal pressure of 60 MN/m^2 and the hoop stress at the inner surface is 156 MN/m^2 (tensile), calculate the hoop stress at the outer surface and the longitudinal stress. The diameters are 120 mm and 80 mm.
(*b*) The pressure at the inside and outside edges of a tube are atmospheric (zero) and 30 MN/m^2 respectively. If the hoop stress at the inside edge is known to be 90 MN/m^2 (compressive) what is its value at the outside edge?
(*c*) At the outside surface of a thick cylinder the circumferential stress is known to be 22 MN/m^2 (tensile) when the internal pressure is 6 MN/m^2. Calculate the circumferential stress at the inside surface and at the point where the radial stress is 3·5 MN/m^2. If the diameters are 200 mm and 250 mm what is the longitudinal stress?

Answer. (*a*) Required hoop stress, 96 MN/m^2; longitudinal stress, 48 MN/m^2; (*b*) 60 MN/m^2; (*c*) Hoop stresses, 28 and 25·5 MN/m^2; longitudinal stress, 11 MN/m^2.

2. Use the formulae derived in Examples 12.2 and 12.3 to solve the following problems. In (*a*), (*b*) and (*e*) draw diagrams showing the variation of the circumferential and radial stresses through the wall of the cylinder.

(*a*) The diameters of a thick-walled tube are 125 mm and 175 mm. Calculate the maximum and minimum hoop tensions due to an internal pressure of 80 bar.
(*b*) Calculate the maximum tensile and compressive hoop stresses in a cylinder which is simultaneously subjected to an internal pressure of 280 bar and an external pressure of 140 bar. The internal and external diameters are 125 mm and 225 mm respectively.
(*c*) Find the minimum external/internal diameter ratio for an internal pressure of 150 bar with a maximum hoop stress of 50 MN/m^2.
(*d*) Find the necessary thickness of a hydraulic main 150 mm internal diameter to withstand a pressure of 70 bar with a maximum hoop stress of 14 MN/m^2.
(*e*) Calculate the safe internal pressure in a cast-iron hydraulic cylinder, 450 mm internal diameter and 250 mm thick, if the safe tensile stress is 15 MN/m^2.

Answer. (*a*) 24·67 and 16·67 MN/m^2; (*b*) 12·5 (tensile) and 1·5 (compressive) MN/m^2; (*c*) 1·363; (*d*) 54·9 mm; (*e*) 79·2 bar.

3. A thick-walled tube with closed ends is subjected to an internal pressure. If the external diameter is $k \times$ internal diameter prove that:

(*a*) The hoop stress at the outer surface is $2/(1 + k^2)$ times that at the inner surface.
(*b*) The hoop stress at the outer surface is twice the longitudinal stress.
(*c*) The ratio of the maximum hoop stress to the value given by thin cylinder theory is $(k^2 + 1)/(k + 1)$ (taking d as the internal diameter in the thin cylinder case).

Use the result (*c*) to show that, when $k = 1{\cdot}1$, thin cylinder theory underestimates the maximum hoop stress by about 5 per cent.

4. Show that the maximum (compressive) hoop stress in a cylinder subjected to an external pressure P is

$$(-)\, 2Pk^2/(k^2 - 1)$$

where k is the external/internal diameter ratio.

The inner tube of a compound cylinder has internal and external diameters of 160 mm and 240 mm respectively. Calculate the external diameter of a second tube to be shrunk on to the first so that, due to the shrinkage, the maximum compressive hoop stress in the inner tube is (numerically) equal to the maximum tensile hoop stress in the outer tube.
Answer. 319 mm.

5. A hollow steel cylinder, 150 mm internal and 200 mm external diameter, is open at the ends and is subjected to an external pressure of 140 bar. Calculate the maximum compressive circumferential stress due to this pressure and the decrease in the internal and external diameters of the cylinder.

$$E = 200\ \text{GN/m}^2 \text{ and } \nu = 0{\cdot}28.$$

Answer. 64 MN/m^2; 0·048 and 0·046 mm.

6. A compound tube has internal and external diameters of 150 mm and 250 mm respectively. It is formed by shrinking one tube on to another of the same material, the common diameter being 200 mm. If the shrinkage allowance is 0·05 mm calculate the radial pressure between the tubes at the common radius ($E = 200$ GN/m^2).

Find also the maximum and minimum hoop stresses in both tubes due to the shrinkage. If the compound tube is subjected to an internal pressure of 700 bar, what are the resultant stresses at the inner and outer surfaces of each tube?

Draw a diagram to show the distribution of the hoop stresses before and after the internal pressure is applied.
Answer. Radial pressure at common radius, 6·15 MN/m^2. Hoop stresses in MN/m^2 (positive are tensile):

r (mm)	75	100 (inner)	100 (outer)	125
Due to shrinkage	−28·1	−22	28	21·9
Due to pressure	148·8	100·9	100·9	78·8
Resultant	120·7	78·9	128·9	100·7

7. Calculate for the compound tube of Example 12·6 the minimum temperature difference between the two tubes to allow of the outer one passing over the inner. Coefficient of expansion, $\alpha = 11 \times 10^{-6}$/°C.

Find also the external diameter of a single tube to withstand the same internal pressure for the same maximum hoop stress (128·9 MN/m^2), the internal diameter being 150 mm.
Answer. 22·7°C; 275·7 mm.

8. A steel cylinder, 100 mm external diameter and 75 mm internal diameter, has another steel cylinder, external diameter 125 mm, shrunk on to it. If the maximum tensile hoop stress induced in the outer cylinder is 60 MN/m^2, find the radial compressive stress between the cylinders and the maximum compressive hoop stress in the inner cylinder.

By how much must the external diameter of the inner tube have exceeded the internal diameter of the outer tube to have produced these stresses?

$E = 200$ GN/m².
Answer. 13·2 MN/m², 60·2 MN/m²; 0·0535 mm.

9. A steel ring 150 mm external diameter and 100 mm internal diameter is shrunk on another steel ring having an internal diameter of 50 mm. The thickness of both rings is 20 mm. If the axial force required to push the rings apart is 56·54 kN and the coefficient of friction for the mating surfaces is 0·2, find the pressure at the mating surfaces and the shrinkage allowance. Also sketch curves showing how the radial stress and the tangential hoop stress vary from the inside to the outside of the compound ring.
$E = 200 \times 10^9$ N/m². (*I.Mech.E.*)
Answer. 45 MN/m²; 0·096 mm. Hoop stresses (σ_H MN/m²) are

r (mm)	25	50 (inner)	50 (outer)	75
σ_H	−120	−75	117	72

10. A steel plug, 75 mm diameter, is forced into a steel ring, 125 mm external diameter and 50 mm wide. From a reading taken by fixing, in a circumferential direction, an electrical resistance strain gauge on the external surface of the ring the strain is found to be $1{\cdot}5 \times 10^{-4}$. Assuming a coefficient of friction 0·2 for the mating surfaces, find the force required to push the plug out of the ring. Also estimate the greatest hoop stress in the ring.

$E = 200$ GN/m². (*I.Mech.E.*)
Answer. 62·8 kN; 56·7 MN/m².

11. A steel cylinder 200 mm external diameter and 150 mm internal diameter has another cylinder 250 mm external diameter shrunk on to it.

If the maximum tensile stress induced in the outer cylinder is 75 MN/m², find the radial compressive stress between the cylinders.

Determine the circumferential stresses at inner and outer diameter of both cylinders and show, by means of a diagram, how these stresses vary with the radius. Calculate the necessary shrinkage allowance at the common surface.
$E = 200$ GN/m². (*U.L.*)
Answer. Radial pressure, 16·5 MN/m². Hoop stresses (σ_H MN/m²) are:

r (mm)	75	100 (inner)	100 (outer)	125
σ_H	−75·25	−58·8	75	58·5

Shrinkage allowance 0·133 mm.

12. A solid circular shaft is subjected to a uniform radial compressive stress. Show, without using thick cylinder formulae, that the radial and circumferential stresses at all radii are equal to the external radial stress.

A steel shaft, originally 100 mm diameter, is subjected to a uniform radial compressive stress of 20 MN/m². Assuming the radial stress remains constant, find the uniform longitudinal stress required to reduce the initial diameter by 0·0125 mm and calculate the alteration in volume for a 150 mm length of shaft.

$E = 200$ GN/m² and Poisson's ratio = 0·304. (*U.L.*)
Answer. 36·4 MN/m² (tensile); 8·24 mm³ (decrease).

13. Find the ratio of thickness to internal diameter for a tube subjected to internal pressure when the ratio of the internal pressure to the greatest circumferential stress is 0·5.

Find the alteration in thickness of metal in such a tube, 200 mm internal diameter, when the internal pressure is 750 bar.

Poisson's ratio = 0·304. E = 200 GN/m². (*U.L.*)
Answer. 0·366; 0·0125 mm (if longitudinal stress = 0).

14. A steel cylindrical plug of 125 mm diameter is forced into a steel sleeve of 200 mm external diameter and 100 mm long. If the greatest circumferential stress in the sleeve is 90 MN/m², find the torque required to turn the plug in the sleeve assuming the coefficient of friction between the plug and the sleeve is 0·2. (*U.L.*)

Answer. 19·36 kN-m.

15. The inner tube of a compound cylinder is 15 cm external and 10 cm internal diameter. It is subjected to an external pressure of 40 MN/m² when the internal pressure is 120 MN/m². Working from first principles, determine the circumferential stress at the external and internal surfaces and find the radial and circumferential stresses at the mean radius. Draw a diagram to show how the radial and circumferential stresses vary with the radius. (*U.L.*)
Answer. Hoop stresses: at inner surface, 168 MN/m² (tensile); at outer surface 88 MN/m² (tensile). At mean radius: radial stress 68 MN/m² (compressive); hoop stress 116 MN/m² (tensile).

16. A steel shaft 40 mm diameter is to be encased in a bronze sleeve 60 mm outside diameter which is to be forced into position and, before forcing on, the inside diameter of the sleeve is smaller than the diameter of the shaft, the difference in these diameters being 0·05 mm.

Find, due to the forcing on

(*a*) the radial pressure between the shaft and the sleeve,
(*b*) the maximum hoop stress in the sleeve,
(*c*) the change in the outside diameter of the sleeve.

For steel, E = 200 GN/m², Poisson's ratio = 0·29.
For bronze, E = 120 GN/m², Poisson's ratio = 0·34. (*U.L.*)
Answer. (*a*) 44·6 MN/m²; (*b*) 115·8 MN/m² (tensile); (*c*) 0·0214 mm (increase).

17. A bronze bush having an outside diameter of 140 mm and an inside diameter of 80 mm is pressed into a recess in a body which is assumed to be perfectly rigid. If the diameter of the recess is 139·96 mm, find the radial pressure produced on the outer surface of the bush and the maximum hoop stress in the bush.

Determine also the change in the inside diameter of the bush.

For bronze take E = 110 GN/m² and Poisson's ratio = 0·35. (*U.L.*)
Answer. 19·4 and 57·6 MN/m², 0·0419 mm (decrease).

18. The four elastic constants are E, G, K and ν where E is Young's modulus of elasticity, G is the shear modulus of elasticity, K is the bulk modulus and ν is Poisson's ratio.

(*a*) State the relationship between (i) E, G and ν and (ii) E, K and ν.
(*b*) A hollow cylinder having internal and external diameters of 90 mm and 150 mm was subjected to an internal pressure of 20 MN/m² when the longitudinal strain was found to be 46×10^{-6}; this strain includes the effect of longitudinal stress. The same longitudinal strain was obtained by subjecting the cylinder to an axial pull of 59·3 kN with no internal pressure. Determine the values of the four elastic constants for the material from which the cylinder was made. (*U.L.*)

Answer. (*a*) (i) $E = 2G(1 + \nu)$. (ii) $E = 3K(1 - 2\nu)$. (*b*) $E = 114$ GN/m², $G = 45$ GN/m², $K = 81{\cdot}5$ GN/m², $\nu = 0{\cdot}267$.

19. State the general formulae for the hoop tension σ_H and the radial tension σ_R in a cylindrical pipe with a thick wall subjected to internal or external radial forces. From these formulae derive a formula for the hoop stress at a distance x from the axis of a hollow cylinder internal radius R_1 and external radius R_2 subjected to an internal pressure p_0.

A pipe of 18 mm external diameter and 6 mm bore carries oil at a pressure of 36 MN/m². If the pipe is free of longitudinal stress, what is the maximum intensity of stress? (*U.L.*)

Answer. $\sigma_H = A + \frac{B}{x^2}$; $\sigma_R = A - \frac{B}{x^2}$ where A and B are constants and x is the distance from the axis;

$$\frac{p_0 R_1^2}{R_2^2 - R_1^2}\left(\frac{R_2^2}{x^2} + 1\right); \ 45 \text{ MN/m}^2$$

20. A thick-walled steel cylinder having an inside diameter of 150 mm is to be subjected to an internal pressure of 40 MN/m². Find to the nearest mm the outside diameter required if the hoop tension in the cylinder wall is not to exceed 120 MN/m².

Using the diameter found above, calculate the actual hoop stress at the inner and outer surfaces of the cylinder and plot a graph showing how the hoop tension varies across the cylinder wall; calculate an intermediate value for this purpose. (*U.L.*)

Answer. 213 mm; 118·8 and 78·8 MN/m².

21. A bronze sleeve having an outside diameter 75 mm is forced on to a steel rod 55 mm diameter, the initial inside diameter of the sleeve being 0·06 mm smaller than the rod diameter. When in service the compound rod is subjected to an external pressure of 20 MN/m² and at the same time to a rise in temperature of 100°C. Determine:

(*a*) the radial pressure between the sleeve and the rod; and

(*b*) the greatest circumferential stress in the sleeve under service conditions.

For steel, $E = 200$ GN/m², $\nu = 0{\cdot}30$, $\alpha = 11 \times 10^{-6}$ per °C

For bronze, $E = 100$ GN/m², $\nu = 0{\cdot}33$, $\alpha = 19 \times 10^{-6}$ per °C. (*U.L.*)

Answer. Stresses in MN/m² (tensile positive).

	(*a*)	(*b*)
due to force fit	(−) 27·2	90·6
due to external pressure	(−) 21·6	−14·7
due to temperature rise	20·0	−66·4
resultant	(−) 28·8	9·5

22. A steel cylinder having inside and outside diameters of 200 mm and 275 mm respectively is shrunk on to another steel cylinder having inside and outside diameters of 150 mm and 200 mm respectively.

Find the necessary radial pressure between the cylinders due to shrinkage only so that when there is an internal pressure of 80 MN/m² the final value of the maximum hoop stress in the inner cylinder is 120 MN/m².

Determine also the magnitude of the maximum hoop stress in the outer cylinder when the internal pressure is acting. (*U.L.*)

Answer. 6·07 MN/m²; 117·5 MN/m².

23. A steel cylinder is required to withstand an internal pressure of 500 bar, the outside diameter is to be 200 mm and the maximum allowable hoop tension is 125 MN/m².

Determine

(*a*) the greatest permissible value for the inside diameter;
(*b*) the change in the inside diameter due to the stresses in the cylinder wall;
(*c*) the maximum shearing stress in the cylinder wall.
$E = 200$ GN/m², $\nu = 0{\cdot}3$. (*U.L.*)

Hint. Note that at radius r the maximum shearing stress is B/r^2 with the usual notation.
Answer. (*a*) 131 mm; (*b*) 0·084 mm (allowing for longitudinal stress); (*c*) 87·5 MN/m².

24. A compound cylinder consists of two steel cylinders and has an outside diameter of 250 mm and an inside diameter of 150 mm; the common diameter between the cylinders is 200 mm. The compound cylinder is subjected to an internal pressure of 70 MN/m². The outer cylinder is shrunk on the inner cylinder and the initial difference between the common diameters is such that, when the internal pressure is applied, the maximum hoop stress in the inner cylinder is equal to the maximum hoop stress in the outer cylinder.

(*a*) Determine the radial pressure between the cylinders due to the shrink fit only.
(*b*) Calculate the initial difference between the common diameters.
(*c*) Make a diagram showing how the hoop stress varies across the cylinder wall when the internal pressure is applied.

For steel take $E = 203$ GN/m². (*U.L.*)
Answer. (*a*) 5·25 MN/m²; (*b*) 0·0420 mm; (*c*)

r (mm)	75	100 (inner)	100 (outer)	125
stress (MN/m²)	124·8	82·1	124·8	97·5

25. A hydraulic cylinder is designed to operate at a pressure of 600 bar with an internal diameter of 80 mm. Determine the minimum wall thickness required if the maximum hoop stress is not to exceed 250 MN/m².

By how much would the external diameter of the cylinder increase under this fully loaded condition? The cylinder is made of steel with $E = 200$ GN/m² and $\nu = 0{\cdot}26$. (*U.L.*)
Answer. 11·1 mm; 0·0845 mm.

26. A hub of 100 mm diameter and 100 mm long is shrunk on to a shaft of 50 mm diameter and is required to transmit a torque between the hub and shaft of 1 kN-m. The coefficient of friction between the hub and shaft is 0·3. The radial stress on the outside of the hub is negligible.

Sketch the stress distribution across the thickness of the hub corresponding to the minimum required interface pressure and give values of the tangential stress at the inner and outer faces of the hub and of the radial stress at the inner face. (*U.L.*)

Answer.

r (mm)	25	50
hoop stress (MN/m²)	14·15	5·66
radial stress (MN/m²)	8·49	0

27. Discuss the assumptions made in deriving the formulae for radial and hoop stress in pressurised "thin" and "thick" cylinders.

A simple thick cylinder is designed to withstand an internal pressure of 20 MN/m². It has an external diameter of 0·6 m and the maximum shear stress is 120 MN/m². Assuming that there is no external pressure determine (i) the thickness of the cylinder and (ii) the maximum hoop stress. If the actual thickness of a cylinder manufactured to this design is 20 per cent greater than the value determined from (i) above and there is an external compressive pressure of 6 MN/m², calculate the maximum internal pressure which the cylinder could withstand if the maximum shear stress is unaltered. (*U.L.*)

Answer. (i) 26·1 mm; (ii) 220 MN/m²; 30·8 MN/m².

28. Show that the maximum intensity of hoop tension in a thin rotating disc is

$$\tfrac{1}{8}\rho\omega^2R^2(3 + \nu)$$

where ρ is the density of the material, R is the radius of the disc, ω its angular velocity and ν is Poisson's ratio for the material. (*U.L.*)

Hint. In the usual equations for stress the constant B must be zero, otherwise the stress at $r = 0$ becomes infinite. (Note that the maximum stress in the solid disc is half that of the disc with a very small central hole, Example 12.10.)

29. A steel disc of 40 cm external diameter is of uniform thickness and has a 5 cm diameter hole at the centre. Find the speed of rotation about an axis perpendicular to the plane of the disc which will produce a maximum tensile stress of 80 MN/m².

Find also the value of the maximum radial stress and the radius at which it occurs.

The radial stress σ_x and the hoop stress σ_y at radius r are given by

$$\sigma_x = A - \frac{B}{r^2} - \tfrac{1}{8}\rho\omega^2r^2(3 + \nu)$$

$$\sigma_y = A + \frac{B}{r^2} - \tfrac{1}{8}\rho\omega^2r^2(1 + 3\nu)$$

where A and B are constants, ρ = density of material, ω = angular velocity, ν = Poisson's ratio.

For steel $\rho = 7{\cdot}83$ Mg/m³ and $\nu = 0{\cdot}3$. (*U.L.*)

Answer. 5226 rev/min: 29·62 MN/m² at 7·07 cm radius.

30. A thin circular steel disc of uniform thickness has an outside diameter of 50 cm with a central hole of 8 cm diameter and it rotates at a speed of 4000 rev/min. Find (*a*) the maximum tensile stress in the disc, (*b*) the maximum shearing stress, and (*c*) the increase in the outside diameter caused by the rotation.

$E = 200$ GN/m²; $\nu = 0{\cdot}28$; density of steel = 7·83 Mg/m³. (*U.L.*)

Hint. (*a*) the maximum tensile stress is the hoop stress at the inner radius; (*b*) the principal stresses at a given point are σ_H and σ_R and the maximum value of the corresponding shearing stress is $\tfrac{1}{2}(\sigma_H - \sigma_R)$, which has its greatest value at the inner radius; (*c*) the increase in diameter is proportional to the circumferential strain.

Answer. (*a*) 70·9 MN/m²; (*b*) 35·45 MN/m²; (*c*) 0·043 mm.

Chapter 13

Strain Energy

In all cases strain energy (or resilience) is denoted by U. It has the units of energy (or work done) which are derived from the product force × distance, i.e. newton metres (denoted by N-m). The name *joule* (denoted by J) is given to the SI unit of energy and the two units may be regarded as alternatives.

Tension or Compression. For a uniform bar, subjected to a stress σ,

$$U = \frac{\sigma^2}{2E} \times \text{(volume of the bar)}$$

where E = Young's modulus.

$$\text{Proof resilience} = \frac{{\sigma_p}^2}{2E} \times \text{(volume of the bar)}$$

where σ_p = proof stress.

$$\text{Modulus of resilience} = \frac{{\sigma_p}^2}{2E}$$

Impact loads. If a mass falls freely on to a bar, beam or spring, stress waves pass through the material and the determination of the maximum stress induced is complicated. Approximate results can be obtained, however, by consideration of energy changes. It is usual to assume that the decrease in potential energy of the falling mass is taken up as strain energy by the material of the bar, beam or spring and that the maximum stress is the same as that for the same energy conditions in a static case.

The maximum stress σ produced by a mass of weight W which falls a distance h before straining a uniform bar, length L and cross-sectional area A, is

$$\sigma = \frac{W}{A}\left(1 + \sqrt{\left[1 + \frac{2AEh}{WL}\right]}\right)$$

If the maximum extension is negligible compared with h then

$$\sigma = \sqrt{\frac{2EWh}{AL}}$$

For a suddenly applied load W,

$$\sigma = \frac{2W}{A}$$

Shear. For a body subjected to a uniform shear stress τ,

$$U = \frac{\tau^2}{2G} \times \text{(volume of the body)}$$

where G = modulus of rigidity.

Bending. For a beam (or cantilever) length L, in which M and EI are the bending moment and flexural rigidity respectively at any distance x from one end

$$U = \int_0^L \frac{M^2}{2EI}\,dx$$

In each of the following cases W is the total load (concentrated or distributed). The results apply to uniform beams only and no account is taken of the strain energy due to shear.

	Type of Load	$U =$
Cantilever	Concentrated at free end	$\frac{W^2L^3}{6EI}$
	Uniformly distributed throughout	$\frac{W^2L^3}{40EI}$
Simply-supported beam	Concentrated at mid-span	$\frac{W^2L^3}{96EI}$
	Uniformly distributed throughout	$\frac{W^2L^3}{240EI}$
Encastré beam	Concentrated at mid-span	$\frac{W^2L^3}{384EI}$
	Uniformly distributed throughout	$\frac{W^2L^3}{720EI}$

Table 4

Torsion. For a hollow shaft, external diameter D and internal diameter d,

$$U = \frac{\tau^2}{4G}\left(\frac{D^2 + d^2}{D^2}\right) \times \text{(volume of the shaft)}$$

where τ = maximum shear stress,

G = modulus of rigidity.

For a solid (circular) shaft,

$$U = \frac{\tau^2}{4G} \times \text{(volume of the shaft)}$$

WORKED EXAMPLES

13.1. Define "strain energy" and derive a formula for it in the case of a uniform bar in tension.

Calculate the strain energy in a bar 3 m long and 50 mm diameter when it is subjected to a tensile load of 200 kN. $E = 206\ \text{GN/m}^2$.

Solution. When a tensile force is applied to a bar it causes an extension in the direction in which it acts. The force, therefore, does work on the bar and, provided the material of the bar is elastic, this work done reappears when the force is removed. The energy stored in a bar when it is subjected to such a force is called *strain energy* (or *resilience*). The work done in compression, shear, bending and torsion can also be stored as strain energy. In all cases, the symbol used for strain energy is U.

Suppose a uniform bar, length L and cross-sectional area A, is subjected to a gradually applied tensile load P. If the limit of proportionality is not exceeded, then the extension x is proportional to P and the graph of load against extension (Fig. 13.1) is a straight line. Hence the strain energy is

U = work done in stretching the bar (which is represented by the area under the load–extension graph)

= *average* force × total extension

$= \frac{1}{2}Px$

If $\sigma = P/A$ (the tensile stress) then $P = \sigma A$.
Also, $x = L \times \text{strain} = L\sigma/E$, where E = Young's modulus.

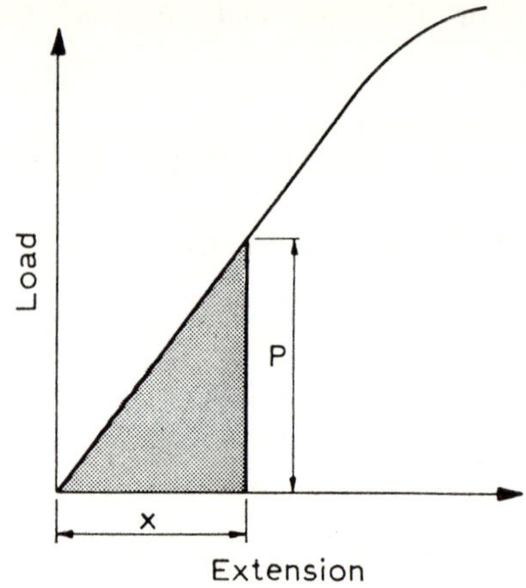

Fig. 13.1

Thus

$$U = \tfrac{1}{2} \times \sigma A \times \frac{L\sigma}{E}$$

$$= \frac{\sigma^2}{2E} \times AL$$

$$= \frac{\sigma^2}{2E} \times \text{(volume of the bar)} \qquad \text{(i)}$$

The strain energy per unit volume is

$$\frac{\sigma^2}{2E} = \tfrac{1}{2} \times \text{stress} \times \text{strain} \qquad \text{(ii)}$$

since strain $= \sigma/E$.

Since strain energy is equal to work done, its units are newton-metres (N-m). The newton-metre has the name *Joule*.

For the bar given in the question,

$$\text{Stress } \sigma = \frac{(200 \times 10^3\ \text{N})}{\frac{1}{4}\pi \times (50 \times 10^{-3}\ \text{m})^2} = \frac{4 \times 200 \times 10^3}{\pi \times 2500 \times 10^{-6}}\ \text{N/m}^2$$

$$= 101{\cdot}9\ \text{MN/m}^2$$

$$\text{Volume} = \tfrac{1}{4}\pi \times (50 \times 10^{-3}\ \text{m})^2 \times 3\ \text{m} = 5{\cdot}89 \times 10^{-3}\ \text{m}^3$$

The strain energy is, therefore,

$$U = \frac{\sigma^2}{2E} \times \text{volume} = \frac{(101{\cdot}9 \times 10^6\ \text{N/m}^2)^2 \times (5{\cdot}89 \times 10^{-3}\ \text{m}^3)}{2 \times (206 \times 10^9\ \text{N/m}^2)}$$

$$= 148{\cdot}4\ \text{N-m (or J)} \qquad (\textit{Ans})$$

13.2. What is meant by the terms "proof resilience" and "modulus of resilience"?

Two bars are subjected to equal, gradually applied tensile loads. One bar is 50 mm diameter throughout and the other, which has the same length, is turned down to a diameter of 25 mm over the middle third of its length, the remainder having a diameter of 50 mm. Compare the strain energies of the two bars assuming that they are of the same material.

Compare also the amounts of energy which the two bars can absorb in simple tension without exceeding a given stress within the limit of proportionality.

Solution. The *proof resilience* is the greatest amount of strain energy which a piece of material can absorb without suffering permanent strain. If σ_p is the maximum stress which the material can withstand without permanent distortion then, provided this stress is uniform,

$$\text{Proof resilience} = \frac{\sigma_p^{\ 2}}{2E} \times (\text{volume of the material})$$

(σ_p is sometimes called the proof stress. In this chapter it can be taken as the limit of proportionality but see also Chapter 16.)

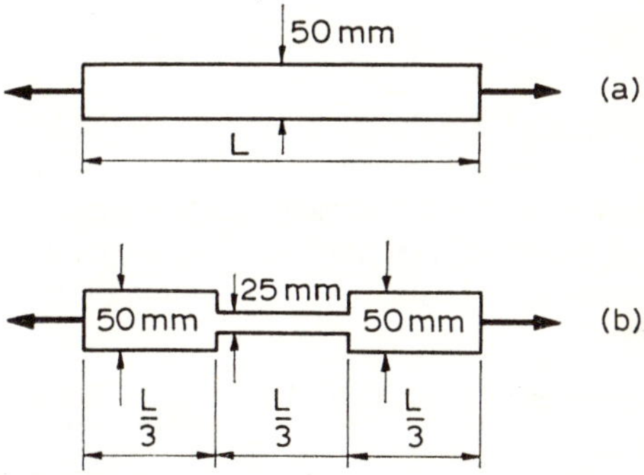

Fig. 13.2

Modulus of resilience is defined as the proof resilience per unit volume and thus equals $\sigma_p^{\ 2}/2E$. This quantity is a mechanical property of the material, e.g. for mild steel $E = 206$ GN/m^2 and $\sigma_p = 300$ MN/m^2 (approximately) and the modulus of resilience is about 220×10^3 N-m/m^3.

The two bars given in the question are shown in Fig. 13.2a and b. Suppose the length is L and that the tensile load applied to each is P.

Working in newton metre units throughout, the strain energies are

For the uniform bar,

$$U = \left(\frac{P}{\frac{1}{4}\pi(0{\cdot}050)^2}\right)^2 \times \frac{1}{2E} \times \tfrac{1}{4}\pi(0{\cdot}050)^2 \times L$$

$$= 800LP^2/\pi E \qquad \text{(i)}$$

For the end portions of the non-uniform bar which are 0·050 m diameter and length $\frac{2}{3}L$ (total) we have, from (i), replacing L by $\frac{2}{3}L$,

$$U = 1600LP^2/3\pi E \qquad \text{(ii)}$$

For the middle portion of the non-uniform bar, which is 0·025 m diameter and length $\frac{1}{3}L$,

$$U = \left(\frac{P}{\frac{1}{4}\pi(0{\cdot}025)^2}\right)^2 \times \frac{1}{2E} \times \tfrac{1}{4}\pi(0{\cdot}025)^2 \times \tfrac{1}{3}L$$

$$= 3200LP^2/3\pi E \qquad \text{(iii)}$$

If the load is the same for both bars, then, from (i), (ii) and (iii),

$$\frac{\text{Resilience of uniform bar}}{\text{Resilience of non-uniform bar}} = \frac{800LP^2/\pi E}{\dfrac{1600LP^2}{3\pi E} + \dfrac{3200LP^2}{3\pi E}}$$

$$= \tfrac{1}{2}$$

Suppose that, for the given stress, P is the load on the non-uniform bar. The maximum (given) stress occurs in the 25 mm diameter portion, and hence for the same stress to be attained in the uniform bar the load required is $(50/25)^2 \times P = 4P$ (the load being proportional to the square of the diameter). Replacing P by $4P$ in (i), the required ratio is

$$\frac{\text{resilience of uniform bar}}{\text{resilience of non-uniform bar}} = \frac{800L(4P)^2/\pi E}{\dfrac{1600LP^2}{3\pi E} + \dfrac{3200LP^2}{3\pi E}}$$

$$= 8$$

13.3. A mass of weight W falls a distance h before beginning to stretch a bar, length L, and cross-sectional area A. Derive expressions for the maximum stress induced in the bar when

(*a*) the maximum extension is negligible compared with h,
(*b*) the maximum extension is of the same order as h.

A bar, 3 m long and 50 mm diameter, hangs vertically and has a collar securely attached at the lower end. Find the maximum stress induced when

(i) a mass of weight 2 kN falls 100 mm on to the collar
(ii) a mass of weight 20 kN falls 10 mm on to the collar.

Take $E = 200\ \text{GN/m}^2$.

Solution. Suppose the mass falls the given distance on to a collar at the lower end of a vertical bar as shown in Fig. 13.3. A small

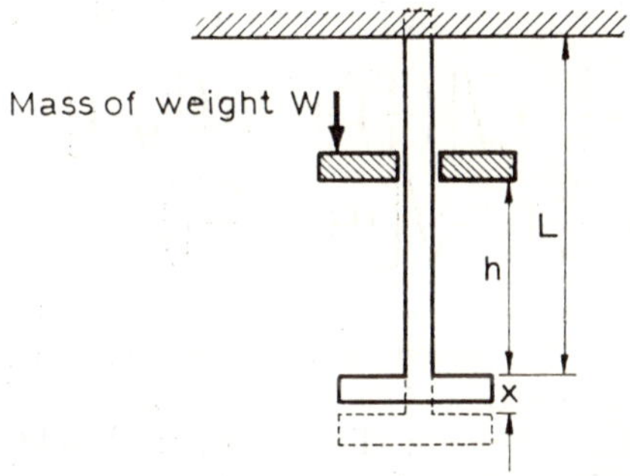

Fig. 13.3

oscillation is set up and, provided the limit of proportionality is not exceeded, the collar will take up the same final position as in the case of a gradually applied load W. The maximum instantaneous extension is much greater than the steady value. Suppose it to be x as shown in the diagram.

Decrease in potential energy of the mass
= strain energy of the bar

$$W(h + x) = \frac{\sigma^2}{2E} \times AL \tag{1}$$

where σ is the maximum stress induced.

(*a*) If x is negligible compared with h then, from (1),

$$Wh = \sigma^2 AL/2E$$

$$\sigma = \sqrt{\frac{2EWh}{AL}} \tag{2}$$

(*b*) In this case x must be expressed in terms of σ,

$$x = \text{strain} \times L = \frac{\text{stress}}{E} \times L = \frac{\sigma L}{E}$$

Substituting in (1),

$$W\left(h + \frac{\sigma L}{E}\right) = \frac{\sigma^2 AL}{2E}$$

$$\left(\frac{AL}{2E}\right)\sigma^2 - \left(\frac{LW}{E}\right)\sigma - Wh = 0$$

Multiplying through by E/AL,

$$\tfrac{1}{2}\sigma^2 - \left(\frac{W}{A}\right)\sigma - \frac{WEh}{AL} = 0$$

This is a quadratic in σ and its solution is

$$\sigma = \frac{W}{A} \pm \sqrt{\left[\left(\frac{W}{A}\right)^2 + 4(\tfrac{1}{2})\left(\frac{WEh}{AL}\right)\right]}$$

$$= \frac{W}{A}\left(1 + \sqrt{\left[1 + \frac{2AEh}{WL}\right]}\right) \tag{3}$$

the positive square root giving the maximum stress.

For the bar given in the question the maximum extension is probably less than 1/500 of the total length since the material is presumably steel with a limit of proportionality of 300 – 400 MN/m². Hence x is less than 6 mm (approx.).

(i) x is negligible compared with h and, therefore, using equation (2) the maximum stress is

$$\sigma = \sqrt{\frac{2 \times (200 \times 10^9\ \text{N/m}^2) \times (2 \times 10^3\ \text{N}) \times (0{\cdot}1\ \text{m})}{\frac{1}{4}\pi \times (0{\cdot}050\ \text{m})^2 \times 3\ \text{m}}}$$

$$= 116{\cdot}5\ \text{MN/m}^2 \qquad (Ans)$$

(ii) Using the "exact" formula (3), maximum stress is

$$\sigma = \frac{20 \times 10^3\ \text{N}}{\frac{1}{4}\pi \times (0{\cdot}050)\ \text{m}^2} \times \left[1 + \sqrt{\left(1 + \frac{2 \times \frac{1}{4}\pi \times (0{\cdot}050\ \text{m})^2 \times (200 \times 10^9\ \text{N/m}^2) \times 0{\cdot}01\ \text{m}}{(20 \times 10^3\ \text{N}) \times 3\text{m}}\right)}\right]$$

$$= 127{\cdot}1\ \text{MN/m}^2 \qquad (Ans)$$

The reader should note that equation (2) would give the same answer for (i) and (ii).

13.4. An unknown mass falls 10 mm on to a collar rigidly attached to the lower end of a vertical bar, 3 m long and 600 mm² in section.

If the maximum instantaneous extension is known to be 1·5 mm, what is the corresponding stress and the value of the unknown mass? $E = 200$ GN/m². Take the weight of 1 kg mass as 9·81 N.

Solution.
Maximum stress

$$= E \times \text{maximum strain}$$
$$= (200 \times 10^9 \text{ N/m}^2) \times \left(\frac{1\cdot5 \times 10^{-3} \text{ m}}{3 \text{ m}}\right)$$
$$= 100 \text{ MN/m}^2 \qquad (Ans)$$

The formula derived in Example 13·3 will not be used since it is simpler to work from first principles when the maximum stress is given.
Resilience of the bar

$$= \frac{\sigma^2}{2E} \times \text{volume}$$
$$= \frac{(100 \times 10^6 \text{ N/m}^2)^2 \times 3 \text{ m} \times (600 \times 10^{-6} \text{ m}^2)}{2 \times (200 \times 10^9 \text{ N/m}^2)}$$
$$= 90 \text{ N-m (or joules)} \qquad \text{(i)}$$

Decrease in potential energy of falling mass

$$= W(h + x)$$
$$= W \text{ N } (0\cdot010 \text{ m} + 0\cdot0015 \text{ m}) = 0\cdot0115W \text{ N-m} \qquad \text{(ii)}$$

where W N is the weight of the unknown mass.
Assuming that all the energy given up by the falling mass is taken up by the bar as strain energy, results (i) and (ii) are equal, i.e.

$$0\cdot0115W = 90 \qquad W = 7830 \text{ N}$$

the mass therefore weighs 7830 N and is

$$7830 \text{ N}/(9\cdot81 \text{ N/kg}) = 798 \text{ kg} \qquad (Ans)$$

13.5. Explain what is meant by the term "suddenly applied load". Calculate the maximum stress and extension produced in a bar 2 m long and 25 mm diameter when it is subjected to a suddenly applied load of 45 kN.

$E = 200$ GN/m².

Solution. A *suddenly applied load* is one in which the full force is applied instantaneously although there is no impact. Suppose, for example, the weight shown in Fig. 13.3 is originally in contact with the collar but is supported independently. If it is then suddenly released the maximum stress induced will be greater than in the case of the same weight gradually applied.

Putting $h = 0$ in equation (3) of Example 13.3 the stress due to a suddenly applied load is

$$\sigma = \frac{W}{A}(1 + \sqrt{1}) = \frac{2W}{A}$$

For the bar given in the present question, the maximum stress is

$$\sigma = \frac{2 \times (45 \times 10^3 \text{ N})}{\frac{1}{4}\pi \times (0{\cdot}025 \text{ m})^2} = 183{\cdot}3 \text{ MN/m}^2 \qquad (Ans)$$

The corresponding extension is

$$L \times \text{strain} = L\sigma/E = \frac{2 \text{ m} \times (183{\cdot}3 \times 10^6 \text{ N/m}^2)}{200 \times 10^9 \text{ N/m}^2}$$

$$= 1{\cdot}833 \text{ mm} \qquad (Ans)$$

13.6. Derive an expression for the strain energy in a material subjected to a uniform shear stress.

Calculate the local strain energy at a point in a material where the shear stress is 80 MN/m².

$G = 28$ GN/m².

Solution. Consider a rectangular block length l, width b, and height h as shown in Fig. 13.4. Suppose that the block is rigidly fixed at its base and a shear force F is gradually applied along the top face as shown. If the resulting deflection in the direction of the force is x,

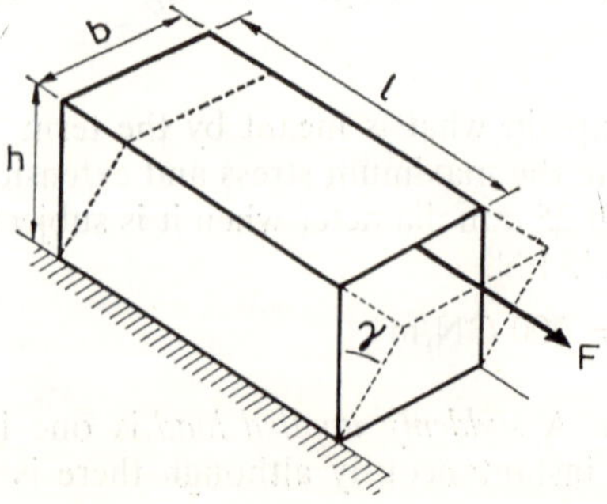

Fig. 13.4

the shear strain is $\gamma = x/h$ (see Example 1.6, Chapter 1). The strain energy is

$$\begin{aligned} U &= \text{work done by the shear force} \\ &= \text{average force} \times \text{deflection} \\ &= \tfrac{1}{2}Fx \end{aligned}$$

Since $F = \tau \times bl$ (where τ = shear stress) and $x = \gamma h$ we have

$$\begin{aligned} U &= \tfrac{1}{2}\tau bl \times \gamma h \\ &= \tfrac{1}{2}\tau \times \frac{\tau^2}{G} \times blh \\ &= \frac{\tau^2}{2G} \times (\text{volume of the block}) \end{aligned}$$

where G = modulus of rigidity = τ/γ.

The shear strain energy per unit volume is $\tau^2/2G$. If τ_p is the proof shear stress then $\tau_p{}^2/2G$ is called the *modulus of shear resilience.*

Substituting the values given in the question, the local strain energy per unit volume is

$$\begin{aligned} \frac{\tau^2}{2G} &= \frac{(80 \times 10^6\ \text{N/m}^2)^2}{2(28 \times 10^9\ \text{N/m}^2)} \\ &= 114{\cdot}3 \times 10^3\ \text{N-m/m}^3 \qquad (Ans) \end{aligned}$$

(Generally, the shear stress is not uniformly distributed over the cross-section and only the local shear strain energy is required.)

13.7. Show that the total strain energy of a beam or cantilever can be expressed as

$$\int_0^L \frac{M^2}{2EI}\,dx$$

and give the meaning of each symbol.

Compare the total strain energies of two equal uniform beams (simply-supported at their ends), one carrying a concentrated central load and one carrying a uniformly distributed load when

(*a*) the total load on each beam is the same,
(*b*) the maximum stress due to bending in each beam is the same.

Solution. Consider a small length δx of a beam as shown in Fig. 13.5. Suppose M is the bending moment for this section of the beam. Then, from the bending equation, the stress in a thin layer distance y

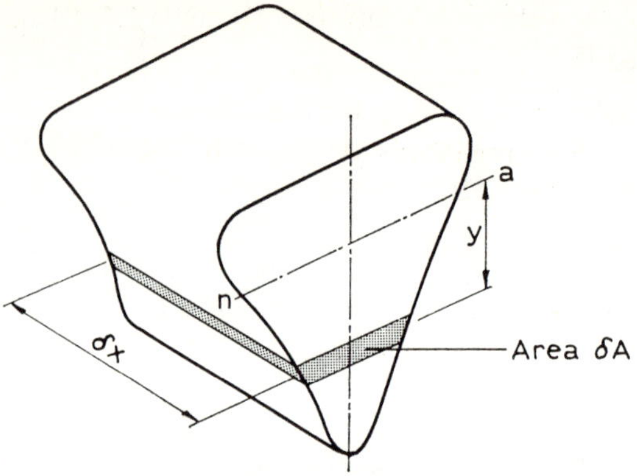

Fig. 13.5

from the neutral axis, as shown, is $\sigma = My/I$, where I is the second moment of area of the cross-section about the neutral axis. If δx is the length of the block then the volume of this layer is $\delta x \delta A$, where δA is its cross-sectional area. Since δx is small the stress can be considered uniform through the layer. Hence the strain energy in the layer is

$$\frac{\sigma^2}{2E} \times (\text{volume of the layer}) = \left(\frac{My}{I}\right)^2 \times \frac{1}{2E} \times \delta x \delta A$$

For the whole block the strain energy is

$$\delta U = \int \left(\frac{My}{I}\right)^2 \times \frac{1}{2E} \times \delta x dA$$
$$= \frac{M^2 \delta x}{2EI^2} \int y^2 dA$$

δx being a constant in this integration. M, E and I are also constants, since δx is small. The integral must be taken over the whole cross-section and is, by definition, the second moment of area of the section (I). Hence

$$\delta U = \frac{M^2 \delta x}{2EI^2} \times I = \frac{M^2 \delta x}{2EI}$$

For a whole beam or cantilever, length L, taking an origin at one end the total strain energy is,

$$U = \int_0^L \frac{M^2}{2EI} dx \qquad \text{(i)}$$

M being the bending moment at any distance x from the origin. L and I are defined above and E is Young's modulus.

N.B. The above theory takes no account of the strain energy due to shear.

For a simply-supported beam, with a central point load the bending moment at any distance x from one end is

$$M = x \times \text{reaction} = Wx/2$$

provided that x is less than $L/2$. Since the bending diagram is symmetrical about mid-span and EI is constant for a uniform beam

$$\begin{aligned} U &= 2\int_0^{L/2} \frac{M^2}{2EI}\,dx = \frac{1}{EI}\int_0^{L/2}\left(\frac{Wx}{2}\right)^2 dx \\ &= \frac{W^2}{4EI}\int_0^{L/2} x^2 dx \\ &= \frac{W^2}{4EI}\left[\frac{x^3}{3}\right]_0^{L/2} \\ &= \frac{W^2L^3}{96EI} \qquad \text{(ii)} \end{aligned}$$

For a uniformly distributed load w/unit length the reaction at each end is $wL/2$ and the bending moment at any distance x from one end is

$$M = \tfrac{1}{2}wLx - \tfrac{1}{2}wx^2 = \tfrac{1}{2}w(Lx - x^2)$$

The total strain energy is thus

$$\begin{aligned} U &= \int_0^L \frac{M^2}{2EI}\,dx = \frac{1}{2EI}\int_0^L \left[\frac{w}{2}(Lx - x^2)\right]^2 dx \\ &= \frac{w^2}{8EI}\int_0^L (L^2x^2 - 2Lx^3 + x^4)dx \\ &= \frac{w^2}{8EI}\left[\frac{L^2x^3}{3} - \frac{Lx^4}{2} + \frac{x^5}{5}\right]_0^L \\ &= \frac{w^2L^5}{240EI} \\ &= \frac{W^2L^3}{240EI} \qquad \text{(iii)} \end{aligned}$$

where $W = wL$, the total load.

(*a*) If the total load is the same for both beams, then from (ii) and (iii), the required ratio is

$$\frac{\text{strain energy for point load case}}{\text{strain energy for distributed load case}} = \left(\frac{W^2L^3}{96EI}\right)\bigg/\left(\frac{W^2L^3}{240EI}\right)$$
$$= 240/96 = 5/2 \quad (Ans)$$

(*b*) Let W be the central point load. For the same maximum bending moment (and hence the same maximum bending stress) the distributed load is $2W$. Replacing W by $2W$ in (iii) the required ratio is

$$\frac{\text{strain energy for point load case}}{\text{strain energy for distributed load case}} = \left[\frac{W^2L^3}{96EI}\right]\bigg/\left[\frac{(2W)^2L^3}{240EI}\right]$$
$$= 240/(96 \times 2^2) = 5/8$$
(*Ans*)

13.8. Explain how the deflection under a single point load can be found by a strain energy method.

A uniform beam, length L is simply supported at its ends and carries a concentrated load W at a point whose distances from the ends are a and b. Find the deflection under the load.

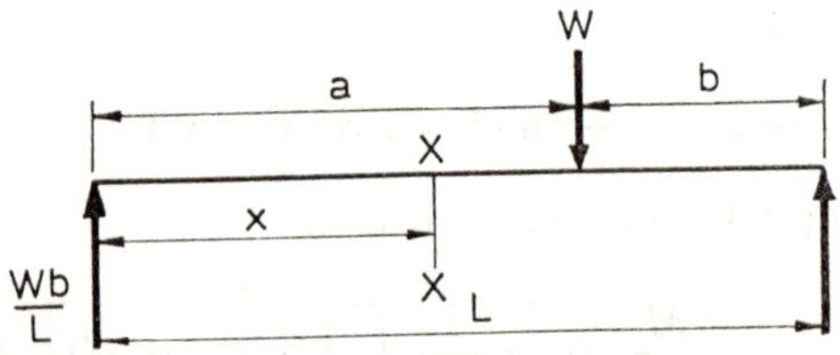

Fig. 13.6

Solution. The given point load is the only external force which does work (since there is no deflection at a support and thus the reactions do no work). If W is a gradually applied concentrated load and δ is the deflection it causes at the point where it acts then the work done is the average force multiplied by the deflection, i.e. $\frac{1}{2}W\delta$.

Since the external work done is stored as strain energy then the total strain energy of the beam can be equated to $\frac{1}{2}W\delta$. The value of δ can thus be found but the method, as given here, is applicable to the deflections at the point of loading only.

Figure 13.6 shows the beam given in the question. Taking the load to be a distance a from the left-hand end, the reaction there is Wb/L. Hence the bending moment at a section distance x from this end is Wbx/L, provided that x is less than a. The strain energy for the portion of the beam between the load and the left-hand end is

$$U = \int_0^a \frac{M^2}{2EI}\,dx = \frac{1}{2EI}\int_0^a \left(\frac{Wbx}{L}\right)^2 dx$$

$$= \frac{W^2b^2}{2L^2EI}\int_0^a x^2 dx$$

$$= \frac{W^2b^2a^3}{6L^2EI} \qquad \text{(i)}$$

Interchanging a and b, the corresponding amount for the portion of the beam to the right of W is

$$U = \frac{W^2a^2b^3}{6L^2EI} \qquad \text{(ii)}$$

Since the external work done equals the total strain energy then, from (i) and (ii),

$$\tfrac{1}{2}W\delta = \frac{W^2b^2a^3}{6L^2EI} + \frac{W^2a^2b^3}{6L^2EI}$$

$$= \frac{W^2a^2b^2\,(a+b)}{6L^2EI} = \frac{W^2a^2b^2}{6LEI}$$

since $a + b = L$.

Dividing each side by $W/2$, the required deflection is

$$\delta = \frac{Wa^2b^2}{3LEI} \qquad \textit{(Ans)}$$

13.9. Explain how the maximum instantaneous stress and deflection produced in a beam by an impact load can be found by the method of the "equivalent static load".

When a load of 20 kN is gradually applied at a certain point in a beam it produces a deflection of 15 mm and a maximum bending stress of 75 MN/m^2. From what height can a load of 5 kN fall on to the beam at this point if the maximum bending stress is to be 150 MN/m^2?

Solution. Suppose a mass of weight W falls a distance h on to a beam, as shown in Fig. 13.7, causing a maximum instantaneous deflection δ at the point where it strikes. The decrease in potential energy of the falling mass is $W(h + \delta)$. Let W_E be the equivalent static load, i.e.

the load that would cause the same deflection δ at the same point when gradually applied. The work done by W_E is $\frac{1}{2}W_E\delta$.

Hence

$$W(h + \delta) = \tfrac{1}{2}W_E\delta \qquad \text{(i)}$$

It is usually necessary to express δ in terms of W_E. (In the case of central loading, for example, $\delta = W_EL^3/48EI$.) If this is done equation (i) becomes a quadratic for W_E. The solution can then be used in calculating any required stress or deflection using the usual equations for static loading.

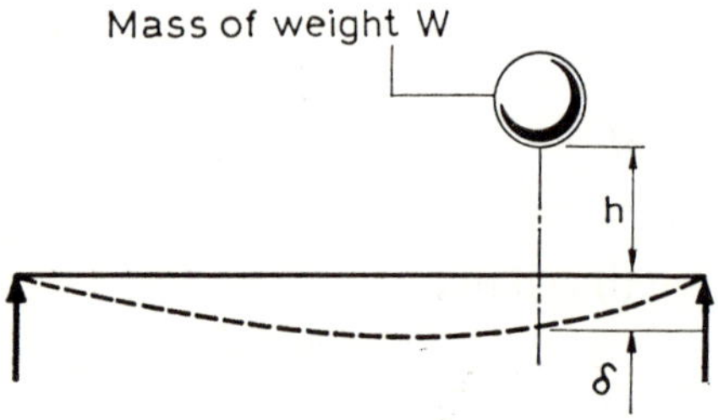

Fig. 13.7

For the conditions given in the question, the required static load for a stress of 150 MN/m² is

$$W_E = \frac{150 \text{ MN/m}^2}{75 \text{ MN/m}^2} \times 20 \text{ kN} = 40 \text{ kN}$$

The corresponding deflection is

$$\delta = \frac{40 \text{ kN}}{20 \text{ kN}} \times 15 \text{ mm} = 30 \text{ mm}$$

since static stresses and deflections are proportional to the loads producing them.

Substituting in (i),

$$5 \text{ kN } (h + 30 \text{ mm}) = \tfrac{1}{2} \times 40 \text{ kN} \times 30 \text{ mm}$$
$$h + 30 \text{ mm} = 120 \text{ mm}$$

and the required height is

$$h = 90 \text{ mm} \qquad (Ans)$$

13.10. A static concentrated load of 10 kN applied to a simply supported beam at mid-span, produces a deflection of 6 mm and a maximum bending stress of 20 MN/m². Calculate the maximum value of the momentary stress produced when a load of 5 kN is allowed to fall through a height of 18 mm on to the beam at the middle of the span. (*U.L.*)

Solution. Let W_E kN be the static load equivalent to the given impact load.

By proportion, the deflection produced by the equivalent static load W_E is

$$\delta = \frac{W_E}{10} \times 6 \text{ mm} = \frac{6W_E}{10} \text{ mm}$$

As in the previous solution,

$$W(h + \delta) = \tfrac{1}{2}W_E\delta$$

or, in kN and mm units,

$$5\left(18 + \frac{6W_E}{10}\right) = \tfrac{1}{2}W_E \frac{6W_E}{10} = \frac{3W_E^2}{10}$$

Rearranging and multiplying through by 10,

$$W_E^2 - 10W_E - 300 = 0$$
$$W_E = \tfrac{1}{2}(1 \pm \sqrt{[100 + 1200]}) = 23{\cdot}03 \text{ kN}$$

taking the positive square root.

Since a static load of 10 kN produces a maximum bending stress of 20 MN/m² then, by proportion, the maximum stress produced in the impact case is

$$\sigma_{max} = \frac{23{\cdot}03}{10} \times 20 \text{ MN/m}^2 = 46{\cdot}1 \text{ MN/m}^2 \qquad (Ans)$$

13.11. Derive an expression for the strain energy in a hollow shaft, external diameter D and internal diameter d, when it transmits a torque which produces a maximum shear stress τ.

A solid shaft, 150 mm diameter, is to be replaced by a hollow shaft of the same material, length, and weight. Calculate the diameters of the latter if its strain energy is to be 20 per cent greater than that of the solid shaft when transmitting torque at the same maximum shear stress.

Solution. Suppose a torque T is gradually applied to the shaft and produces a twist θ (radians) in a length L. From the torsion equation

$T/J = G\theta/L = \tau/r$ and, if τ is the maximum shear stress, $r = D/2$,

$$T = \frac{\tau}{r} J = \frac{2\tau}{D} \frac{\pi}{32} (D^4 - d^4) \qquad \text{(i)}$$

$$\theta = \tau L/rG = 2\tau L/DG \qquad \text{(ii)}$$

where G is the modulus of rigidity.

The strain energy for the shaft is, using (i) and (ii),

$$\begin{aligned} U &= \text{work done by the torque} \\ &= \text{average torque} \times \text{angle of twist} \\ &= \tfrac{1}{2} T\theta \\ &= \frac{1}{2}\left[\frac{2\tau}{D}\frac{\pi}{32}(D^4 - d^4)\right] \times \left[\frac{2\tau L}{DG}\right] \\ &= \frac{\tau^2}{16G} \frac{\pi(D^4 - d^4)L}{D^2} \\ &= \frac{\tau^2}{4G} \frac{(D^2 + d^2)}{D^2} \frac{\pi(D^2 - d^2)L}{4} \\ &= \frac{\tau^2}{4G} \frac{(D^2 + d^2)}{D^2} \text{(volume of the shaft)} \qquad \text{(iii)} \end{aligned}$$

Thus the *average* strain energy per unit volume is $\frac{\tau^2}{4G} \frac{(D^2 + d^2)}{D^2}$ but the *local* strain energy per unit volume is not constant since the shear stress varies across the section. It was shown in Example 13.6 that for a *uniform* shear stress the strain energy is $(\tau^2/2G) \times$ (volume). Hence the maximum value of $(D^2 + d^2)/D^2$ is 2, which corresponds to the case of a tube with an infinitely thin wall, so that $D = d$.

For a solid shaft, $d = 0$, and, from (iii), the total strain energy is

$$U = \frac{\tau^2}{4G} \times \text{(volume of the shaft)} \qquad \text{(iv)}$$

Let D and d be the external and internal diameters of the required hollow shaft. Since the length, weight, and material is the same for the solid and hollow shafts then the volumes and cross-sectional areas of the two shafts are equal.

Hence, area of hollow shaft = area of solid shaft, or

$$\tfrac{1}{4}\pi(D^2 - d^2) = \tfrac{1}{4}\pi \times (150 \text{ mm})^2$$

$$D^2 - d^2 = 22\,500 \text{ mm}^2 \qquad \text{(v)}$$

Also, for the required ratio of strain energies, we have, from (iii) and (iv),

Strain energy for hollow shaft = 1·2 × strain energy for solid shaft

$$\frac{\tau^2}{4G}\frac{(D^2+d^2)}{D^2} \times (\text{volume}) = 1{\cdot}2 \times \frac{\tau^2}{4G} \times (\text{volume})$$

$$D^2 + d^2 = 1{\cdot}2D^2$$

$$D^2 = 5d^2 \qquad \text{(vi)}$$

Solving the simultaneous equations (v) and (vi),

$$d = 75 \quad \text{and} \quad D = 168$$

The required diameters are 168 mm external and 75 mm internal.

13.12. If σ_1 and σ_2 are the principal stresses, at a point in a material in a two-dimensional stress system, derive an expression for the strain energy, per unit volume, in terms of σ_1, σ_2, Poisson's ratio and Young's modulus.

Find the strain energy in N-m, stored in a steel bar 250 mm long, and of cross-section 25 mm by 6 mm when it is subjected simultaneously to an axial pull of 18 kN and to a compressive stress of 60 MN/m² on its narrow edges.

(For steel, E = 200 GN/m² and Poisson's ratio = 0·28.)

(I.Mech.E.)

Solution. From equation (ii) of Example 13.1, the strain energy, per unit volume, due to a single direct stress is

$$\frac{\sigma^2}{2E} = \tfrac{1}{2} \times \text{stress} \times \text{strain} \qquad \text{(i)}$$

For simple stresses the strain is σ/E where E = Young's modulus, but in complex stress systems the principal stresses and principal strains must be used. Hence, due to σ_1 the strain energy, per unit volume, is

$$\tfrac{1}{2} \times \sigma_1 \times \left(\frac{\sigma_1}{E} - \frac{\nu\sigma_2}{E}\right) = \frac{{\sigma_1}^2}{2E} - \frac{\nu\sigma_1\sigma_2}{2E} \qquad \text{(ii)}$$

(where ν = Poisson's ratio), since the strain in the direction in which σ_1 acts is $\dfrac{\sigma_1}{E} - \dfrac{\nu\sigma_2}{E}$.

Similarly, the strain energy, per unit volume, due to σ_2 is

$$\frac{\sigma_2^2}{2E} - \frac{\nu\sigma_2\sigma_1}{2E} \qquad \text{(iii)}$$

From (ii) and (iii), the total strain energy per unit volume is

$$\left(\frac{\sigma_1^2}{2E} - \frac{\nu\sigma_1\sigma_2}{2E}\right) + \left(\frac{\sigma_2^2}{2E} - \frac{\nu\sigma_2\sigma_1}{2E}\right) = \frac{1}{2E}(\sigma_1^2 + \sigma_2^2 - 2\nu\sigma_1\sigma_2) \qquad \text{(iv)}$$

For the steel bar in the question, the principal stresses are

$$\sigma_1 \text{ (corresponding to the 18 kN load)} = \frac{18 \text{ kN}}{25 \text{ mm} \times 6 \text{ mm}}$$
$$= 120 \text{ MN/m}^2$$

and

$$\sigma_2 = -60 \text{ MN/m}^2$$

taking compressive stresses as negative.

From (iv), the total strain energy of the bar is

$$U = \frac{1}{2E}(\sigma_1^2 + \sigma_2^2 - 2\nu\sigma_1\sigma_2) \times \text{volume of the bar}$$

$$= \frac{1}{2 \times 200 \times 10^9}[(120 \times 10^6)^2 + (-60 \times 10^6)^2 - 2 \times 0{\cdot}28 \times (120 \times 10^6)(-60 \times 10^6)] \times [250 \times 25 \times 6 \times 10^{-9}]$$

$$= \frac{(60 \times 10^6)^2 \times 10^{-9}}{400 \times 10^9} \times [2^2 + (-1)^2 - 2 \times 0{\cdot}28 \times 2 \times (-1)] \times 37\,500$$

$$= 2{\cdot}065 \text{ N-m} \qquad (Ans)$$

PROBLEMS

1. Find the total strain energy of a mild steel bar, 25 mm diameter and 2 m long, when it is subjected to a steady tensile load of 60 kN. $E = 200$ GN/m^2.
Answer. 36·7 N-m (or J).

2. Two round bars have the same length L. One is diameter d for a length $L/3$ and diameter $2d$ for a length $2L/3$. The other is diameter $2d$ for a length $2L/3$ and diameter $3d$ for a length $L/3$. Compare the amounts of energy stored in the bars when they are subjected to equal tensile loads, assuming that they are of the same material.

Compare also the amounts of energy that can be absorbed by the two bars in simple tension without exceeding a given stress within the limit of proportionality.

Answer. For same load, $\frac{\text{resilience of first bar}}{\text{resilience of second bar}} = \frac{27}{11}$; for same stress, ratio is 27/176.

3. Calculate the modulus of resilience in tension and in shear for a steel having the following properties.

Limits of proportionality: in tension, 325 MN/m²; in shear, 240 MN/m².

$E = 206$ GN/m²; $G = 80$ GN/m².

Answer. In tension, 256 kN-m/m³; in shear, 360 kN-m/m³ (or kJ/m³).

4. A solid shaft carries a flywheel of mass 150 kg and having a radius of gyration of 0·5 m. The shaft is rotating at a steady speed of 60 rev/min when it is suddenly fixed at a point 4 m from the flywheel. Calculate the shaft diameter if the maximum instantaneous shear stress produced is 150 MN/m². Assume that the kinetic energy of the flywheel is taken up as torsional strain energy by the shaft. Neglect the inertia of the shaft.

$G = 80$ GN/m².

Answer. 57·9 mm.

5. A bar, 1·2 m long and 25 mm diameter hangs vertically and has a collar rigidly attached at the lower end. When a mass of 500 kg is gradually lowered on to the collar, the extension is 0·05 mm. Find the maximum instantaneous stress induced in the bar when this mass falls a distance 2·5 mm on to the collar.

Answer. 110·4 MN/m².

6. Find the maximum height from which a mass of 2500 kg may fall on to a column 100 mm diameter and 600 mm high, if the stress is limited to 100 MN/m². $E = 200$ GN/m². If you use a formula, prove it.

Answer. 4·8 mm.

7. Derive, by strain energy methods, an expression for the deflection at the mid-point of a simply-supported beam carrying a central point load.

A simply-supported beam is of symmetrical section 250 mm deep ($I = 10000$ cm⁴) and spans 4 m. From what height can a mass of 2000 kg fall on to the mid-point of the beam without causing a maximum instantaneous bending stress greater than 120 MN/m²? $E = 200$ GN/m².

Answer. 9·26 mm.

8. Explain the meaning of the term "strain energy" and state the type of problems for which the conception may be used. A cantilever has its section tapered so that the 2nd moment of area I varies from I_0 at the support to zero at the free end, the value of I at a distance x from the free end being $I_0 x/l$ where l is the cantilever length. A load W is applied at the free end. By how much does the free end displace, assuming the material to be homogeneous and of Young's modulus E? Strain energy methods should be used. (*R.Ae.S.*)

Answer. $Wl^3/2EI_0$.

9. Prove that the strain energy stored in a bent beam is

$$\int (M^2/2EI)dx$$

where M is the bending moment at any section distant x from the origin, E is the modulus of elasticity of the material at the section, I is the relevant moment of inertia at the section.

A load of 30 kN is applied gradually to the centre of a simply supported beam of 3 m span. The section has a moment of inertia of 1500 cm^4, and E may be taken as 200 GN/m^2. Find the amount of strain energy stored in the beam, and hence find the deflection under the load. *(I.Struct.E.)*

Answer. 84·4 N-m; 5·63 m.

10. Obtain an expression for the resilience of a solid circular bar of diameter d and length L when subjected to a uniform tensile stress σ_t.

A bar is 60 mm diameter and 1 m long. Determine the stress produced in it by a mass of weight 1·8 kN falling 80 mm before commencing to stretch the bar. (E = 200 GN/m^2). *(I.Struct.E.)*

Answer. $\pi\sigma_t^2 d^2 L/8E$; 143·4 MN/m^2.

11. Explain what you understand by the term "strain energy", and derive from first principles an expression giving the strain energy in a tension bar of length L and cross-sectional area A when under a load F.

A tension bar 3 m long is made up of two parts: 1·8 m of its length has a cross-sectional area of 12 cm^2, while the remaining 1·2 m has a cross-sectional area of 24 cm^2. An axial load of 80 kN is gradually applied.

Find the total strain energy produced in the bar and compare this value with that obtained in a uniform bar of the same length *and having the same volume* when under the same load.

E = 200 GN/m^2. *(I.Struct.E.)*

Answer. $\dfrac{F^2L}{2AE}$; 32 N-m; $\dfrac{\text{strain energy in given bar}}{\text{strain energy in uniform bar}} = \dfrac{28}{25}$

12. A rod of material for which Young's modulus is E has a cross-sectional area a and a length l. Obtain an expression for the strain energy stored in it when elastically stressed by a gradually applied direct load P.

A bolt 500 mm long has a diameter of 125 mm for one-half of its length and of 112 mm for the other half. It may, on occasion, be subjected to a maximum tensile force of 1 MN. Determine the total extension of the bolt and the strain energy in each part. Young's modulus, 200 GN/m^2. *(I.Mech.E.)*

Answer. $P^2l/2aE$; extension = 0·229 mm; strain energy = 50·9 N-m (125 mm diameter part) and 63·5 N-m (112 mm diameter part).

13. Two shafts, one of steel and the other of phosphor-bronze, are of the same length and are subjected to equal torques. If the steel shaft is 10 mm diameter find the diameter of the phosphor-bronze shaft so that it will store the same amount of strain energy, per unit volume, as the steel shaft. Also determine the ratio of the maximum shear stresses induced in the two shafts. Take the modulus of rigidity for phosphor-bronze as 48 GN/m^2, and for steel as 80 GN/m^2. *(I.Mech.E.)*

Answer. Required diameter, 10·8 mm; $\dfrac{\tau\ (\text{steel})}{\tau\ (\text{bronze})} = 1{\cdot}265.$

14. A hollow shaft, subjected to a pure torque, attains a maximum shearing stress τ. Given that the strain energy stored per unit volume is $\tau^2/3G$, where G is the modulus of rigidity, calculate the ratio of the shaft diameters. Determine the actual diameters for such a shaft required to transmit 4 MN at 110 rev/min

with uniform torque when the energy stored is 20000 N-m/m^3 of material; take $G = 80$ MN/m^2.
Answer. $\sqrt{3}$; 306 and 177 mm.

15. A hollow shaft 200 mm external diameter and 125 mm internal diameter transmits 1·5 MW at a speed of 150 rev/min. Calculate the shearing stress at the inner and outer surfaces of the shaft and the strain energy per metre length. G (steel) $= 80 \times 10^9$ N/m^2. *(U.L.)*
Answer. 44·9 and 71·8 MN/m^2; 428 N-m.

16. What is meant by "resilience"?

A bar of steel, l m in length and d m square in cross-section, is subjected to a uniform bending moment *(a)* in a plane parallel to one of the sides of the bar and *(b)* in the plane of a diagonal. The maximum fibre stress is σ N/m^2 within the elastic range. Derive an expression of the elastic strain energy in each case. *(U.L.)*

Answer. *(a)* $\dfrac{\sigma^2 d^2 l}{6E}$ N-m; *(b)* $\dfrac{\sigma^2 d^2 l}{12E}$ N-m

17. Compare the strain energy of a beam, simply supported at the ends and loaded with a uniformly distributed load, with that of the same beam centrally loaded and having the same value of the maximum bending stress. *(U.L.)*

Answer. $\dfrac{U \text{ for point load}}{U \text{ for distributed load}} = \dfrac{5}{8}$

18. A concentrated load W gradually applied to a horizontal beam, simply supported at the ends, produces a deflection y at the load point. If this load falls through a distance h, before making contact with the beam, find an expression, in terms of h and y, for the maximum deflection at the load point. Neglect loss of energy at impact.

In a given beam a concentrated load W, gradually applied, produces a deflection of 5 mm at the load point and a maximum bending stress of 60 MN/m^2. Find the greatest height from which a load $0{\cdot}1W$ can be dropped without exceeding the elastic limit stress of 270 MN/m^2. *(U.L.)*
Answer. Maximum deflection $= y + \sqrt{(y^2 + 2yh)}$; 484 mm.

19. The shearing stress at the surface of a solid steel shaft is 75 MN/m^2 when subjected to the action of a torque. When used as a beam, simply supported at the ends and centrally loaded the greatest bending stress at the surface of the shaft is 120 MN/m^2. Compare the strain energies of the shaft for the two conditions of loading.

Is this ratio affected if the shaft is hollow and the stresses remain unchanged? ($E = 210 \times 10^9$ N/m^2; $G = 80{\cdot}5 \times 10^9$ N/m^2). *(U.L.)*

Answer. $\dfrac{\text{strain energy in torsion}}{\text{strain energy in bending}} = 6{\cdot}11$; ratio is unaffected if shaft is hollow.

20. A hollow steel shaft has an external diameter of 225 mm and it was found that, when running at 125 rev/min and transmitting 2·2 MW, the angle of twist per metre length of shaft was 0·5°. If G is assumed to be 80×10^9 N/m^2, estimate the internal diameter of the shaft and the maximum shearing stress.

If the total strain energy U is expressed as $U = u \times$ volume of shaft, find the

value of u; find also the strain energy per m^3 of material at the external surface of the shaft. (U.L.)
Answer. 103 mm; 78·5 MN/m^2; 23·3 and 38·6 kN-m/m^3.

21. By an extension of the method used in Example 13.12 show that at a point in a material subject to a three dimensional stress system the strain energy, per unit volume, is

$$\frac{1}{2E}[\sigma_1^2 + \sigma_2^2 + \sigma_3^2 - 2\nu(\sigma_1\sigma_2 + \sigma_2\sigma_3 + \sigma_3\sigma_1)]$$

where σ_1, σ_2 and σ_3 are the principal stresses. Suppose σ is the mean of the three principal stresses, i.e. $\sigma = \frac{1}{3}(\sigma_1 + \sigma_2 + \sigma_3)$, so that the principal stresses can be replaced by two systems as follows:

(*a*) a stress σ in each direction which produces a change in volume but no change in shape.

(*b*) perpendicular stresses $(\sigma_1 - \sigma)$, $(\sigma_2 - \sigma)$ and $(\sigma_3 - \sigma)$ in the directions of σ_1, σ_2 and σ_3 which produce a change in shape but no change in volume.

Show that the strain energies due to these systems are respectively.

(*a*) $$\frac{3\sigma^2}{2E}(1 - 2\nu)$$

(*b*) $$\frac{1}{12G}[(\sigma_1 - \sigma_2)^2 + (\sigma_2 - \sigma_3)^2 + (\sigma_3 - \sigma_1)^2]$$

Check that the sum of these results equals the expression found above.

22. A rolled steel beam is simply supported on a span of 6 m. A gradually applied load of 6 kN acting at mid-span produces a deflection of 1·08 mm at mid-span. Determine the value of EI, the flexural constant for the beam, stating the units in which it is expressed.

If the load of 6 kN is dropped on to the beam at mid-span, falling freely through a distance of 20 mm before striking the beam, determine the maximum deflection produced. (U.L.)
Answer. 25×10^6 N-m^2; 7·74 mm.

23. A vertical steel rod, of 25 mm diameter, checks the fall on its end of a weight of 2·25 kN which drops through a distance of 4 mm before it strikes the rod. Find the shortest length of rod which will bear the impact if the stress is not to exceed 136 MN/m^2. $E = 200$ GN/m^2.

Verify that the length found is the *least* possible length. (U.L.)
Answer. 425 mm.

24. A beam of uniform section and span l is firmly fixed at the ends. Show that, due to a central load P, the maximum deflection is $Pl^3/192EI$.

A steel bar of rectangular section 40 mm wide and 10 mm deep is arranged in a horizontal position as a beam with firmly fixed ends and of span 0·9 m. A load of 140 N is allowed to fall freely on to the beam at mid-span. Find the height above the beam from which the load must fall in order to produce in the beam a maximum stress of 140 MN/m^2. Take $E = 200$ GN/m^2.

Describe briefly the behaviour of the beam immediately after impact. (U.L.)
Answer. 9·28 mm.

25. A piece of steel is subjected to the following system of two-dimensional stress, namely, two tensile stresses of 90 and 45 MN/m^2 on planes at right angles to each other and a shear stress of 30 MN/m^2 on the same planes.

Working from first principles, find

(i) the principal stresses;
(ii) the principal strains;
(iii) the strain energy per unit volume.

If the tensile stress of 45 MN/m^2 is reversed in sign, find the new value to which the 90 MN/m^2 stress must be changed, assuming it to remain tensile, so that the strain energy remains the same, and find the new principal strains.

Take $E = 200$ GN/m^2, $G = 80$ GN/m^2, Poisson's ratio $= \frac{1}{4}$. (*U.L.*)

Answer. (i) 105 and 30 MN/m^2 (tensile); (ii) 487×10^{-6} and $18{\cdot}5 \times 10^{-6}$; (iii) 25·9 kN-m/$m^3$; stress must be changed to 67·5 MN/m^2 (tensile); 441×10^{-6} and -356×10^{-6}.

26. A steel ring of rectangular cross-section 8 mm wide by 5 mm thick has a mean diameter of 300 mm. A narrow radial saw cut is made and tangential separating forces of 5 N each are applied at the cut in the plane of the ring.

Determine the additional separation due to these forces.

Modulus of elasticity = 200 GN/m^2. (*U.L.*)

Hint. Use bending strain energy. Show that, at an angular position θ from the saw cut, $M = WR(1 - \cos\theta)$ where W = one force and R = radius. Putting $dx = Rd\theta$ and integrating between 0 and 2π total strain energy $= 3\pi W^2R^3/2EI$.

Answer. 9·54 mm (if $b = 8$ mm) or 3·73 mm (if $b = 5$ mm).

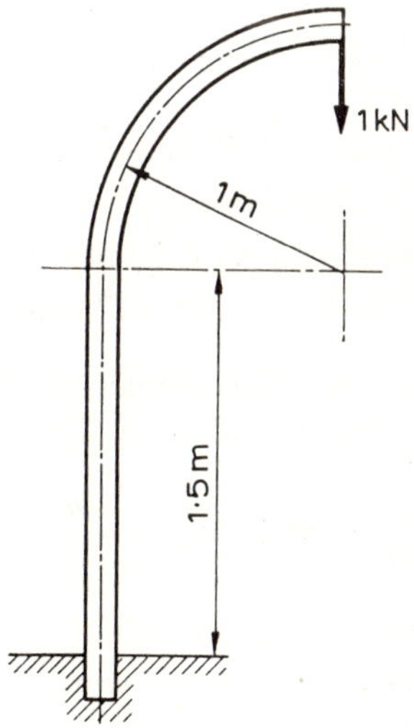

Fig. 13.8

27. A steel bar 60 mm diameter is bent to the shape shown in Fig. 13·8 and the lower end is firmly fixed in the ground in a vertical position. A load of 1 kN is applied at the free end. Calculate the vertical deflection at the free end. $E =$ 200 GN/m^2. (*U.L.*)

Hint. Use bending strain energy as in the previous question.

Answer. 18·0 mm.

Chapter 14

Springs

The *stiffness* of a spring is the load per unit deflection (or torque per unit angle of twist).

Close-coiled Helical Spring

(*a*) Due to an axial load W, the deflection and maximum shear stress are (for circular wire)

$$\delta = \frac{8WND^3}{Gd^4} \quad \text{and} \quad \tau_{max} = \frac{8WD}{\pi d^3}$$

where D = mean coil diameter,
d = wire diameter,
G = modulus of rigidity,
N = number of "free" coils.

(*b*) Due to an axial couple M, the angle of twist (radians) and maximum bending stress are (for all wire sections)

$$\phi = \frac{ML}{EI} \quad \text{and} \quad \sigma_{max} = \frac{M}{Z}$$

where E = Young's modulus,
L = total length of wire,
I = second moment of area of wire section,
Z = section modulus for wire.

Semi-elliptic Leaf (or Laminated) Spring. The deflection and maximum stress due to a central load W are

$$\delta = \frac{3WL^3}{8Enbt^3} \quad \text{and} \quad \sigma_{max} = \frac{3WL}{2nbt^2}$$

where L = length of the spring,
b = breadth of each plate,
t = thickness of each plate,
n = number of plates,
E = Young's modulus.

Quarter-elliptic Leaf (or Laminated) Spring. The deflection and maximum stress due to an end load W are

$$\delta = \frac{6WL^3}{Enbt^3} \qquad \text{and} \qquad \sigma_{max} = \frac{6WL}{nbt^2}$$

the symbols having the same meanings as in the semi-elliptic case.

Flat Spiral Spring. Due to a couple M, the rotation of the spindle and maximum stress are

$$\phi = \frac{ML}{EI} \qquad \text{and} \qquad \sigma_{max} = \frac{2M}{Z}$$

where L = total length of spring
E = Young's modulus
I = second moment of area of the section
Z = modulus of the section.

Table 5 gives the resilience per unit volume for each type of spring. In all cases the average value is given, i.e. total resilience/total

Type of Spring	Resilience per Unit Volume
Bar in tension or compression.	$\frac{\sigma_{max}^2}{2E}$
Simply-supported beam centrally loaded or a cantilever with an end load; rectangular cross-section.	$\frac{\sigma_{max}^2}{18E}$
Solid circular bar in torsion or helical spring of round wire subjected to axial load.	$\frac{\tau_{max}^2}{4G}$
Helical spring of round wire subjected to axial couple.	$\frac{\sigma_{max}^2}{8E}$
Semi- or quarter-elliptic leaf spring.	$\frac{\sigma_{max}^2}{6E}$
Flat spiral spring, rectangular cross-section.	$\frac{\sigma_{max}^2}{24E}$

Table 5

volume. The maximum stress in each case is σ_{max} (direct) or τ_{max} (shear).

WORKED EXAMPLES

14.1. What is a "spring"? Define the terms "proof load", "proof stress", and "proof resilience" as applied to springs.

Calculate the total strain energy of a rectangular bar, 2·5 m long, 6 cm wide, and 1 cm thick when

(*a*) it is subjected to a steady tensile load,
(*b*) it is subjected to a uniform bending moment,
(*c*) it is simply supported at the ends and carries a central point load.

The maximum stress in each case is 200 MN/m² and the bending takes place with the neutral axis parallel to the longer edge of the section. $E = 200 \times 10^9$ N/m².

Solution. A *spring* is a device in which the material is so arranged that it can suffer considerable changes in form without being permanently distorted. The purpose of a spring is to absorb energy and restore it as required. Although straight bars in bending, torsion or tension are used as springs, there are two groups which are especially useful in practice: those consisting of a length of wire wound into a helix or spiral, and those consisting of several approximately flat plates.

The *proof load* is the greatest load which the spring can carry without suffering permanent distortion.

The *proof stress* is the maximum stress occurring in the spring when it is subjected to the proof load.

The *proof resilience* is the strain energy of the spring when it is subjected to the proof load.

(*a*) The total strain energy is, by equation (i), Example 13.1 of Chapter 13,

$$U = \frac{\sigma_{max}^2}{2E} \times \text{volume of the bar}$$

$$= \frac{(200 \times 10^6 \text{ N/m}^2)^2}{2 \times (200 \times 10^9 \text{ N/m}^2)} \times 2{\cdot}5 \text{ m} \times 0{\cdot}06 \text{ m} \times 0{\cdot}01 \text{ m}$$

$$= 150 \text{ N-m (or J)} \qquad (Ans)$$

(*b*) The appropriate second moment of area is

$$I = \frac{bd^3}{12} = \frac{(6\text{ cm}) \times (1\text{ cm})^3}{12} = \tfrac{1}{2}\text{ cm}^4 = (\tfrac{1}{2} \times 10^{-8})\text{ m}^4$$

The bending moment for the given maximum stress is

$$M = \frac{\sigma_{max}}{y} I = \frac{(200 \times 10^6\text{ N/m}^2)}{(\tfrac{1}{2} \times 10^{-2}\text{ m})} \times (\tfrac{1}{2} \times 10^{-8}\text{ m}^4)$$
$$= 200\text{ N-m}$$

From equation (i), Example 13.7 of Chapter 13, the total strain energy (for a uniform bending moment) is

$$U = \frac{M^2 L}{2EI} = \frac{(200\text{ N-m})^2 \times 2{\cdot}5\text{ m}}{2 \times (200 \times 10^9\text{ N/m}^2) \times (\tfrac{1}{2} \times 10^{-8}\text{ m}^4)}$$
$$= 50\text{ N-m (or J)} \qquad (Ans)$$

(*c*) The maximum bending moment is again 200 N-m and since, for a central point load, this equals $WL/4$ the maximum central load is

$$W = \frac{4}{L} \times M = \frac{4}{2{\cdot}5\text{ m}} \times 200\text{ N-m}$$
$$= 320\text{ N}$$

Using formula (ii), Example 13.7 of Chapter 13, the total strain energy is

$$U = \frac{W^2 L^3}{96EI} = \frac{(320\text{ N})^2 \times (2{\cdot}5\text{ m})^3}{96 \times (200 \times 10^9\text{ N/m}^2) \times (0{\cdot}5 \times 10^{-8}\text{ m}^4)}$$
$$= 16\tfrac{2}{3}\text{ N-m (or J)} \qquad (Ans)$$

14.2. Derive an expression for the maximum shear stress in a close-coiled helical spring, mean coil diameter D, wire diameter d, subjected to an axial load W.

A close-coiled helical spring is to carry a load of 100 N and the mean coil diameter is to be eight times the wire diameter. Calculate these diameters if the maximum shear stress is to be 100 MN/m².

Solution. Figure 14.1a shows a helical (or cylindrical spiral) spring in which the coils are regarded as being very flat. At Fig. 14.1b one coil is shown. The load W produces a twisting moment at any cross-section (e.g. at A) of (approximately)

$$T = W \times \text{radius of coils} = \tfrac{1}{2} WD \qquad \text{(i)}$$

Assuming that the torsion equation applies at a section such as A the maximum shear stress is given by

$$\tau_{max} = \frac{T}{J} \times r = \frac{\frac{1}{2}WD \times \frac{1}{2}d}{\frac{1}{32}\pi d^4}$$

$$= \frac{8WD}{\pi d^3} \qquad \text{(ii)}$$

(Since the torsion equation was established for straight bars, equation (ii) is approximate only, the error depending on the ratio D/d.)

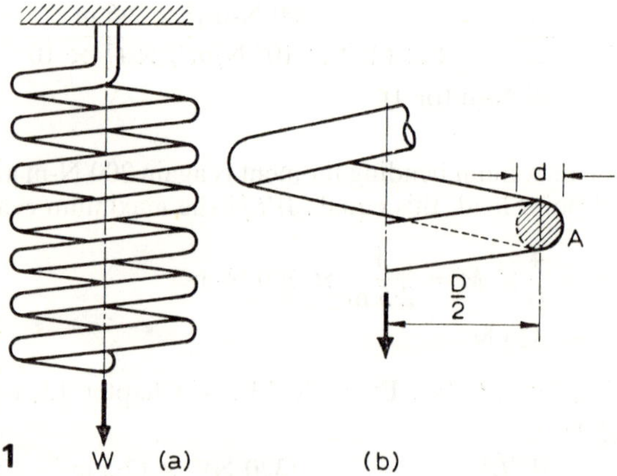

Fig. 14.1

With the values given in the question,

$$\tau_{max} = 100 \text{ MN/m}^2, \; W = 100 \text{ N and } D = 8d$$

Thus, from (ii),

$$100 \times 10^6 = 8 \times 100 \times 8d/\pi d^3$$

$$d^2 = \frac{64}{\pi} \times 10^{-6} \qquad d = 4{\cdot}51 \times 10^{-3} \text{ m}$$

Wire diameter = 4·51 mm *(Ans)*

Mean coil diameter = 8 × 4·51 = 36·1 mm *(Ans)*

14.3. Assuming without proof the formulae for the strength and stiffness of a round bar in torsion, derive an expression for the extension of a close-coiled helical spring in terms of the load W, the mean coil diameter D, the diameter of the wire section d, the number of active coils N, and the modulus of rigidity G.

The following data apply to two close-coiled helical steel springs.

Spring	N	D	d	Axial Length Uncompressed
A	8	100 mm	6 mm	70 mm
B	10	75 mm	5 mm	80 mm

Spring B is placed inside A and both are compressed between a pair of parallel plates until the distance between the plates measures 60 mm. Calculate (*a*) the load applied to the plates, and (*b*) the maximum intensity of shear stress in each spring. $G = 80 \times 10^9$ N/m². (*I.Mech.E.*)

Solution. Referring to equation (i) of Example 14.2, the twisting moment at any section of the wire is $WD/2$.

Assuming the torsion equation $T/J = G\theta/L = \tau/r$, the twist in a small length δL of the wire at this section is

$$\delta\theta = \frac{T\delta L}{GJ} = \frac{WD\delta L}{2GJ} \qquad \text{(i)}$$

Consider the weight to be at the end of a radial arm; the distance it moves, due to this twist, is

$$\delta\theta \times \text{radius of the coil} = \frac{WD\delta L}{2GJ} \times \frac{D}{2}$$

$$= \frac{WD^2\delta L}{4GJ} \qquad \text{(ii)}$$

Since the conditions are the same for each section of the wire, the total deflection of the spring is, from (ii),

$$\delta = \frac{WD^2L}{4GJ} \qquad \text{(iii)}$$

where L = total length of wire in the coil.

If N is the number of active (or "free") coils the effective length of wire is

$$L = N \times \text{circumference of the coils}$$

$$= \pi ND \qquad \text{(iv)}$$

Substituting from (iv) in (iii) and putting $J = \pi d^4/32$, the total deflection is

$$\delta = \frac{WD^2(\pi ND)}{4G(\pi d^4/32)} = \frac{8WND^3}{Gd^4} \tag{v}$$

Rearranging (v), $W = Gd^4\delta/8ND^3$ and since springs A and B are compressed 10 mm and 20 mm respectively, the total load on the parallel plates is (working in newtons and metres)

$$\begin{aligned} W &= \left(\frac{Gd^4\delta}{8ND^3}\right) \text{ for A} + \left(\frac{Gd^4\delta}{8ND^3}\right) \text{ for B} \\ &= \frac{80 \times 10^9}{8}\left[\frac{(0{\cdot}006)^4 \times (0{\cdot}01)}{8 \times (0{\cdot}1)^3} + \frac{(0{\cdot}005)^4 \times (0{\cdot}02)}{10 \times (0{\cdot}075)^3}\right] \\ &= 45{\cdot}9 \text{ N} \end{aligned}$$

(*Ans*)

Substituting the value of W found from (v) above in equation (ii) of Example 14.2, the maximum shear stress for each spring is

$$\tau_{max} = \frac{8D}{\pi d^3}\left(\frac{Gd^4\delta}{8ND^3}\right) = \frac{Gd\delta}{\pi ND^2} \tag{vi}$$

Using (vi) the maximum shear stresses are

For spring A,

$$\begin{aligned} \tau_{max} &= \frac{(80 \times 10^9) \times 0{\cdot}006 \times 0{\cdot}01}{\pi \times 8 \times (0{\cdot}1)^2} \\ &= 19{\cdot}1 \text{ MN/m}^2 \end{aligned}$$

(*Ans*)

For spring B,

$$\begin{aligned} \tau_{max} &= \frac{(80 \times 10^9) \times 0{\cdot}005 \times 0{\cdot}02}{\pi \times 10 \times (0{\cdot}075)^2} \\ &= 45{\cdot}3 \text{ MN/m}^2 \end{aligned}$$

(*Ans*)

14.4. Use strain energy methods to find for a close-coiled helical spring

(*a*) the deflection due to an axial load W, and
(*b*) the angular twist due to an axial couple M. The spring is made of wire diameter d and the mean coil diameter is D.

Find the mass of such a spring which would absorb the energy of a truck of mass 10 tonne and moving at 1 m/s if

(i) the spring is compressed by the impact,
(ii) the spring is "wound up" by the impact.

Working stresses: 350 MN/m^2 (bending) and 280 MN/m^2 (torsion). $E = 200 \times 10^9$ N/m^2 and $G = 80 \times 10^9$ N/m^2. The density of the spring material is 7·83 Mg/m^3.

Solution. (*a*) The twisting moment at every section of the wire is $T = WD/2$ as shown in Example 14.2. Also, the total angle of twist for the whole spring is, from the torsion equation,

$$\theta = \frac{TL}{GJ} = \frac{\pi TND}{G(\pi d^4/32)} = \frac{32TND}{Gd^4} \qquad \text{(i)}$$

since $L = \pi ND$, N being the number of coils.

Using (i), the total strain energy of the spring is

$$U = \tfrac{1}{2}T\theta = \tfrac{1}{2}T\left(\frac{32TND}{Gd^4}\right) = \frac{16T^2ND}{Gd^4} \qquad \text{(ii)}$$

Substituting for T and equating the work done by W to the strain energy

$$\tfrac{1}{2}W\delta = \frac{16(\tfrac{1}{2}WD)^2ND}{Gd^4}$$

and the deflection is

$$\delta = \frac{8WND^3}{Gd^4} \qquad \text{(iii)}$$

(*b*) At every section of the wire there is a bending moment equal to the applied couple M. Assuming that bending theory is applicable, the strain energy is (since M is a constant)

$$U = \frac{M^2L}{2EI} \qquad \text{(iv)}$$

where L is the total length of wire.

Equating the work done by M to the bending strain energy and substituting for L and I,

$$\tfrac{1}{2}M\phi = \frac{M^2L}{2EI}$$

and the angle of twist of one end of the spring relative to the other is

$$\phi = \frac{ML}{EI} = \frac{M(\pi ND)}{E(\pi d^4/64)} = \frac{64MND}{Ed^4} \tag{v}$$

For the truck in question, the kinetic energy to be absorbed is given by

$$KE = \tfrac{1}{2}mv^2 \tag{vi}$$

where m = mass of truck,
v = velocity of truck.

Since $m = 10$ tonne $= 10 \times 10^3$ kg, we have from (vi)

$$\begin{aligned} KE &= \tfrac{1}{2} \times (10 \times 10^3 \text{ kg}) \times (1 \text{ m/s})^2 \\ &= 5000 \text{ kg-m}^2/\text{s}^2 \\ &= 5000 \text{ N-m (or J)} \end{aligned}$$

(i) If the spring absorbs the energy in direct compression then the "wire" of the spring is in torsion. Hence from equation (iv), Example 13.11 of Chapter 13 the strain energy is

$$U = \frac{\tau_{max}^2}{4G} \times \text{volume of the spring}$$

Equating the strain energy to the kinetic energy and substituting for τ_{max} (the maximum shear stress) and G,

$$\frac{\tau_{max}^2}{4G} \times (\text{volume of the spring}) = \text{KE of truck}$$

and volume of the spring is

$$\begin{aligned} &\frac{(\text{KE of truck}) \times 4G}{\tau_{max}^2} \\ &= \frac{5000 \text{ N-m} \times 4 \times (80 \times 10^9 \text{ N/m}^2)}{(280 \times 10^6 \text{ N/m})^2} \\ &= 1/49 \text{ m}^3 \end{aligned}$$

Hence, the mass of the spring is

$$\text{Volume} \times \text{density} = \frac{1}{49} \text{ m}^3 \times 7{\cdot}83 \text{ Mg/m}^3$$

$$= 0{\cdot}160 \times 10^6 \text{ Mg} = 160 \text{ kg} \qquad (Ans)$$

(ii) If σ_{max} is the maximum bending stress in the wire, $M = \sigma_{max}I/y$. Substituting in (iv), the strain energy is

$$U = \frac{(\sigma_{max}\, I/y)^2 L}{2EI} = \frac{\sigma_{max}{}^2 IL}{2y^2 E} = \frac{\sigma_{max}{}^2(\pi d^4/64)L}{2(\frac{1}{2}d)^2 E}$$

$$= \frac{\sigma_{max}{}^2}{8E} \times (\tfrac{1}{4}\pi d^2 L)$$

$$= \frac{\sigma_{max}{}^2}{8E} \times (\text{volume of the spring}) \qquad \text{(vii)}$$

Equating this strain energy to the kinetic energy,

$$\frac{\sigma_{max}{}^2}{8E} \times (\text{volume of the spring}) = \text{KE of truck}$$

and the mass of the spring is

$$\text{Volume} \times \text{density}$$

$$= \left[\frac{\text{KE of truck} \times 8E}{\sigma_{max}{}^2}\right] \times \text{density}$$

$$= \frac{(5000 \text{ N-m}) \times 8 \times (200 \times 10^9 \text{ N/m}^2)}{(350 \times 10^6 \text{ N/m}^2)^2} \times 7{\cdot}83 \text{ Mg/m}^3$$

$$= 511 \text{ kg} \qquad (Ans)$$

14.5. Derive an expression for the maximum bending stress in a semi-elliptic laminated spring consisting of n plates, each breadth b and thickness t, when it is subjected to a central load W. Find also the strain energy stored in the spring in terms of its volume.

How many plates, each 65 mm wide and 6 mm thick, would be required for such a spring 1 m long, if it is to carry a central load of 1·7 kN, the maximum bending stress being 140 MN/m²?

Solution. A semi-elliptic laminated spring is shown diagrammatically in Fig. 14.2. It is also known as a leaf or carriage spring and consists of a number of plates or leaves strapped together. Assuming that each plate is free to slide relative to its neighbours (i.e. neglecting friction between the plates) then, at any section, each plate acts as a separate beam.

Each plate overlaps its neighbour by an amount $L/2n$ at each end. Hence, any section XX, where there are x complete plates, is a distance $xL/2n$ from the end of the spring. The bending moment at XX is, therefore,

$$\frac{W}{2} \times \frac{xL}{2n} = \frac{WLx}{4n} \qquad \text{(i)}$$

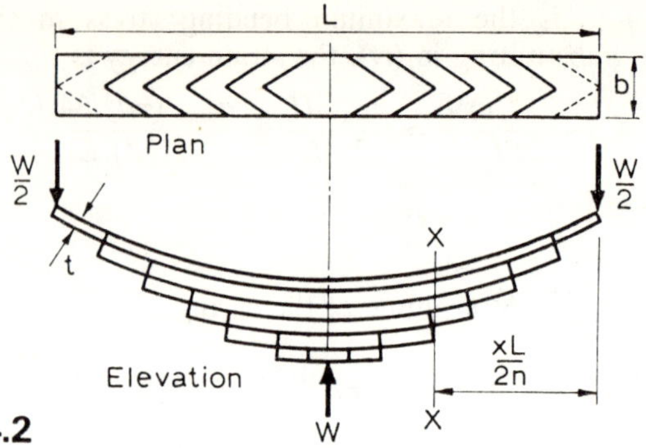

Fig. 14.2

Since all the plates at XX have the same section and are constrained to bend to the same radius of curvature then the bending moment is shared equally between them. The bending moment for each plate at XX is

$$M = \frac{1}{x} \times \frac{WLx}{4n} = \frac{WL}{4n} \tag{ii}$$

Since the bending moment (ii) is independent of x it applies to all sections such as XX. In practice, to make the bending moment more uniform everywhere, the ends of the separate plates (except the longest) are often cut away as shown in Fig. 14.2.

In this case, the maximum stress in each plate (at any section) is, from the bending equation,

$$\sigma_{max} = \frac{M}{I} \times y_{max}$$

The bending moment is given by (ii), $I = bt^3/12$ and $y_{max} = \frac{1}{2}t$. Thus the maximum bending stress is

$$\sigma_{max} = \frac{(WL/4n)}{(bt^3/12)} \times \tfrac{1}{2}t = \frac{3WL}{2nbt^2} \tag{iii}$$

The strain energy stored in a beam carrying a uniform bending moment M is $(M^2/2EI) \times$ (length of the beam). For the given spring, substituting from (ii), the total strain energy is

$$U = \left(\frac{WL}{4n}\right)^2 \times \frac{1}{2EI} \times \text{(total length of all the leaves)}$$

$$= \left(\frac{\sigma_{max}bt^2}{6}\right)^2 \times \frac{1}{2EI} \times \text{(total length of all the leaves)}$$

since. from (iii), $WL/2n = \frac{1}{3}\sigma_{max}bt^2$.

Substituting $I = bt^3/12$, the total strain energy is

$$U = \frac{(\sigma_{max}{}^2)b^2t^4}{36} \times \frac{1}{2E(bt^3/12)} \times \text{(total length of leaves)}$$

$$= \frac{\sigma_{max}{}^2}{6E} \times bt \times \text{(total length of leaves)}$$

$$= \frac{\sigma_{max}{}^2}{6E} \times \text{(volume of the spring)} \qquad \text{(iv)}$$

For the values given in the question, the number of plates required is, from (iii),

$$n = \frac{3WL}{2\sigma_{max}bt^2}$$

$$= \frac{3 \times 1700 \text{ N} \times 1 \text{ m}}{2 \times (140 \times 10^6 \text{ N/m}^2) \times 0{\cdot}065 \text{ m} \times (0{\cdot}006 \text{ m})^2}$$

$$= 7{\cdot}8$$

Taking the next highest whole number, 8 plates are required. (*Ans*)

14.6. A laminated spring, 900 mm long, is made up of plates each 50 mm wide and 6 mm thick. If the bending stress in the plates is limited to 80 MN/m^2, how many plates are required in order that the spring may carry a central point load of 800 N? If $E = 200 \times 10^9$ N/m^2, what is the deflection under this load? Deduce the formula used for deflection.

Solution. The formula will be proved first. Assuming equation (ii), of Example 14.5 (though this should be derived in an examination answer) the bending moment on each plate at each section is $WL/4n$, where W is the central load, L the total length of the spring, and n is the number of plates. Since the plates are held in contact, the central deflection for each is the same. Considering the longest plate, the differential equation of flexure is

$$EI\frac{d^2y}{dx^2} = -M = -\frac{WL}{4n}$$

Integrating once, the slope at any point is given by

$$EI\frac{dy}{dx} = -\frac{WLx}{4n} + A \qquad \text{(i)}$$

If the origin is taken at the left-hand end, $dy/dx = 0$ when $x = \frac{1}{2}L$. Using this condition, we have from (i),

$$0 = -\frac{WL}{4n}\left(\frac{L}{2}\right) + A \quad \text{or} \quad A = \frac{WL^2}{8n}$$

Substituting for A in (i) and integrating again, the deflection at any point is given by

$$EIy = -\frac{WLx^2}{8n} + \frac{WL^2x}{8n} \qquad \text{(ii)}$$

the constant of integration being zero since $y = 0$ when $x = 0$.

The deflection at mid-span is, from (ii), putting $x = \frac{1}{2}L$,

$$\delta = \frac{1}{EI}\left[-\frac{WL}{8n}\left(\frac{L}{2}\right)^2 + \frac{WL^2}{8n}\left(\frac{L}{2}\right)\right]$$

$$= \frac{WL^3}{32nEI} \qquad \text{(iii)}$$

where I is the second moment of area for *one* plate.

If b is the breadth and t the thickness of each plate, $I = bt^3/12$, and, substituting in (iii), the central deflection is

$$\delta = \frac{WL^3}{32nE(bt^3/12)} = \frac{3WL^3}{8nEbt^3} \qquad \text{(iv)}$$

This result can also be obtained by the method of Example 5.1, Chapter 5, since the plate bends to an arc of a circle (M/I being constant).

For the spring given in the question, the number of plates required is, (by equation (iii) of Example 14.5),

$$n = \frac{3WL}{2\sigma_{max}bt^2}$$

$$= \frac{3 \times (800\ \text{N}) \times (0{\cdot}9\ \text{m})}{2 \times (80 \times 10^6\ \text{N/m}^2) \times (0{\cdot}05\ \text{m}) \times (0{\cdot}006\ \text{m})^2}$$

$$= 7{\cdot}5$$

Taking the next highest whole number, 8 plates are required. *(Ans)*

In calculating the deflection, n must be taken as 8.

From (iv), the deflection under the given load is

$$\delta = \frac{3 \times (800\ \text{N}) \times (0{\cdot}9\ \text{m})^3}{8 \times 8 \times (200 \times 10^9\ \text{N/m}^2) \times (0{\cdot}05\ \text{m}) \times (0{\cdot}006\ \text{m})^3}$$

$$= 12{\cdot}66\ \text{mm} \qquad \textit{(Ans)}$$

14.7. A leaf spring of the semi-elliptic type has 10 plates each 75 mm wide and 10 mm thick. The length of the spring is 1·25 m and the plates are of steel having a proof stress (bending) of 600 MN/m². To what radius should the plates be initially bent?

From what height can a mass of 45 kg fall on to the centre of the spring if the maximum stress produced is to be one-half of the proof stress?

$E = 200$ GN/m² and the gravitational acceleration is 9·81 m/s².

Solution. Leaf springs are usually given an initial curvature opposite to that produced by the load. The radius to which the plates are bent is so chosen that the spring just straightens under the proof load. In the present example, applying the bending equation to one plate, the required radius is

$$R = \frac{Ey}{\sigma} = \frac{E}{\sigma_{max}}\left(\frac{t}{2}\right)$$

$$= \frac{(200 \times 10^9 \text{ N/m}^2) \times (0{\cdot}005 \text{ m})}{(600 \times 10^6 \text{ N/m}^2)}$$

$$= 1{\cdot}67 \text{ m} \qquad (Ans)$$

Let W_E be the equivalent static load which would produce the same maximum stress and deflection as the falling mass. In the usual notation,

$$\sigma_{max} = \frac{3W_E L}{2nbt^2}$$

$$W_E = \frac{2nbt^2\sigma_{max}}{3L}$$

$$= \frac{2 \times 10 \times (0{\cdot}075 \text{ m}) \times (0{\cdot}01 \text{ m})^2 \times (300 \times 10^6 \text{ N/m}^2)}{3 \times (1{\cdot}25 \text{ m})}$$

$$= 12 \text{ kN}$$

The corresponding deflection is

$$\delta = \frac{3W_E L^3}{8nEbt^3}$$

$$= \frac{3 \times (12 \times 10^3 \text{ N}) \times (1{\cdot}25 \text{ m})^3}{8 \times 10 \times (200 \times 10^9 \text{ N/m}^2) \times (0{\cdot}075 \text{ m}) \times (0{\cdot}01 \text{ m})^3}$$

$$= 58{\cdot}6 \text{ mm}$$

The weight of the falling mass is

$$W = mg = 45 \text{ kg} \times 9{\cdot}81 \text{ m/s}^2 = 441 \text{ N}$$

Equating the decrease of potential energy of the falling mass to the strain energy of the spring (as in Example 13.9, Chapter 13)

$$W(h + \delta) = \tfrac{1}{2}W_E\delta$$

Hence the height from which the mass may fall is

$$h = \delta\left(\frac{W_E}{2W} - 1\right) = 58{\cdot}6\ \text{mm}\left(\frac{12000}{2 \times 441} - 1\right)$$

$$= 738\ \text{mm} \qquad (Ans)$$

14.8. Obtain expressions for the maximum stress and deflection of a quarter-elliptic laminated spring in terms of the load W, the plate width b and thickness t, and number of plates n.

A spring of this type, 900 mm long, is to be made of steel plates which are available in multiples of 1 mm for thickness and 5 mm for width. The spring is to carry a load of 7·5 kN and the end deflection must not exceed 100 mm. The bending stress must not be greater than 300 MN/m².

Find suitable values for the size and number of plates to be used, taking the width as five times the thickness.

$E = 200\ \text{GN/m}^2$.

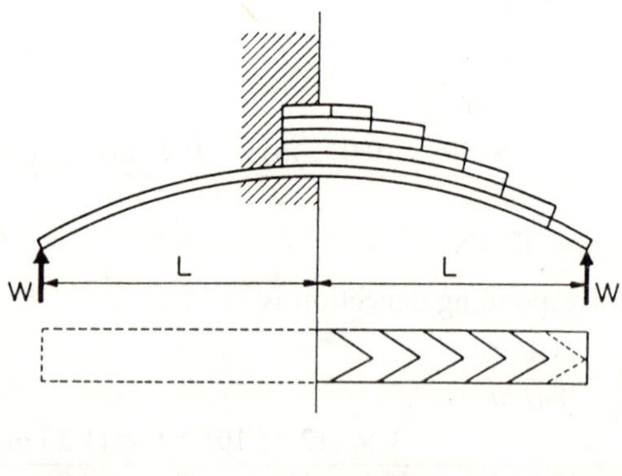

Fig. 14.3

Solution. A quarter-elliptic or cantilever laminated spring, length L, is shown diagrammatically in Fig. 14.3. The load W is carried at the free end and it is clear that the maximum stress and deflection are equal to those of a semi-elliptic spring, length $2L$, carrying a central load $2W$ (and a reaction W at each end).

Hence, replacing W by $2W$ and L by $2L$ in equation (iii) of Example 14.5, the maximum stress for the quarter-elliptic spring is

$$\sigma_{max} = \frac{3(2W)(2L)}{2nbt^2} = \frac{6WL}{nbt^2} \tag{i}$$

Similarly, making the same adjustment to equation (iv) of Example 14.6, the deflection is

$$\delta = \frac{3(2W)(2L)^3}{8nEbt^3} = \frac{6WL^3}{nEbt^3} \tag{ii}$$

(The formulae of Examples 14.5 and 14.6 should not be assumed in an "examination proof".)

For the spring given in the question, we have, dividing (i) by (ii),

$$\frac{\sigma_{max}}{\delta} = \frac{6WL}{nbt^2} \bigg/ \frac{6WL^3}{nEbt^3} = \frac{Et}{L^2}$$

or

$$t = \frac{\sigma_{max}L^2}{\delta E} = \frac{(300 \times 10^6 \text{ N/m}^2) \times (0{\cdot}9 \text{ m})^2}{(0{\cdot}1 \text{ m}) \times (200 \times 10^9 \text{ N/m}^2)}$$
$$= 12{\cdot}15 \text{ mm} \tag{iii}$$

From (iii), the size of plate required is 13 mm thick and 65 mm wide. Using this size, the number of plates required is, from (i),

$$n = \frac{6WL}{\sigma_{max}bt^2} = \frac{6 \times 7500 \text{ N} \times 0{\cdot}9 \text{ m}}{(300 \times 10^6 \text{ N/m}^2) \times 0{\cdot}065 \text{ m} \times (0{\cdot}013 \text{ m})^2}$$
$$= 12{\cdot}25 \tag{iv}$$

Also, from (ii), the number is

$$n = \frac{6WL^3}{\delta Ebt^3}$$
$$= \frac{6 \times 7500 \text{ N} \times (0{\cdot}9 \text{ m})^3}{0{\cdot}1 \text{ m} \times (200 \times 10^9 \text{ N/m}^2) \times 0{\cdot}065 \text{ m} \times (0{\cdot}013 \text{ m})^3}$$
$$= 11{\cdot}5 \tag{v}$$

From results (iv) and (v) the number of plates required to satisfy both conditions is 13. It is possible in certain problems, where the number of plates is small and the multiples of plate sizes are small, that in taking the next highest whole number for n the plate size can be reduced without the maximum deflection and stress being exceeded. In the present case, however, it will be found that 13 plates, each 60 mm × 12 mm, will not satisfy the conditions of the problem. Hence, 13 plates, each 65 mm × 13 mm, are required.

14.9. A flat spiral spring has a total length of wire L. At the centre it is attached to a spindle and the other end is pinned. Obtain an expression for the angle through which the spindle turns when a couple M is applied to it. Find also the strain energy of the spring in terms of its volume and the maximum bending stress when it is made of wire of rectangular cross-section.

Calculate the length of steel strip 25 mm broad and 1 mm thick required for such a spring if one complete turn of the spindle is to produce a maximum bending stress of 150 MN/m². $E = 200$ GN/m².

What is the strain energy of the spring for this condition?

Solution. Figure 14.4 shows a flat spiral spring one end being fastened (pin-jointed) at C and the other being attached to a central spindle.

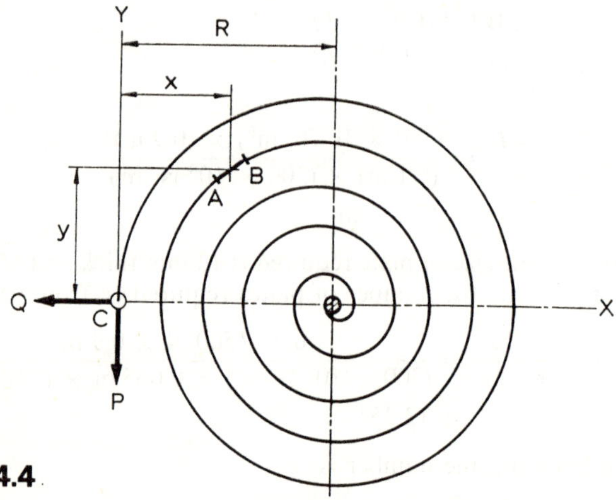

Fig. 14.4

Take coordinate axes through C such that the x-axis passes through the spindle and let Q and P be the components of the reaction at C in the x and y directions respectively due to a couple M applied to the spindle.

Consider a small portion of the spring AB whose length is δL. The bending moment on this portion is $(Px - Qy)$ and, from the relationships established in Chapter 5, the change in slope (δi) over the length AB is,

$$\delta i = \frac{(\text{bending moment}) \times (\text{length})}{EI}$$

$$= \frac{1}{EI} \times (Px - Qy) \times \delta L$$

The total angle turned through, ϕ radians, is the summation of all such elements. Thus

$$\phi = \int \frac{1}{EI}(Px - Qy)dL$$

$$= \frac{1}{EI}(P \int x\, dL - Q \int y\, dL) \qquad \text{(i)}$$

The integrals in (i) are to be taken over the whole length of the spring and are the first moments of this length about the y and x axes respectively. Thus

$$\int x\, dL = \bar{x} \times L \quad \text{and} \quad \int y\, dL = \bar{y} \times L$$

where $\bar{x}$ and $\bar{y}$ are the centroidal distances of the spring's profile from these axes. The centroid of the spring is approximately at the centre of the spindle and, assuming this to be *exactly* so, we have $\bar{y} = 0$ and $\bar{x} = R$ (the distance from C to the spindle). Thus from (i)

$$\phi = \frac{1}{EI} \times PR \times L$$

Taking moments about the spindle, M (the applied couple) $= PR$ and thus the angle turned through is

$$\phi = \frac{ML}{EI} \qquad \text{(ii)}$$

The maximum bending moment occurs at D on the opposite side of the spindle to C (see Fig. 14.4) and is approximately

$$M_{max} = 2R \times P = 2M$$

If the wire is rectangular in section, breadth b and thickness d the maximum stress is

$$\sigma_{max} = \frac{M_{max}}{Z} = \frac{2M}{\frac{1}{6}bd^2} = \frac{12M}{bd^2} \qquad \text{(iii)}$$

From (iii) $M = bd^2\sigma_{max}/12$ and using (ii) the strain energy is

$$U = \text{average couple} \times \text{angle turned through}$$

$$= \tfrac{1}{2}M\phi = \tfrac{1}{2}M\left(\frac{ML}{EI}\right)$$

$$= \frac{M^2L}{2EI} = \tfrac{1}{2}\left(\frac{bd^2\sigma_{max}}{12}\right)^2 \times \frac{L}{E \times bd^3/12}$$

$$= \frac{\sigma_{max}^2}{24E} \times bdL$$

$$= \frac{\sigma_{max}^2}{24E} \times \text{volume of the spring}$$

With the figures given in the question we have, from (iii),

$$M = \frac{bd^2\sigma_{max}}{12}$$
$$= \tfrac{1}{12}[(25 \times 10^{-3}\ \text{m}) \times (1 \times 10^{-3}\ \text{m})^2 \times (150 \times 10^6\ \text{N/m}^2)]$$
$$= 0{\cdot}3125\ \text{N-m}$$

For one complete turn of the spindle $\phi = 2\pi$ radians and, from (ii),

$$L = \frac{\phi EI}{M} = \frac{2\pi \times (200 \times 10^9\ \text{N/m}^2)}{0{\cdot}3125\ \text{N-m}} \times \left(\frac{25 \times 1^3}{12} \times 10^{-12}\ \text{m}^4\right)$$
$$= 8{\cdot}38\ \text{m}$$

The strain energy can be calculated in two ways. Either

$$U = \frac{{\sigma_{max}}^2}{24E} \times \text{volume of the spring}$$
$$= \frac{(150 \times 10^6\ \text{N/m}^2)^2}{24 \times (200 \times 10^9\ \text{N/m}^2)} \times (25 \times 10^{-3}\ \text{m}) \times (1 \times 10^{-3}\ \text{m}) \times (8{\cdot}38\ \text{m})$$
$$= 0{\cdot}982\ \text{N-m (or J)} \qquad (Ans)$$

or

$$U = \tfrac{1}{2}M\phi$$
$$= \tfrac{1}{2} \times 0{\cdot}3125 \times 2\pi$$
$$= 0{\cdot}982\ \text{N-m (or J)} \qquad (Ans)$$

PROBLEMS

1. Assuming, without proof, the formulae for the strength and stiffness of a shaft subject to pure torsion, show that the torsional resilience of a shaft is equal to ${\tau_{max}}^2/4G$ times the volume of the shaft where τ_{max} is the maximum shear stress and G the modulus of rigidity. A spring with a small helix angle and 10 active coils is required to carry a load of 9 kN when compressed axially by 18 mm. Find the mean coil diameter and the diameter of spring section for a maximum shearing stress of 400 MN/m² ($G = 80 \times 10^9$ N/m²). *(I.Mech.E.)*
Answer. Coil diameter, 38·7 mm; wire diameter, 13·04 mm.

2. Derive a formula for the strain energy, in terms of the maximum bending stress, the volume of the spring and E, stored in a closely-coiled helical spring when subjected to an axial twisting moment.

A spring with 8 free coils of mean diameter 60 mm and 6 mm × 6 mm wire section, is used as a flexible coupling for a light direct drive between a motor and a machine. If the bending stress in the spring section is limited to 80 MN/m², find the greatest horse-power which may be transmitted at 1500 rev/min. How much strain energy is stored in the spring under the above conditions? (E = 200 GN/m²). (*I.Mech.E.*)

Answer. $U = \frac{\sigma_{max}^2}{8E} \times$ volume (for round wire) and $\frac{\sigma_{max}^2}{6E} \times$ volume (for rectangular wire); 452 W; 0·290 N-m.

3. The particulars of the close-coiled helical springs, shown in the assembly in Fig. 14.5, are as follows:

Spring	Length uncompressed	Number of free coils	Mean coil diameter	Diameter of section
A	200 mm	12	100 mm	10 mm
B	175 mm	15	75 mm	6 mm

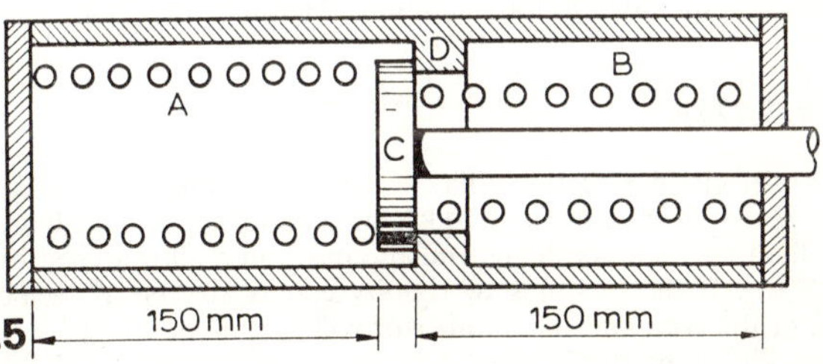

Fig. 14.5

Find the pressure exerted by the disc C on the shoulder D. What force must be applied to the rod to hold the disc 12·5 mm from the shoulder? Also find the shear stresses, induced in the springs, when the disc is in this latter position. (G = 80 × 10⁹ N/m²). (*I.Mech.E.*)
Answer. 348 N; 487 N; τ_{max} (in A) = 132·6 MN/m²; τ_{max} (in B) = 30·2 MN/m².

4. A laminated spring of the semi-elliptic type has a span of 675 mm and is built of leaves 8 mm thick by 45 mm wide. How many leaves would be required to carry a central load of 4·4 kN without the stress in the steel exceeding 220 MN/m²? What will be the deflection at the centre due to this load? Any formulae used should be derived. ($E = 200 \times 10^9$ N/m²). (*I.Mech.E.*)
Answer. 8 leaves; 13·77 mm.

5. In the steel cantilever and spring arrangement, shown in Fig. 14.6, an increasing load W is applied to deflect the cantilever and compress the spring. When W is zero there is a gap x between the top of the spring and the underside of the beam and when W is equal to 16 N the downward movement of the free end of the cantilever is 6 mm ($x < 6$ mm). If the close-coiled helical spring has 16 active coils, a mean coil diameter of 75 mm, and diameter of wire section 6 mm, find the value of x and the load on the cantilever when the gap is just closed. $E = 200 \times 10^9$ N/m²; $G = 80 \times 10^9$ N/m². (*I.Mech.E.*)
Answer. x = 1·57 mm; 1·97 N.

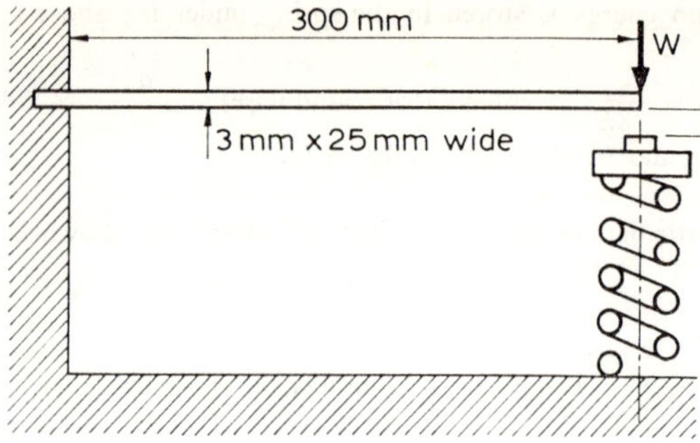

Fig. 14.6

6. Prove that the energy stored per unit volume of a compressed helical spring made of round wire is $\tau^2/4G$, where τ is the maximum shearing stress in the wire and G is the modulus of rigidity.

Determine the mass of such a spring which requires a force of 9 kN to produce a compression of 140 mm, the maximum shearing stress being 450 MN/m^2.

Density of material, 7·83 Mg/m^3.

Modulus of rigidity, 82×10^9 N/m^2. (*U.L.*)

Answer. 7·99 kg.

7. Wire of circular section (dia. = 8 mm) is available to make a close-coiled helical spring of a stiffness of 20 kN/m. The spring will not be required to carry more than 280 N static load, and the maximum allowable stress in the wire is 80 MN/m^2. If the modulus of rigidity equals 80 GN/m^2, determine the mean radius of the coils and the number of coils required. (*U.L.*)

Answer. 28·7 mm; 10·82 coils.

8. A close-coiled helical steel spring is required having a sliding fit over a rod of 28 mm diameter. The spring is to carry a maximum axial load of 120 N and the deflection at this load is to be 20 mm; the shearing stress must not exceed 200 MN/m^2.

Determine the diameter of the wire required, the mean diameter of the coil and the number of turns necessary. Take G = 80 GN/m^2. (*U.L.*)

Answer. 3·15 mm (root of cubic equation); 31·15 mm; 5·43 coils.

9. Make a sketch of a leaf spring showing the shape to which the ends of the plates should be made and give the reason for doing this.

A leaf spring which carries a central load of 9 kN consists of plates each 76 mm wide and 8 mm thick. If the length of the spring is 900 mm, determine the least number of plates required if the maximum stress due to bending is limited to 220 MN/m^2 and the maximum deflection must not exceed 32 mm. Take E = 200 GN/m^2.

Find for the number of plates obtained, the actual values of the maximum stress and maximum deflection and also the radius to which the plates should be formed if they are to straighten under the given load. (*U.L.*)

Answer. 12 plates; 208·2 MN/m^2; 26·3 mm; 3·84 m.

10. A laminated steel spring, simply supported at the ends and centrally loaded, with a span of 750 mm, is required to carry a proof load of 8 kN, and the central deflection is not to exceed 50 mm. The bending stress must not be greater than 400 MN/m². Plates are available in multiples of 1 mm for thickness and in multiples of 4 mm for width.

Determine suitable values for the thickness, width and number of plates and the radius to which the plates should be formed. Assume the width to be ten times the thickness. ($E = 200$ GN/m².) (*U.L.*)

Answer. 11 plates, each 6 mm × 60 mm; initial radius 1·584 m.

11. A close-coiled helical spring is to have a stiffness of 8·75 N per cm of compression, a maximum load of 40 N, and a maximum shearing stress of 125 MN/m². The solid length of the spring (i.e. when the coils are touching) is to be 46 mm. Find the diameter of the wire, the mean radius of the coils and the number of coils required.

Modulus of rigidity = 40 GN/m². (*U.L.*)

Answer. 2·70 mm; 12·13 mm; 17.

12. A close-coiled helical spring of circular section extends unit length when subjected to an axial load W, and there is an angular rotation of 1 radian when a torque T is independently applied, about the axis of the spring. If Poisson's ratio is ν and the mean diameter of the coil is D, show that

$$T/W = \tfrac{1}{4}D^2(1 + \nu)$$

Determine Poisson's ratio for the material of a spring of 75 mm mean diameter of coils, if 240 N extends the spring 130 mm and a torque of 3·4 N-m produces an angular rotation of 60°. (*U.L.*)

Answer. $\nu = 0{\cdot}25$.

13. A composite spring has two close-coiled helical steel springs in series; each spring has a mean coil diameter of 8 times the diameter of its wire. One spring has 20 coils and wire diameter 2·5 mm. Find the diameter of the wire in the other if it has 15 coils and the stiffness of the composite spring is 1·25 kN/m.

Find the greatest axial load that can be applied to the spring and the corresponding extension for a maximum shearing stress of 300 MN/m². ($G = 80 \times 10^9$ N/m².) (*U.L.*)

Answer. $d = 1{\cdot}96$ mm; 56·6 N; 45·3 mm.

14. A carriage spring, centrally loaded and simply supported at the ends, has 10 steel plates, each 50 mm wide by 6 mm thick. If the longest plate is 750 mm long, find the initial radius of curvature of the plates when the greatest bending stress is 150 MN/m² and the plates are finally straight. Neglecting the loss of energy at impact, determine the greatest height from which a mass weighing 200 N may be dropped centrally on the spring without exceeding the limiting bending stress of 150 MN/m². ($E = 200$ GN/m².) (*U.L.*)

Answer. 4 m; 87·9 mm.

15. A quarter-elliptic, i.e. cantilever, leaf spring has a length of 500 mm and consists of plates each 50 mm wide and 6 mm thick. Find the least number of plates which can be used if the deflection under a gradually applied load of 2 kN is not to exceed 70 mm.

If instead of being gradually applied, the load of 2 kN falls a distance of 6 mm on to the undeflected spring, find the maximum deflection and stress produced. ($E = 200$ GN/m².) (*U.L.*)

Answer. 10 plates; 144·6 mm; 694 MN/m².

16. Derive the expression for the strain energy of a carriage spring in terms of the maximum bending stress, the volume of the spring and the elastic modulus E. The spring is supported at the ends and loaded at the centre, and the ends of the plates are pointed in the usual manner.

A carriage spring, as above, is to be constructed using plates 60 mm wide and 6 mm thick and is to carry a suddenly applied central load of 4 kN. The maximum values of stress and deflection produced by this load are not to exceed 420 MN/m^2 and 50 mm respectively. Determine the length of the spring and the number of plates required. $E = 200$ GN/m^2. (*U.L.*)

Answer. $\dfrac{\sigma_{max}^2}{2E} \times$ volume; 756 mm; 10 plates.

17. A flat spiral spring is made of steel 12 mm broad and 0·5 mm thick. The end at the greatest radius is attached to a fixed point and the other end to a spindle. The length of the steel strip is 6 m. Determine:

(*a*) the maximum turning moment which can be applied to the spindle if the stress in the strip is not to exceed 550 MN/m^2;
(*b*) the number of turns required to be given to the spindle;
(*c*) the energy then stored in the spring. ($E = 200$ GN/m^2). (*U.L.*)

Answer. (*a*) 0·1375 N-m. (*b*) 5·25 turns. (*c*) 2·27 N-m (or J).

18. An instrument control spring is made of phosphor-bronze 1 mm wide and 0·1 mm radial thickness. It is to be formed into a flat spiral spring pinned at the outer end and at the inner end to the collet on the instrument arbor.

Calculate the necessary length of spring so that a torque of 450 dyne-cm will cause a rotation of 90 degrees.

E for phosphor-bronze $= 113$ GN/m^2.
(Note that 1 newton $= 10^5$ dynes). (*U.L.*)

Answer. 328·7 mm.

19. A flat spiral spring is formed from a strip of material of rectangular section 12 mm wide and 0·8 mm thick and of total length 3·6 m. It is pinned at its outer end and a winding couple of 0·25 N-m is applied at the central spindle. Find (*a*) the maximum stress and (*b*) the strain energy stored in the material.

Work from first principles. Take $E = 200$ GN/m^2. (*U.L.*)

Answer. (*a*) 390·6 MN/m^2; (*b*) 1·098 N-m.

Chapter 15

Shearing Stresses in Beams

The vertical shear stress in a beam at a height h from the neutral axis is

$$\tau = \frac{F}{Ib_0}\int_h^Y yb\,dy = \frac{F}{Ib_0}A\bar{y}$$

where F = total shear force on the cross-section considered,
I = second moment (or moment of inertia) of the area of the cross-section about the neutral axis,
b_0 = breadth of cross-section at the height h,
b = breadth of cross-section at height y above the neutral axis,
Y = distance from neutral axis to most strained fibre,
A = area of cross-section of beam above height h,
$\bar{y}$ = distance of centroid of the area A from the neutral axis.

For a rectangular cross-section τ (maximum) $= \frac{3}{2}\tau$ (mean)
For a circular cross-section, τ (maximum) $= \frac{4}{3}\tau$ (mean)

The shearing deflection over a short length δx of a beam or cantilever of rectangular section (width b and depth d) is

$$\frac{6F}{5Gbd}\delta x$$

where F is the local shearing force and G is the modulus of rigidity. If L is the total length of such a beam or cantilever and W the total load then the maximum deflections due to shearing strain are,

Cantilever with end point load, $\dfrac{6WL}{5Gbd}$

Cantilever with uniformly distributed load, $\dfrac{3WL}{5Gbd}$

Beam, simply supported at its ends with central point load, $\dfrac{3WL}{10Gbd}$

Beam, simply supported at its ends with uniformly distributed load, $\dfrac{3WL}{20Gbd}$

WORKED EXAMPLES

15.1. Working from first principles, derive a formula for calculating the intensity of shear stress at any point in the cross-section of a beam subjected to a transverse shearing force.

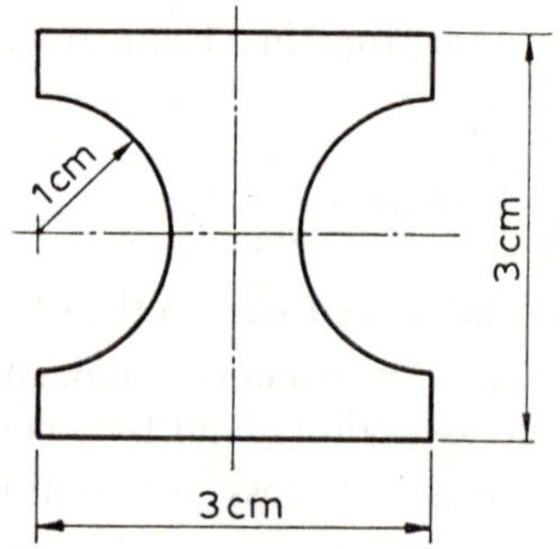

Fig. 15.1

Find the ratio of the maximum shear stress to the mean shear stress for the beam section shown in Fig. 15.1. (The centroid of a semicircle is $4R/3\pi$ from its centre.) *(I.Mech.E.)*

Solution. Consider a length δx of the beam as shown in Fig. 15.2a.

Let M and $M + \delta M$ be the bending moments at the left- and

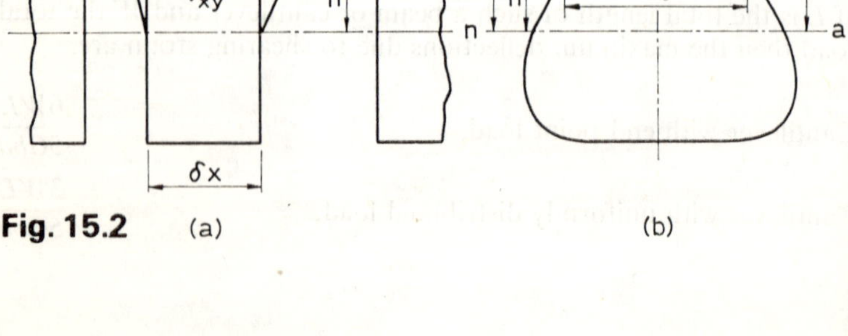

Fig. 15.2 (a) (b)

right-hand ends of this portion respectively. From the bending equation, the longitudinal stress on a small area, distance y from the neutral axis (Fig. 15.2b) at the left-hand end of this portion is

$$\sigma = \frac{My}{I}$$

where I is the second moment of area of the cross-section. Hence the force on a slice width b, thickness δy and length δx pushing from left to right is

$$\sigma b \delta y = \frac{Myb\delta y}{I}$$

Similarly, from the conditions at the right-hand end there is a force pushing from right to left of

$$\frac{(M + \delta M)yb\delta y}{I}$$

Hence there is a resultant force from right to left on the slice of

$$\frac{(M + \delta M)yb\delta y}{I} - \frac{Myb\delta y}{I} = \frac{\delta Myb\delta y}{I} \qquad \text{(i)}$$

Although the sum of all forces, such as (i), for the whole section is zero (there being no resultant longitudinal force) there is a resultant force if only part of the cross-section is considered. Consider the equilibrium of the block ABCD (Fig. 15.2a) whose end section is PQR (Fig. 15.2b).

Due to the bending moments there is a longitudinal force on this block given by the sum of all terms such as (i) for slices above AB. Since there can be no shear stress at a free surface such as CD, then the equilibrium of ABCD can only be maintained by a shear force along the face AB.

Assuming that the shear stress is uniform laterally and longitudinally, let its value be τ (in the direction shown). Then the total shear force acting on the block (from left to right) is

$$\tau \times \text{AB} \times \text{PR} = \tau \delta x b_0 \qquad \text{(ii)}$$

where b_0 is the breadth of the section at AB.

For the equilibrium of ABCD the force given by (ii) must equal the sum of all forces such as (i) for the slices above AB. Hence

$$\tau \delta x b_0 = \sum \frac{\delta Myb\delta y}{I} = \frac{\delta M}{I} \sum yb\delta y \qquad \text{(iii)}$$

δM and I being constants for any one section and the summation being considered over the whole area PQR.

From (iii),

$$\tau = \frac{\delta M}{\delta x} \frac{1}{Ib_0} \sum yb\delta y$$

or, in the calculus notation,

$$\tau = \frac{dM}{dx} \frac{1}{Ib_0} \int_h^Y ybdy$$

$$= \frac{F}{Ib_0} \int_h^Y ybdy \qquad \text{(iv)}$$

since $dM/dx = F$, the total shearing force (equation (ii), Example 2.8 of Chapter 2).

h = height of AB (and PR) above the neutral axis.

Y = distance from neutral axis to CD (or Q).

The integral $\int_h^Y ybdy$ is the first moment of the area PQR about the neutral axis. Hence, if A is the area PQR and $\bar{y}$ is the distance of its centroid from the neutral axis, (iv) can be rewritten as

$$\tau = \frac{F}{Ib_0} A\bar{y} \qquad \text{(v)}$$

This longitudinal shearing stress is accompanied by a vertical (complementary) shearing stress of equal magnitude, and hence equations (iv) and (v) are formulae for the vertical shearing stress at any height h from the neutral axis.

For the section given, the maximum shearing stress will occur at the neutral axis, since b_0 is a minimum there and $A\bar{y}$ has its maximum value there.

I for the section

$$= (I \text{ for a square } 3 \text{ cm} \times 3 \text{ cm}) - (I \text{ for a circle radius } 1 \text{ cm})$$

$$= \frac{3 \text{ cm} \times (3 \text{ cm})^3}{12} - \frac{\pi \times (2 \text{ cm})^4}{64}$$

$$= 5{\cdot}965 \text{ cm}^4$$

Using the centroid position given in the question,

$A\bar{y}$ = (first moment of a rectangle 3 cm × $1\frac{1}{2}$ cm about longer edge) − (first moment of a semicircle, radius 1 cm, about bounding diameter)

$$= (3\text{ cm} \times 1\tfrac{1}{2}\text{ cm} \times \tfrac{3}{4}\text{ cm}) - \left[\tfrac{1}{8}\pi \times (2\text{ cm})^2 \times \frac{4 \times 1\text{ cm}}{3\pi}\right]$$

$$= 2\tfrac{17}{24}\text{ cm}^3$$

Hence, the maximum shear stress is

$$\tau_{max} = \frac{F}{Ib_0}A\bar{y} = \frac{F \times 2\frac{17}{24}\text{ cm}^3}{5{\cdot}965\text{ cm}^4 \times 1\text{ cm}}$$

$$= 0{\cdot}4545F\text{ N/cm}^2 = 4{\cdot}545F\text{ kN/m}^2$$

F being in newtons.

The mean shear stress is

$$\tau_{mean} = \frac{F}{\text{total area}} = F/[(3\text{ cm} \times 3\text{ cm}) - \tfrac{1}{4}\pi \times (2\text{ cm})^2]$$

$$= 0{\cdot}1707F\text{ N/cm}^2 = 1{\cdot}707F\text{ kN/m}^2$$

The required ratio is

$$\frac{\tau_{max}}{\tau_{mean}} = \frac{4{\cdot}545F}{1{\cdot}707F} = 2{\cdot}664 \qquad (Ans)$$

15.2. Figure 15.3 shows the vertical cross-section of a beam which at this section is subjected to a vertical shearing force of 8 kN. Plot to a suitable scale a curve showing how the intensity of shear stress

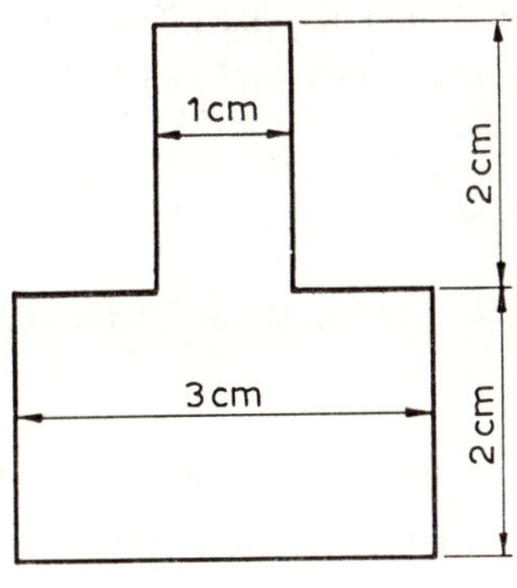

Fig. 15.3

varies throughout the depth of the section. What is the ratio of the maximum shear stress to the mean shear stress? (*I.Mech.E.*)

Solution. The neutral axis is $1\frac{1}{2}$ cm from the bottom edge and the second moment of area of the section is $8\frac{2}{3}$ cm^4.

Considering sections at each $\frac{1}{2}$ cm of the depth as shown in Fig. 15.4, we have from equation (v) of Example 15.1,

TABLE 6

Section	A (cm^2)	$\bar{y}$ (cm)	b_0 (cm)	$\tau = \frac{F}{Ib_0} A\bar{y}$ (MN/m^2)
AA	0	—	—	0
BB	$\frac{1}{2}$	$2\frac{1}{4}$	1	270/26 = 10·39
CC	1	2	1	240/13 = 18·46
DD	$1\frac{1}{2}$	$1\frac{3}{4}$	1	630/26 = 24·23
EE (b_0 = 1 cm)	2	$1\frac{1}{2}$	1	360/13 = 27·69
FF (b_0 = 3 cm)	2	$1\frac{1}{2}$	3	120/13 = 9·23
n.a. (neutral axis)	$4\frac{1}{2}$	$\frac{3}{4}$	3	270/26 = 10·39
GG	3	1	3	120/13 = 9·23
HH	$1\frac{1}{2}$	$1\frac{1}{4}$	3	150/26 = 5·77
JJ	0	—	—	0

The last four cases have been calculated by considering the areas between the sections and the bottom edge.

Two values (at EE and FF) have been calculated for the "junction" between the top and bottom portions since theoretically there is a sudden change in the shear stress intensity at this point. Since there can be no shear stress along the top edge of the bottom portion (except where it joins the top portion) the theory fails to give the correct values at such sections. A similar case occurs with the underside of a flange or the inside faces of a hollow beam. In practice, there is a lateral variation of shear stress.

The values obtained are plotted in Fig. 15.4 alongside the section view.

The mean shearing stress for the section is

$$\tau_{mean} = \frac{F}{\text{total area}} = \frac{8 \text{ kN}}{(3 \text{ cm} \times 2 \text{ cm}) + (1 \text{ cm} \times 2 \text{ cm})}$$

$$= 1 \text{ kN/cm}^2 = 10 \text{ MN/m}^2$$

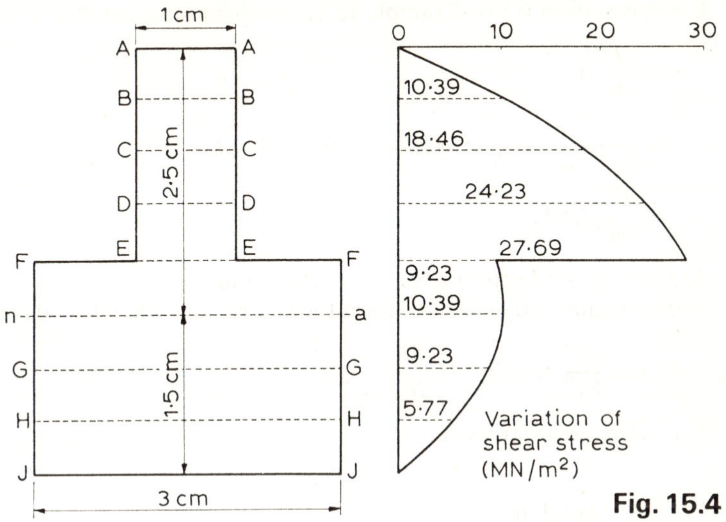

Fig. 15.4

The maximum shear stress clearly occurs at EE and hence the required ratio is

$$\frac{\tau_{max}}{\tau_{mean}} = \frac{27{\cdot}69\ \mathrm{MN/m^2}}{10\ \mathrm{MN/m^2}} = 2{\cdot}769 \qquad (Ans)$$

15.3. Establish formulae for the shear stress at a distance h from the neutral axis at a section of a beam where the total shear force is F, if the cross-section of the beam is

(*a*) a rectangle, width b and depth d,
(*b*) a circle, diameter D.

Find, in each case, the ratio of the maximum to the mean shear stress and the distance from the neutral axis at which the local shear stress equals the mean shear stress.

Solution.

(*a*) Let BB (Fig. 15.5) be the required distance h from the neutral axis. Then the area of the section above BB is

$$A = b(\tfrac{1}{2}d - h)$$

The distance of its centroid from the neutral axis is

$$\bar{y} = \tfrac{1}{2}(\tfrac{1}{2}d + h)$$

Using equation (v) of Example 15.1, the shear stress at XX is

$$\tau = \frac{F}{Ib_0} A\bar{y}$$

$$= \frac{F}{(bd^3/12) \times b} \times b(\tfrac{1}{2}d - h) \times \tfrac{1}{2}(\tfrac{1}{2}d + h)$$

$$= \frac{6F}{bd^3} [(\tfrac{1}{2}d)^2 - h^2] \qquad \text{(i)}$$

At the top and bottom edges $h^2 = (\frac{1}{2}d)^2$ and $\tau = 0$.
The maximum value of τ occurs at the neutral axis when $h = 0$ and

$$\tau_{max} = \frac{6F}{bd^3} \times (\tfrac{1}{2}d)^2 = \frac{3F}{2bd}$$

The mean shear stress is

$$\tau_{mean} = \frac{F}{\text{total area}} = \frac{F}{bd}$$

Hence the required ratio is

$$\frac{\tau_{max}}{\tau_{mean}} = \left(\frac{3F}{2bd}\right) \Big/ \left(\frac{F}{bd}\right) = \frac{3}{2} \qquad (Ans)$$

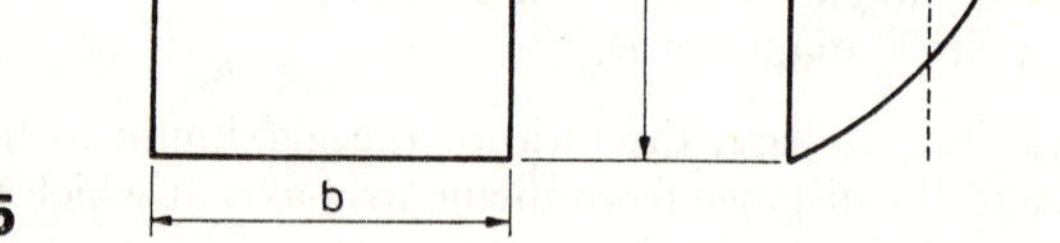

Fig. 15.5

The local shear stress (τ) equals the mean (τ_{mean}) where

$$\frac{6F}{bd^3} [(\tfrac{1}{2}d)^2 - h^2] = \frac{F}{bd}$$

$$\tfrac{1}{4}d^2 - h^2 = \tfrac{1}{6}d^2 \qquad h^2 = d^2/12$$

$$h = d/\sqrt{12} = 0{\cdot}2887d \qquad (Ans)$$

From (i) the variation of τ is parabolic and is shown in Fig. 15.5.

(*b*) Since there are no simple formulae for the area and centroid position of a segment of a circle, equation (iv) of Example 15.1 will be used. Let BB (Fig. 15.6) be the required distance h above the neutral axis. The integration is most easily effected by expressing b and y in terms of the angle θ shown in Fig. 15.6.

$$y = \tfrac{1}{2}D \sin \theta \qquad \text{hence } \frac{dy}{d\theta} = \tfrac{1}{2}D \cos \theta$$

Therefore, dy may be replaced by $\frac{1}{2}D \cos \theta \, d\theta$ if the limits of integration are also written in terms of angles. Let θ_1 be the value of θ when $y = h$. At the top of the section where $y = \frac{1}{2}D$, $\theta = \frac{1}{2}\pi$. Also,

$$b = 2 \times \tfrac{1}{2}D \cos \theta = D \cos \theta \quad \text{and} \quad b_0 = D \cos \theta_1$$

Substituting in equation (iv) of Example 15.1, the shear stress at the required height is

$$\begin{aligned}
\tau &= \frac{F}{Ib_0} \int_h^Y ybdy \\
&= \frac{F}{(\pi D^4/64) D \cos \theta_1} \int_{\theta_1}^{\pi/2} (\tfrac{1}{2}D \sin \theta) \times (D \cos \theta) \tfrac{1}{2}D \cos \theta \, d\theta \\
&= \frac{16F}{\pi D^2 \cos \theta_1} \int_{\theta_1}^{\pi/2} \cos^2 \theta \sin \theta \, d\theta \\
&= \frac{16F}{\pi D^2 \cos \theta_1} \int_{\theta_1}^{\pi/2} (-\cos^2 \theta) \, d(\cos \theta) \\
&= \frac{16F}{\pi D^2 \cos \theta_1} \left[-\frac{\cos^3 \theta}{3} \right]_{\theta_1}^{\pi/2} \\
&= \frac{16F \cos^3 \theta_1}{3\pi D^2 \cos \theta_1} \\
&= \frac{16F}{3\pi D^2} (1 - \sin^2 \theta_1) \\
&= \frac{16F}{3\pi D^2} \left[1 - \left(\frac{2h}{D} \right)^2 \right] \qquad \text{(ii)}
\end{aligned}$$

(since $h = \frac{1}{2}D \sin \theta_1$).

The graph given by (ii) is a parabola as shown in Fig. 15.6. The maximum value of τ occurs at the neutral axis where $h = 0$, and is

$$\tau_{max} = \frac{16F}{3\pi D^2}$$

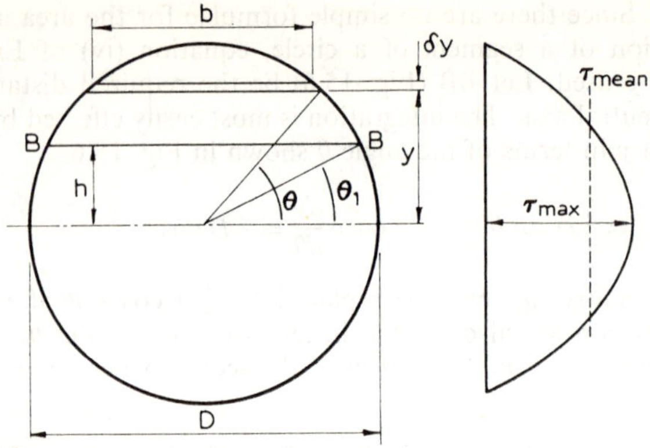

Fig. 15.6

The shear stress is

$$\tau_{mean} = \frac{F}{\text{total area}} = \frac{4F}{\pi D^2}$$

Hence the required ratio is

$$\frac{\tau_{max}}{\tau_{mean}} = \left(\frac{16F}{3\pi D^2}\right) \bigg/ \left(\frac{4F}{\pi D^2}\right) = \frac{4}{3} \qquad (Ans)$$

The local shear stress (τ) equals the mean (τ mean) where

$$\frac{16F}{3\pi D^2}\left[1 - \left(\frac{2h}{D}\right)^2\right] = \frac{4F}{\pi D^2}$$

$$1 - \left(\frac{2h}{D}\right)^2 = \frac{3}{4}$$

$$\left(\frac{2h}{D}\right) = \frac{1}{2}$$

$$h = \tfrac{1}{4}D \qquad (Ans)$$

15.4. A beam of I-section, 20 cm deep and 8 cm wide, has flanges 1 cm thick and web $\frac{1}{2}$ cm thick. It carries at one section a shearing force of 30 kN. Calculate the shear stress in the flange at $\frac{1}{2}$ cm and 1 cm from the outside edges.

Find an expression for the shear stress in the web at any distance h cm from the neutral axis, and draw a diagram showing the variation of shear stress across the section.

What percentage of the total shear force is carried by the web? What is the intensity of shear stress in the web, assuming that the web carries the whole shearing force uniformly distributed?

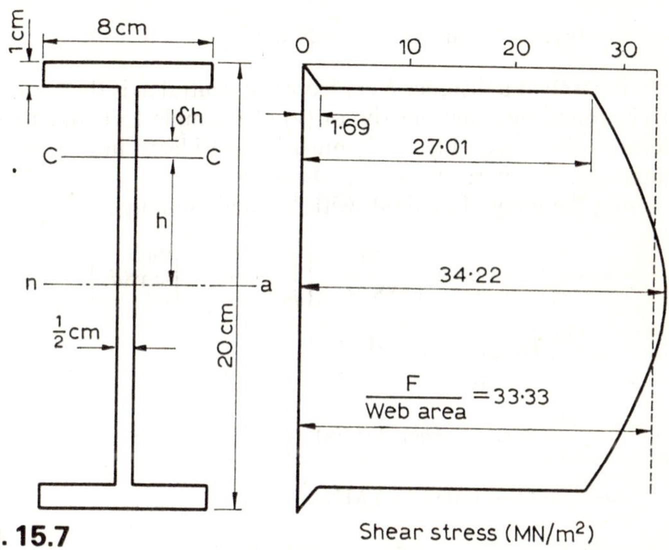

Fig. 15.7

Solution. The reader should verify that the second moment of area about the neutral axis is $I = 1688 \text{ cm}^4$.

The section is shown in Fig. 15.7 and the shear stress at any point in the flange can be found from equation (v) of Example 15.1.

At ½ cm from outside edge,

$$A = \tfrac{1}{2} \text{ cm} \times 8 \text{ cm} = 4 \text{ cm}^2$$
$$\bar{y} = 10 \text{ cm} - \tfrac{1}{4} \text{ cm} = 9\tfrac{3}{4} \text{ cm}$$
$$b_0 = 8 \text{ cm}$$

and hence

$$\tau = \frac{F}{Ib_0} A\bar{y} = \frac{30 \text{ kN} \times 4 \text{ cm}^2 \times 9\tfrac{3}{4} \text{ cm}}{1688 \text{ cm}^4 \times 8 \text{ cm}}$$
$$= 0{\cdot}0866 \text{ kN/cm}^2 = 0{\cdot}866 \text{ MN/m}^2 \qquad (Ans)$$

At 1 cm from outside edge

$$A = 1 \text{ cm} \times 8 \text{ cm} = 8 \text{ cm}^2$$
$$\bar{y} = 10 \text{ cm} - \tfrac{1}{2} \text{ cm} = 9\tfrac{1}{2} \text{ cm}$$
$$b_0 = 8 \text{ cm}$$

Hence

$$\tau = \frac{F}{Ib_0} A\bar{y} = \frac{30 \text{ kN} \times 8 \text{ cm}^2 \times 9\tfrac{1}{2} \text{ cm}}{1688 \text{ cm}^4 \times 8 \text{ cm}}$$

$$= 0{\cdot}169 \text{ kN/cm}^2 = 1{\cdot}69 \text{ MN/m}^2$$

In the web at a distance h cm from the neutral axis the shear stress can be found by equations (iv) or (v) of Example 15.1. In either case the area above CC (Fig. 15.7) must be considered in two parts, the flange and that part of the web above CC.

Using the integral method, with kN and cm units,

$$\tau = \frac{F}{Ib_0}\int_h^Y ybdy = \frac{30}{1688 \times \frac{1}{2}}\left\{\int_h^9 y\tfrac{1}{2}dy + \int_9^{10} y8dy\right\}$$

$$= \frac{30}{844}\left\{\left[\frac{y^2}{4}\right]_h^9 + \left[4y^2\right]_9^{10}\right\}$$

$$= \frac{30}{844}\left[(20\tfrac{1}{4} - \tfrac{1}{4}h^2) + (400 - 324)\right]$$

$$= (34{\cdot}22 - 0{\cdot}0889h^2) \text{ MN/m}^2$$

Using the first moment of area method,

$$A\bar{y}\ (\text{total}) = A\bar{y}\ (\text{flange}) + A\bar{y}\ (\text{that part of the web above CC})$$

$$= [8 \times 1 \times 9\tfrac{1}{2}] + [\tfrac{1}{2}(9 - h) \times \tfrac{1}{2}(9 + h)]$$

$$= 76 + \tfrac{1}{4}(81 - h^2) = 96\tfrac{1}{4} - \tfrac{1}{4}h^2$$

$$\text{Hence } \tau = \frac{F}{Ib_0} A\bar{y} = \frac{30}{1688 \times \frac{1}{2}}(96\tfrac{1}{4} - \tfrac{1}{4}h^2) \text{ kN/cm}^2$$

$$= (34{\cdot}22 - 0{\cdot}0889h^2) \text{ MN/m}^2$$

Using this equation for τ the following values are obtained,

h cm	0	± 3	± 6	± 9
τ MN/m^2	34·22	33·42	31·02	27·01

A diagram showing the variation of shear stress across the section is given in Fig. 15.7.

The shear *force* carried by an element of the web, thickness δh (at CC) as shown in Fig. 15.7, is $\tau \times \delta h \times b$. Using the expression for τ obtained above and working again in kN and cm units, the total shear force carried by the web is

$$\int_{-9}^{+9} \tau b dh = 2\int_{0}^{9} (3{\cdot}422 - 0{\cdot}008\,89h^2)\tfrac{1}{2} dh$$
$$= \left[3{\cdot}422h - 0{\cdot}008\,89\left(\frac{h^3}{3}\right)\right]_0^9$$
$$= 28{\cdot}64 \text{ kN}$$

Hence, the percentage of the total shear force carried by the web is

$$\frac{28{\cdot}64}{30} \times 100 \quad \text{i.e. } 95{\cdot}5 \text{ per cent}$$

If the total shear force were uniformly distributed over the web the shear stress would be

$$\tau = \frac{30 \text{ kN}}{18 \text{ cm} \times \frac{1}{2} \text{ cm}} = 3{\cdot}333 \text{ kN/cm}^2 = 33{\cdot}33 \text{ MN/m}^2 \qquad (\textit{Ans})$$

(This result is a good approximation to the shear stress at all points in the web and, in practice, it is usual to assume that the web carries the total shear force uniformly distributed.)

15.5. A horizontal beam of I section 10 cm deep has flanges 6 cm by 0·7 cm and a web 0·36 cm thick. The transverse loading is such that at a certain section there is a bending moment of 3 kN-m together with a vertical shearing force. If the maximum principal stress is limited to 80 MN/m² what is the value of the shearing force? (*I.Mech.E.*)

Solution. The reader should verify that the second moment of area of the section is $I = 201{\cdot}1 \text{ cm}^4$. The bending stress varies (linearly) from zero at the neutral axis to a maximum at the outside edge of the flanges. The shear stress variation is similar to that shown in Fig. 15.7. The shear stress is small in the flange where the bending stress is greatest and is a maximum at the neutral axis where the bending stress is zero. The maximum principal stress occurs in the web at the point where it joins the flange, since the bending stress there is nearly equal to that at the outside edge and the shear stress is little short of the value at the neutral axis.

If F kN is the vertical shearing force on the section the shear stress in the web at this point (4·3 cm from the neutral axis) is

$$\tau = \frac{F}{Ib_0} A\bar{y} = \frac{F\,\text{kN}}{201{\cdot}1\ \text{cm}^4 \times 0{\cdot}36\ \text{cm}} \times$$

$$[6\ \text{cm} \times 0{\cdot}7\ \text{cm} \times (5 - 0{\cdot}36)\ \text{cm}]$$

$$= 0{\cdot}2698F\ \text{kN/cm}^2 = 2{\cdot}698F\ \text{MN/m}^2$$

The bending stress at the same point is

$$\sigma = \frac{My}{I} = \frac{3\ \text{kN-m} \times 4{\cdot}3\ \text{cm}}{201{\cdot}1\ \text{cm}^4} = 6{\cdot}415\ \text{kN/cm}^2$$

$$= 64{\cdot}15\ \text{MN/m}^2$$

With the usual relationship for principal stresses

$$(\sigma - \sigma_x)(\sigma - \sigma_y) = \tau_{xy}{}^2$$

In the present case,

the principal stress $\sigma = 120\ \text{MN/m}^2$
the bending stress $\sigma_x = 64{\cdot}15\ \text{MN/m}^2$,
$\sigma_y = 0$ and $\tau_{xy} = (2{\cdot}698F)\ \text{MN/m}^2$. Hence

$$(80 - 64{\cdot}15)(80 - 0) = (2{\cdot}698F)^2$$

from which

$$F^2 = 174{\cdot}2 \qquad F = 13{\cdot}2$$

The shearing force is therefore 13·2 kN. (*Ans*)

15.6. The beam of Example 15.4 is simply supported over a span of 6 m. Find, for a central concentrated load of 50 kN the maximum deflection

(*a*) due to shearing, assuming the web carries the whole shearing force uniformly distributed
(*b*) due to bending.

Take $E = 200\ \text{GN/m}^2$ and $G = 80\ \text{GN/m}^2$.

Solution. (*a*) Figure 15.8 shows the type of distortion due to shear assuming uniform shear stress in the web,
Shear stress in web is

$$\tau = \frac{\text{shear force}}{\text{web area}} = \frac{25\ \text{kN}}{18\ \text{cm} \times \frac{1}{2}\ \text{cm}} = 27{\cdot}78\ \text{MN/m}^2$$

and the corresponding shear strain is

$$\gamma = \frac{\tau}{G} = \frac{27{\cdot}78 \text{ MN/m}^2}{80 \text{ GN/m}^2} = 0{\cdot}000\,3473$$

Hence, the mid-span deflection due to shear is

$$\gamma \times 3 \text{ m} = 0{\cdot}000\,3473 \times 3$$
$$= 0{\cdot}001\,04 \text{ m} \quad \text{or} \quad 1{\cdot}04 \text{ mm}$$

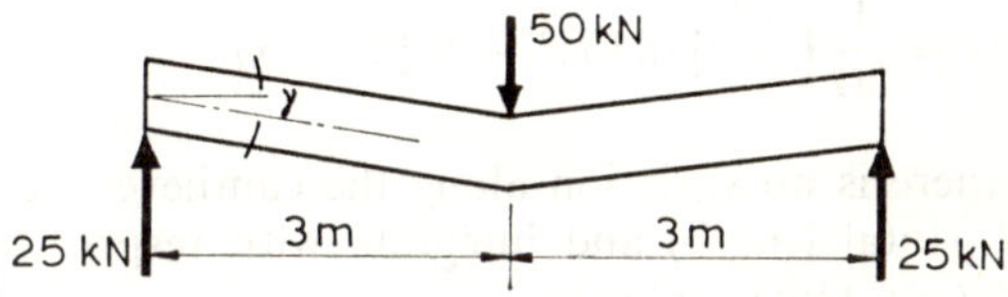

Fig. 15.8

(*b*) With the usual formula, the deflection due to bending is

$$\delta = \frac{WL^3}{48EI} = \frac{50 \text{ kN} \times (6 \text{ m})^3}{48 \times 200 \text{ GN/m}^2 \times 1\,688 \text{ cm}^4}$$
$$= 0{\cdot}066\,74 \text{ m} \quad \text{or} \quad 66{\cdot}74 \text{ mm} \qquad (\textit{Ans})$$

(In most practical beams and cantilevers the shear deflection is small compared with that due to bending.)

15.7. Obtain expressions for the deflection at the free end of a cantilever, of rectangular cross-section, due to shearing strain when it carries

(*a*) a concentrated load at the free end
(*b*) a uniformly distributed load

Calculate the ratio (shear deflection)/(bending deflection) in each case if $E = 2{\cdot}5G$ and the length of the cantilever is 8 times the depth.

Solution. The required results can be derived by strain energy methods. Referring to Fig. 15.2, consider a small slice length δx and width b_0 (represented by AB and PR) and of thickness δh. Its volume is $b_0 \delta x \delta h$ and if τ is the shear stress then the corresponding shear strain energy δU is given by

$$\delta U = \frac{\tau^2}{2G} \times \text{volume} = \frac{\tau^2}{2G}\, b_0 \delta x \delta h \qquad \text{(i)}$$

and, in general, τ varies with both x and h.

(*a*) If the cantilever carries a concentrated load W at the free end then the shearing force F is constant along its length and $F = W$. Thus, from equation (i), of Example 15.3 we have, for a rectangular cross-section,

$$\tau = \frac{6W}{bd^3}[(\tfrac{1}{2}d)^2 - h^2]$$

Substituting this result in (i) above, gives

$$\delta U = \frac{1}{2G}\left(\frac{6W}{bd^3}\right)^2 [(\tfrac{1}{2}d)^2 - h^2]^2 b_0 \delta x \delta h$$

Since there is no variation along the cantilever we may replace δx with L (total length) and integrate with respect to h. Thus since $b_0 = b$ (constant) we have

$$U = \frac{L}{2G}\int_{-d/2}^{+d/2}\left(\frac{6W}{bd^3}\right)^2 [(\tfrac{1}{2}d)^2 - h^2]^2 b\,dh$$

$$= \frac{18LW^2}{Gbd^6}\left[\frac{d^4h}{16} - \frac{d^2h^3}{6} + \frac{h^5}{5}\right]_{-d/2}^{+d/2}$$

$$= \frac{3W^2L}{5Gbd}$$

If δ is the deflection at the free end due to shearing strain then the work done by the load W in producing this deflection is $\tfrac{1}{2}W\delta$. Equating this to the total internal shear strain energy gives

$$\tfrac{1}{2}W\delta = \frac{3W^2L}{5Gbd} \quad \text{or} \quad \delta = \frac{6WL}{5Gbd} \qquad \text{(ii)} \quad (Ans)$$

The deflection due to bending is $WL^3/3EI$ and with $E = 2{\cdot}5G$ and $L = 8d$ (as given in the question) the required ratio is

$$\frac{\delta\,(\text{shearing})}{\delta\,(\text{bending})} = \left(\frac{6WL}{5Gbd}\right)\bigg/\left(\frac{WL^3}{3EI}\right) = \frac{18EI}{5GbdL^2}$$

$$= \frac{18E}{5GbdL^2}\left(\frac{bd^3}{12}\right) = \frac{3E}{10G}\left(\frac{d}{L}\right)^2$$

$$= \frac{3}{10} \times 2{\cdot}5 \times \left(\frac{1}{8}\right)^2 = \frac{3}{256} \qquad (Ans)$$

(*b*) In this case there is a variation of shearing force along the cantilever. Consider a small portion, length δx, distance x from the fixed end as shown in Fig. 15.9. This element can be considered as a cantilever with an end load equal to the local shearing force F. The

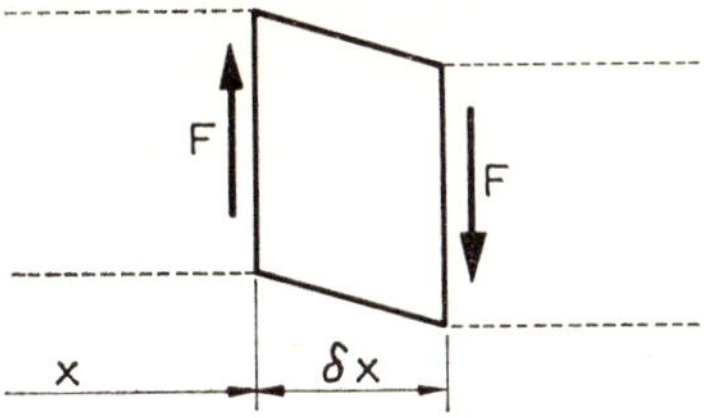

Fig. 15.9

shearing deflection of this element can be obtained from expression (ii) above replacing W by F and L by δx. The deflection of the element is therefore $(6F/5Gbd)\delta x$ and, for the whole cantilever, the deflection

$$\delta = \int_0^L \frac{6F}{5Gbd} dx \qquad \text{(iii)}$$

For a uniformly distributed load w per unit length the shearing force at distance x from the fixed end is $F = w(L - x)$ and (iii) becomes

$$\delta = \frac{6}{5Gbd} \int_0^L w(L - x)dx$$

$$= \frac{6w}{5Gbd}\left[Lx - \tfrac{1}{2}x^2\right]_0^L$$

$$= \frac{3wL^2}{5Gbd} = \frac{3WL}{5Gbd} \qquad (Ans)$$

where $W = wL$, the total load.

On comparison with the bending deflection $(WL^3/8EI)$ we have

$$\frac{\delta\text{ (shearing)}}{\delta\text{ (bending)}} = \left(\frac{3WL}{5Gbd}\right) \Big/ \left(\frac{WL^3}{8EI}\right)$$

$$= \frac{24EI}{5GbdL^2} = \frac{2E}{5G}\left(\frac{d}{L}\right)^2$$

If $E = 2{\cdot}5G$ and $L = 8d$, this gives a ratio of 1/64. (*Ans*)

(As in the previous section the shearing deflection is small compared with that due to bending.)

15.8. Figure 15.10 shows the section of a horizontal cantilever which carries a vertical end load W. Determine the position of W relative to the edge AC in order that the cantilever shall not be subjected to torsion. Work from first principles. (*U.L.*)

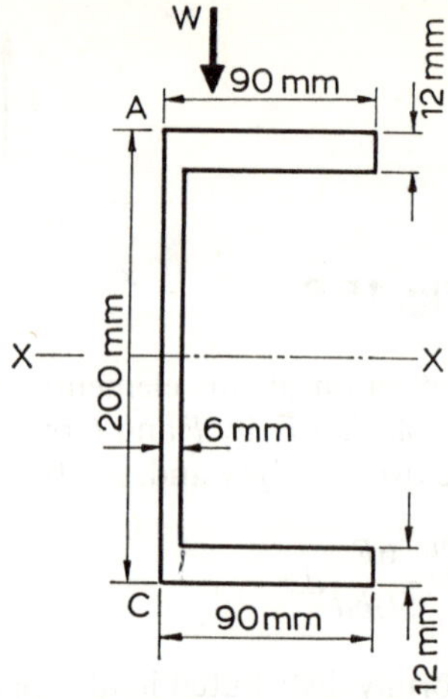

Fig. 15.10

Solution. In a beam or cantilever of open section such as a T, I or channel section it is found (see Example 15.4) that the vertical web carries nearly all the shearing force and the vertical shearing stresses in the flanges are small. However, horizontal shearing stresses occur in the flanges and torsion may occur.

Suppose, in Fig. 15.11a, PQRS is a small element of the flange of a beam or cantilever, length δx and end area A. Let M and $M + \delta M$ be the bending moments at the sections SR and PQ respectively. Then if $\bar{y}$ is the distance of the centroid of the area A from XX, there is a net longitudinal force on the element due to the bending moments of

$$\frac{\delta M \bar{y}}{I} \times A \tag{i}$$

As shown in Fig. 15.11b this can only be resisted by a shearing stress τ acting along the face PS (since there can be no shearing stress on a free surface such as QR). If t is the flange thickness at P this force is

$$\tau \times \text{area} = \tau \times t\delta x \tag{ii}$$

Equating (i) and (ii) we obtain

$$\tau = \frac{A\bar{y}}{It} \times \frac{\delta M}{\delta x}$$

In the limiting case $dM/dx = F$, the shearing force, and

$$\tau = \frac{FA\bar{y}}{It} \tag{iii}$$

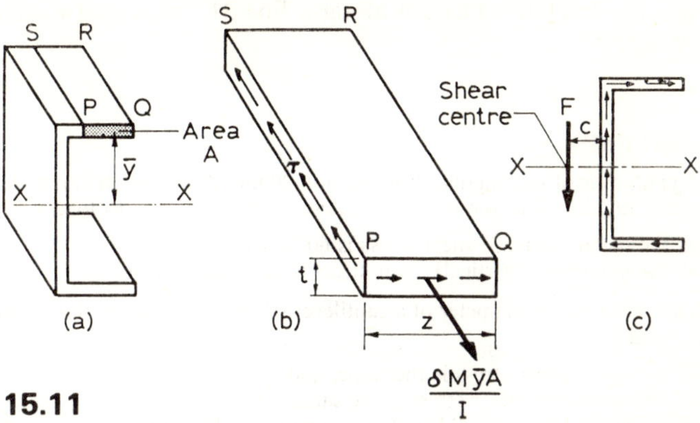

Fig. 15.11

Since complementary shear stresses are equal there is a horizontal shear stress τ in the flange at P and the general direction of the shear stress on the section is everywhere parallel to the boundary as shown in Fig. 15.11c.

In a channel section the resulting forces in the flanges form a couple and torsion will occur unless an equal and opposite couple is produced by the external force and the shear force in the web. The line along which the external force must act to prevent torsion is independent of the magnitude of the force and its point of intersection with XX is known as the *shear centre*.

For the section shown in Fig. 15.10 and working in mm units, $I_{xx} = 21{\cdot}8 \times 10^6$ mm⁴. At a point in the flange such as P, distance z from the edge Q, the area $A = tz$, $\bar{y} = 94$ mm and, from (iii),

$$\tau = \frac{F \times tx \times 94}{21{\cdot}8 \times 10^6 t} = 4{\cdot}31 \times 10^{-6} Fz$$

By integration the total force in one flange is

$$\int_0^{84} \tau t \, dz = \int_0^{84} 4{\cdot}31 \times 10^{-6} Fz \times 12 \, dz$$

$$= 51{\cdot}72 \times 10^{-6} F \left[\frac{z^2}{2}\right]_0^{84}$$

$$= 0{\cdot}1825F$$

There is an equal and opposite force in the lower flange and these two forces are 188 mm apart. If the shear centre is distance c from the centre-line of the web, see Fig. 15.11c, then, equating the moments of the couples,

$$Fc = 0{\cdot}1825F \times 188 \qquad c = 34{\cdot}3 \text{ mm}$$

Hence the load W must act along a line 31·3 mm (outwards) from the edge AC. (*Ans*)

PROBLEMS

1. A cantilever of rectangular cross-section 90 mm deep and 30 mm wide carries a concentrated load at the free end. Calculate the value of this load if

(*a*) the mean (vertical) shear stress is 60 MN/m², and
(*b*) the maximum (vertical) shear stress is to be 60 MN/m².

What must be the diameter of a cantilever of circular cross-section to carry the same load if

(*c*) the mean shear stress is the same, and
(*d*) the maximum shear stress is the same?

Answer. (*a*) 162 kN; (*b*) 108 kN; (*c*) 58·6 mm; (*d*) 55·3 mm.

2. Show that, according to the usual theory, the ratio of the maximum shear stress to the mean shear stress on a section of a hollow circular beam is $4(D^2 + Dd + d^2)/3(D^2 + d^2)$, where D = external diameter and d = internal diameter.

Hence show that for all values of D/d, the ratio lies between $\frac{4}{3}$ and 2. For what value of D/d is the ratio $\frac{5}{3}$?

Answer. 3·732.

3. A bar of square section is used as a beam so that the plane of bending is parallel to a diagonal. The side of the square is 40 mm and the shearing force at one section is 16 kN.

Calculate

(*a*) the mean shear stress,
(*b*) the shear stress at the neutral axis,
(*c*) the magnitude and position of the maximum shear stress.

Draw a diagram showing the variation of shear stress across the section.

Answer. (*a*) 10 MN/m²; (*b*) 10 MN/m²; (*c*) 11·25 MN/m² at 7·071 mm from neutral axis. The variation is shown in Fig. 15.12.

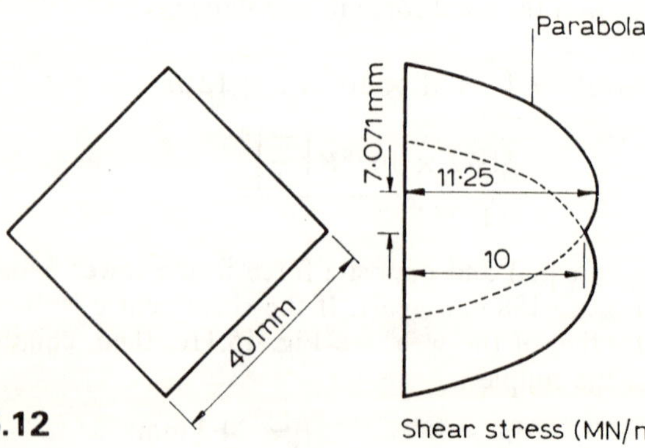

Fig. 15.12

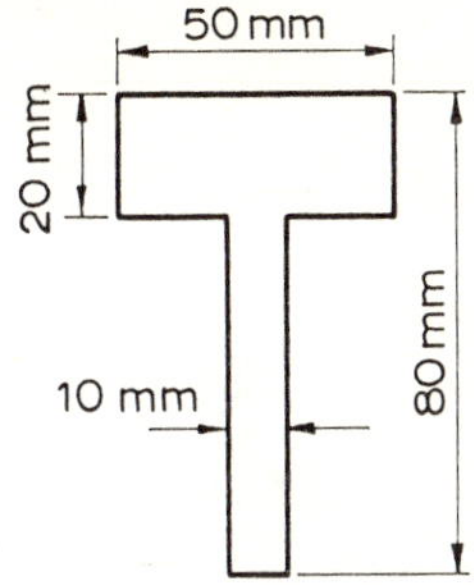

Fig. 15.13

4. A beam has the section shown in Fig. 15.13 and is used with the cross-piece horizontal and uppermost. It carries at one section a total (vertical) shear force of 15 kN. Draw to scale a diagram showing the distribution of shear stress across this section. Prove the formula giving the shear stress at any section.
Answer. At h mm from the top edge the shear stress τ MN/m² is given by

h	0	10	20 (flange)	20 (web)	25	30	40	60	80
τ	0	3·69	5·53	27·67	27·89	27·67	25·81	16·60	0

5. A beam of I section, 6 m long, has an overall depth of 254 mm, the flanges are 152 mm wide and 25·4 mm thick and the web thickness is 12·7 mm ($I = 110{\cdot}4 \times 10^6$ mm^4). If the beam carries a uniformly distributed load of 400 kN total and is simply supported at its ends, calculate the maximum (vertical) shear stresses in the web and flanges at a section 1·2 m from one end. Find also the percentage of the total shear force at any section carried by the web.
Answer. Maximum shear stress in web, 43·38 MN/m²; in flange, 3·156 MN/m². Web carries 89·23 per cent.

6. The cantilever bracket, shown in Fig. 15.14 is pulled with force P, equal to 14·14 kN, inclined at 45° to the horizontal. Find, for the point A,

(*a*) the intensity of shear stress,
(*b*) the intensity of normal stress on the cross-section, and
(*c*) the principal stresses. (*I.Mech.E.*)

Hint. Resolve P into its vertical and horizontal components. The vertical component gives a bending moment and shearing force. The horizontal component is an "eccentric load".
Answer. (*a*) 16·41 MN/m²; (*b*) 221·9 MN/m²; (*c*) 223·2 MN/m² (compressive) and 1·20 MN/m² (tensile).

7. If the section (b m × d m) of a rectangular beam is subjected to a shearing force of V N, show that the shear stress at s metres above the neutral axis is

$$\frac{6V}{bd^3}\left\{\left(\frac{d}{2}\right)^2 - s^2\right\} \text{ N/m}^2$$

Determine the principal stresses (at 0·013 m down from the upper face) at the built-in end of a cantilever of length 0·3 m and cross-section $b = 0{\cdot}02$ m, $d =$

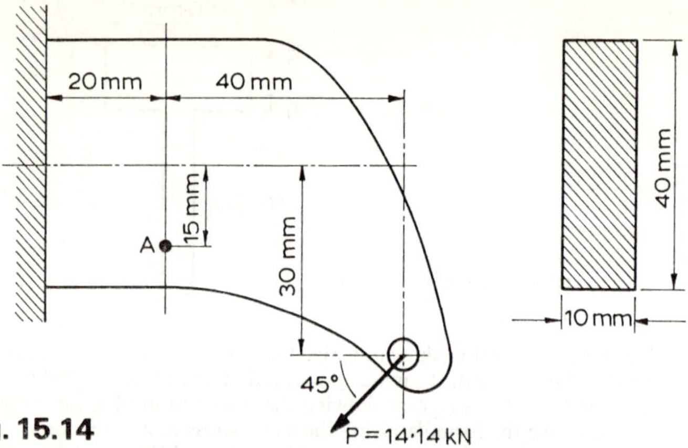

Fig. 15.14

0·04 m. The cantilever carries 500 N at its free end and 250 N at 0·15 m from the free end. (*U.L.*)
Answer. 12·42 MN/m² (tensile); 0·12 MN/m² (compressive).

8. A bar of hexagonal cross-section and of side length 10 mm is used as a cantilever, one diagonal being horizontal. A load hung from the bar subjects it to a shearing force of 2 kN. Plot the shear stress distribution diagram. (*U.L.*)
Answer. The shear stress τ MN/m² at h mm from the neutral axis is given by

h	0	±1·68	±2	±4	±6	±8	±8·66
τ	9·24	9·69 (max)	9·67	8·76	6·31	1·96	0

9. A steel cantilever of I-section 20 cm deep and 8 cm wide has flanges 1 cm thick and web ½ cm thick. It is 2 m long and carries at the free end a concentrated load of 10 kN.

Calculate, neglecting the self-weight of the cantilever,

(*a*) the maximum (vertical) shear stress in the web,
(*b*) the deflection at the free end due to bending,
(*c*) the deflection at the free end due to shear, assuming the web carries the whole of the shear force uniformly distributed.

$E = 200 \times 10^9$ N/m²; $G = 80 \times 10^9$ N/m².
Answer. (*a*) 11·4 MN/m²; (*b*) 8·0 mm; (*c*) 0·278 mm.

10. A compound girder has a top flange 6 cm wide by 1 cm thick, the bottom flange being similarly 12 cm by 1 cm, and the web is ½ cm thick, the girder being 50 cm deep overall. At a certain section the shearing force is 100 kN in a direction parallel to the web.

Find the position and value of the maximum shearing stress, and the values of the shearing stresses at the top and bottom of the web; find also the amount of the shearing force taken by the web.

Prove briefly any formula used. (*U.L.*)
Answer. 285 mm from top edge; 47·9 MN/m²; 22·6 and 33·8 MN/m²; 98·56 kN.

11. A beam of rectangular cross-section ($b \times d$) is simply supported over a span L. It carries a total load W uniformly distributed over the span. Obtain an expression for the deflection at mid-span due to shear.

If, in such a beam, the length is 20 times the depth and $E = 2{\cdot}5G$ find the ratio of the deflection at mid-span due to shear to that due to bending.

Answer. $3WL/20Gbd$; 3/500.

12. For a given cantilever of rectangular cross-section, length l, depth d and carrying a concentrated load at the free end, show that

$$\frac{\text{deflection due to shearing strain}}{\text{bending deflection}} = \text{constant} \times (d/l)^2$$

and find the value of the constant for a steel cantilever. Hence find the least value of l/d if the deflection due to shearing strain is not to exceed 1 per cent of the total deflection.

For steel $E = 203$ GN/m², $G = 78$ GN/m². (*U.L.*)

Answer. 0·781; 8·836.

13. Derive an expression for the deflection due to shear in a cantilever of rectangular section, loaded with a concentrated load at the free end. The usual parabolic distribution of shearing stress over the cross-section may be assumed.

A cantilever of length L, rectangular in section, of depth d and breadth b, carries two point loads, each W, one at the free end and the other half-way along its length. If E/G is 2·5 find the ratio of d to L, for which the deflection at the end due to shear will be 1/100 of that due to bending. (*U.L.*)

Answer. $6WL/5Gbd$; 0·108.

14. An I-beam has flanges 8 cm wide and 1 cm thick and a web ½ cm thick. The overall depth is 18 cm.

When loaded as a beam there is, at a certain section, a bending moment M and a shearing force S.

At a point on this section 6 cm below the neutral layer, the strains are measured along the three directions OA, OB and OC as indicated in Fig. 15.15.

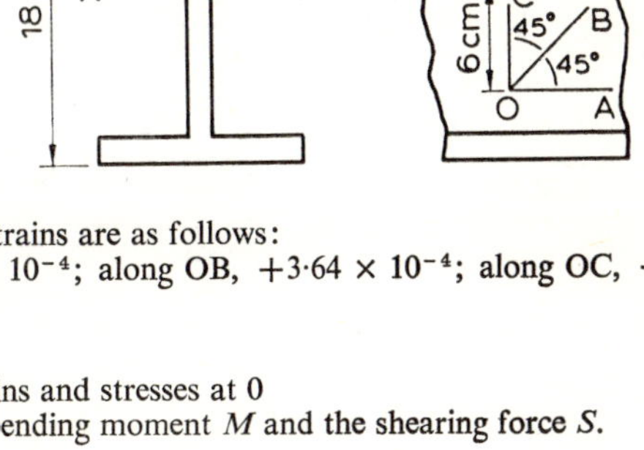

Fig. 15.15

The values of these strains are as follows:

Along OA, $+3{\cdot}42 \times 10^{-4}$; along OB, $+3{\cdot}64 \times 10^{-4}$; along OC, $-1{\cdot}14 \times 10^{-4}$.

Determine

(*a*) the principal strains and stresses at 0
(*b*) the value of the bending moment M and the shearing force S.

Take I_{xx} for the beam section 1328 cm^4; $E = 208$ GN/m^2 and Poisson's ratio as 0·3. (*U.L.*)

Answer. (*a*) $+4{\cdot}52$ and $-2{\cdot}24 \times 10^{-4}$; $+88{\cdot}0$ and $-20{\cdot}3$ MN/m^2; (*b*) 13·67 kN-m; 374 N.

15. Define the "shear centre" of a beam.

A beam of channel section carries a vertical load and is supported so that the two flanges are horizontal. The flanges and web have equal thicknesses which are small compared to the depth of the web (D) and the width of the flanges (B).

Show by working from first principles that the shear centre is at a distance $3B^2/(6B + D)$ from the web.

Assumptions must be clearly stated, where they are made. (*U.L.*)

16. A timber beam of uniform rectangular section has a length of 840 mm; the breadth of the beam is 30 mm and the depth is 70 mm. The beam is simply supported at its ends and it carries two loads 240 mm apart each W kN symmetrically situated at 300 mm from the ends of the beam.

Determine the magnitudes of the loads for a maximum stress due to bending of 7 MN/m^2 and find the maximum deflection of the beam due to both bending and shear. Formulae for deflection due to bending and shear must be derived, but in the case of shear the general formula for the distribution of shearing stress over the depth of the beam may be assumed.

For timber take $E = 10$ GN/m^2 and $G = 0{\cdot}6$ GN/m^2. (*U.L.*)

Answer. 1·143 kN; 2·93 mm; 0·33 mm.

Chapter 16

Mechanical Properties and Theories of Failure

For a tensile test specimen of a given material, gauge length L and original cross-sectional area A, the extension after fracture (or *elongation*) e is given by

$$\text{Elongation } e = c\sqrt{A} + bL$$

$$\text{percentage elongation} = 100\left(\frac{c\sqrt{A}}{L} + b\right)$$

where b and c are Unwin's constants for the material.

The parabolic relationship for limiting fatigue range is

$$r = r_0 - m\sigma^2_{(average)}$$

where r_0 = limiting fatigue range for zero mean stress,
r = limiting fatigue range for a mean stress $\sigma_{(average)}$,
m = an experimental constant.

If σ_1, σ_2 and σ_3 are the principal stresses (in descending order of magnitude) in a three-dimensional complex stress system and σ is the equivalent stress in simple tension, then

(*a*) Greatest principal stress theory:

$$\sigma_1 = \sigma$$

(*b*) Greatest principal strain theory:

$$\sigma_1 - \nu(\sigma_2 + \sigma_3) = \sigma$$

(*c*) Maximum shear stress theory:

$$\sigma_1 - \sigma_3 = \sigma$$

(*d*) Total strain energy theory:

$$\sigma_1^2 + \sigma_2^2 + \sigma_3^2 - 2\nu(\sigma_1\sigma_2 + \sigma_2\sigma_3 + \sigma_3\sigma_1) = \sigma^2$$

(*e*) Mises-Hencky (shear strain energy) theory:

$$(\sigma_1 - \sigma_2)^2 + (\sigma_2 - \sigma_3)^2 + (\sigma_3 - \sigma_1)^2 = 2\sigma^2$$

WORKED EXAMPLES

16.1. Distinguish between (*a*) *elasticity* and *plasticity* (*b*) *ductility* and *brittleness.*

Describe the phenomena observed during a tensile test on a mild steel specimen. Sketch the graph of load against extension, indicating points of interest.

Solution. (*a*) *Elasticity* is the property of a material by which the deformation disappears when the load producing it is removed. Many materials behave in this way up to a certain stress called the *elastic limit.* Unless otherwise stated, it is also assumed that Hooke's law applies, and the stress is proportional to the corresponding strain. This is, for practical purposes, true of many metals up to the *limit of proportionality* stress and it is found by experiment that these two limits are almost identical.

Plasticity is the property by which some permanent deformation remains when the load is removed.

(*b*) A *ductile* material is one which can be drawn out in tension to a smaller cross-section. It is therefore capable of considerable distortion without fracture. A *brittle* material is one which fractures with little previous distortion.

A material may fall into different categories depending on the stress and temperature conditions to which it is subjected.

In a tensile test a specimen (usually a round or flat bar) is gradually pulled in a testing machine until it fractures. The central portion of the specimen should be of uniform section and two gauge points are set on it. The original distance between these points is called the *gauge length.* Values of the extension (or *elongation*) over this length and the corresponding loads are recorded at frequent intervals during the test.

A typical load–extension diagram for mild steel in tension is shown in Fig. 16.1. The scale and proportions of the diagram depend on the exact composition of the material, the dimensions of the testpiece and its previous history. At the beginning of the test the extensions are small but proportional to the corresponding loads and this elastic stage is generally redrawn to an enlarged scale as shown. These small extensions are measured with instruments called *extensometers.*

At A, the limit of proportionality, the graph departs from a straight line. From B, the *yield point*, to C there is a considerable increase in extension even with a slightly reduced load. From C to D the specimen is ductile and there is a general reduction in cross-section. From D to E a considerable increase in extension is obtained with little increase in load. The specimen is now plastic and at E, the point of maximum load, a neck or waist appears at one point of the specimen, Fig. 16.4b. This becomes more pronounced so that further extension is obtained with a reducing load and finally the specimen fractures at the neck.

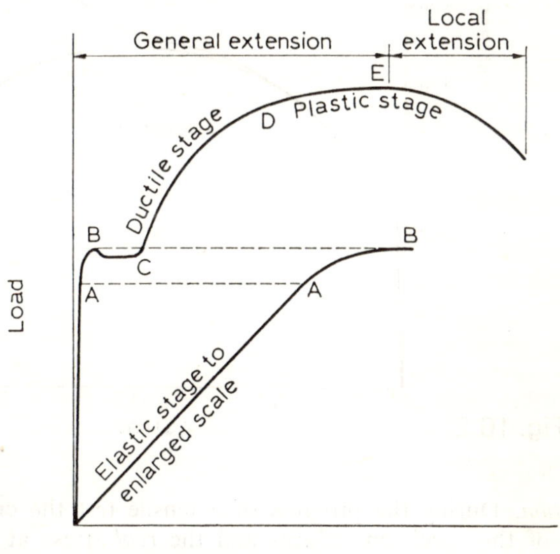

Fig. 16.1

16.2. Distinguish between *real* and *nominal* stress. Define

(*a*) upper and lower yield stress, (*b*) tensile strength,
(*c*) percentage elongation after fracture and (*d*) percentage reduction of area.

Determine the upper yield stress, tensile strength, percentage elongation and reduction of area at fracture, and nominal and real stress at fracture for a specimen which gave the following results:

Gauge length, 80 mm; original diameter, 15·96 mm; load at yield point, 60 kN; maximum load, 90 kN, load at fracture, 65 kN; distance between gauge points at fracture, 108 mm; minimum diameter at fracture, 9·20 mm.

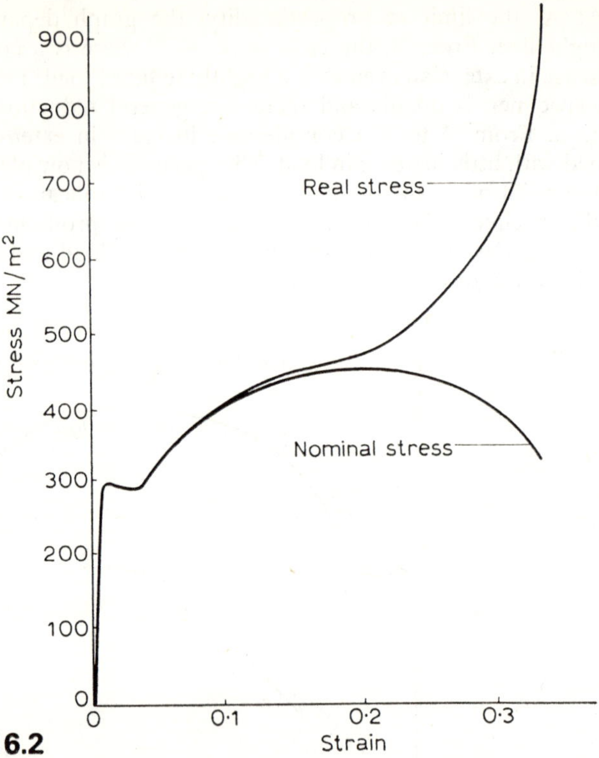

Fig. 16.2

Solution. During the progress of a tensile test the cross-sectional area of the specimen reduces and the *real* stress at any stage is the load divided by the corresponding minimum area. The *nominal* stress is based on the original area throughout and a graph of nominal stress is similar in shape to the load diagram (see Fig. 16.2). The difference between real and nominal stress is small in the early stages of a tensile test but increases rapidly in the later stages as shown in Fig. 16.2.

Precise definitions of the terms given in the question will be found in British Standard 18: 1971 *Tensile Testing of Metals*. The following shorter descriptions are sufficient at this stage.

(*a*) At point B on Fig. 16.1 the specimen continues to stretch at a constant, or even reduced, load. This effect is known as *yielding*. *Upper yield stress* is the stress at B, immediately before yield begins. *Lower yield stress* is the lowest stress at which yielding occurs.

(*b*) *Tensile strength* is the maximum load divided by the original cross-sectional area. It is thus the greatest *nominal* stress attained during the test. (It is sometimes called *ultimate tensile stress*).

(*c*) If L_0 is the original gauge length and L_u is the distance between the gauge points of the fractured test-piece, the *percentage elongation* after fracture is

$$100(L_u - L_0)/L_0$$

(*d*) If S_0 is the original cross-sectional area and S_u is the minimum cross-sectional area of the fractured test-piece, the *percentage reduction of area* is

$$100(S_0 - S_u)/S_0$$

With the above definitions and the data given in the question, the following results are obtained

$$S_0 = \tfrac{1}{4}\pi(15{\cdot}96\ \text{mm})^2 = 200\ \text{mm}^2$$
$$S_u = \tfrac{1}{4}\pi(9{\cdot}20\ \text{mm})^2 = 66{\cdot}5\ \text{mm}^2$$

(*a*) Upper yield stress

$$= 60\ \text{kN}/200\ \text{mm}^2 = (60 \times 10^3\ \text{N})/(200 \times 10^{-6}\ \text{m}^2)$$
$$= 300\ \text{MN/m}^2 \qquad (Ans)$$

(*b*) Tensile strength

$$= 90\ \text{kN}/200\ \text{mm}^2 = (90 \times 10^3\ \text{N})/(200 \times 10^{-6}\ \text{m}^2)$$
$$= 450\ \text{MN/m}^2 \qquad (Ans)$$

(*c*) Percentage elongation

$$= \frac{108\ \text{mm} - 80\ \text{mm}}{80\ \text{mm}} \times 100$$
$$= 35{\cdot}0 \qquad (Ans)$$

(*d*) Percentage reduction of area

$$= \left(\frac{200\ \text{mm}^2 - 66{\cdot}5\ \text{mm}^2}{200\ \text{mm}^2}\right) \times 100$$
$$= 66{\cdot}8$$

(*e*) Nominal stress at fracture

$$= \frac{65\ \text{kN}}{200\ \text{mm}^2} = \frac{65 \times 10^3\ \text{N}}{200 \times 10^{-6}\ \text{m}^2}$$
$$= 325\ \text{MN/m}^2 \qquad (Ans)$$

Real stress at fracture

$$= \frac{65 \text{ kN}}{66{\cdot}5 \text{ mm}^2} = \frac{65 \times 10^3 \text{ N}}{66{\cdot}5 \times 10^{-6} \text{ m}^2}$$

$$= 978 \text{ MN/m}^2 \qquad (Ans)$$

16.3. The following results were obtained during the first part of a tensile test on an aluminium specimen, gauge length 69 mm, diameter 13·82 mm.

Load (kN)	0	5·0	7·5	10·0	12·5	15·0
Extension (mm)	0	0·0329	0·0493	0·0658	0·0960	0·1852

Plot a load–extension diagram. Define the following terms and find their values for this material:

(*a*) Young's modulus of elasticity,
(*b*) limit of proportionality stress,
(*c*) 0·1 per cent proof stress under load.

Solution. Figure 16.3 is the required graph. Non-ferrous metals do not show a yield point as defined in the previous solution and, in such cases, it is usual to specify a proof stress. Definitions (*a*) and (*c*) are based on those in the British Standard publication mentioned in the previous solution.

(*a*) *Young's Modulus of Elasticity.* The value of the increase in stress divided by the corresponding increase in strain for the straight portion of the stress–strain (or load–extension) curve.
(*b*) *Limit of Proportionality stress.* The stress at which the strain (or extension) ceases to be proportional to the corresponding stress (or load).
(*c*) *Proof stress.* The stress at which a non-proportional extension, equal to a specified percentage of the original gauge length, occurs.

The proof stress is determined from a stress–strain (or load–extension) curve on which a line is drawn parallel to the straight portion of the curve and distant from it by an amount representing an increase of strain equal to the required percentage strain: the point at which the line cuts the curve gives the proof stress (see Fig. 16.3).

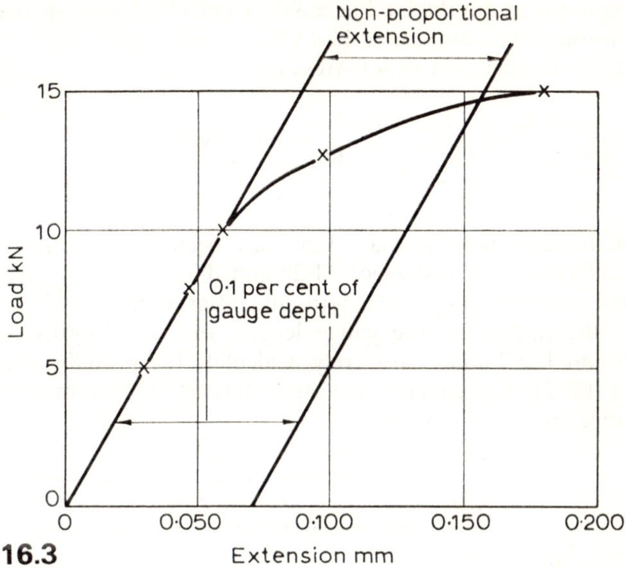

Fig. 16.3

In specifying or describing a proof stress the required percentage strain shall be quoted, e.g. 0·1 per cent proof stress

In the present example, original area $= \frac{1}{4}\pi(13{\cdot}82 \text{ mm})^2$
$= 150 \text{ mm}^2$

(*a*) The extension is 0·0658 mm per 10 kN of load.

$$\text{Increase in stress} = \frac{10 \text{ kN}}{150 \text{ mm}^2} = \frac{10 \times 10^3 \text{ N}}{150 \times 10^{-6} \text{ m}^2} = 66{\cdot}7 \text{ MN/m}^2$$

Corresponding increase in strain is

$$\frac{0{\cdot}0658 \text{ mm}}{69 \text{ mm}} = 0{\cdot}000954$$

Young's modulus of elasticity is

$$\frac{66{\cdot}7 \times 10^6 \text{ N/m}^2}{0{\cdot}000954} = 70 \times 10^9 \text{ N/m}^2 = 70 \text{ MN/m}^2 \qquad (Ans)$$

(*b*) Load at limit of proportionality = 10 kN
Limit of proportionality stress is

$$\frac{10 \text{ kN}}{150 \text{ mm}^2} = \frac{10 \times 10^3 \text{ N}}{150 \times 10^{-6} \text{ m}^2} = 66{\cdot}7 \text{ MN/m}^2 \qquad (Ans)$$

(*c*) 0·1 per cent of the gauge length is 0·069 mm.

From the graph the load necessary to produce a non-proportional extension of this amount is 14·4 kN.

Hence 0·1 per cent proof stress is

$$\frac{14{\cdot}4\ \text{kN}}{150\ \text{mm}^2} = \frac{14{\cdot}4 \times 10^3\ \text{N}}{150 \times 10^{-6}\ \text{m}^2} = 96\ \text{MN/m}^2 \qquad (Ans)$$

16.4. Explain how Barba's Law has been derived from experimental results. A test piece 11·28 mm diameter was marked with 50 mm and 100 mm gauge lengths and, after fracture, which occurred near the middle of the gauge length, the two lengths measured 65·6 and 126·2 mm respectively. Calculate the probable elongation for a standard specimen of the same material, 22·56 mm diameter on a gauge length of 113 mm. *(U.L.)*

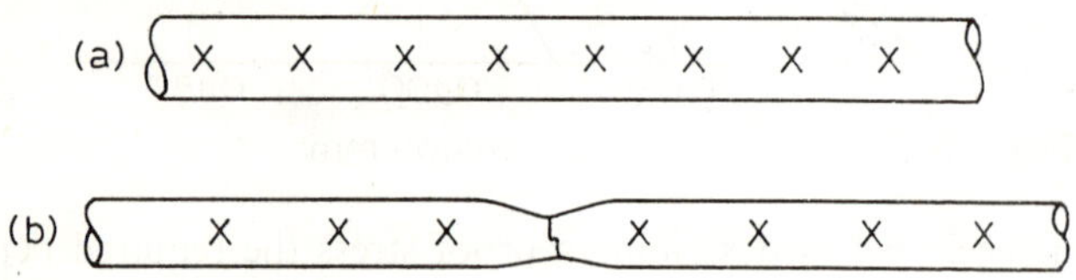

Fig. 16.4

Solution. Suppose a tensile test specimen (Fig. 16.4) is marked with several equal intervals before testing, as at (*a*). Up to the maximum load, point E in Fig. 16.1, the extension is nearly the same for each section. If, after fracture occurs, the two parts of the specimen are fitted together, Fig. 16.4b, it is found that the extension of the section which includes the fracture is greatly increased whereas the extension of the other sections is almost the same as at the maximum load. The extension at fracture (or elongation) e can therefore be considered in two parts, a general extension which is proportional to the gauge length L and a local extension due to the necking. Thus

$$e = a + bL \qquad \text{(i)}$$

where a and b are constants for bars of the same material and cross-section.

If bars of the same material but of different cross-section are tested, it is found that the local extension a is proportional to the square root of the cross-sectional area A. Hence, from (i),

$$e = c\sqrt{A} + bL \qquad \text{(ii)}$$

where c and b depend only on the material. These are known as Unwin's constants.

For round bars $A = \frac{1}{4}\pi d^2$ and (ii) becomes

$$e = c'd + bL \tag{iii}$$

where d is the diameter and $c' = c\sqrt{(\frac{1}{4}\pi)}$.

From (ii) and (iii)

$$\text{Percentage elongation} = \frac{e}{L} \times 100 = 100\left(\frac{c\sqrt{A}}{L} + b\right) \tag{iv}$$

$$= 100\left(\frac{c'd}{L} + b\right) \tag{v}$$

From (iv) and (v) it follows that geometrically similar test pieces of the same material [constant $(\sqrt{A})/L$ or constant d/L] have the same *percentage* elongation. This is known as *Barba's Law*.

The specimens mentioned in the question are of circular cross-section and equation (iii) is the most convenient.

Working in mm units, we obtain for the 50 mm gauge lengths

$$(65{\cdot}6 - 50) = (c' \times 11{\cdot}28) + (b \times 50) \tag{vi}$$

and, for the 100 mm gauge length

$$(126{\cdot}2 - 100) = (c' \times 11{\cdot}28) + (b \times 100) \tag{vii}$$

Solving the simultaneous equations (vi) and (vii)

$$b = 0{\cdot}212 \qquad c' = 0{\cdot}443$$

Hence, for the standard specimen the probable elongation is

$$e = (0{\cdot}443 \times 22{\cdot}56) + (0{\cdot}212 \times 113)$$
$$= 33{\cdot}9 \text{ mm} \qquad (\textit{Ans})$$

16.5. Explain the terms *fatigue* and *limiting fatigue range* as applied to engineering materials.

Give a relation connecting the limiting fatigue range and the mean stress during the cycle.

A certain steel has an ultimate strength of 540 MN/m^2. The limiting range for alternating stress is $\pm$180 MN/m^2. Estimate the probable safe maximum stress for an unlimited number of cycles if the minimum stress is 120 MN/m^2.

Solution. Experiments show that a material may fail at a stress considerably lower than its ultimate strength in a normal tensile test if this stress is repeated a large number of times. The term *fatigue* is used for the effects on a material of repeated cycles of stress.

If the limits of stress during the cycle are of the same sign, e.g. both tensile, the stress is said to be *fluctuating*. If the lower limit is zero, the term *repeated* stress is used. *Reverse* (or *alternating*) stress implies limits which are numerically equal but opposite in sign.

As the range of stress during the cycle is reduced the number of applications required to cause failure is increased. In the case of steels it is found that, for given mean stress, there is a limiting range within which failure does not occur, however many cycles are applied. This is called the *limiting fatigue range* and experiments show that it is approximately equal to the range which the material can withstand for 10 million cycles. For some non-ferrous materials, however, it is not yet certain that such limiting ranges exist and failures after hundreds of millions of cycles have been reported.

The following parabolic relation is an empirical one suggested by Gerber on the basis of experiments by Wöhler

$$r = r_0 - m\sigma^2_{(average)} \qquad \text{(i)}$$

where r_0 = limiting fatigue range for zero mean stress (alternating stress)

r = limiting fatigue range for a mean stress $\sigma_{(average)}$

m = an experimental constant.

Using the data of the question and working in MN/m^2

$$r_0 = 2 \times 180 = 360$$

when σ_{av} reaches the ultimate tensile stress the range must be zero. Thus $r = 0$ when $\sigma_{av} = 540$ and substituting these values in (i),

$$0 = 360 - m \times 540^2 \qquad m = 1/810$$

Let σ_{max} and σ_{min} be the upper and lower limits of stress during the cycle. Then

$$\sigma_{min} = 120$$

$$r = \sigma_{max} - \sigma_{min} = \sigma_{max} - 120$$

$$\sigma_{av} = \tfrac{1}{2}(\sigma_{max} + \sigma_{min}) = \tfrac{1}{2}(\sigma_{max} + 120)$$

Substituting these results in (i),

$$\sigma_{max} - 120 = 360 - \frac{1}{810} \times [\tfrac{1}{2}(\sigma_{max} + 120)]^2$$

or, rearranging,

$$[\tfrac{1}{2}(\sigma_{max} + 120)]^2 + 810(\sigma_{max} - 120) - 360 \times 810 = 0$$

$$\tfrac{1}{2}\sigma_{max}{}^2 + 870\sigma_{max} - 385200 = 0$$

This is a quadratic equation for σ_{max}, the roots being 398 and -3878. Taking the positive root the probable safe maximum stress is 398 MN/m². *(Ans)*

16.6. Explain briefly the various theories which have been put forward to obtain a criterion of failure under conditions of complex stress.

At a point in the wall of a thin steel tube there are perpendicular stresses of 40 MN/m² and 20 MN/m², both tensile. Calculate the equivalent stress in simple tension according to the maximum principal strain theory and the (total) strain energy theory.

Poisson's ratio = 0·28.

Solution. Failure here means elastic breakdown and the onset of permanent strain. Let σ be the stress at which this occurs in simple tension (for practical purposes the limit of proportionality may be used). Suppose in the three-dimensional complex stress system on the same material, the principal stresses are σ_1, σ_2 and σ_3 in descending order, tensions being positive.

The main theories of the conditions for elastic failure are as follows:

(*a*) *Greatest principal stress theory* (*Rankine*). This states that failure occurs when the greatest principal stress reaches a critical value. For the complex stress case, this stress is σ_1. In simple tension it is σ. Hence the criterion is

$$\sigma_1 = \sigma \qquad \text{(i)}$$

(*b*) *Greatest principal strain theory* (*St. Venant*). This states that the relevant quantity is the greatest principal strain. In the complex stress case this is

$$\frac{\sigma_1}{E} - \frac{\nu\sigma_2}{E} - \frac{\nu\sigma_3}{E}$$

(see page 289) and in simple tension it is σ/E. On equating these strains, we have

$$\sigma_1 - \nu(\sigma_2 + \sigma_3) = \sigma \qquad \text{(ii)}$$

(*c*) *Maximum shear stress theory* (*Coulomb, Guest*). The maximum shear stress on an interface is half the difference of the corresponding principal stresses (see page 277, Problem 7). This is $\frac{1}{2}(\sigma_1 - \sigma_3)$ for the complex stress system and $\frac{1}{2}(\sigma - 0)$ in simple tension. Thus, by this theory

$$\sigma_1 - \sigma_3 = \sigma \qquad \text{(iii)}$$

(*d*) *Total strain energy theory* (*Beltrami*). By extending the theory given on pages 375–6 to three-dimensional conditions the total strain energy is

$$\frac{1}{2E}\left[\sigma_1^2 + \sigma_2^2 + \sigma_3^2 - 2\nu(\sigma_1\sigma_2 + \sigma_2\sigma_3 + \sigma_3\sigma_1)\right]$$

In simple tension the strain energy is $\sigma^2/2E$ and this leads to the relationship

$$\sigma_1^2 + \sigma_2^2 + \sigma_3^2 - 2\nu(\sigma_1\sigma_2 + \sigma_2\sigma_3 + \sigma_3\sigma_1) = \sigma^2 \qquad \text{(iv)}$$

(*e*) *Mises-Hencky* (*shear strain energy*) *theory*. This states that the relevant quantity is

$$(\sigma_1 - \sigma_2)^2 + (\sigma_2 - \sigma_3)^2 + (\sigma_3 - \sigma_1)^2$$

and it can be shown that this expression represents the shear strain energy. In simple tension the principal stresses are σ, 0 and 0, and the corresponding expression is therefore $2\sigma^2$.

Hence the criterion is

$$(\sigma_1 - \sigma_2)^2 + (\sigma_2 - \sigma_3)^2 + (\sigma_3 - \sigma_1)^2 = 2\sigma^2 \qquad \text{(v)}$$

In the present example (working in MN/m^2) we have

$$\sigma_1 = 40 \qquad \sigma_2 = 20 \qquad \sigma_3 = 0$$

Principal strain theory. From (ii), the equivalent stress in simple tension is

$$\sigma = \sigma_1 - \nu(\sigma_2 + \sigma_3) = 40 - 0{\cdot}28(20 + 0)$$
$$= 34 \text{ MN/m}^2$$

Total strain energy theory. From (iv)

$$\sigma^2 = \sigma_1^2 + \sigma_2^2 + \sigma_3^2 - 2\nu(\sigma_1\sigma_2 + \sigma_2\sigma_3 + \sigma_3\sigma_1)$$
$$= 40^2 + 20^2 + 0 - 2 \times 0{\cdot}28(40 \times 20 + 20 \times 0 + 0 \times 40)$$
$$= 1552$$
$$\sigma = 39{\cdot}4 \text{ MN/m}^2 \qquad (\textit{Ans})$$

16.7. Discuss briefly the merits of the various theories of elastic failure.

A certain steel has a proportionality limit of 270 MN/m² in simple tension. Under a certain two-dimensional stress system the principal stresses are 105 MN/m² (tensile) and 30 MN/m² (compressive). Calculate the factor of safety according to

(*a*) the maximum shear stress theory,

(*b*) the Mises-Hencky theory.

Solution. The greatest principal stress theory is reasonably correct for brittle materials such as cast iron. The greatest principal strain theory is of very little value. The maximum shear stress theory is widely used for ductile materials, particularly in the design of shafts subjected to combined bending and torsion.

Some experimental results on ductile materials support the (total) strain energy theory but more are in agreement with the Mises-Hencky criterion, which is widely regarded as the most reliable basis for design.

Working in MN/m², the question gives

$$\sigma_1 = 105 \qquad \sigma_2 = 0 \qquad \sigma_3 = -30$$

(*a*) By the maximum shear stress theory, the equivalent single tensile stress is

$$\sigma = \sigma_1 - \sigma_3 = 105 - (-30) = 135$$

Factor of safety = 270/135 = 2 (*Ans*)

(*b*) Mises-Hencky theory.

$$\begin{aligned} 2\sigma^2 &= (\sigma_1 - \sigma_2)^2 + (\sigma_2 - \sigma_3)^2 + (\sigma_3 - \sigma_1)^2 \\ &= (150 - 0)^2 + (0 + 30)^2 + (-30 - 105)^2 \\ &= 30\,150 \\ \sigma &= 122{\cdot}8 \end{aligned}$$

Factor of safety = 270/122·8 = 2·2 (*Ans*)

16.8. Explain the term "equivalent bending moment" as used in connection with shafting subjected to a bending moment combined with a twisting moment.

For a shaft of solid circular section subjected to a bending moment M combined with a twisting moment T, deduce the formula for the equivalent bending moment M_E in terms of M, T and if necessary, Poisson's ratio ν, to correspond with each of the following hypotheses of elastic failure: (*a*) maximum principal stress; (*b*) maximum shearing stress; (*c*) maximum strain energy.

For a shaft 10 cm diameter made of steel for which the limiting stress in simple tension is 120 MN/m^2 draw to scale a graph for each of the above theories showing the limits within which combined values of M and T must occur according to each of the hypotheses. $\nu = 0{\cdot}286$. (*U.L.*)

Solution. The equivalent bending moment (M_E) is that which, if acting alone, would produce the same value of the quantity, used for the criterion of failure, as the given bending moment (M) and twisting moment (T) in combination.

For a circular shaft $J = 2I$ and, if d is the external diameter, the stresses at the surface are

$$\text{Bending stress } \sigma_x = My/I = Md/2I$$

$$\text{Shear stress } \tau_{xy} = Tr/J = Td/2J = Td/4I$$

This is a case of combined stress in two dimensions in which $\sigma_y = 0$. The principal stresses (σ_1 and σ_2) are the roots of the equation,

$$(\sigma - \sigma_x)(\sigma - \sigma_y) = \tau_{xy}{}^2$$

$$\sigma^2 - \sigma\left(\frac{Md}{2I}\right) - \left(\frac{Td}{4I}\right)^2 = 0$$

and the principal stresses are

$$\sigma_1 \text{ and } \sigma_2 = \frac{d}{4I}[M \pm \sqrt{(M^2 + T^2)}] \qquad \text{(i)}$$

The third principal stress is zero and falls between these two values. For a bending moment M_E acting alone the maximum direct stress is

$$\sigma = M_E y/I = M_E d/2I \qquad \text{(ii)}$$

(*a*) *Maximum principal stress.* For this criterion $\sigma = \sigma_1$ and

$$\frac{M_E d}{2I} = \frac{d}{4I}[M + \sqrt{(M^2 + T^2)}]$$

$$M_E = \tfrac{1}{2}[M + \sqrt{(M^2 + T^2)}]$$

(*b*) *Maximum shearing stress.* In the combined case the maximum shearing stress is $\frac{1}{2}(\sigma_1 - \sigma_2)$ and in the equivalent bending moment case it is $\frac{1}{2}(\sigma - 0)$. Thus from (i) and (ii)

$$\frac{1}{2}\left(\frac{M_E d}{2I}\right) = \frac{1}{2}\left(\frac{d}{4I}\right) \times \{[M + \sqrt{(M^2 + T^2)}] - [M - \sqrt{(M^2 + T^2)}]\}$$

$$M_E = \sqrt{(M^2 + T^2)}$$

(*c*) *Maximum strain energy.* Putting $\sigma_3 = 0$ in equation (iv) of Example 16.6 we have

$$\left(\frac{M_E d}{2I}\right)^2 = \left\{\frac{d}{4I}[M + \sqrt{(M^2 + T^2)}]\right\}^2 + \left\{\frac{d}{4I}[M - \sqrt{(M^2 + T^2)}]\right\}^2 - 2\nu\frac{d^2}{16I^2}[M + \sqrt{(M^2 + T^2)}][M - \sqrt{(M^2 + T^2)}]$$

from which

$$M_E^2 = M^2 + \tfrac{1}{2}T^2(1 + \nu)$$

With the numerical data given in the question

$$M_E = \sigma I/y = \sigma \times \pi d^3/32$$
$$= \tfrac{1}{32}[120\ \text{MN/m}^2 \times \pi \times (10\ \text{cm})^3] = 11{\cdot}8\ \text{kN-m}$$

Taking values of M from 0 to 11·8 kN-m and using the three criteria mentioned the maximum values of T are as follows:

M kN-m		0	2	4	6	8	10	11·8
T kN-m	(*a*)	23·6	21·5	19·2	16·5	13·4	9·2	0
	(*b*)	11·8	11·6	11·1	10·2	8·7	6·3	0
	(*c*)	14·7	14·5	13·8	12·7	10·8	7·9	0

Figure 16·5 illustrates the results and pairs of values of M and T must lie within each boundary to satisfy the corresponding criterion.

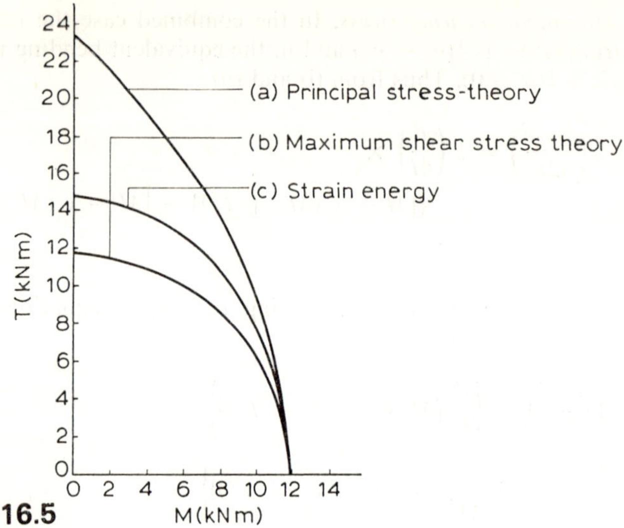

Fig. 16.5

PROBLEMS

1. The following results were obtained during a tensile test on mild steel. Plot these values and find Young's modulus of elasticity and the limit of proportionality stress.

Load in kN	0	20	40	60	80	90	100	110
Extensometer reading	0	27·4	55·1	82·5	109·8	124·5	141·0	165·2

Diameter of test piece 22·56 mm; gauge length, 113 mm; 1 division of the extensometer corresponds to an extension of 10^{-3} mm.
Answer. E = 206 GN/m^2; limit of proportionality stress = 225 MN/m^2.

2. The tensile test of Problem No. 1 was continued until the specimen fractured. The maximum load was 166 kN and the final breaking load was 118 kN. The minimum (final) diameter was 13·8 mm and the total extension was 36·9 mm. Calculate

(i) ultimate tensile stress,
(ii) real and nominal stress at fracture,
(iii) percentage elongation and percentage reduction of area.

Answer. (i) 415 MN/m^2; (ii) 789 and 295 MN/m^2; (iii) 32·6 and 62·6.

3. Two test pieces were cut from a piece of mild steel plate and, when tested, gave the following results—

Specimen	Cross-section	Gauge length	Percentage elongation
1	50 mm × 6 mm	150 mm	26·2
2	75 mm × 6 mm	100 mm	33·2

What is the probable percentage elongation for a piece of the same plate 60 mm by 6 mm when tested on a gauge length of 200 mm?
Answer. 24·7.

4. In a tensile test on a 19 mm mild steel bar the gauge length of 200 mm was divided into eight equal sections. After fracture these were found to measure (in order) 32·5, 34·0, 34·2, 44·5, 33·0, 32·7, 33·0, 32·2 mm. Plot a graph showing how the elongation varies with gauge length, the fracture being included as centrally as possible.

Estimate the percentage elongation of a specimen of the same material, 250 mm long and 25·4 mm diameter.
Answer. 38·5.

5. A certain steel was found to have an ultimate strength of 450 MN/m^2 when tested in simple tension. In addition, a specimen of the same material failed after about 10^7 cycles with a repeated stress of 270 MN/m^2, i.e. a range of 270 MN/m^2 with a mean stress of 135 MN/m^2.

Estimate the limiting fatigue range for zero mean stress.
Answer. 297 MN/m^2.

6. A mild steel test-piece 40 mm wide and 7 mm thick is tested to destruction in a tensile testing machine. The extensions of a 200 mm gauge length, for different values of the load, are recorded below—

Load (kN)	20	40	60	80	100
Extension (mm)	0·070	0·140	0·210	0·279	0·448
Load (kN)	87	100	120	125	90
Extension (mm)	3·7	5·1	9·6	28·4	50

The final load given is the breaking load.

Plot load–extension diagrams, one to cover the complete test, and another showing the elastic stage to a larger scale of extensions.

From your diagrams determine

(i) the modulus of elasticity,
(ii) the yield point stress,
(iii) the ultimate tensile strength,
(iv) the percentage elongation. (*I.Struct.E.*)

Answer. (i) 205 × 10^9 N/m^2; (ii) 311 × 10^6 N/m^2; (iii) 446 × 10^6 N/m^2; (iv) 25.

7. The following results were obtained from a tensile test on a strip of a certain alloy, 12·5 mm wide, 0·6 mm thick, and of gauge length 50 mm.

Load (kN)	0	0·45	0·90	1·35	1·6	1·8
Extension (mm)	0	0·0443	0·0886	0·133	0·155	0·181
Load (kN)	1·9	2·0	2·1	2·2	2·3	2·7
Extension (mm)	0·198	0·219	0·246	0·281	0·332	0·645
Load (kN)	3·0	3·1	The last load given			
Extension (mm)	1·05	1·94	is the breaking load			

Plot diagrams of load on a base of extension, to suitable scales, and determine

(*a*) Young's modulus.
(*b*) The 0·1 per cent proof stress.
(*c*) The ultimate tensile stress. (*I.Struct.E.*)

Answers. (*a*) $67{\cdot}6 \times 10^9$ N/m^2; (*b*) 290 MN/m^2; (*c*) 413 MN/m^2.

8. (*a*) Explain why it is necessary, when quoting values of percentage elongation, to state the gauge-length and cross-sectional area of the tensile test-piece used.

(*b*) A test-piece of diameter 25·4 mm and gauge-length 200 mm, was made of a certain mild steel, and gave 26·1 per cent elongation. A second test-piece of diameter 14 mm and gauge length 50 mm, of the same material, gave 36·2 per cent elongation. Calculate the probable elongation percentage of a test-piece of this material, 7·98 mm diameter and 80 mm gauge-length. (*I.Struct.E.*)

Answer. 26·7.

9. In a tensile test of a mild steel specimen, 25 mm wide and 10 mm thick, of gauge length 100 mm, the following readings were recorded:

Load (kN)	16	32	48	64	68	72	76
Elongation (mm)	0·032	0·064	0·096	0·128	0·137	0·147	0·173
Load (kN)	79	76·8	83·7	103·8	111	112·8	108
Elongation (mm)	0·605	1·815	2·42	7·25	12·0	16·8	22·0
Load (kN)	96	92·7					
Elongation (mm)	24·0	Fracture					

Plot separate load/extension diagrams, for the elastic stage and for the plastic stage, and determine—

(*a*) Young's modulus.
(*b*) Elastic limit stress.
(*c*) Yield point stress.
(*d*) Ultimate tensile stress.
(*e*) Percentage elongation. (*I.Struct.E.*)

Answer. (*a*) 197 GN/m^2; (*b*) 270 MN/m^2; (*c*) 315 MN/m^2; (*d*) 451 MN/m^2; (*e*) 24·4.

10. Describe briefly any machine used for endurance tests on a material. Sketch the arrangement of the specimen and the method of loading.

What relation has been suggested between the limiting fatigue range of stress and the mean stress during the cycle?

A certain steel has an ultimate strength of 450 MN/m^2. For zero mean stress its limiting fatigue range is 270 MN/m^2. Estimate the limiting fatigue range for a mean stress of 150 MN/m^2. (*U.L.*)
Answer. 240 MN/m^2. (The descriptive part of the solution is not covered in this book.)

11. Show that the strain energy per unit volume in a two dimensional system is given by the formula

$$U = \frac{1}{2E}(\sigma_1^2 + \sigma_2^2 - 2\nu\sigma_1\sigma_2)$$

in which σ_1 and σ_2 are the principal stresses. Hence show in the case of combined bending and torsion

$$U = \frac{1}{2E}(\sigma^2 + 2[1 + \nu]\tau^2)$$

in which σ is the maximum bending stress, τ the maximum shear stress and ν is Poisson's ratio.

A 50 mm diameter mild steel shaft when subjected to pure torsion ceases to be elastic when the torque reaches 4·2 kN-m. A similar shaft is subjected to a torque of 2·5 kN-m and a bending moment M kN-m. If maximum strain energy is the criterion of elastic failure find the value of M. Poisson's ratio $\nu = 0{\cdot}28$. (*I.Mech.E.*)
Answer. 27 kN-m.

12. A horizontal circular shaft of a diameter d and diametral moment of inertia I is subjected to a bending moment $M \cos\theta$ in a vertical plane and to an axial twisting moment $M \sin\theta$. Show that the principal stresses at the ends of a vertical diameter are $\frac{1}{2}Mk(\cos\theta \pm 1)$, where $k = d/2I$.

If strain energy is the criterion of failure, show that

$$S = \frac{S_0\sqrt{2}}{\sqrt{[\cos^2\theta(1 - \nu) + (1 + \nu)]}}$$

where

S = maximum shearing stress,
S_0 = maximum shearing stress in the special case when $\theta = 0$,
ν = Poisson's ratio. (*U.L*).

13. Show that for a material subjected to two principal stresses, σ_1 and σ_2, the strain energy per unit volume of material is equal to

$$\frac{1}{2E}(\sigma_1^2 + \sigma_2^2 - 2\nu\sigma_1\sigma_2)$$

A thin-walled steel tube, of internal diameter 150 mm, closed at its ends, is subjected to an internal fluid pressure of 3 MN/m^2. Find the thickness of the tube if the criterion of failure is the maximum strain energy. Assume a factor of

safety of 4 and take the elastic limit in pure tension as 300 MN/m². Poisson's ratio $\nu = 0{\cdot}28$. (*I.Mech.E.*)
Answer. 2·96 mm (assuming safety factor is ratio of stresses), 1·48 mm (assuming safety factor is ratio of energies).

14. For a certain material subjected to plane stress it is assumed that the criterion of elastic failure is the shear strain energy per unit volume. By considering co-ordinates relative to two axes at 45° to the principal axes show that the limiting values of the two principal stresses can be represented by an ellipse having semi-diameters $\sigma_e\sqrt{2}$ and $\sigma_e\sqrt{\frac{2}{3}}$ where σ_e is the equivalent simple tension. Hence show that for a given value of the major principal stress the elastic factor of safety is greatest when the minor principal stress is half the major, both stresses being of the same sign. (*U.L.*)

15. Derive a formula for the shear strain-energy in an element of material subjected to principal stresses σ_x and σ_y with the third principal stress zero.

A circular shaft 100 mm diameter is subjected to combined bending and twisting moments, the bending moment being three times the twisting moment.

If the direct tension yield-point of the material is 360 MN/m² and the factor of safety on yield is to be 4, calculate the allowable twisting moment by the three following theories of elastic failure:

(*a*) maximum principal stress theory,
(*b*) maximum shearing stress theory,
(*c*) maximum shear strain-energy theory. (*U.L.*)

Answer. (*a*) 2·86, (*b*) 2·79, (*c*) 2·83 kN-m.

16. A cast iron cylinder has outside and inside diameters of 20 cm and 12 cm. If the ultimate tensile strength of the cast iron is 150 MN/m² and its Poisson's ratio is $\frac{1}{4}$, find, according to each of the following theories of failure, the internal pressure which would cause rupture:

(*a*) Maximum principal stress theory.
(*b*) Maximum principal strain theory.
(*c*) Maximum strain-energy theory.

Assume no longitudinal stress in the cylinder. Which of the results obtained do you consider should be applied to this case? (*U.L.*)
Answer. (*a*) 70·6 MN/m², (*b*) 63·2 MN/m², (*c*) 58·5 MN/m²; (*a*) should be applied.

17. A sample of steel is tested (*a*) by direct tension of a solid bar and (*b*) by submitting a cantilevered circular tube to a load at the free end causing both torsion and bending. The limit of proportionality was found in case (*a*) to be a direct tensile stress of 262 MN/m² and in case (*b*) to be at a bending stress of 124 MN/m² together with a shearing stress of 117 MN/m².

Examine these results and state whether they are consistent with any of the following theories of elastic failure: (i) maximum principal stress theory, (ii) maximum shearing stress theory, (iii) maximum strain energy theory.

Assume reasonable values for any constants not given. (*U.L.*)
Answer. Consistent with (ii).

18. A 50 mm diameter shaft is made of a material which in a direct tension test gave elastic failure at 340 MN/m². Poisson's ratio for the material is 0·3.

Estimate the torque which will just cause failure in the shaft when applied in addition to a bending moment of 3 500 N-m taking as the criterion of failure

(i) the maximum principal stress,
(ii) the maximum strain energy.

Principal stress and strain energy formulae may be used without derivation; equivalent bending moment formulae may be used but they should be derived. (*U.L.*)

Answer. (i) 3·35 kN-m, (ii) 2·78 kN-m.

19. Show that the energy stored in an elastic body subject to the three principal stresses σ_1, σ_2 and σ_3 is given by:

$$U = \frac{1}{2E}[\sigma_1^2 + \sigma_2^2 + \sigma_3^2 - 2\nu(\sigma_1\sigma_2 + \sigma_2\sigma_3 + \sigma_3\sigma_1)]/\text{unit volume}$$

The principal stresses in a material, which fails elastically at 300 MN/m^2 when subject to uniaxial tensions, are such that $\sigma_1 = 2\sigma_2$ and $\sigma_3 = 0$. Poisson's ratio for the material is 0·3. Find the value of σ_1, corresponding to elastic failure using the total strain energy as a criterion. (*U.L.*)
Answer. 308 MN/m^2.

20. A horizontal shaft of 8 cm diameter projects from a bearing and, in addition to the torque transmitted, the shaft carries a vertical load of 8 kN at 30 cm from the bearing. If the safe stress for the material, as determined in a simple tension test, is 150 MN/m^2, find the safe torque to which the shaft may be subjected using as the criterion (*a*) the maximum shearing stress, (*b*) the maximum strain energy. Poisson's ratio = 0·29.

A formula for the principal stress may be used without proof but if a formula for either equivalent torque or equivalent bending moment is used such formula should be derived. (*U.L.*)
Answer. (*a*) 7·15 N-m; (*b*) 8·9 kN-m.

Chapter 17

Special Bending and Torsion Problems

This chapter contains examples whose solutions require a knowledge of methods rather than formulae. Four types of problems are considered: flat circular plates, unsymmetrical bending, the bending of curved bars and deformation beyond the yield point.

For a *symmetrically loaded circular plate* the differential equation of flexure is

$$\frac{d}{dr}\left[\frac{1}{r}\frac{d(r\theta)}{dr}\right] = -\frac{F}{D}$$

where F = shearing force per unit circumferential length at radius r,
$\theta = dy/dr$, y being the deflection at radius r,
$D = Et^3/12(1 - \nu^2)$, the flexural rigidity, where E = Young's modulus of elasticity, t = plate thickness and ν = Poisson's ratio.

If F is expressed in terms of r, the equation can be integrated twice to find θ, the slope, and again to find y, the deflection.

The bending moment M in a radial plane per unit circumferential length is given by

$$M = D\left(\frac{d\theta}{dr} + \frac{\nu\theta}{r}\right)$$

In *unsymmetrical bending* the bending stress at a given point in the section is the resultant of the stresses due to bending about the two principal axes. The determination of the position of the principal axes and the values of I (the second moment of area) about them is discussed in Appendix 1.

In *heavily curved beams* the neutral axis does not pass through the centroid of the section. If e is the distance of the neutral axis from the centroid (positive when the neutral axis is between the centroid and the centre of curvature)

$$e = \frac{1}{A}\int \frac{y^2}{R_0 + y}\, dA$$

where A = area of the cross-section

y = distance from the neutral axis, positive outwards

R_0 = initial radius of the neutral surface.

The bending stress σ is given by

$$\sigma = \frac{My}{Ae(R_0 + y)}$$

where M is the bending moment on the section.

In problems on *plastic deformation* it is assumed that straining beyond the yield point takes place at a constant stress. When the section of a beam reaches a fully-plastic state, a "plastic hinge" is formed. The ratio of the moment of resistance under this condition to that at which yielding commences is called the "shape factor".

WORKED EXAMPLES

17.1. A thin flat plate of length L, breadth B, and thickness t is built in along its edges and subjected to a normal loading of w per unit area. From consideration of the central deflection show that the maximum bending stress at the centre will be given approximately by

$$\frac{wB^2L^4}{4t^2(L^4 + B^4)} \qquad (U.L.)$$

Solution. First consider bending in the plane parallel to the length L as shown in Fig. 17.1 and let w_1 be the corresponding part of the normal loading. The deflection and stress due to this loading are the same as for a built-in beam, length L and rectangular cross-section,

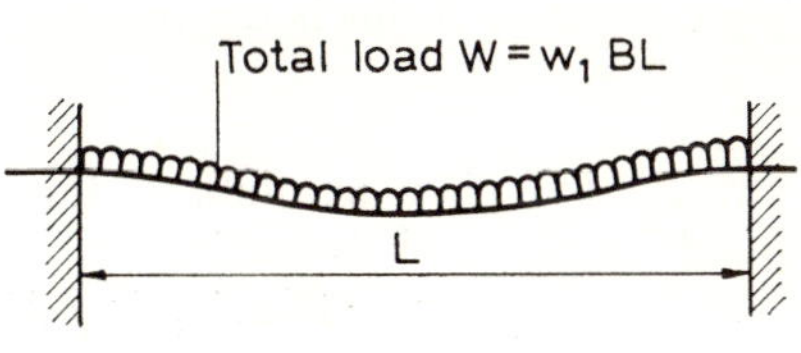

Fig. 17.1

width B and depth t. The total load $W = w_1BL$ and, for the cross-section, $I = Bt^3/12$. Thus the central deflection is

$$\delta = \frac{WL^3}{384EI} = \frac{(w_1BL)L^3 \times 12}{384EBt^3}$$

$$= \frac{w_1L^4}{32Et^3} \qquad \text{(i)}$$

For bending in the plane parallel to the breadth B the corresponding load is $(w - w_1)$ and the central deflection is

$$\delta = \frac{(w - w_1)B^4}{32Et^3} \qquad \text{(ii)}$$

Equating (i) and (ii),

$$w_1L^4 = (w - w_1)B^4$$

$$w_1 = \frac{wB^4}{L^4 + B^4} \qquad \text{(iii)}$$

The bending moment at mid-span is $WL/24$ (see chapter 6) and the corresponding maximum stress is

$$\sigma = \frac{My}{I} = \frac{(WL/24)(t/2)}{(Bt^3/12)} = \frac{WL}{4Bt^2}$$

Substituting from (iii)

$$W = w_1BL = \left(\frac{wB^4}{L^4 + B^4}\right) BL$$

$$\sigma = \frac{wL^2B^4}{4t^2(L^4 + B^4)}$$

Interchanging B and L, the corresponding stress for bending in the perpendicular plane is

$$\sigma = \frac{wB^2L^4}{4t^2(L^4 + B^4)}$$

Since L is greater than B this is the required maximum stress.

17.2. A circular flat plate of uniform thickness is supported uniformly round its rim and is subjected on one side to a normal pressure which on any concentric ring is uniformly distributed. Considering a diametral section of the plate derive the differential equation

$$\frac{d^2\theta}{dr^2} + \frac{1}{r}\frac{d\theta}{dr} - \frac{\theta}{r^2} + \frac{S}{D} = 0$$

where θ = slope at any radius r,

S = shearing force per unit length of arc at radius r,

D = flexural rigidity of plate

$= Et^3/12(1 - \nu^2)$

where t = thickness of plate and ν = Poisson's ratio. *(U.L.)*

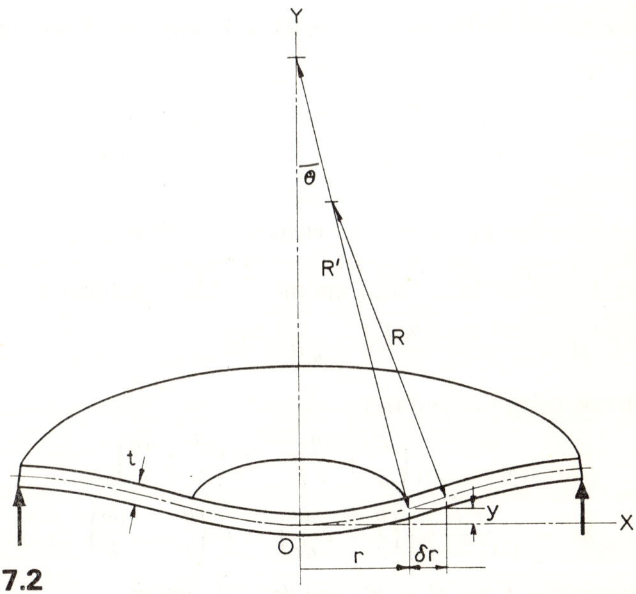

Fig. 17.2

Solution. Take the origin O at the centre of the deflected plate as shown in Fig. 17.2 with axes OX and OY in the plane of a diametral section. Let y be the deflection at radius r so that the slope in this plane $\theta = dy/dr$. Bending occurs not only in the XOY plane but also in a plane at right angles to it. Let the radii of curvature in these planes (at radius r) be R and R'. Then, as shown in Chapter 5,

$$\frac{1}{R} = \frac{d^2y}{dx^2} = \frac{d\theta}{dr} \qquad \text{(i)}$$

Since the loading, and hence the distortion, are symmetrical a circumferential "section", originally cylindrical, becomes part of a cone whose apex is on the axis OY. Hence the radius of curvature R' is given by $r = R'\theta$ and

$$1/R' = \theta/r \tag{ii}$$

Using the relationship between strain and radius of curvature (see Chapter 3) and equations (i) and (ii) we have, at the surface of the plate,

$$\epsilon = \frac{\frac{1}{2}t}{R} = \tfrac{1}{2}t\frac{d\theta}{dr}$$

$$\epsilon' = \frac{\frac{1}{2}t}{R'} = \tfrac{1}{2}t\frac{\theta}{r}$$

The stresses in the same directions at the same point (see Chapter 10) are

$$\sigma = \frac{E}{1-\nu^2}(\epsilon + \nu\epsilon') = \frac{tE}{2(1-\nu^2)}\left(\frac{d\theta}{dr} + \frac{\nu\theta}{r}\right) \tag{iii}$$

$$\sigma' = \frac{E}{1-\nu^2}(\epsilon' + \nu\epsilon) = \frac{tE}{2(1-\nu^2)}\left(\frac{\theta}{r} + \nu\frac{d\theta}{dr}\right) \tag{iv}$$

Consider the bending of an element of unit length, the maximum stress being σ. The distance from the neutral axis to the surface is $t/2$, $I = (1 \times t^3)/12$ and, from the bending equation $\sigma/y = M/I$,

$$M = \frac{\sigma \times \frac{1}{12}(1 \times t^3)}{\frac{1}{2}t} = \tfrac{1}{6}\sigma t^2$$

Hence, using (iii) and (iv)

$$M = \frac{t^3E}{12(1-\nu^2)}\left(\frac{d\theta}{dr} + \frac{\nu\theta}{r}\right) = D\left(\frac{d\theta}{dr} + \frac{\nu\theta}{r}\right) \tag{v}$$

$$M' = \frac{t^3E}{12(1-\nu^2)}\left(\frac{\theta}{r} + \nu\frac{d\theta}{dr}\right) = D\left(\frac{\theta}{r} + \nu\frac{d\theta}{dr}\right) \tag{vi}$$

where $D = t^3E/12(1-\nu^2)$, the *flexural rigidity*.

Figure 17.3 shows a small element of the plate which subtends a small angle ϕ at the centre. M and $M + \delta M$ are the moments per unit length at radii r and $r + \delta r$ in the diametral plane, S and $S + \delta S$ are the corresponding shearing forces per unit length. M' is the moment per unit length in the perpendicular plane and the component of the moment $M'\delta r$ in the central plane of the element through the centre is

$$M'\delta r \cos(90^\circ - \tfrac{1}{2}\phi) = M'\delta r \sin\tfrac{1}{2}\phi = \tfrac{1}{2}M'\delta r\phi$$

since ϕ is a small angle.

Taking moments in this plane about a point at radius $(r + \delta r)$ we have

$$(M + \delta M)\phi(r + \delta r) - M\phi r - \tfrac{1}{2}(2M'\delta r\phi) + S\phi r\delta r = 0$$

Cancelling through by ϕ and neglecting products of small quantities, we obtain

$$M\delta r + r\delta M - M'\delta r + Sr\delta r = 0$$

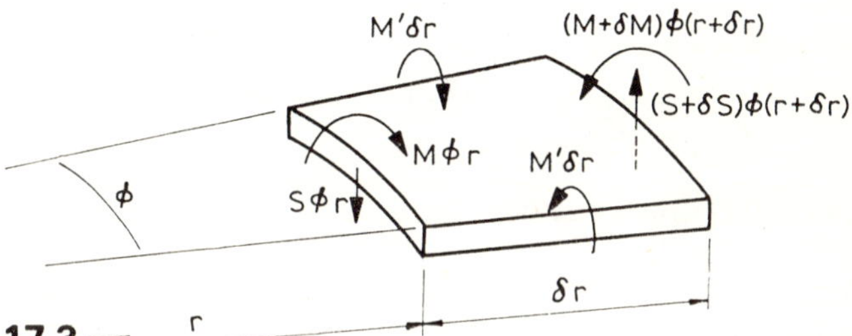

Fig. 17.3

or dividing by δr and taking the limiting case

$$M + r\frac{dM}{dr} - M' + Sr = 0$$

Substituting from (v) and (vi),

$$D\left(\frac{d\theta}{dr} + \frac{\nu\theta}{r}\right) + rD\left[\frac{d^2\theta}{dr^2} + \nu\left(\frac{r\,d\theta/dr - \theta}{r^2}\right)\right] - D\left(\frac{\theta}{r} + \nu\frac{d\theta}{dr}\right) + Sr = 0$$

or, dividing through by rD and re-arranging,

$$\frac{d^2\theta}{dr^2} + \frac{1}{r}\frac{d\theta}{dr} - \frac{\theta}{r^2} + \frac{S}{D} = 0$$

as required.

17.3. A diaphragm of uniform thickness is 10 cm diameter. It is freely supported at its edge and loaded on one face with a uniform pressure of 20 kN/m^2.

If the maximum stress is limited to 100 MN/m^2 find the necessary thickness of the diaphragm and determine also the maximum deflection produced.

At any radius r, the bending moment per unit length of arc is

$$M = D\left(\frac{d\theta}{dr} + \nu\frac{\theta}{r}\right)$$

where θ is the slope at radius r.

$$\frac{d}{dr}\left[\frac{1}{r}\frac{d(r\theta)}{dr}\right] + \frac{S}{D} = 0$$

where $S =$ the shearing force per unit length of arc at radius r
and $D =$ the flexural rigidity of the plate
$= Et^3/12(1 - \nu^2)$

$E = 200\ \text{GN/m}^2$; $\nu = 0{\cdot}28$ (*U.L.*)

Solution. The second equation given in the question is equivalent to the relationship derived in the previous solution. Working in symbols, let w per unit area be the uniform pressure. Then at radius r,

Total shear force around circumference
= total load within circumference

$$2\pi r S = w\pi r^2$$

$$S = \tfrac{1}{2}wr$$

Substituting in the given equation

$$\frac{d}{dr}\left[\frac{1}{r}\frac{d(r\theta)}{dr}\right] = -\frac{wr}{2D}$$

and, on integrating once,

$$\frac{1}{r}\frac{d(r\theta)}{dr} = -\frac{wr^2}{4D} + A$$

$$\frac{d(r\theta)}{dr} = -\frac{wr^3}{4D} + Ar$$

where A is a constant. On integrating again,

$$r\theta = -\frac{wr^4}{16D} + \tfrac{1}{2}Ar^2 + B$$

$$\theta = -\frac{wr^3}{16D} + \tfrac{1}{2}Ar + \frac{B}{r} \qquad \text{(i)}$$

where B is a second constant. But $\theta = dy/dr$ and at $r = 0$ the slope $= 0$. Hence B must be zero and

$$\frac{dy}{dr} = -\frac{wr^3}{16D} + \tfrac{1}{2}Ar \qquad \text{(ii)}$$

from which

$$y = -\frac{wr^4}{64D} + \tfrac{1}{4}Ar^2 + C \qquad \text{(iii)}$$

where C is a third constant of integration.

But $y = 0$ when $r = 0$ and therefore $C = 0$. From (i),

$$\frac{d\theta}{dr} = -\frac{3wr^2}{16D} + \tfrac{1}{2}A$$

and, substituting in the expression for M given in the question

$$M = D\left(-\frac{3wr^2}{16D} + \tfrac{1}{2}A - \frac{\nu wr^2}{16D} + \tfrac{1}{2}\nu A\right)$$

$$= \tfrac{1}{2}AD(1 + \nu) - \tfrac{1}{16}wr^2(3 + \nu) \qquad \text{(iv)}$$

Since the plate is freely supported at its edge $M = 0$ when $r = R$, the external radius. Thus putting $M = 0$ and $r = R$ in (iv),

$$\tfrac{1}{2}AD(1 + \nu) = \tfrac{1}{16}wR^2(3 + \nu)$$

$$A = \frac{wR^2}{8D}\left(\frac{3 + \nu}{1 + \nu}\right) \qquad \text{(v)}$$

Using this value of A and putting $r = 0$ for maximum bending moment we obtain, from (iv),

$$M_{max} = \tfrac{1}{16}wR^2(3 + \nu)$$

This result is the bending moment per unit length so putting $I = \tfrac{1}{12}(1 \times t^3)$ and y (distance from neutral axis) $= \tfrac{1}{2}t$ the maximum stress is

$$\sigma = \frac{My}{I} = \frac{\tfrac{1}{16}wR^2(3 + \nu) \times \tfrac{1}{2}t}{\tfrac{1}{12}(1 \times t^3)} = \frac{3wR^2(3 + \nu)}{8t^2}$$

Re-arranging and substituting the numerical data given in the question,

$$t^2 = \frac{3wR^2(3 + \nu)}{8\sigma}$$

$$= \frac{3 \times (20 \times 10^3\ \text{N/m}^2) \times (0{\cdot}05\ \text{m})^2 \times (3 + 0{\cdot}28)}{8 \times (100 \times 10^6\ \text{N/m}^2)}$$

$$= 6{\cdot}15 \times 10^{-7}\ \text{m}^2$$

The necessary thickness t is therefore

$7{\cdot}84 \times 10^{-4}$ m or $0{\cdot}784$ mm (*Ans*)

Since the origin of co-ordinates is taken at the centre of the *deflected* plate the maximum deflection is obtained by putting $r = R$. Hence, from (iii), with the value of A given by (v)

$$y_{max} = -\frac{wR^4}{64D} + \frac{wR^2}{8D}\left(\frac{3+\nu}{1+\nu}\right) \times \frac{R^2}{4}$$

$$= \frac{wR^4}{64D}\left(\frac{5+\nu}{1+\nu}\right)$$

Substituting the figures given in the question

$$D = \frac{Et^3}{12(1-\nu^2)} = \frac{(200 \times 10^9\ \text{N/m}^2) \times (7{\cdot}84 \times 10^{-4}\ \text{m})^3}{12\,(1 - 0{\cdot}28^2)}$$

$$= 8{\cdot}71\ \text{N-m}$$

and the maximum deflection

$$y_{max} = \frac{wR^4}{64D}\left(\frac{5+\nu}{1+\nu}\right) = \frac{(20 \times 10^3\ \text{N/m}^2) \times (0{\cdot}05\ \text{m})^4}{64 \times (8{\cdot}71\ \text{N-m})} \times \left(\frac{5+0{\cdot}28}{1+0{\cdot}28}\right)$$

$= 0{\cdot}925 \times 10^{-3}$ m or $0{\cdot}925$ mm (*Ans*)

17.4. A uniform circular plate of radius a and thickness t is clamped along its edge in such a manner that the slope and deflection of the plate at the edge are zero. The plate carries a downward central load P on the top surface. The strains at the lower surface of the plate, at radius r, in the radial and circumferential directions are given by

$$\epsilon_r = \tfrac{1}{2}t\frac{d\phi}{dr} \quad \text{and} \quad \epsilon_\theta = \frac{t\phi}{2r} \quad \text{respectively}$$

where ϕ is the slope of the surface of the plate in the radial direction.

With Young's modulus $= E$ and Poisson's ratio $= \nu$ obtain expressions for the stresses, σ_r and σ_θ, at the lower surface of the plate, in the radial and circumferential directions respectively.

Obtain a general expression for the deflection of the plate, and show that the maximum deflection is given by

$Pa^2/16\pi D$

where D is the flexural rigidity.

The following relationship may be used:

$$\frac{d}{dr}\left[\frac{1}{r}\frac{d}{dr}(r\phi)\right] = \pm\frac{Q}{D}$$

where Q is the shear force/circumferential length at radius r (the sign in front of Q/D depends on the sign convention adopted).

Show that as $r \to 0$ so $\sigma_r \to \infty$ and $\sigma_\theta \to \infty$. Comment on the fact that the maximum deflection is finite although the stresses near the centre approach infinity. (*U.L.*)

Solution. At radius r the total shear force on a circumferential ring is P the central load. Hence the shear force per circumferential length is $Q = P/2\pi r$ and substituting in the differential equation

$$\frac{d}{dr}\left[\frac{1}{r}\frac{d}{dr}(r\phi)\right] = -\frac{Q}{D} = -\frac{P}{2\pi r D}$$

Integrating,

$$\frac{1}{r}\frac{d}{dr}(r\phi) = -\frac{P}{2\pi D}\log r + A \qquad \text{(N.B. logs are to base e)}$$

$$\frac{d}{dr}(r\phi) = -\frac{Pr}{2\pi D}\log r + Ar$$

where A is a constant.

Integrating again and noting that the integral of $r \log r$ is $\frac{1}{2}r^2$ $(\log r - \frac{1}{2})$, we have

$$r\phi = -\frac{Pr^2}{4\pi D}(\log r - \tfrac{1}{2}) + \tfrac{1}{2}Ar^2 + B$$

where B is a second constant. At $r = 0$, the slope $\phi = 0$ and $r^2 \log r$ tends to zero. Thus $B = 0$ and

$$\phi = -\frac{Pr}{4\pi D}(\log r - \tfrac{1}{2}) + \tfrac{1}{2}Ar \qquad \text{(i)}$$

Also, since the edge is clamped, $\phi = 0$ when $r = a$ and

$$A = \frac{P}{2\pi D}(\log a - \tfrac{1}{2})$$

Substituting this result in (i), the slope is

$$\frac{dy}{dr} = \phi = -\frac{Pr}{4\pi D}(\log r - \tfrac{1}{2}) + \frac{Pr}{4\pi D}(\log a - \tfrac{1}{2})$$

$$= \frac{Pr}{4\pi D}(\log a - \log r) \qquad \text{(ii)}$$

Integrating to obtain the deflection we have

$$y = \frac{Pr^2}{8\pi D}\log a - \frac{Pr^2}{8\pi D}(\log r - \tfrac{1}{2}) + C$$

where C is a further constant. If the origin is taken at the centre of the *deflected* plate, $y = 0$ when $r = 0$ and hence $C = 0$. Thus the deflection equation is

$$y = \frac{Pr^2}{8\pi D}(\log a - \log r + \tfrac{1}{2})$$

$$= \frac{Pr^2}{8\pi D}\left(\log\frac{a}{r} + \tfrac{1}{2}\right)$$

Putting $r = a$, the maximum deflection is

$$y_{max} = \frac{Pa^2}{8\pi D}\left(\log\frac{a}{a} + \tfrac{1}{2}\right) = \frac{Pa^2}{16\pi D}$$

as required.

From (ii), $\phi = \dfrac{Pr}{4\pi D}\log\dfrac{a}{r}$ and by differentiation

$$\frac{d\phi}{dr} = \frac{P}{4\pi D}\left(\log\frac{a}{r} - 1\right)$$

Substituting in the expressions given in the question the strains are

$$\epsilon_r = \frac{Pt}{8\pi D}\left(\log\frac{a}{r} - 1\right) \quad \text{and} \quad \epsilon_\theta = \frac{Pt}{8\pi D}\log\frac{a}{r}$$

The corresponding stresses are given by:

$$\sigma_r = \frac{E}{1 - \nu^2}(\epsilon_r + \nu\epsilon_\theta)$$

$$= \left(\frac{E}{1 - \nu^2}\right)\left(\frac{Pt}{8\pi D}\right)\left(\log\frac{a}{r} - 1 + \nu\log\frac{a}{r}\right)$$

which, since $D = Et^3/12(1 - \nu^2)$, becomes

$$\sigma_r = \frac{3P}{2\pi t^2}\left[(1 + \nu)\log\frac{a}{r} - 1\right]$$

and, similarly,

$$\sigma_\theta = \frac{3P}{2\pi t^2}\left[(1 + \nu)\log\frac{a}{r} - \nu\right]$$

From the form of these expressions it can be seen that as $r \to 0$, $\log a/r \to \infty$ and the stresses tend to infinity.

In practice the load P cannot be applied at a point and a more elaborate analysis, (see Problem 4, p. 481) allowing for P being distributed over a small area, would lead to finite stresses. That the maximum deflection is finite arises from the fact that deflection is the integral sum of the strains throughout the plate and this can remain finite even though the strains become infinite at one point.

17.5. A cantilever consists of an 80 mm × 80 mm × 10 mm angle with the top face AB horizontal (Fig. 17.4). It carries a load of 2 kN at a distance of 1 m from the fixed end, the line of action of the load passing through the centroid of the section and inclined at 30° to the vertical.

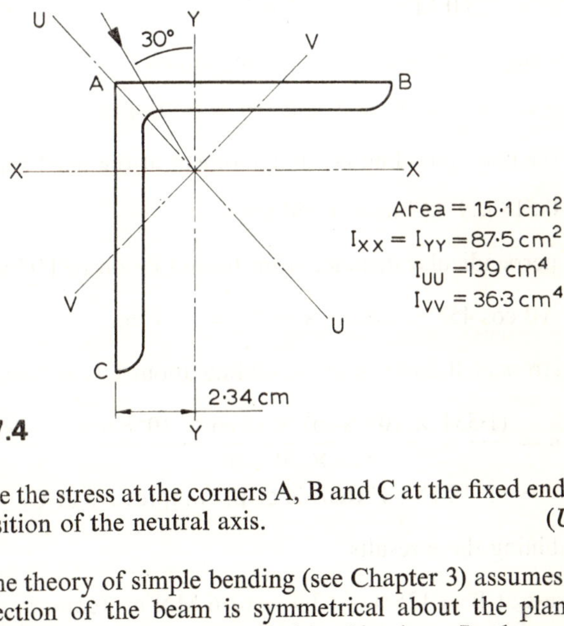

Fig. 17.4

Determine the stress at the corners A, B and C at the fixed end and also the position of the neutral axis. (*U.L.*)

Solution. The theory of simple bending (see Chapter 3) assumes that the cross-section of the beam is symmetrical about the plane of bending and that the loads are applied in this plane. In the present example the principal axes of the cross-section are UU (an axis of symmetry) and VV which is perpendicular to it. These axes make angles of 45° with XX and YY.

The component of the load in the plane UU is 2 kN × cos 15° = 2 kN × 0·966 = 1·932 kN and the corresponding bending moment at the fixed end is M = 1·932 kN × 1 m = 1·932 kN-m.

The distance OA = 2·34 cm × sec 45° = 2·34 × 1·414 = 3·31 cm and, considering bending about VV the stress at A is

$$\sigma = \frac{M \times OA}{I_{VV}} = \frac{(1{\cdot}932 \times 10^3 \text{ N-m}) \times (3{\cdot}31 \times 10^{-2} \text{ m})}{36{\cdot}3 \times 10^{-8} \text{ m}^4}$$

$$= 176 \text{ MN/m}^2 \text{ (tensile)} \qquad (Ans)$$

There is no stress at A for bending about UU.

The perpendicular distance of B and C from the axis VV is

$$(\text{AB} \cos 45° - \text{AO}) = (8{\cdot}0 \text{ cm} \times 0{\cdot}707 - 3{\cdot}31 \text{ cm}) = 2{\cdot}35 \text{ cm}$$

and thus, for bending about VV the stress at B and C is

$$\sigma = \frac{(1{\cdot}932 \times 10^3 \text{ N-m}) \times (2{\cdot}35 \times 10^{-2} \text{ m})}{36{\cdot}3 \times 10^{-8} \text{ m}^4}$$

$$= 125{\cdot}0 \text{ MN/m}^2 \text{ (compressive)}$$

The component of the load in the plane VV is

$$2 \text{ kN} \times \cos 75° = 2 \times 0{\cdot}259 = 0{\cdot}518 \text{ kN}$$

and the corresponding bending moment at the fixed end is

$$0{\cdot}518 \text{ kN} \times 3 \text{ m} = 1{\cdot}554 \text{ kN-m}.$$

The perpendicular distance from B (and C) on to UU is

$$\text{AB} \cos 45° = 8{\cdot}0 \text{ cm} \times 0{\cdot}707 = 5{\cdot}66 \text{ cm}$$

The stress at B and C due to bending about UU is therefore

$$\sigma = \frac{(1{\cdot}554 \times 10^3 \text{ N-m}) \times (5{\cdot}66 \times 10^{-2} \text{ m})}{139 \times 10^{-8} \text{m}^4}$$

$$= 63{\cdot}4 \text{ MN/m}^2 \text{ (tensile at B, compressive at C)}$$

Combining these results,

Stress at B = 125·0 − 63·4 = 61·6 MN/m² (compressive) (*Ans*)

Stress at C = 125·0 + 63·4 = 188·4 MN/m² (compressive) (*Ans*)

The neutral axis passes through 0; let α be the angle (measured anti-clockwise) which it makes with VV. If P is a point on the neutral axis having coordinates (u, v) relative to the axes VV and UU then $u = \text{OP} \sin \alpha$ and $v = \text{OP} \cos \alpha$. A moment M applied in the plane of the original load has components $M \cos 15°$ and $M \sin 15°$ about the axes VV and UU respectively. Since the resultant stress at P is zero we have

$$\frac{M \cos 15^\circ \times \mathrm{OP} \sin \alpha}{I_{VV}} + \frac{M \sin 15^\circ \times \mathrm{OP} \cos \alpha}{I_{UU}} = 0$$

or

$$\tan \alpha = -\frac{I_{VV}}{I_{UU}} \tan 15^\circ = -\frac{36{\cdot}3}{139} \tan 15^\circ = -0{\cdot}070$$

$$\alpha = -4^\circ\, 0'$$

The neutral axis is therefore in a direction of 4° 0′ clockwise from VV. (*Ans*)

17.6. Taking as a starting point the elementary knowledge that $\sigma/y = E/R$, in pure bending, show that the stress at a point (x, y) in a cross-section subjected to a direct load W and bending moments M_{XX} and M_{YY} about axes XX and YY which are mutually perpendicular but not principal axes is given by

$$\frac{W}{A} + \frac{(M_{XX}I_{YY} - M_{YY}I_{XY})y + (M_{YY}I_{XX} - M_{XX}I_{XY})x}{I_{XX}I_{YY} - I^2{}_{XY}}$$

where A, I_{XX}, I_{YY} and I_{XY} are respectively the area, the two second moments of area and the product of inertia with respect to XX and YY, both of which pass through the centre of area of the section. A thin sheet of thickness t is bent to form an equal angle section of legs equal to a. It carries as a *simply* supported beam of length L a uniform load of intensity w, the legs being vertical and horizontal. Find the maximum bending stress induced. (*U.L.*)

Solution. Suppose (Fig. 17.5) the principal axes UU and VV make an angle θ with the given axes XX and YY. Let the point (x, y) have co-ordinates (u, v) relative to these axes.

The moment M_{XX} has components $M_{XX} \cos \theta$ and $M_{XX} \sin \theta$ about the axes UU and VV respectively and thus the stress at (x, y) due to M_{XX} is

$$\sigma = \frac{M_{XX} \cos \theta \,.\, v}{I_{UU}} + \frac{M_{XX} \sin \theta \,.\, u}{I_{VV}}$$

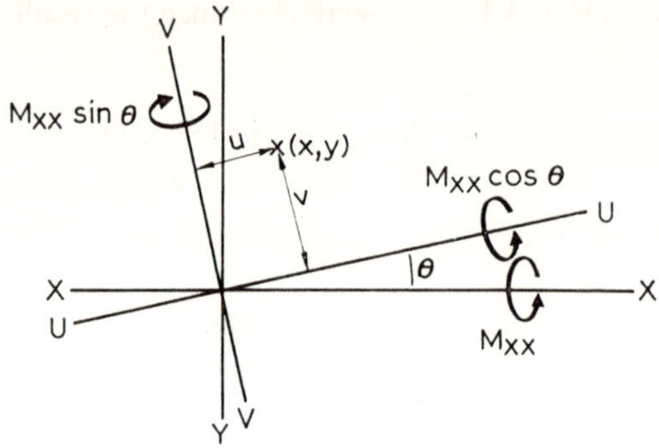

Fig. 17.5

As shown in Appendix 1, $u = x\cos\theta + y\sin\theta$ and $v = y\cos\theta - x\sin\theta$. Thus the stress becomes

$$\sigma = M_{XX}\left[\frac{\cos\theta\,(y\cos\theta - x\sin\theta)}{I_{UU}} + \frac{\sin\theta\,(x\cos\theta + y\sin\theta)}{I_{VV}}\right]$$

$$= \frac{M_{XX}}{I_{UU}I_{VV}}\left[\begin{matrix} x(I_{UU}\sin\theta\cos\theta - I_{VV}\sin\theta\cos\theta) \\ +\,y(I_{UU}\sin^2\theta + I_{VV}\cos^2\theta) \end{matrix}\right]$$

$$= \frac{M_{XX}}{I_{UU}I_{VV}}\left[\begin{matrix} \tfrac{1}{2}x\sin 2\theta(I_{UU} - I_{VV}) + \tfrac{1}{2}y(I_{UU} + I_{VV}) \\ +\,\tfrac{1}{2}y\cos 2\theta(I_{VV} - I_{UU}) \end{matrix}\right]$$

But, from equations (xi) and (xv) of Appendix 1,

$$I_{UU} + I_{VV} = I_{XX} + I_{YY}$$

$$I_{VV} - I_{UU} = 2I_{XY}\operatorname{cosec} 2\theta$$

Thus:

$$\sigma = \frac{M_{XX}}{I_{UU}I_{VV}}\left[-xI_{XY} + \tfrac{1}{2}y(I_{XX} + I_{YY}) + yI_{XY}\cot 2\theta\right]$$

Again, from equations (viii) and (xiv) of Appendix 1,

$$\cot 2\theta = \frac{I_{YY} - I_{XX}}{2I_{XY}}$$

$$I_{UU}I_{VV} = I_{XX}I_{YY} - I^2{}_{XY}$$

The stress due to M_{XX} is therefore

$$\sigma = \frac{M_{XX}}{I_{XX}I_{YY} - I_{XY}^2} \times \left[-xI_{XY} + \tfrac{1}{2}y(I_{XX} + I_{YY}) + yI_{XY}\left(\frac{I_{YY} - I_{XX}}{2I_{XY}}\right)\right]$$

$$= \frac{M_{XX}(-xI_{XY} + yI_{YY})}{I_{XX}I_{YY} - I_{XY}^2}$$

Similarly the stress due to M_{YY} is

$$\sigma = \frac{M_{YY}(-yI_{XY} + xI_{XX})}{I_{XX}I_{YY} - I^2_{XY}}$$

Adding these results and allowing for the direct stress W/A, the resultant stress is

$$\sigma = \frac{W}{A} + \frac{(M_{XX}I_{YY} - M_{YY}I_{XY})y + (M_{YY}I_{XX} - M_{XX}I_{XY})x}{I_{XX}I_{YY} - I_{XY}^2}$$

as required.

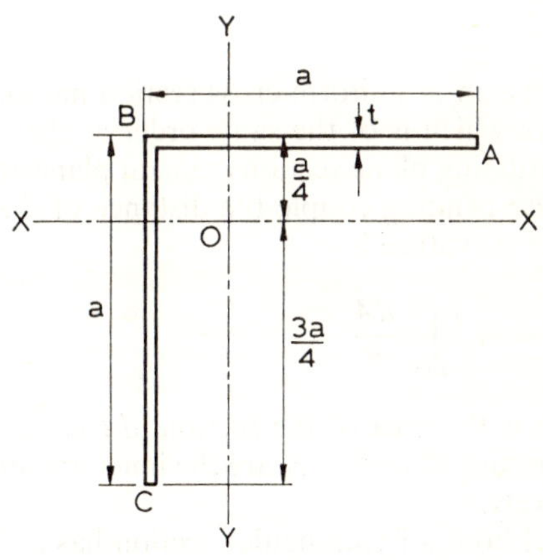

Fig. 17.6

Let ABC be the given angle section as shown in Fig. 17.6. The centroid of the section O is $a/4$ from each leg. Using the theorem of parallel axes

$$I_{YY} = I_{XX} = (I \text{ for AB})_{XX} + (I \text{ for BC})_{XX}$$

$$= at \times (\tfrac{1}{4}a)^2 + \frac{ta^3}{12} + at \times (\tfrac{1}{4}a)^2$$

$$= 5a^3t/24$$

In calculating the product moment each leg can be considered as concentrated at its own centroid. Hence:

$$I_{XY} = 2 \times at \times \tfrac{1}{4}a \times \tfrac{1}{4}a = \tfrac{1}{8}a^3t$$

For the given loading, $M_{XX} = wL^2/8$ and $M_{YY} = 0$. The maximum stress occurs at one of the corners A, B or C. Considering C, $x = -a/4$ and $y = -3a/4$. Using the result derived in the first part of the solution

$$\sigma = \frac{(wL^2/8) \times (5a^3t/24) \times (-3a/4) + (-wL^2/8) \times (a^3t/8) \times (-a/4)}{(5a^3t/24) \times (5a^3t/24) - (a^3t/8)^2}$$

$$= \left(\frac{wL^2}{8a^2t}\right)\frac{(-\frac{5}{24} \times \frac{3}{4}) + (\frac{1}{8} \times \frac{1}{4})}{(\frac{5}{24})^2 - (\frac{1}{8})^2}$$

$$= (-)\, 9wL^2/16a^2t$$

The reader should check that the stresses at the other extremities A and B are numerically smaller and this result is therefore the maximum bending stress induced.

17.7. A beam of uniform cross-section has an initial mean radius of curvature which is of the same order as the radial depth of the section. Assuming plane sections remain plane after bending prove that for a pure bending couple the distance of the neutral axis from the centre of curvature is

$$R_n = A \Big/ \int_{r_1}^{r_2} \frac{dA}{R}$$

where A is the area of the section, dA is the area of an elementary strip at radius R, and r_1, r_2 are the inner and outer radii of the section respectively.

A steel hook of rectangular section has an inner radius of 50 mm and an outer radius of 100 mm. The width is 25 mm. A bending moment of 1 kN-m is applied to the section, tending to open the hook. Calculate the maximum tensile and compressive stresses.

(*U.L.*)

Solution. In deriving the theory of simple bending (Chapter 3) it was assumed that the beam was initially straight. The results obtained can be applied to curved beams provided that the initial radius of curvature is large compared with the cross-sectional dimensions.

For heavily curved beams, however, a new analysis is required. In particular the neutral axis no longer passes through the centroid of the cross-section nor is the stress proportional to the distance from the neutral axis.

Suppose, in Fig. 17.7a, P and Q represent the intersections of the neutral plane with two cross-sections which initially intersect at O.

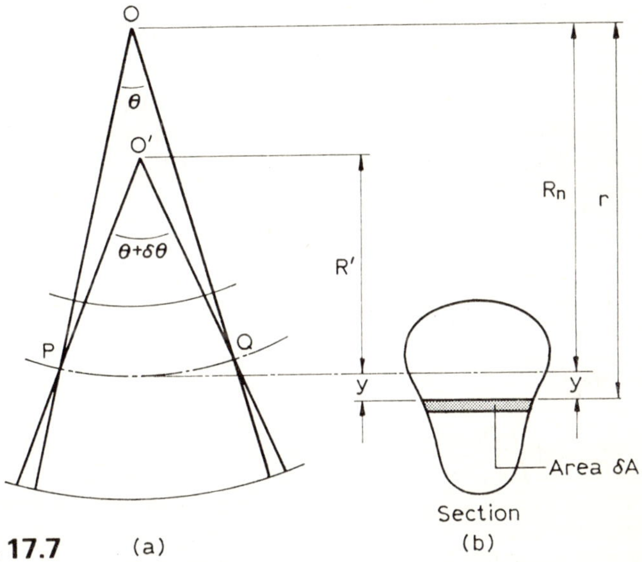

Fig. 17.7

Let O′ be the intersection of these planes after bending. Suppose R_n and R' are the radii of curvature of the neutral plane before and after bending. Then

$$PQ = R_n\theta = R'(\theta + \delta\theta) \qquad \text{(i)}$$

If δA is an element of area, distance y below the neutral axis, Fig. 17.7b, the longitudinal strain on the element is

$$\frac{\text{Increase in length}}{\text{Original length}} = \frac{(R' + y)(\theta + \delta\theta) - (R_n + y)\theta}{(R_n + y)\theta}$$
$$= \frac{R'(\theta + \delta\theta) + y\theta + y\delta\theta - R_n\theta - y\theta}{(R_n + y)\theta}$$

and, using (i) this becomes

$$\text{Strain} = \frac{y\delta\theta}{(R_n + y)\theta}$$

and the corresponding stress

$$\sigma = \frac{Ey\delta\theta}{(R_n + y)\theta} \tag{ii}$$

Integrating over the whole cross-section,

$$\text{Total force} = \frac{E\delta\theta}{\theta}\int_{r_1}^{r_2} \frac{y}{R_n + y}\, dA = \frac{E\delta\theta}{\theta}\int_{r_1}^{r_2} \frac{r - R_n}{r}\, dA \tag{iii}$$

where r is the radius from the original centre of curvature O to the element δA. In pure bending the net force on a cross-section is zero and the integrals in (iii) are therefore zero. Thus

$$\int_{r_1}^{r_2}\left(1 - \frac{R_n}{r}\right) dA = 0$$

$$\int_{r_1}^{r_2} dA = \int_{r_1}^{r_2} \frac{R_n}{r}\, dA$$

The integral on the left is the total area A, R_n is a constant and thus

$$R_n = A \bigg/ \int_{r_1}^{r_2} \frac{dA}{r}$$

as required.

Similarly the total moment about the neutral axis is

$$M = \frac{E\delta\theta}{\theta}\int_{r_1}^{r_2} \frac{y^2}{R_n + y}\, dA \tag{iv}$$

and this integral may be written in two parts thus:

$$\int_{r_1}^{r_2} \frac{(y^2 + R_n y) - R_n y}{R_n + y}\, dA = \int_{r_1}^{r_2} y\,.\,dA - R_n \int_{r_1}^{r_2} \frac{y}{R_n + y}\, dA$$

The first of these integrals is the first moment of the cross-section about the neutral axis. It can be written as Ae where e is the distance of the centroid from the neutral axis. As shown by (iii) the second integral is proportional to the net force on the section and is therefore zero. Equation (iv) therefore becomes

$$M = \frac{E\delta\theta}{\theta} \times Ae$$

and from (ii)

$$E\frac{\delta\theta}{\theta} = \sigma(R_n + y)/y$$

$$M = \frac{\sigma(R_n + y)}{y} Ae$$

$$\frac{\sigma}{y} = \frac{M}{Ae(R_n + y)} \quad \text{(v)}$$

Working in mm units, we have, for the numerical part of the question:

$$A = 50 \times 25 = 1250,\ r_1 = 50,\ r_2 = 100 \text{ and}$$

$$\delta A = 25 \times \delta r \text{ Hence}$$

$$R_n = \frac{A}{\int_{r_1}^{r_2} \frac{dA}{r}} = \frac{1250}{\int_{50}^{100} \frac{25}{r}\,dr}$$

$$= \frac{50}{\left[\log_e r\right]_{50}^{100}}$$

$$= \frac{50}{\log_e 2} = \frac{50}{0{\cdot}69315} = 72{\cdot}13 \text{ (mm)}$$

The radius of the central axis of the section is 75 mm and hence the neutral axis is 2·87 mm above the centroid in the sense of Fig. 17.8. Since this result is obtained as the small difference of two large quantities, R_n must be calculated with appropriate accuracy, e.g. using five-figure tables.

At the inner surface $y = (-)$ 22·13 mm and, with $M = (-)$ 1000 kN-mm (M is in the opposite sense to that assumed in the theory) the stress there, from (v), is

$$\sigma = \frac{My}{Ae(R_n + y)} = \frac{(-)\,1000 \times 10^3 \text{ N-mm} \times (-)\,22{\cdot}13 \text{ mm}}{1250 \text{ mm}^2 \times 2{\cdot}87 \text{ mm} \times (72{\cdot}13 - 22{\cdot}13) \text{ mm}}$$

$$= 123 \text{ N/mm}^2 \quad \text{or} \quad \text{MN/m}^2 \text{ (tensile)}$$

Similarly at the outer surface $y = 27{\cdot}87$ mm and

$$\sigma = (-)\,77{\cdot}7 \text{ N/mm}^2 \quad \text{or} \quad \text{MN/m}^2 \text{ (compressive)} \qquad (Ans)$$

Figure 17·8 shows the variation of stress across the section.

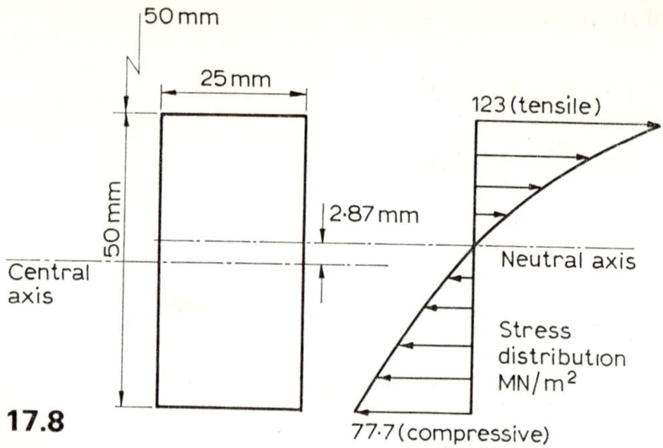

Fig. 17.8

17.8. Show that, for a curved bar of rectangular cross-section, the distance of the neutral axis from the centroid of the cross-section expressed as a first approximation is

$$e = R\left[1 - \frac{1}{1+\frac{1}{3}(d/2R)^2}\right]$$

where R is the radius of curvature of the centre line of the bar and d is the radial depth of the cross-section.

A crane hook of rectangular cross-section carries a bending moment of 2 kN-m; the radius of the centre line of the hook is 16 cm, the depth of the cross-section is 8 cm and the breadth is 4 cm.

Determine the magnitude of the maximum stress in the cross-section and indicate where it occurs. (*U.L.*)

Solution. In the notation of the previous example $r_1 = R - \frac{1}{2}d$ and $r_2 = R + \frac{1}{2}d$. Hence the distance of the neutral axis from the original centre of curvature is

$$R_n = \frac{bd}{\displaystyle\int_{R-\frac{1}{2}d}^{R+\frac{1}{2}d} \frac{d}{r}\,dr}$$

$$= \frac{d}{\log_e \dfrac{R+\frac{1}{2}d}{R-\frac{1}{2}d}} = \frac{d}{\log_e \dfrac{1+(d/2R)}{1-(d/2R)}}$$

From the expansion of log (1 + x) we obtain

$$\log_e\left(\frac{1+x}{1-x}\right) = (x - \tfrac{1}{2}x^2 + \tfrac{1}{3}x^3 - \tfrac{1}{4}x^4 \ldots) - (-x - \tfrac{1}{2}x^2 - \tfrac{1}{3}x^3 - \tfrac{1}{4}x^4 \ldots)$$
$$= 2(x + \tfrac{1}{3}x^3 + \tfrac{1}{5}x^5 \ldots)$$

Putting $x = d/2R$ the position of the neutral axis is given by

$$R_n = \frac{d}{2\left[\dfrac{d}{2R} + \dfrac{1}{3}\left(\dfrac{d}{2R}\right)^3 + \dfrac{1}{5}\left(\dfrac{d}{2R}\right)^5 \cdots\right]}$$

and the distance between the neutral axis and central axis is

$$e = R - R_n$$
$$= R\left\{1 - \frac{d}{2R\left[\dfrac{d}{2R} + \frac{1}{3}\left(\dfrac{d}{2R}\right)^3 + \frac{1}{5}\left(\dfrac{d}{2R}\right)^5 + \ldots\right]}\right\}$$
$$= R\left\{1 - \frac{1}{1 + \frac{1}{3}\left(\dfrac{d}{2R}\right)^2 + \frac{1}{5}\left(\dfrac{d}{2R}\right)^4 + \ldots}\right\}$$
$$= R\left\{1 - \frac{1}{1 + \frac{1}{3}\left(\dfrac{d}{2R}\right)^2}\right\}$$

as a first approximation.

This result can be developed further if required by expanding the fraction as a series. Thus

$$\left[1 + \tfrac{1}{3}\left(\frac{d}{2R}\right)^2 + \tfrac{1}{5}\left(\frac{d}{2R}\right)^4\right]^{-1}$$
$$= 1 - \left[\tfrac{1}{3}\left(\frac{d}{2R}\right)^2 + \tfrac{1}{5}\left(\frac{d}{2R}\right)^4\right] + \left[\tfrac{1}{3}\left(\frac{d}{2R}\right)^2 + \tfrac{1}{5}\left(\frac{d}{2R}\right)^4\right]^2$$
$$= 1 - \frac{d^2}{12R^2} - \frac{d^4}{80R^4} + \frac{d^4}{144R^4} \ldots$$
$$= 1 - \frac{d^2}{12R^2} - \frac{d^4}{180R^4} \cdots$$
$$e = R\left[1 - \left(1 - \frac{d^2}{12R^2} - \frac{d^4}{180R^4} - \ldots\right)\right]$$
$$= \frac{d^2}{12R}\left(1 + \frac{d^2}{15R^2} + \ldots\right)$$

The approximation $e \simeq d^2/12R$ is sufficient for most examples. Using it in the numerical part of the question

$$e = \frac{d^2}{12R} = \frac{(8\text{ cm})^2}{12 \times (16\text{ cm})} = \tfrac{1}{3}\text{ cm}$$

The maximum stress occurs at the inner surface for which $y = -(4 - \frac{1}{3}) = -3\frac{2}{3}$ cm. The radius to the neutral axis is $(16 - \frac{1}{3}) = 15\frac{2}{3}$ cm and the area $A = 8 \times 4 = 32\text{ cm}^2$. Thus, the maximum stress is

$$\begin{aligned}\sigma &= \frac{My}{Ae(R_n + y)}\\ &= \frac{2\text{ kN-m} \times (-3\frac{2}{3})\text{ cm}}{32\text{ cm}^2 \times \frac{1}{3}\text{ cm} \times 12\text{ cm}}\\ &= 57{\cdot}3\text{ MN/m}^2 \qquad (Ans)\end{aligned}$$

17.9. A rectangular steel beam, width b and depth d, is simply supported over a span L and a load W is gradually applied at mid-span. The steel follows a linear stress–strain law up to a yield stress σ_Y; at this constant stress considerable plastic deformation occurs. It may be assumed that the properties of the steel are the same in tension and compression.

Calculate the value of W at which

(*a*) yielding commences,
(*b*) yielding penetrates half-way to the neutral axis,
(*c*) a plastic hinge is formed.

What is the value of the shape factor for a rectangular section?

Solution. The stress–strain relationship described in the question is illustrated by Fig. 17.9 and should be compared with the behaviour of mild steel in a tension test as described on pages 430-1 (see Fig. 16.1). It ignores any drop in stress at yield and assumes that the material undergoes considerable strain at constant stress.

In applying this relationship to the bending of beams it is further assumed that plane cross-sections remain plane and that the same conditions apply in compression as in tension.

(*a*) Up to the load at which yielding commences the simple theory of bending applies and the distribution of stress over the section is shown in Fig. 17.10a. For the cross-section $I = bd^3/12$, the distance from the neutral axis to the edge of the section is $\frac{1}{2}d$ and, from the

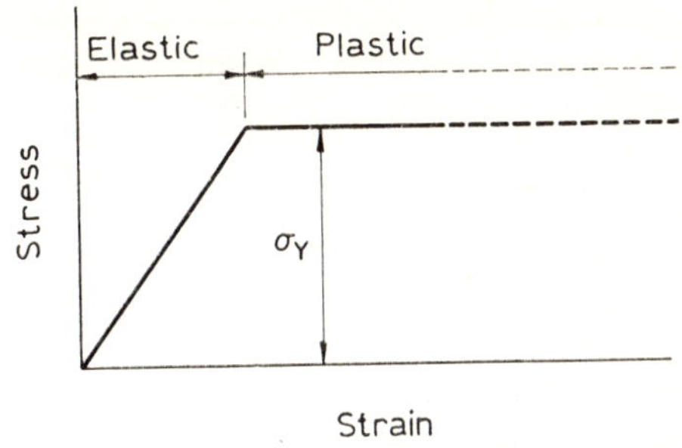

Fig. 17.9

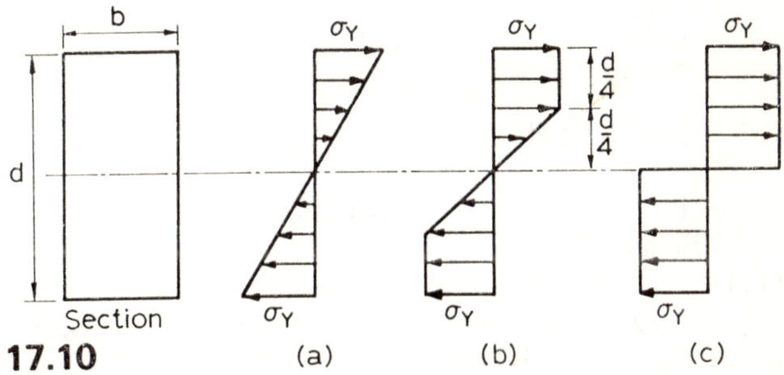

Fig. 17.10

bending equation, the moment of resistance when yielding commences is

$$M = \frac{\sigma I}{y} = \frac{\sigma_Y \times (bd^3/12)}{\frac{1}{2}d} = \tfrac{1}{6}\sigma_Y bd^2$$

The maximum bending moment for a centrally-loaded beam simply-supported at its ends is $\frac{1}{4}WL$ and hence

$$\tfrac{1}{4}WL = \tfrac{1}{6}\sigma_Y bd^2 \quad \text{from which } W = 2\sigma_Y bd^2/3L \qquad (Ans)$$

(*b*) If the load is increased yielding penetrates from the extreme fibres towards the neutral axis. Although the outer fibres are strained more and more the maximum stress remains at the value σ_Y and the stress distribution is of the form shown in Fig. 17.10b. The central portion of the cross-section remains elastic and its moment of resistance can be calculated using simple bending theory and a reduced depth. If yielding penetrates half way to the neutral axis, $y = \frac{1}{2}d$. The corresponding bending moment is

$$M\ (\text{elastic}) = \frac{\sigma_Y \frac{1}{12} b (\frac{1}{2}d)^3}{\frac{1}{4}d} = \frac{\sigma_Y bd^2}{24}$$

For the plastic portion the stress σ_Y is constant and there are forces, each $\sigma_Y \times b \times \frac{1}{4}d$, acting at distances $\frac{3}{8}d$ from the neutral axis. These produce a moment

$$M\,(\text{plastic}) = 2\sigma_Y \times \tfrac{1}{4}bd \times \tfrac{3}{8}d = \frac{3\sigma_Y bd^2}{16}$$

Thus, total moment of resistance is

$$\begin{aligned} M &= M\,(\text{elastic}) + M\,(\text{plastic}) \\ &= \sigma_Y bd^2/24 + 3\sigma_Y bd^2/16 \\ &= 11\sigma_Y bd^2/48 \end{aligned}$$

The corresponding load is

$$W = 11\sigma_Y bd^2/12L \qquad (Ans)$$

(*c*) If the load is further increased plastic deformation continues until yielding penetrates to the neutral axis as indicated in Fig. 17.10c. (A very small elastic region remains near the neutral axis but its contribution to the moment of resistance may be neglected).

The corresponding forces above and below the neutral axis are each $\sigma_Y \times b \times \frac{1}{2}d$ and act at distances $\frac{1}{4}d$ from it. Thus the moment of resistance is

$$M = 2 \times \sigma_Y \times \tfrac{1}{2}bd \times \tfrac{1}{4}d = \tfrac{1}{4}\sigma_Y bd^2$$

and the corresponding load is

$$W = \sigma_Y bd^2/L \qquad (Ans)$$

At this load a plastic "hinge" is formed at mid-span and any further increase would cause collapse through the rotation of the halves of the beam about its mid-point.

Shape factor is a property of the cross-section and is defined as the ratio of the moment of resistance in the fully plastic state to that at which yielding commences.

Thus, using the present results,

$$\text{shape factor} = \frac{\text{moment in case } (c)}{\text{moment in case } (a)}$$

$$= \frac{\sigma_Y bd^2}{4} \Big/ \frac{\sigma_Y bd^2}{6} = \frac{3}{2} \qquad (Ans)$$

17.10. A fixed ended beam of uniform section is ten metres long and subject to a uniformly distributed load applied over the entire length of the beam. The elastic moment of resistance of the cross-section of the beam is 100 kN-m and the shape factor is 1·10.

Find the maximum value of the load which may be applied without exceeding elastic conditions and compare it with the value obtainable immediately prior to total plastic collapse. (*U.L.*)

Solution. The bending moment diagram for the elastic condition is shown in Fig. 17.11a. As shown in Chapter 6 the maximum value occurs at the supports and equals $WL/12$. With the value given in the question

$$WL/12 = 100 \text{ kN-m}$$

$$W = 100 \text{ kN-m} \times 12/10 \text{ m}$$

$$= 120 \text{ kN total or } 12 \text{ kN/m} \qquad (Ans)$$

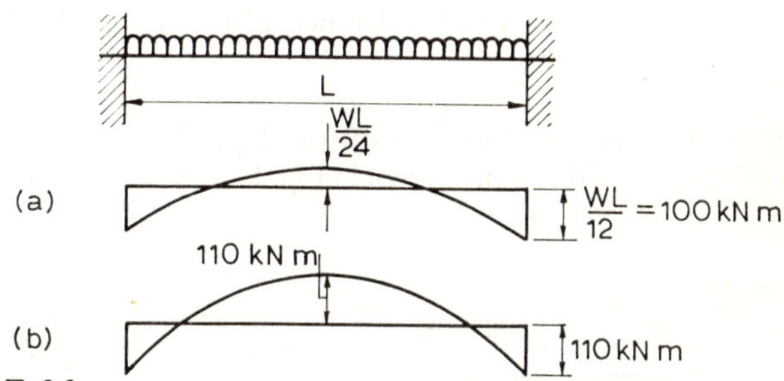

Fig. 17.11

The shape factor for the section is 1·10 and therefore a plastic hinge is formed when the bending moment reaches 110 kN-m. This occurs first at the ends but the beam can withstand further loading because it then behaves as if it were simply supported. The bending moment at the ends remains at 110 kN-m but that at mid-span increases until it reaches the same value. At this stage a plastic hinge is formed at mid-span and the beam can withstand no further load. The bending moment diagram for this condition is shown in Fig. 17.11b. By symmetry the end reactions are each $\frac{1}{2}W$ and, taking moments at mid-span,

$$110 = \tfrac{1}{2}W \times \tfrac{1}{2}L - \tfrac{1}{2}W \times \tfrac{1}{4}L - 110$$

from which $WL/8 = 220$, and

$$W = 220 \times 8/10 = 176 \text{ kN total or } 17{\cdot}6 \text{ kN/m} \qquad (Ans)$$

Hence the ratio required is

$$\frac{\text{Maximum load under elastic conditions}}{\text{Load at point of plastic collapse}} = \frac{120}{176} = \frac{15}{22} \qquad (Ans)$$

17.11. A beam has a T cross-section and is made by welding together two plates, the cross-sections of which are rectangular, with dimensions 160 mm × 10 mm. The beam is subjected to a bending moment of such magnitude that yielding occurs over the lower 40 mm of the web although the yield stress is not developed at the top surface of the flange. The yield stress of 250 MN/m^2 may be assumed to be constant over the area which has yielded while over the remainder of the section the stress is proportional to the distance from the neutral axis. Under these conditions determine the position of the neutral axis and the moment of resistance of the section.

Compare this moment of resistance with that obtained when the section is fully plastic. *(U.L.)*

Solution. The cross-section is shown in Fig. 17.12a and the distribution of stress is of the form shown in Fig. 17.12b. The neutral axis does

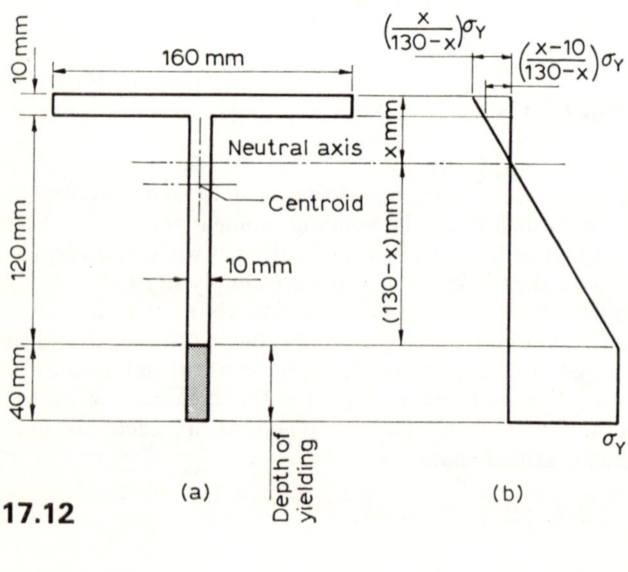

Fig. 17.12

not pass through the centroid of the section. Let its distance from the top edge of the flange be x mm. Then by proportion,

$$\text{Stress at top of flange} = \left(\frac{x}{130 - x}\right)\sigma_Y$$

$$\text{Stress at bottom edge of flange} = \left(\frac{x - 10}{130 - x}\right)\sigma_Y$$

$$\text{Average stress in flange} = \left(\frac{x - 5}{130 - x}\right)\sigma_Y$$

Working in millimetres, the forces in the four parts of the cross-section are as follows:

Flange:

$$\left(\frac{x - 5}{130 - x}\right)\sigma_Y \times 160 \times 10 = 1\,600\sigma_Y\left(\frac{x - 5}{130 - x}\right)$$

Web above n.a.:

$$\tfrac{1}{2}\left(\frac{x - 10}{130 - x}\right)\sigma_Y \times (x - 10) \times 10 = 5\sigma_Y\frac{(x - 10)^2}{(130 - x)}$$

Elastic portion of web below n.a.:

$$\tfrac{1}{2}\sigma_Y \times (130 - x) \times 10 = 5\sigma_Y(130 - x)$$

Plastic portion of web:

$$\sigma_Y \times 40 \times 10 = 400\sigma_Y$$

Since the resultant force on the section is zero, the total forces above and below the n.a. are equal. Hence:

$$1\,600\sigma_Y\left(\frac{x - 5}{130 - x}\right) + 5\sigma_Y\frac{(x - 10)^2}{(130 - x)} = 5\sigma_Y(130 - x) + 400\sigma_Y$$

Multiplying through by $(130 - x)/5\sigma_Y$ and collecting terms,

$$640x = 28\,800 \qquad x = 45$$

The neutral axis is therefore 45 mm from the top edge. (For comparison, the centroid is 47·5 mm from the top edge.) (*Ans*)

With this result the forces for the four portions of the cross-section become

Flange:

$$(1600 \times 10^{-6}\ \text{m}^2) \times (250 \times 10^6\ \text{N/m}^2) \times \left(\frac{45 - 5}{130 - 45}\right)$$
$$= 188{\cdot}2\ \text{kN}$$

Web above n.a.:

$$(5 \times 10^{-3}\ \text{m}) \times (250 \times 10^6\ \text{N/m}^2) \times \frac{(45 - 10)^2}{(130 - 45)} \times 10^{-3}\ \text{m}$$
$$= 18{\cdot}0\ \text{kN}$$

Elastic portion of web below n.a.:

$$(5 \times 10^{-3}\ \text{m}) \times (250 \times 10^6\ \text{N/m}^2) \times (130 - 45) \times 10^{-3}\ \text{m}$$
$$= 106{\cdot}3\ \text{kN}$$

Plastic portion of web:

$$(400 \times 10^{-6}\ \text{m}^2) \times (250 \times 10^6\ \text{N/m}^2) = 100{\cdot}0\ \text{kN}$$

The centroid of a trapezium whose parallel sides a and b are distance h apart, is

$$(a + 2b)h/3(a + b)$$

from the side a and hence the distance from the n.a. at which the flange force effectively acts is 40·2 mm.

The corresponding distances for the other three forces are 23·3 mm, 56·7 mm and 105 mm respectively. The moment of resistance of the section is, taking moments about the n.a.,

$$\begin{aligned} M = {} & (188{\cdot}2 \times 10^3\ \text{N}) \times (40{\cdot}2 \times 10^{-3}\ \text{m}) \\ & + (18{\cdot}0 \times 10^3\ \text{N}) \times (23{\cdot}3 \times 10^{-3}\ \text{m}) \\ & + (106{\cdot}3 \times 10^3\ \text{N}) \times (57{\cdot}7 \times 10^{-3}\ \text{m}) \\ & + (100 \times 10^3\ \text{N}) \times (105 \times 10^{-3}\ \text{m}) \\ = {} & 24{\cdot}6\ \text{kN-m} \end{aligned}$$

(*Ans*)

When the section is fully plastic the areas above and below the n.a. are equal since the resultant force is zero. In the present case the flange and web are equal in area and the n.a. is located at the junction between them.

The forces in the flange and web are each (σ_Y × area), i.e.

$$(250 \times 10^6\ \text{N/m}^2) \times (160 \times 10 \times 10^{-6}\ \text{m}^2) = 400\ \text{kN}$$

This distance between them is 85 mm. The moment of resistance is therefore

$$M = (400 \times 10^3 \text{ N}) \times (85 \times 10^{-3} \text{ m}) = 34 \text{ kN-m}$$

The ratio of the moments of resistance for the two conditions is therefore 24·6/34·0 = 0·724 (*Ans*)

17.12. A solid shaft 1 m long and 40 mm diameter is made of a material with a yield stress in shear of 150 MN/m². The elastic modulus of rigidity is 90 GN/m². Determine

(*a*) The total angle of twist in radians and the torque in kN-m when the material of the shaft just reaches its yield stress, and
(*b*) the torque in kN-m required to increase the angle of twist to twice that at yield.

Prove any formulae used. (*U.L.*)

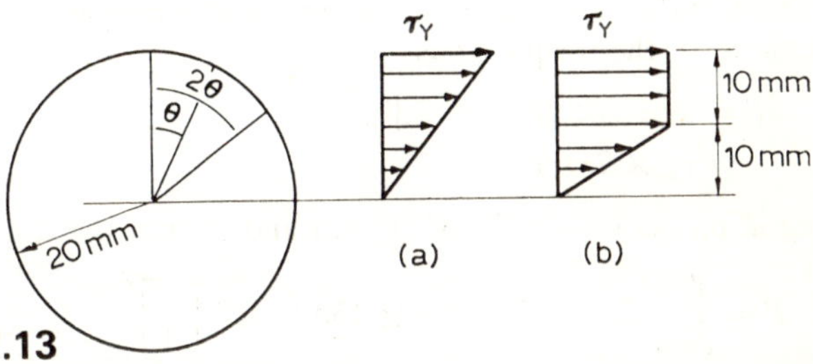

Fig. 17.13

Solution. (*a*) The stress distribution for the elastic condition is shown in Fig. 17.13a. The proof of the torsion equation for this case is given in Chapter 4 where it is shown that

$$\tau/r = T/J = G\theta/L$$

and, with the figures given in the equation,

$$\theta = \tau L/Gr$$

$$= \frac{(150 \times 10^6 \text{ N/m}^2) \times (1 \text{ m})}{(90 \times 10^9 \text{ N/m}^2) \times (20 \times 10^{-3} \text{ m})}$$

$$= \tfrac{1}{12} \text{ rad}$$ (*Ans*)

The polar second moment

$$J = \pi d^4/32 = \pi \times (40 \times 10^{-3} \text{ m})^4/32$$
$$= 2{\cdot}51 \times 10^{-7} \text{ m}^4$$

Hence the torque is

$$T = \tau J/r = \frac{(150 \times 10^6\ \text{N/m}^2) \times (2{\cdot}51 \times 10^{-7}\ \text{m}^4)}{20 \times 10^{-3}\ \text{m}}$$

$$= 1{\cdot}88\ \text{kN-m} \qquad (Ans)$$

(*b*) If the angle of twist is doubled the strain at each radius of the cross-section is doubled and the corresponding distribution of stress is shown in Fig. 17.13b. The central part of the shaft, up to a radius of 10 mm, remains elastic and the usual theory can be applied.

$$J = \tfrac{1}{32}\pi \times (20 \times 10^{-3}\ \text{m})^4 = 1{\cdot}57 \times 10^{-8}\ \text{m}^4$$

and the elastic torque is

$$T = \tau J/r = \frac{(150 \times 10^6\ \text{N/m}^2) \times (1{\cdot}57 \times 10^{-8}\ \text{m}^4)}{10 \times 10^{-3}\ \text{m}}$$

$$= 0{\cdot}235\ \text{kN-m}$$

For the outer part of the shaft where yielding occurs the stress is constant at the yield value τ_Y. For an annular ring at radius r, radial thickness δr, the torque δT is

$$\delta T = \text{stress} \times \text{area} \times \text{radius}$$

$$= \tau_Y \times 2\pi r \delta r \times r$$

Integrating between radii of 10 mm and 20 mm the total torque is

$$T = \int_{0{\cdot}01}^{0{\cdot}02} 2\pi r^2 \tau_Y dr = 2\pi \times 150 \times 10^6 \left[\frac{r^3}{3}\right]_{0{\cdot}01}^{0{\cdot}02}$$

$$= 700\pi\ \text{N-m} = 2{\cdot}20\ \text{kN-m}$$

Hence the torque required to increase the angle of twist to twice that at yield is $0{\cdot}235 + 2{\cdot}20 = 2{\cdot}44$ kN-m. (*Ans*)

PROBLEMS

1. Obtain an expression for the central deflection of the plate in Example 17.1. Determine also the maximum bending stress at the centre and the central deflection due to a central load W using similar assumptions.

Answer. $\dfrac{wL^4B^4}{32Et^3(L^4 + B^4)}$; $\dfrac{3WBL^3}{4t^2(L^4 + B^4)}$; $\dfrac{WL^3B^3}{16Et^3(L^4 + B^4)}$

2. The equation giving the deflected shape of a circular plate of uniform thickness t under loading which is normal to the plane containing the rim and symmetrical with respect to its centre, is

$$\frac{d}{dr}\left[\frac{1}{r}\frac{d}{dr}(r\phi)\right] = \frac{F}{D}$$

where ϕ is the slope of any radial line at radius r, F is the corresponding shearing force per unit length of arc, and D is the "flexural stiffness",

$$Et^3/12(1 - \nu^2)$$

Explain briefly the main assumptions made in deriving this equation.

A diaphragm of light alloy is 15 cm diameter, 1 mm thick and firmly clamped at its edges before loading. Calculate and plot a curve giving the deflected shape of the diaphragm under a uniform pressure of 15 kN/m². Take $E = 70$ GN/m² and $\nu = 0{\cdot}3$. (*U.L.*)

Answer.

r (cm)	0	1·5	3	4·5	6	7·5
deflection (mm)	1·157	1·006	0·6829	0·3404	0·0907	0

3. Obtain an expression for the maximum deflection of a circular plate clamped at its outer radius R and subjected to a uniform pressure p on the whole surface.

Compare this deflection with the maximum deflection of a beam having a rectangular section of unit width and thickness t, carrying a load p per unit length on a span $2R$ and having its ends fixed.

Answer. $\dfrac{3pR^4}{16Et^3}(1 - \nu^2)$; $\dfrac{\text{plate deflection}}{\text{beam deflection}} = \tfrac{3}{8}(1 - \nu^2)$

4. A thin, uniform circular flat plate of radius R is freely supported at its periphery and loaded in the centre with a load P which may be assumed as uniformly distributed over a small circular area of radius r_0.

The following equations apply to this case:

$$\left.\begin{aligned} &\text{(i)}\ \theta/r = A + \frac{B}{r^2} - \frac{kr^2}{4r_0^2} \\ &\text{(ii)}\ d\theta/dr = A - \frac{B}{r^2} - \frac{3kr^2}{4r_0^2} \end{aligned}\right\} \text{ when } 0 < r < r_0$$

$$\left.\begin{aligned} &\text{(iii)}\ \theta/r = C + \frac{D}{r^2} - k \log_e r \\ &\text{(iv)}\ d\theta/dr = C - \frac{D}{r^2} - k(\log_e r + 1) \end{aligned}\right\} \text{ when } r_0 < r < R$$

where θ is the slope at radius r; A, B, C and D are constants depending on boundary conditions and

$$k = 3(1 - \nu^2)P/\pi Et^3$$

where ν is Poisson's ratio, E is Young's modulus and t is the thickness.

Evaluate the constants A, B, C and D and hence (say by using equation (iii)) deduce that as $r_0 \rightarrow 0$ the central deflection tends to

$$\frac{3(1 - \nu)(3 + \nu)}{4\pi Et^3} PR^2 \qquad (U.L.)$$

Answer. $A = k\left[\log_e \dfrac{R}{r_0} + \dfrac{1}{1 + \nu} - \dfrac{r_0^2}{4R^2}\left(\dfrac{1 - \nu}{1 + \nu}\right)\right]$; $B = 0$;

$C = k\left[\log_e R + \dfrac{1}{1 + \nu} - \dfrac{r_0^2}{4R^2}\left(\dfrac{1 - \nu}{1 + \nu}\right)\right]$; $D = -\tfrac{1}{4}kr_0^2$

5. A pressure vessel is fitted with a circular manhole of 0·6 m diameter, the cover of which is made of a plate 25 mm thick. Assuming that the cover is rigidly clamped around the edge of the manhole, determine the maximum allowable pressure in the vessel if the maximum principal strain in the cover plate must not exceed that produced by a simple direct stress of 150 MN/m². Poisson's ratio, ν, for the material of the cover is 0·3.

At any radius r, the bending moment per unit length of arc is

$$M_1 = D\left(\frac{d\theta}{dr} + \nu\frac{\theta}{r}\right)$$

and the bending moment per unit length of radius is:

$$M_2 = D\left(\frac{\theta}{r} + \nu\frac{d\theta}{dr}\right)$$

where θ is the slope at radius r and

$$\frac{d}{dr}\left[\frac{1}{r}\frac{d(r\theta)}{dr}\right] + \frac{S}{D} = 0$$

where S = shearing force per unit length of arc at radius r, and the flexural rigidity, D, of the plate is $Et^3/[12(1-\nu^2)]$. (*U.L.*)
Answer. 381·6 kN/m².

6. Dry sand of bulk density ρ is poured on to a uniform disc of radius a, forming a right conical mound of base radius a and height $a/2$. The disc is of thickness t and is fully built-in round its periphery. Young's modulus for the disc material is E and Poisson's ratio is ν. It may be assumed that the deflection of the disc does not affect the distribution of load.

Given that

$$\frac{d}{dr}\left[\frac{1}{r}\frac{d}{dr}(r\phi)\right] = \pm\frac{Q}{D}$$

where Q = shear force/circumferential length at radius r (the sign depends on the convention adopted),
D = flexural rigidity = $Et^3/12(1-\nu^2)$
ϕ = slope of the surface of the disc in the radial direction

show that

(i) $Q = (3ar - 2r^2)\rho g/12$
(ii) the maximum deflection of the disc is of the form $\alpha\rho g a^5/D$ and state the value of the constant α, and
(iii) the maximum radial stress at the edge of the disc is of the form $\beta\rho g a^3/t^2$, and state the value of the constant β. (*U.L.*)

Answer. (ii) $\alpha = 43/9600$; (iii) $\beta = 29/120$.

7. A circular disc of uniform thickness t is clamped around its periphery and also to a central plunger as shown in Fig. 17.14. The effective inner and outer radii are 2·5 cm and 7·5 cm respectively and t = 1·5 mm. The plunger exerts a maximum axial force P.

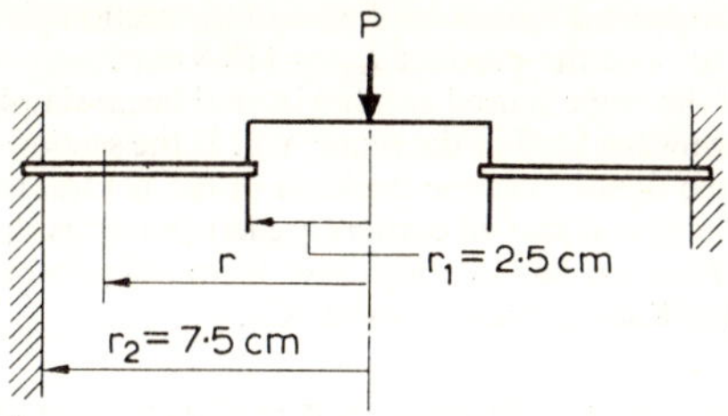

Fig. 17.14

At any radius r, the general relationship between the slope θ and the radius is given by

$$\frac{d}{dr}\left[\frac{1}{r}\frac{d(r\theta)}{dr}\right] + \frac{S}{D} = 0$$

where S is the shearing force per unit length of arc at radius r and

$$D = Et^3/12(1 - \nu^2)$$

Show that the deflection y at radius r is given by

$$y = -\frac{Pr^2}{8\pi D}(\log_e r - 1) + \tfrac{1}{4}C_1 r^2 + C_2 \log_e r + C_3$$

and find the maximum deflection for P = 500 N, assuming E = 200 GN/m² and Poisson's ratio ν = 0·28. *(U.L.)*
Answer. 0·97 mm.

8. Figure 17.15 shows the position of the centroid of a 12 cm × 9 cm × 1 cm angle. Determine the positions of the principal axes of the section and the values

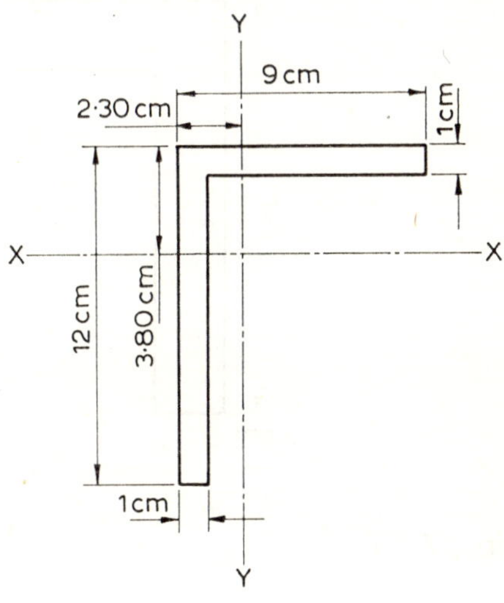

Fig. 17.15

of the principal second moments of area of the section given that $I_{xx} = 290{\cdot}0$ cm^4, $I_{yy} = 140{\cdot}9$ cm^4 and the product $I_{xy} = 118{\cdot}8$ cm^4.

A length of the angle is used as a horizontal beam simply supported at the ends, to carry a downward load in the plane YY. If the section of the beam is arranged as shown in the figure, find the position of the neutral axis.

Assume throughout that all corners are left square as shown in the figure. (*U.L.*)

Answer. 355·7 cm^4 at 28° 55′ clockwise from XX; 75·2 cm^4 at 28° 55′ clockwise from YY; 67° 44′ anticlockwise from XX.

9. A 80 mm × 8 mm × 8 mm angle is used as a beam simply supported at each end over a span of 2 m with one leg of the section horizontal and the other vertically upwards.

It is loaded at the centre of the span with a vertical load which may be assumed to pass through the centroid of the section.

The principal second moments of area for the section are 115 cm^4 and 29·8 cm^4. The distance of the centroid from the outside edge is 2·26 cm and the toe has a radius of 0·53 cm.

Find the position of the neutral axis and calculate the safe load if the maximum stress is not to exceed 120 MN/m^2. Graphical constructions may be used. (*U.L.*)

Hint. The toe radius is given since the point of maximum distance from the VV axis lies on the arc of the toe.

Answer. 30°30′ anticlockwise from horizontal; 2·56 kN.

10. The angle shown in Fig. 17.16 is used as a beam to span 3 m and to carry a load of 10 kN at mid-span, the load being applied along the direction YY. The second moments of area are $I_{xx} = 492$ cm^4 and $I_{yy} = 172$ cm^4. Find the product of inertia and hence the principal axes and the principal second moments of area of the section.

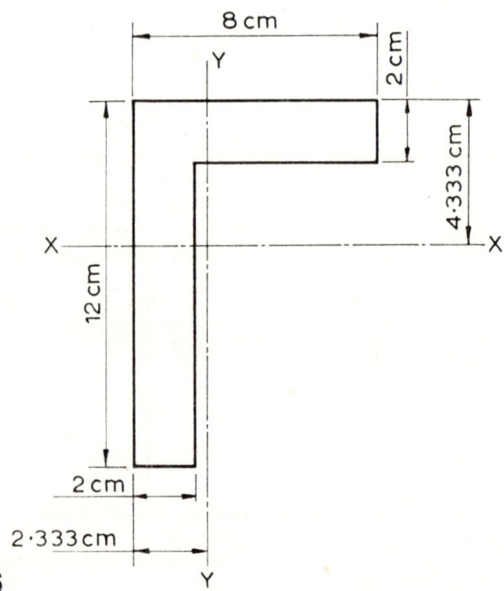

Fig. 17.16

Calculate the neutral axis for the given loading and determine the magnitude of the greatest tensile and compressive stresses.

A construction by the Mohr circle is permissible. (*U.L.*)

Answer. 160 cm^4; principal axes are inclined $22\frac{1}{2}°$ clockwise, to XX and YY; principal second moments are 558·2 and 105·8 cm^4; neutral axis is 42·9 anti-clockwise from XX; 161 MN/m^2 (tensile), 142 MN/m^2 (compressive).

11. A section of a beam, unsymmetrical as shown in Fig. 17.17 is subjected to a bending moment of 2·5 kN-m acting in the horizontal plane through the centroid of the section. Determine the magnitude of the stress produced at the bottom right-hand corner. (*U.L.*)

Answer. 52·8 MN/m^2.

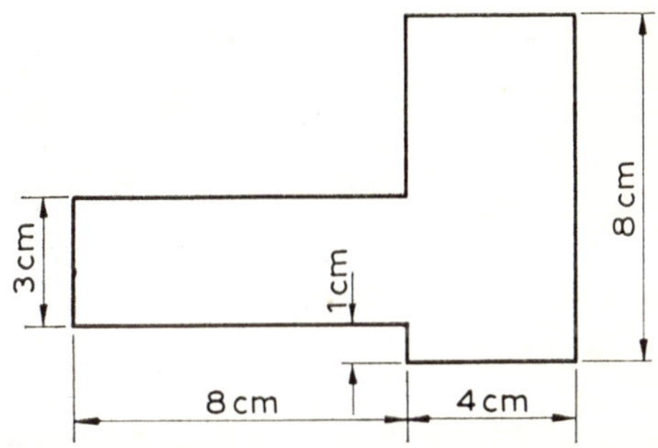

Fig. 17.17

12. A curved beam of rectangular section, initially unstressed, is subjected to a bending moment of 1·5 kN-m which tends to straighten the bar. The section is 4 cm wide by 5 cm deep in the plane of bending and the mean radius of curvature is 10 cm. Find the position of the neutral axis and the magnitudes of the greatest bending stresses and draw a diagram to show approximately how the stress varies across the section. (*U.L.*)

Answer. 2·04 mm from central axis; 112·6 and 79·5 MN/m^2.

13. A curved beam has a rectangular section; the depth of the beam is 5 cm and the radius to which it is curved is 12·5 cm at the mean depth. The beam is subjected at a certain section to a bending moment which would produce a maximum bending stress of 90 MN/m^2 if the beam had been straight. Determine the percentage increase in the maximum stress due to the curvature. (*U.L.*)

Answer. 18·4 per cent.

14. A curved beam has a rectangular section ABCD the depth AB being 9 cm and the width BC 3 cm as shown in Fig. 17.18. The radius of curvature at the mean

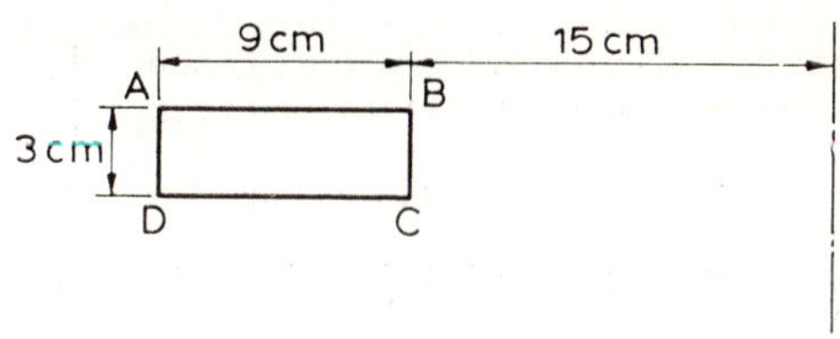

Fig. 17.18

depth is 19½ cm, BC being the side where the radius is 15 cm. There is a normal force P acting on the section producing a uniform compressive stress on ABCD and a bending moment M producing compressive stress at BC and tensile stress at AD. The stress at BC was found to be 75 MN/m² compression and that at AD 15 MN/m² tension. Determine the magnitudes of P and M. Work from first principles. (*U.L.*)
Answer. 142·8 kN; 8·35 kN-m.

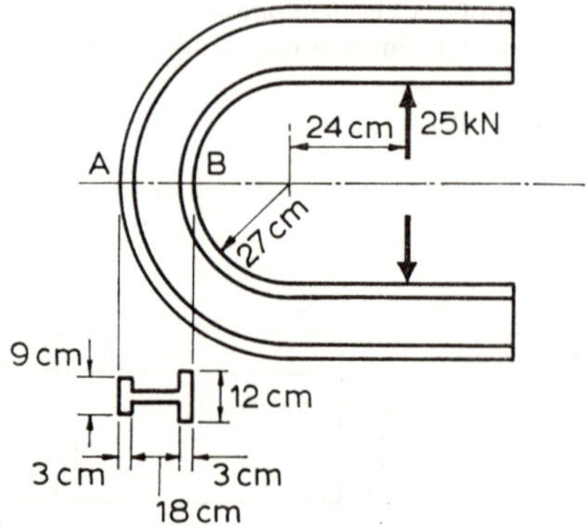

Fig. 17.19

15. Figure 17.19 shows a machine frame and the load to which it is subjected. Determine the stresses acting across the section AB at A and B. Make a diagram showing the distribution of stress across the section. Flange and web thickness 3 cm. (*U.L.*)
Answer. σ_A = 12·46 MN/m² (compression); σ_B = 19·73 MN/m² (tension).

16. A bar having a rectangular cross-section 9 cm by 6 cm is bent to a semi-circular form, the plane of the semi-circle being that of the 9 cm dimension; the inner and outer radii of curvature are 13½ cm and 22½ cm respectively. The half-ring stands on a frictionless horizontal surface with the plane of the semi-circle vertical. A vertical load of 80 kN is applied to the mid-point of the outer semi-circle, this point being mid-way between the two points of support which are 36 cm apart.

Find the maximum tensile and compressive stresses due to bending at a section near the point of application of the load. The depth of the beam is to be considered large compared with the radius of curvature. (*U.L.*)
Answer. 111·1 MN/m² (tensile); 78·6 MN/m² (compressive).

17. A steel member of rectangular section, breadth b = 3 cm and depth d = 8 cm is formed to a U-shape (Fig. 17.20) and subjected to a diametral load W = 22 kN as shown in Fig. 17.20. Find the greatest value for the mean radius R if the maximum allowable stress is 120 MN/m². Denoting the distance of the neutral axis for the bending stresses from the centroid as e, you may assume that $e = d^2/12R$.

Make a diagram showing the distribution of normal stress across the section AB. (*U.L.*)
Answer. R = 96 mm; σ_B = 120 MN/m² (tensile); σ_A = 53·08 MN/m² (compressive).

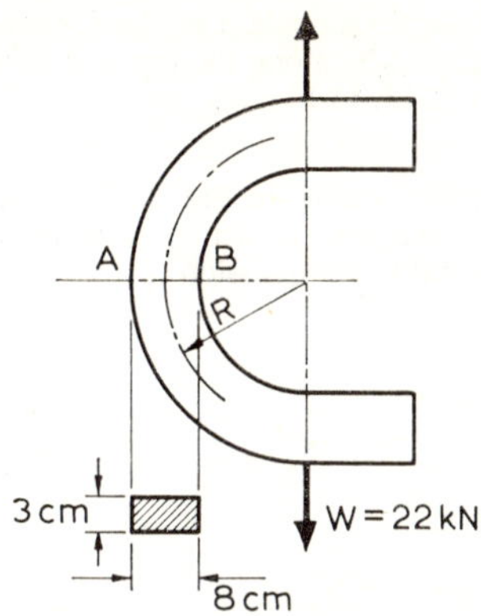

Fig. 17.20

18. A steel bar of rectangular section 8 cm × 3 cm is used as a simply supported beam on a span of 1·2 m and loaded at mid-span. If the yield stress is 275 MN/m^2 and the long edges of the section are vertical find the load when yielding first occurs.

Assuming that a further increase in load causes yielding to spread in towards the neutral axis with the stress in the yielded part remaining constant at 275 MN/m^2, determine the load required to cause yielding for a depth of 1 cm at the top and bottom of the section at mid-span, and find the length of beam over which yielding at the top and bottom faces will have occurred. (*U.L.*)
Answer. 29·3 kN; 35·75 kN; 0·215 m.

19. A rectangular steel beam AB 25 mm wide by 12 mm deep is placed symmetrically on two knife-edges C and D, 0·5 m apart, and loaded by applying equal weights at the ends A and B. The steel follows a linear stress–strain law (E = 200 GN/m^2) up to a yield stress of 280 MN/m^2; at this constant stress considerable plastic deformation occurs. It may be assumed that the properties of the steel are the same in tension and compression.

Calculate the bending moment on the central part of the beam CD when yielding commences, and the deflection of the centre relative to the supports.

If the loads are increased until yielding penetrates half-way to the neutral axis calculate the new value of the bending moment and the corresponding deflection. (*U.L.*)

Answer. 168 N-m; 7·29 mm; 231 N-m; 14·6 mm.

20. A steel beam of I-section has an overall depth of 250 mm and width of flange is 150 mm; the flanges are 25 mm thick and the web 10 mm thick. The beam is 5 m long and rests on two supports equidistant from the ends and 3 m apart; the beam carries a point load W at each end. Consider the two loads to be gradually increased but kept equal and assume the yield stress of the material to be 270 MN/m^2. Determine

(*a*) the magnitude of W when yielding of the material first occurs, stating clearly where the yield stress is reached
(*b*) the magnitude of W when yielding just extends through the thickness of the flanges, and
(*c*) the deflection of the beam at mid-span relative to the supports for the two conditions (*a*) and (*b*).

State the assumptions made. For elastic material take $E = 200$ GN/m². (*U.L.*)
Answer. (*a*) 220·3 kN, along the top and bottom surfaces between supports, (*b*) 245·8 kN, (*c*) 12·15, 15·19 mm.

21. The T-section shown in Fig. 17·21 is subject to a moment M represented vectorially in the figure and acting in the plane of symmetry of the section which puts the edge AB in compression.

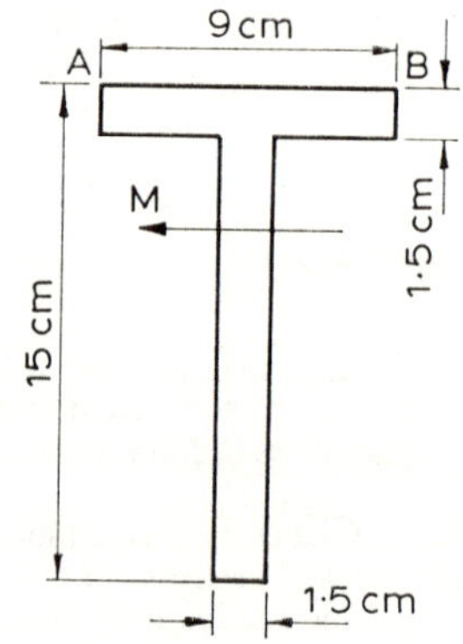

Fig. 17.21

Find the value of the fully plastic moment of resistance of this section and compare it with that moment which first produces yield in the outermost fibres. The yield stress for the material of this section in compression and tension is 210 MN/m². (*U.L.*)
Answer. 29·2 kN-m; 16·5 kN-m.

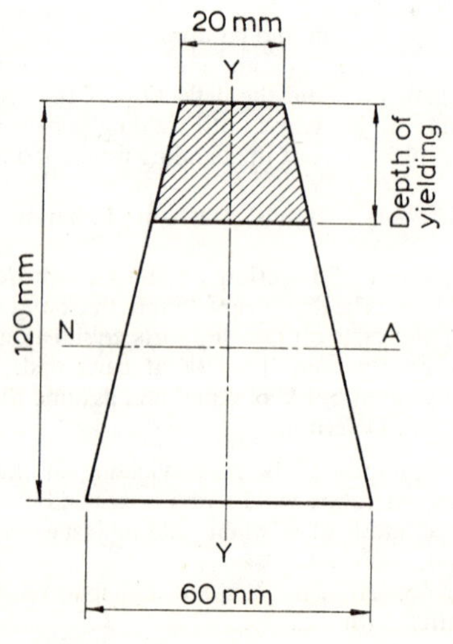

Fig. 17.22

22. A steel beam, subjected to a bending moment in the plane YY, has a section symmetrical about YY as shown in Fig. 17·22.

The bending moment is such that yielding has occurred in the top portion and the yield stress σ_y is just reached at the bottom. Assuming that, when yielding has occurred, the stress in the material remains constant at the value σ_y and that yielding occurs in tension and compression at the same stress, show that the neutral axis is 48·82 mm above the bottom edge and determine the magnitude of the bending moment in terms of σ_y. (*U.L.*)

Answer. ($101{\cdot}4\sigma_y$) N-m where σ_y is in MN/m^2.

23. Making the usual assumptions for the plastic deformation of beams, determine the shape factor for the following sections

(*a*) circle;
(*b*) regular hexagon in which one of the diagonals is the neutral axis;
(*c*) the T-section of Example 17.11.

Answer. (*a*) 1·7, (*b*) 1·6, (*c*) 1·81.

24. A rectangular beam 80 mm deep by 40 mm wide is made of a material for which E in compression is 1·5 times E in tension. Find (*a*) the position of the neutral axis; (*b*) the moment of resistance when the maximum tensile stress is 88 MN/m^2; (*c*) the corresponding maximum compressive stress.

If the beam is freely supported on a span of 2·5 m and carries a uniformly distributed load which induces these maximum stresses, find the deflection at the centre of the span. Take E in tension as 60 GN/m^2. (*U.L.*)

Hint. Work from first principles and assume plane sections remain plane.

Answer. (*a*) 4 mm from centroidal axis; (*b*) 4·135 kN-m; (*c*) 108 MN/m^2; 21·7 mm.

25. A steel shaft 8 cm diameter is solid for a certain distance from one end but hollow for the remainder of its length with an inside diameter of 4 cm.

If a pure torque is transmitted from one end of the shaft to the other of such a magnitude that yielding just occurs at the surface of the solid part of the shaft find the depth of yielding in the hollow part of the shaft and the ratio of the angles of twist per unit length for the two parts of the shaft.

State any assumptions made in arriving at your results. (*U.L.*)

Answer. 5·8 mm; 1·476.

26. A case-hardened shaft is 30 mm diameter with a case depth of 2 mm.

Assuming the case remains perfectly elastic up to its failing stress in shear of 300 MN/m^2 and that the inner core becomes perfectly plastic at a shearing stress of 180 MN/m^2 calculate (*a*) the torque to cause elastic failure in torsion in the case and (*b*) the angle of twist per metre length at failure.

State clearly any assumptions made and prove any formula required to deal with plastic conditions. $G = 80$ GN/m^2 for all the material while elastic. (*U.L.*)

Answer. 1·868 kN-m; $\frac{1}{4}$ rad.

27. A hollow cylinder has a bore diameter one half of its outside diameter. The cylinder is made from a material whose stress–strain behaviour is initially elastic and subsequently perfectly plastic. Show that the ratio of axially-applied torques $T_1:T_2:T_3$ is equal to 1:1·18:1·24 where

T_1 = the maximum torque to which the cylinder may be subjected without causing plastic deformation,

T_2 = the torque that causes plastic flow to a depth of one-half the thickness of the cylinder wall,

T_3 = the minimum torque to cause plastic flow throughout the entire section of the cylinder.

Assume that the shear strain is directly proportional to the radius. *(U.L.)*

Appendix 1

Moments of Area

First and second moments of area are used in problems concerning beams and shafts, and the main results are summarized here.

Consider a plane area, as shown in Fig. A.1, and any axis PP in the same plane (which may lie outside the area).

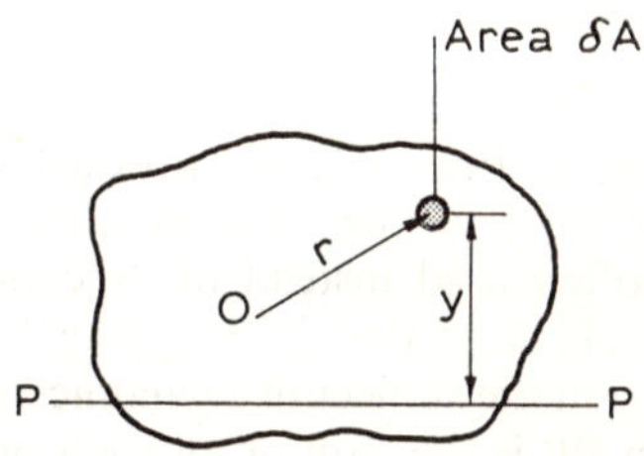

Fig. A.1

Let δA be an element of area, distance y from this axis. Then the product $y\delta A$ is called the *first moment of area* of this element about PP. The total first moment of area about PP is the sum of all such products and, in the calculus notation,

$$\text{Total first moment} = \int y dA \qquad \text{(i)}$$

the integration extending over the whole area, and y being considered positive in one direction (e.g. upwards) and negative in the other. The value of the total first moment depends upon the position of PP and, if a series of axes each parallel to PP are considered, there is one about which the total first moment is zero. This axis is called a *central axis*. For any given area there is one point, called the *centroid*, such that any axis passing through it is a central axis. The position of the centroid corresponds to the centre of gravity of a uniform thin plate of the same shape. The centroid of a figure having two (or more) axes of symmetry is the point of intersection of these axes. The areas and centroid positions for figures occurring in bending moment diagrams are shown in Fig. A.2. For the purposes of calculating the first moment of any figure about any axis the total area can be considered as concentrated at the centroid. This property is used in finding the centroid position of unsymmetrical beam sections. (See Chapter 3, Examples 3.6 and 3.7.)

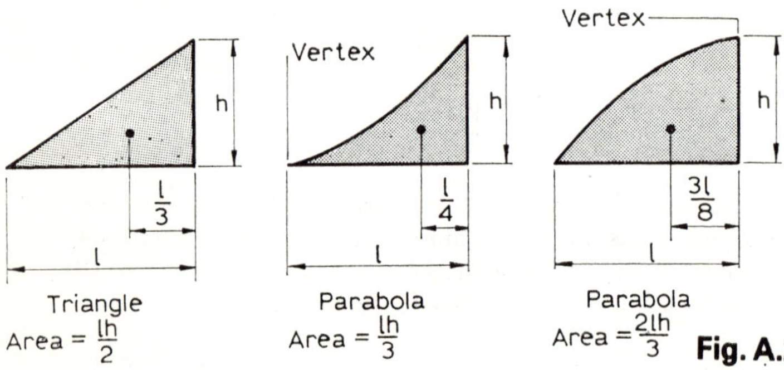

Fig. A.2

Referring to Fig. A.1, the product of $y^2\delta A$ is called the *second moment of area* of the element about PP. (The term *moment of inertia* is often used instead of "second moment of area", but the reader should note that its meaning here is somewhat different to its meaning in the subject of Dynamics.) The total second moment of area about PP is the sum of all such products, and is denoted by I. In the calculus notation,

$$\text{Total second moment } I = \int y^2 dA \qquad \text{(ii)}$$

the integration again extending over the whole area. Again, the value of this integral depends upon the position of the axis but it cannot be zero whatever the position of PP. For a family of parallel axes, however, it is a minimum for the one which passes through the centroid (i.e. the central axis). It is shown in Chapter 3, Example 3.3, that for simple bending the neutral axis is a central axis. The values of I for common geometrical shapes about horizontal neutral axes are given in Fig. A.3. Other useful second moments of area are (using the same notation):

For a rectangle about one edge (length b),

$$I = \frac{bd^3}{3}$$

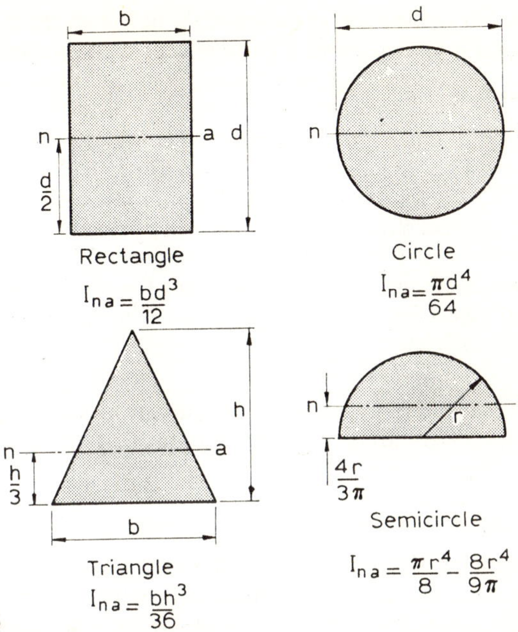

Fig. A.3

For a triangle about the base,

$$I = \frac{bh^3}{12}$$

For a semicircle about the bounding diameter,

$$I = \frac{\pi d^4}{128} = \frac{\pi r^4}{8}$$

If a series of central axes (making different angles to a given datum) are considered it can be proved that I has its maximum and minimum values about two axes which are mutually perpendicular. These are called the *principal axes* and the corresponding values of I are the *principal second moments of area* (or principal moments of inertia). Axes of symmetry are principal axes. If I has the same value about two axes of symmetry (as in the cases of a square, circle, equilateral triangle, regular hexagon, etc.) then it has this value for all other central axes. (This result is used in Chapter 3, Example 3.9.)

Parallel-axis Theorem

It is often necessary to calculate the second moment of area about an axis which does not pass through the centroid. This can be done as follows. If I is the second moment about the required axis and I_0 is the second moment about a parallel central axis, then

$$I = I_0 + Ah^2 \qquad \text{(iii)}$$

where A is the area of the figure and h is the distance between the two axes. This theorem can be used for calculating the principal values of I for figures made up of common geometrical shapes (see Chapter 3, Nos. 6 and 7). Also, the principal values of I for standard beam sections are given in various engineers' hand-books, and, by the above theorem, the required values for built-up sections can be obtained (see Chapter 3, No. 12; also Chapter 7, No. 10).

Radius of Gyration

If I is the second moment of area about any axis of a figure having an area A then the radius of gyration about that axis is defined as

$$k = \sqrt{\frac{I}{A}} \qquad \text{(iv)}$$

Rearranging (iv) we have $I = Ak^2$ and k can be regarded as the distance from the axis at which the entire area may be concentrated to give the same second moment of area (I). If k and k_0 are the radii of gyration for a figure about a non-central and parallel central axis respectively, then the theorem of parallel axes may be stated as

$$k^2 = k_0^2 + h^2 \qquad \text{(v)}$$

This follows from (iii) since $I = Ak^2$ and $I_0 = Ak_0^2$

Polar Second Moment of Area

Consider the point O in Fig. A.1. Then, if r is the distance from O to the element δA, the polar second moment of area (or "*polar moment of inertia*") of the figure about O is

$$J = \int r^2 dA$$

the integration extending over the whole area. Formulae for J in the case of solid and hollow circular sections are given in Chapter 4, page 111. No other cases are considered in this book.

Product moment

Suppose (Fig. A.4) XX and YY are perpendicular axes and δA is an element of area having co-ordinates (x, y). The second moments of area of the complete figure about these axes are

$$I_{XX} = \int y^2 dA \quad \text{and} \quad I_{YY} = \int x^2 dA$$

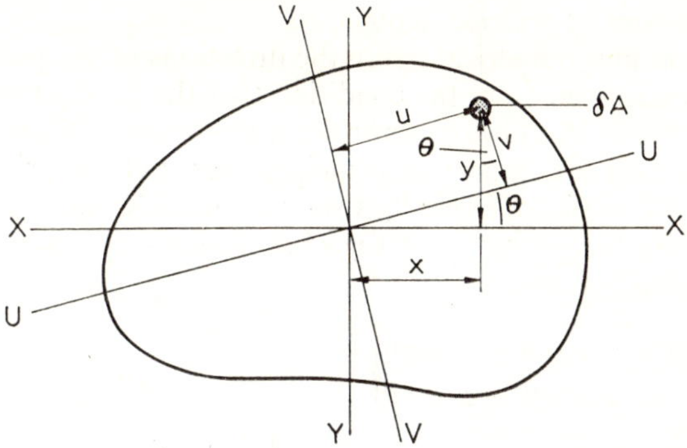

Fig. A.4

The product moment (or product of inertia) is defined as

$$I_{XY} = \int xy dA \qquad \text{(vi)}$$

For axes of symmetry the product moment is zero since x and y follow the usual sign conventions and corresponding parts of the figure on opposite sides of the axes cancel each other.

The theorem of parallel axes used in calculating second moments has its counterpart in product moments. Suppose, Fig. A.5, XX and

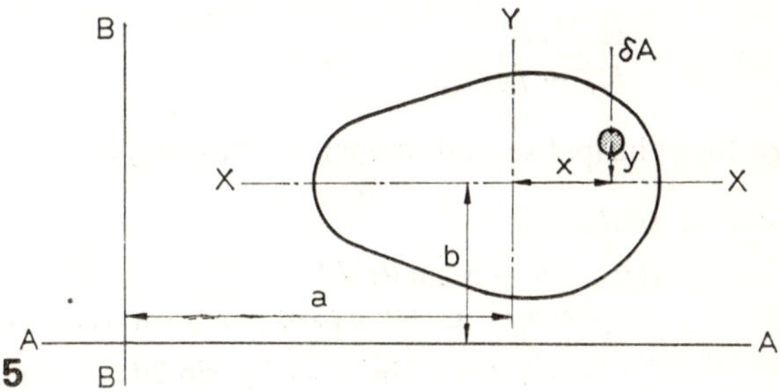

Fig. A.5

YY are principal axes of a given area which intersect at O and AA and BB are parallel axes distances b and a respectively from them. Then the product moment for axes AA and BB is

$$I_{AB} = \int (x + a)(y + b) dA$$
$$= \int xy dA + a \int y dA + b \int x dA + ab \int dA$$

Since O is the centroid and XX, YY are principal axes the first three integrals in this expression are all zero. Thus

$$I_{AB} = ab \times \text{(total area of the figure)}$$

Directions of principal axes

For an unsymmetrical figure the directions of the principal axes can be determined from the condition that the product moment is zero. Let O (Fig. A.4) be the centroid of the figure, XX and YY perpendicular axes which are not principal axes. Let the principal axes UU and VV make an angle θ as shown with XX and YY.

From the geometry of Fig. A.4 the coordinates of δA relative to UU and VV are

$$\left.\begin{aligned} u &= x\cos\theta + y\sin\theta \\ v &= y\cos\theta - x\sin\theta \end{aligned}\right\} \qquad \text{(vii)}$$

The product moment for UU and VV is therefore

$$\begin{aligned} I_{UV} &= \int uv\,dA \\ &= \int(x\cos\theta + y\sin\theta)(y\cos\theta - x\sin\theta)dA \\ &= \cos\theta\sin\theta[\int y^2 dA - \int x^2 dA] \\ &\qquad + (\cos^2\theta - \sin^2\theta)\int xy\,dA \\ &= \tfrac{1}{2}\sin 2\theta(I_{XX} - I_{YY}) + \cos 2\theta . I_{XY} \end{aligned}$$

For *UU* and *VV* to be principal axes, $I_{UV} = 0$ and, thus, equating the last result to zero,

$$\tan 2\theta = \frac{2I_{XY}}{I_{YY} - I_{XX}} \qquad \text{(viii)}$$

Hence the principal second moments are, using (vii),

$$\begin{aligned} I_{UU} &= \int v^2 dA \\ &= \int(y\cos\theta - x\sin\theta)^2 dA \\ &= \cos^2\theta\int y^2 dA + \sin^2\theta\int x^2 dA - 2\sin\theta\cos\theta\int xy\,dA \\ &= I_{XX}\cos^2\theta + I_{YY}\sin^2\theta - I_{XY}\sin 2\theta \qquad \text{(ix)} \end{aligned}$$

and similarly

$$I_{VV} = I_{XX}\sin^2\theta + I_{YY}\cos^2\theta + I_{XY}\sin 2\theta \qquad \text{(x)}$$

Adding (ix) and (x)

$$I_{UU} + I_{VV} = I_{XX} + I_{YY} \qquad \text{(xi)}$$

a result which can also be obtained by noting that the sum of the second moments for a pair of perpendicular axes is equal to the polar second moment *J*, since $r^2 = x^2 + y^2 = u^2 + v^2$.

Using the identities $\cos^2\theta \equiv \frac{1}{2}(\cos 2\theta + 1)$ and $\sin^2\theta \equiv \frac{1}{2}(1 - \cos 2\theta)$ together with (viii), equation (ix) becomes

$$\begin{aligned} I_{UU} &= \tfrac{1}{2}(I_{XX} + I_{YY}) + \tfrac{1}{2}(I_{XX} - I_{YY})\cos 2\theta - I_{XY}\sin 2\theta \\ &= \tfrac{1}{2}(I_{XX} + I_{YY}) - I_{XY}\frac{\cos 2\theta}{\tan 2\theta} - I_{XY}\sin 2\theta \\ &= \tfrac{1}{2}(I_{XX} + I_{YY}) - I_{XY}\left(\frac{\cos^2 2\theta + \sin^2 2\theta}{\sin 2\theta}\right) \\ &= \tfrac{1}{2}(I_{XX} + I_{YY}) - I_{XY}\operatorname{cosec} 2\theta \qquad \text{(xii)} \end{aligned}$$

similarly from (x)

$$I_{VV} = \tfrac{1}{2}(I_{XX} + I_{YY}) + I_{XY}\operatorname{cosec} 2\theta \qquad \text{(xiii)}$$

Multiplying these results together, noting that $\operatorname{cosec}^2 2\theta \equiv 1 + \cot^2 2\theta$, and again using (viii), we have

$$\begin{aligned} I_{UU}I_{VV} &= \tfrac{1}{4}(I_{XX} + I_{YY})^2 - I^2{}_{XY}\operatorname{cosec}^2 2\theta \\ &= \tfrac{1}{4}(I_{XX} + I_{YY})^2 - I^2{}_{XY}(1 + \cot^2 2\theta) \\ &= \tfrac{1}{4}(I_{XX} + I_{YY})^2 - I^2{}_{XY}\left(\frac{I_{YY} - I_{XX}}{2I_{XY}}\right)^2 - I^2{}_{XY} \\ &= \tfrac{1}{4}[(I_{XX} + I_{YY})^2 - (I_{YY} - I_{XX})^2] - I^2{}_{XY} \\ &= I_{XX}I_{YY} - I^2{}_{XY} \qquad \text{(xiv)} \end{aligned}$$

Subtracting (xii) from (xiii)

$$I_{VV} - I_{UU} = 2I_{XY}\operatorname{cosec} 2\theta \qquad \text{(xv)}$$

Application of Mohr circle

The form of equations (xii) and (xiii) suggests that the Mohr circle (see Chapters 9 and 10) can be used to determine I_{UU} and I_{VV}. If I_{XX} and I_{YY} are set out along the horizontal axis, Fig. A.6, and I_{XY} is

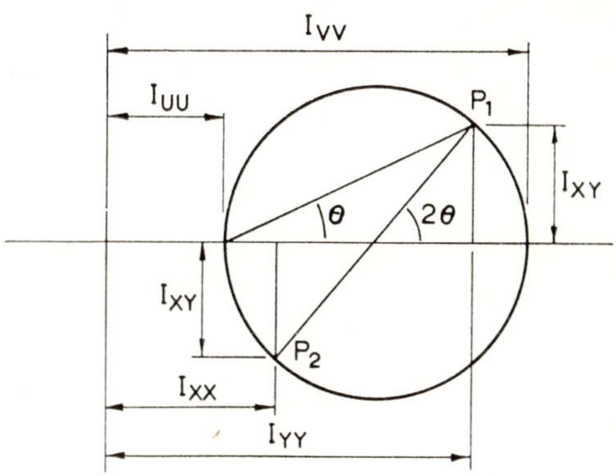

Fig. A.6

drawn as an ordinate at each point then the circle passing through P_1 and P_2 intersects the horizontal axis at points which represent I_{UU} and I_{VV}. The radius of the circle is I_{XY} cosec 2θ and the details of the proof are left to the reader.

The inclination of the principal axes to XX and YY can also be determined by measuring θ on the diagram.

Formulae for principal second moments

By analogy with the relationships for principal stresses (see equation (i) on page 258) the principal second moments I_{UU} and I_{VV} are the roots of the following quadratic equation for I

$$(I - I_{XX})(I - I_{YY}) = I^2{}_{XY}$$

from which

$$I_{UU} \text{ and } I_{VV} = \tfrac{1}{2}(I_{XX} + I_{YY}) \pm \tfrac{1}{2}\sqrt{[(I_{XX} - I_{YY})^2 + 4I^2{}_{XY}]}$$

The same results can be obtained from the geometry of Fig. A.6 or analytically, by eliminating θ from equations (viii), (xii) and (xiii) above.

Index